Lecture Notes in Artificial Intelligence 2838

Edited by J. G. Carbonell and J. Siekmann

Subseries of Lecture Notes in Computer Science

T0190028

Springer
Berlin
Heidelberg
New York
Hong Kong
London
Milan
Paris
Tokyo

Nada Lavrač Dragan Gamberger
Ljupčo Todorovski Hendrik Blockeel (Eds.)

Knowledge Discovery in Databases: PKDD 2003

7th European Conference on Principles and Practice of
Knowledge Discovery in Databases
Cavtat-Dubrovnik, Croatia, September 22-26, 2003
Proceedings

 Springer

Series Editors
Jaime G. Carbonell, Carnegie Mellon University, Pittsburgh, PA, USA
Jörg Siekmann, University of Saarland, Saarbrücken, Germany

Volume Editors

Nada Lavrač
Ljupčo Todorovski
Jožef Stefan Institute, Dept. of Intelligent Systems
Jamova 39, 1000 Ljubljana, Slovenia
E-mail:{Nada.Lavrac/Ljupco.Todorovski}@ijs.si

Dragan Gamberger
Rudjer Bošković Institute
Bijenička 54, 10000 Zagreb, Croatia
E-mail: Dragan.Gamberger@irb.hr

Hendrik Blockeel
Katholieke Universiteit Leuven, Dept. of Computer Science
Celestijnenlaan 200A, 3001 Leuven, Belgium
E-mail: Hendrik.Blockeel@cs.kuleuven.ac.be

Cataloging-in-Publication Data applied for

A catalog record for this book is available from the Library of Congress

Bibliographic information published by Die Deutsche Bibliothek
Die Deutsche Bibliothek lists this publication in the Deutsche Nationalbibliographie;
detailed bibliographic data is available in the Internet at <http://dnb.ddb.de>.

CR Subject Classification (1998): I.2, H.2, J.1, H.3, G.3, I.7, F.4.1

ISSN 0302-9743
ISBN 3-540-20085-1 Springer-Verlag Berlin Heidelberg New York

Springer-Verlag Berlin Heidelberg New York,
a member of BertelsmannSpringer Science+Business Media GmbH

http://www.springer.de

© Springer-Verlag Berlin Heidelberg 2003
Printed in Germany

Typesetting: Camera-ready by author, data conversion by Olgun Computergrafik
Printed on acid-free paper   SPIN: 10955635   06/3142   5 4 3 2 1 0

Preface

The proceedings of ECML/PKDD 2003 are published in two volumes: the *Proceedings of the 14th European Conference on Machine Learning* (LNAI 2837) and the *Proceedings of the 7th European Conference on Principles and Practice of Knowledge Discovery in Databases* (LNAI 2838). The two conferences were held on September 22–26, 2003 in Cavtat, a small tourist town in the vicinity of Dubrovnik, Croatia.

As machine learning and knowledge discovery are two highly related fields, the co-location of both conferences is beneficial for both research communities. In Cavtat, ECML and PKDD were co-located for the third time in a row, following the successful co-location of the two European conferences in Freiburg (2001) and Helsinki (2002). The co-location of ECML 2003 and PKDD 2003 resulted in a joint program for the two conferences, including paper presentations, invited talks, tutorials, and workshops.

Out of 332 submitted papers, 40 were accepted for publication in the ECML 2003 proceedings, and 40 were accepted for publication in the PKDD 2003 proceedings. All the submitted papers were reviewed by three referees. In addition to submitted papers, the conference program consisted of four invited talks, four tutorials, seven workshops, two tutorials combined with a workshop, and a discovery challenge.

We wish to express our gratitude to

- the authors of submitted papers,
- the program committee members, for thorough and timely paper evaluation,
- invited speakers Pieter Adriaans, Leo Breiman, Christos Faloutsos, and Donald B. Rubin,
- tutorial and workshop chairs Stefan Kramer, Luis Torgo, and Luc Dehaspe,
- local and technical organization committee members,
- advisory board members Luc De Raedt, Tapio Elomaa, Peter Flach, Heikki Mannila, Arno Siebes, and Hannu Toivonen,
- awards and grants committee members Dunja Mladenić, Rob Holte, and Michael May,
- Richard van der Stadt for the development of CyberChair which was used to support the paper submission and evaluation process,
- Alfred Hofmann of Springer-Verlag for co-operation in publishing the proceedings, and finally
- we gratefully acknowledge the financial support of the Croatian Ministry of Science and Technology, Slovenian Ministry of Education, Science, and Sports, and the Knowledge Discovery Network of Excellence (KDNet). KDNet also sponsored the student grants and best paper awards, while Kluwer Academic Publishers (the Machine Learning Journal) awarded a prize for the best student paper.

We hope and trust that the week in Cavtat in late September 2003 will be remembered as a fruitful, challenging, and enjoyable scientific and social event.

June 2003   Nada Lavrač
                                                      Dragan Gamberger
                                                      Hendrik Blockeel
                                                      Ljupčo Todorovski

ECML/PKDD 2003 Organization

Executive Committee

Program Chairs:	Nada Lavrač (Jožef Stefan Institute, Slovenia) ECML and PKDD chair
	Dragan Gamberger (Rudjer Bošković Institute, Croatia) ECML and PKDD co-chair
	Hendrik Blockeel (Katholieke Universiteit Leuven, Belgium) ECML co-chair
	Ljupčo Todorovski (Jožef Stefan Institute, Slovenia) PKDD co-chair
Tutorial and Workshop Chair:	Stefan Kramer (Technische Universität München, Germany)
Workshop Co-chair:	Luis Torgo (University of Porto, Portugal)
Tutorial Co-chair:	Luc Dehaspe (PharmaDM, Belgium)
Challenge Chair:	Petr Berka (University of Economics, Prague, Czech Republic)
Advisory Board:	Luc De Raedt (Albert-Ludwigs University Freiburg, Germany)
	Tapio Elomaa (University of Helsinki, Finland)
	Peter Flach (University of Bristol, UK)
	Heikki Mannila (Helsinki Institute for Information Technology, Finland)
	Arno Siebes (Utrecht University, The Netherlands)
	Hannu Toivonen (University of Helsinki, Finland)
Awards and Grants Committee:	Dunja Mladenić (Jožef Stefan Institute, Slovenia)
	Rob Holte (University of Alberta, Canada)
	Michael May (Fraunhofer AIS, Germany)
	Hendrik Blockeel (Katholieke Universiteit Leuven, Belgium)
Local Chairs:	Dragan Gamberger, Tomislav Šmuc (Rudjer Bošković Institute)
Organization Committee:	Darek Krzywania, Celine Vens, Jan Struyf (Katholieke Universiteit Leuven, Belgium),
	Damjan Demšar, Branko Kavšek, Milica Bauer, Bernard Ženko, Peter Ljubič (Jožef Stefan Institute, Slovenia),
	Mirna Benat (Rudjer Bošković Institute),
	Dalibor Ivušić (The Polytechnic of Dubrovnik, Croatia),
	Zdenko Sonicki (University of Zagreb, Croatia)

ECML 2003 Program Committee

H. Blockeel, Belgium
A. van den Bosch, The Netherlands
H. Boström, Sweden
I. Bratko, Slovenia
P. Brazdil, Portugal
W. Buntine, Finland
M. Craven, USA
N. Cristianini, USA
J. Cussens, UK
W. Daelemans, Belgium
L. Dehaspe, Belgium
L. De Raedt, Germany
S. Džeroski, Slovenia
T. Elomaa, Finland
F. Esposito, Italy
B. Filipič, Slovenia
P. Flach, UK
J. Fürnkranz, Austria
J. Gama, Portugal
D. Gamberger, Croatia
J.-G. Ganascia, France
L. Getoor, USA
H. Hirsh, USA
T. Hofmann, USA
T. Horvath, Germany
T. Joachims, USA
D. Kazakov, UK
R. Khardon, USA
Y. Kodratoff, France
I. Kononenko, Slovenia
S. Kramer, Germany
M. Kubat, USA
S. Kwek, USA
N. Lavrač, Slovenia
C. Ling, Canada
R. López de Màntaras, Spain

D. Malerba, Italy
H. Mannila, Finland
S. Matwin, Canada
J. del R. Millán, Switzerland
D. Mladenić, Slovenia
K. Morik, Germany
H. Motoda, Japan
R. Nock, France
D. Page, USA
G. Paliouras, Greece
B. Pfahringer, New Zealand
E. Plaza, Spain
J. Rousu, Finland
C. Rouveirol, France
L. Saitta, Italy
T. Scheffer, Germany
M. Sebag, France
J. Shawe-Taylor, UK
A. Siebes, The Netherlands
D. Sleeman, UK
R.H. Sloan, USA
M. van Someren, The Netherlands
P. Stone, USA
J. Suykens, Belgium
H. Tirri, Finland
L. Todorovski, Slovenia
L. Torgo, Portugal
P. Turney, Canada
P. Vitanyi, The Netherlands
S.M. Weiss, USA
G. Widmer, Austria
M. Wiering, The Netherlands
R. Wirth, Germany
S. Wrobel, Germany
T. Zeugmann, Germany
B. Zupan, Slovenia

PKDD 2003 Program Committee

H. Ahonen-Myka, Finland
E. Baralis, Italy
R. Bellazzi, Italy
M.R. Berthold, USA
H. Blockeel, Belgium
M. Bohanec, Slovenia
J.F. Boulicaut, France
B. Crémilleux, France
L. Dehaspe, Belgium
L. De Raedt, Germany
S. Džeroski, Slovenia
T. Elomaa, Finland
M. Ester, Canada
A. Feelders, The Netherlands
R. Feldman, Israel
P. Flach, UK
E. Frank, New Zealand
A. Freitas, UK
J. Fürnkranz, Austria
D. Gamberger, Croatia
F. Giannotti, Italy
C. Giraud-Carrier, Switzerland
M. Grobelnik, Slovenia
H.J. Hamilton, Canada
J. Han, USA
R. Hilderman, Canada
H. Hirsh, USA
S.J. Hong, USA
F. Höppner, Germany
S. Kaski, Finland
J.-U. Kietz, Switzerland
R.D. King, UK
W. Kloesgen, Germany
Y. Kodratoff, France
J.N. Kok, The Netherlands
S. Kramer, Germany
N. Lavrač, Slovenia
G. Manco, Italy

H. Mannila, Finland
S. Matwin, Canada
M. May, Germany
D. Mladenić, Slovenia
S. Morishita, Japan
H. Motoda, Japan
G. Nakhaeizadeh, Germany
C. Nédellec, France
D. Page, USA
Z.W. Ras, USA
J. Rauch, Czech Republic
G. Ritschard, Switzerland
M. Sebag, France
F. Sebastiani, Italy
M. Sebban, France
A. Siebes, The Netherlands
A. Skowron, Poland
M. van Someren, The Netherlands
M. Spiliopoulou, Germany
N. Spyratos, France
R. Stolle, USA
E. Suzuki, Japan
A. Tan, Singapore
L. Todorovski, Slovenia
H. Toivonen, Finland
L. Torgo, Portugal
S. Tsumoto, Japan
A. Unwin, Germany
K. Wang, Canada
L. Wehenkel, Belgium
D. Wettschereck, Germany
G. Widmer, Austria
R. Wirth, Germany
S. Wrobel, Germany
M.J. Zaki, USA
D.A. Zighed, France
B. Zupan, Slovenia

ECML/PKDD 2003 Additional Reviewers

F. Aiolli
A. Amrani
A. Appice
E. Armengol
I. Autio
J. Azé
I. Azzini
M. Baglioni
A. Banerjee
T.M.A. Basile
M. Bendou
M. Berardi
G. Beslon
M. Bevk
A. Blumenstock
D. Bojadžiev
M. Borth
J. Brank
P. Brockhausen
M. Ceci
E. Cesario
S. Chiusano
J. Clech
A. Cornuéjols
J. Costa
T. Curk
M. Degemmis
D. Demšar
J. Demšar
M. Denecker
N. Di Mauro
K. Driessens
T. Erjavec
T. Euler
N. Fanizzi
S. Ferilli
M. Fernandes
D. Finton
S. Flesca
J. Franke
F. Furfaro
T. Gärtner
P. Garza

L. Geng
A. Giacometti
T. Giorgino
B. Goethals
M. Grabert
E. Gyftodimos
W. Hämäläinen
A. Habrard
M. Hall
S. Hoche
E. Hüllermeier
L. Jacobs
A. Jakulin
T.Y. Jen
B. Jeudy
A. Jorge
R.J. Jun
P. Juvan
M. Kääriäinen
K. Karimi
K. Kersting
J. Kindermann
S. Kiritchenko
W. Kosters
I. Koychev
M. Kukar
S. Lallich
C. Larizza
D. Laurent
G. Leban
S.D. Lee
G. Legrand
E. Leopold
J. Leskovec
O. Licchelli
J.T. Lindgren
F.A. Lisi
T. Malinen
O. Matte-Tailliez
A. Mazzanti
P. Medas
R. Meo
T. Mielikäinen

H.S. Nguyen
S. Nijssen
A. Nowé
M. Ohtani
S. Ontañón
R. Ortale
M. Ould Abdel Vetah
G. Paaß
I. Palmisano
J. Peltonen
L. Peña
D. Pedreschi
G. Petasis
J. Petrak
V. Phan Luong
D. Pierrakos
U. Rückert
S. Rüping
J. Ramon
S. Ray
C. Rigotti
F. Rioult
M. Robnik-Šikonja
M. Roche
B. Rosenfeld
A. Sadikov
T. Saito
E. Savia
C. Savu-Krohn
G. Schmidberger
M. Scholz
A.K. Seewald
J. Sese
G. Sigletos
T. Silander
D. Slezak
C. Soares
D. Sonntag
H.-M. Suchier
B. Sudha
P. Synak
A. Tagarelli
Y. Tzitzikas

R. Vilalta   M. Wurst   I. Zogalis
D. Vladušič   R.J. Yan   W. Zou
X. Wang   X. Yan   M. Žnidaršič
A. Wojna   H. Yao   B. Ženko
J. Wróblewski   X. Yin

ECML/PKDD 2003 Tutorials

KD Standards
Sarabjot S. Anand, Marko Grobelnik, and Dietrich Wettschereck

Data Mining and Machine Learning in Time Series Databases
Eamonn Keogh

Exploratory Analysis of Spatial Data and Decision Making Using Interactive
Maps and Linked Dynamic Displays
Natalia Andrienko and Gennady Andrienko

Music Data Mining
Darrell Conklin

ECML/PKDD 2003 Workshops

First European Web Mining Forum
*Bettina Berendt, Andreas Hotho, Dunja Mladenić, Maarten van Someren, Myra
Spiliopoulou, and Gerd Stumme*

Multimedia Discovery and Mining
Dunja Mladenić and Gerhard Paaß

Data Mining and Text Mining in Bioinformatics
Tobias Scheffer and Ulf Leser

Knowledge Discovery in Inductive Databases
*Jean-François Boulicaut, Sašo Džeroski, Mika Klemettinen, Rosa Meo, and Luc
De Raedt*

Graph, Tree, and Sequence Mining
Luc De Raedt and Takashi Washio

Probabilistic Graphical Models for Classification
Pedro Larrañaga, Jose A. Lozano, Jose M. Peña, and Iñaki Inza

Parallel and Distributed Computing for Machine Learning
Rui Camacho and Ashwin Srinivasan

Discovery Challenge: A Collaborative Effort in Knowledge Discovery
from Databases
Petr Berka, Jan Rauch, and Shusaku Tsumoto

ECML/PKDD 2003 Joint Tutorials-Workshops

Learning Context-Free Grammars
Colin de la Higuera, Jose Oncina, Pieter Adriaans, Menno van Zaanen

Adaptive Text Extraction and Mining
Fabio Ciravegna, Nicholas Kushmerick

Table of Contents

Invited Papers

Contributed Papers

From Knowledge-Based to Skill-Based Systems: Sailing as a Machine Learning Challenge

Pieter Adriaans

FNWI / ILLC
University of Amsterdam
Plantage Muidergracht 24
1018 TV Amsterdam
The Netherlands
pietera@science.uva.nl
http://turing.wins.uva.nl/~pietera/ALS/

Abstract. This paper describes the Robosail project. It started in 1997 with the aim to build a self-learning auto pilot for a single handed sailing yacht. The goal was to make an adaptive system that would help a single handed sailor to go faster on average in a race. Presently, after five years of development and a number of sea trials, we have a commercial system available (www.robosail.com). It is a hybrid system using agent technology, machine learning, data mining and rule-based reasoning. Apart from describing the system we try to generalize our findings, and argue that sailing is an interesting paradigm for a class of hybrid systems that one could call Skill-based Systems.

1 Introduction

Sailing is a difficult sport that requires a lot of training and expert knowledge [1],[9],[6]. Recently the co-operation of crews on a boat has been studied in the domain of cognitive psychology [4]. In this paper we describe the Robosail system that aims at the development of self-learning steering systems for racing yachts [8]. We defend the view that this task is an example of what one could call skill-based systems. The connection between verbal reports of experts performing a certain task and the implementation of ML for those task is an interesting emerging research domain [3],[2], [7]. The system was tested in several real-life race events and is currently commercially available.

2 The Task

Modern single-handed sailing started its history with the organization of the first Observer Single-Handed Transatlantic Race (OSTAR) in 1960. Since that time the sport has known a tremendous development and is the source of many innovations in sailing. A single-handed skipper can only attend the helm for about 20% of his time. The rest is divided between boat-handling, navigation,

N. Lavrač et al. (Eds.): PKDD 2003, LNAI 2838, pp. 1–8, 2003.
© Springer-Verlag Berlin Heidelberg 2003

preparing meals, doing repairs and sleeping. All single-handed races allow the skippers to use some kind of autopilot. In its simplest form such an autopilot is attached to a flux-gate compass and it can only maintain a compass course. More sophisticated autopilots use a variety of sensors (wind, heel, global positioning system etc.) to steer the boat optimally. In all races the use of engines to propel the boat and of electrical winches to operate the sails is forbidden. All boat-handling except steering is to be done with manual power only.

It is clear that a single-handed sailor will be less efficient than a full crew. Given the fact that a single-handed yacht operates on autopilot for more than 80 % of the time a slightly more efficient autopilot would already make a yacht more competitive. In a transatlantic crossing a skipper will alter course maybe once or twice a day based on meteorological data and information and from various other sources like the positions of the competitors. From an economic point of view the automatization of this task has no top priority. It is the optimization of the handling of the helm from second to second that offers the biggest opportunity for improvement. The task to be optimized is then: *steer the ship as fast as possible in a certain direction and give the skipper optimal support in terms of advice on boat-handling, early warnings, alerts etc.*

3   Introduction

Our initial approach to the limited task of maintaining the course of a vessel was to conceive it as a *pure machine learning task*. At any given moment the boat would be in a certain region of a complex state-space defined by the array of sensor inputs. There was a limited set of actions defined in terms of a force exercised on the rudder, and there was a reward defined in terms of the overall speed of the boat. Fairly soon it became clear that it was not possible to solve the problem in terms of simple optimization of a system in a state-space:

- There is no neutral theory-free description of the system. A sailing yacht is a system that exists on the border between two media with strong non-linear behavior, wind and water. The interaction between these media and the boat should ideally be modelled in terms of complex differential equations. A finite set of sensors will never be able to give enough information to analyze the system in all of its relevant aspects. A careful selection of sensors given economical, energy management and other practical constraint is necessary. In order to make this selection one needs a theory about what to measure.
- Furthermore, given the complexity of the mathematical description, there is no guarantee that the system will know regions of relative stability in which it can be controlled efficiently. The only indication we have that efficient control is possible is the fact that human experts do the task well, and the best guess as to select which sensors is the informal judgement of experts on the sort of information they need to perform the task. The array of sensors that 'describes' the system is in essence already anthropomorphic.
- Establishing the correct granularity of the measurements is a problem. Wind and wave information typically comes with the frequency of at least 10 hz.

But hidden in these signals are other concepts that exist only on a different timescale eg. gusts (above 10 seconds), veering (10 minutes) and sea-state (hours). A careful analysis of sensor information involved in sailing shows that sensors and the concepts that can be measured with them cluster in different time-frames (hundreds of seconds, minutes, hours). This is a strong indication for a modular architecture. The fact that at each level decisions of a different nature have to be taken strongly suggest an architecture that consists of a hierarchy of agents that operate in different time-frames: lower agents have a higher measurement granularity, higher agents a lower one.

– Even when a careful selection of sensors is made and an adequate agent-architecture is in place the convergence of the learning algorithms is a problem. Tabula rasa learning is in the context of sailing impossible. One has to start with a rough rule-based system that operates the boat reasonably well and use ML techniques to optimize the system.

In the end we developed a hybrid agent based system. It merges traditional AI techniques like rule based reasoning with more recent methods developed in the ML community. Essential for this kind of systems is the link between expert concepts that have a fuzzy nature and learning algorithms. A simple example of an expert rule in the Robosail system is: *If you sail close-hauled then luff in a gust.* This rule contains the concepts 'close-hauled', 'gust' and 'luff'. The system contains agents that represent these concepts:

– Course agent: If the apparent wind angle is between A and B then you sail close hauled
– Gust agent: If the average apparent wind increases by a factor D more than E seconds then there is a gust
– Luff agent: Steer Z degrees windward.

The related learning methodology then is:

– Task: Learn optimal values for A,B,C,D,E,Z
– Start with expert estimates then
– Optimize using ML techniques

This form of *symbol grounding* is an emerging area of research interest that seems to be of vital importance to the kind of skill-based systems like Robosail [8], [2], [7], [5].

4   The System

The main systems contains four agents: Skipper, Navigator, Watchman and Helmsman. These roles are more or less modelled after the task division on a modern racing yacht [9]. Each agent lives in a different time frame. the agents are ordered in a subsumption hierarchy. The skipper is intended to take strategical decisions with a time interval of say 3 to 6 hours. He has to take into account weather patterns, currents, seastate, chart info etc. Currently this process is only partly automated. It results in the determination of a waypoint, i.e. a location on the map where we want to arrive as soon as possible. The navigator and the watchman have the responsibility to get to the waypoint. The navigator deals with the more tactic aspects of this process. He knows the so-called polar diagrams of the boat and its behavior in various sea states. He also has a number of agents at his disposal that help him to asses the state of the ship: do we carry too much sail, is there too much current, is our trim correct etc. The reasoning of the navigator results in a compass course. This course could change within minutes. The watchman is responsible for keeping this course with optimal velocity in the constantly changing environment (waves, wind shifts etc.). He gives commands to the helmsman, whose only responsibility it is to make and execute plans to get and keep the rudder in certain positions in time.

The AI solution: a hybrid agent based approach

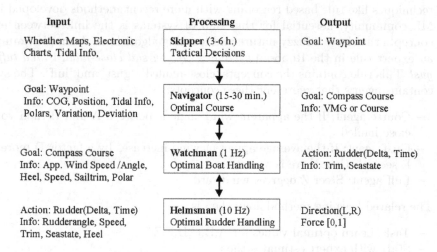

Fig. 1. The hierarchy of main agents

There are a number of core variables: log speed, apparent wind speed and angle, rudder angle, compass course, current position, course on ground and speed on ground. These are loaded into the kernel system. Apart from these core variables there are a number of other sensors that give information. Amongst others: canting angle mast, swivel angle keel, heel sideways, heel fore-aft, depth, sea state, wave direction, acceleration in various directions. Others will activate agents that warn for certain undesirable situations (i.e. depth, temperature of the water). Others are for the moment only used for human inspection (i.e. radar images). For each sensor we have to consider the balance between the contribution to speed and safety of the boat and the negative aspects like energy consumption, weight, increased complexity of the system.

Hybrid Architecture

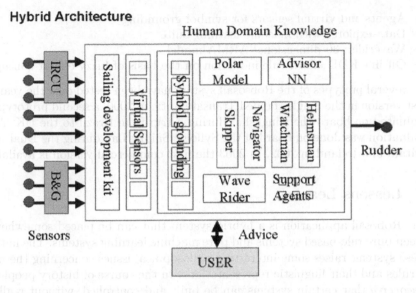

Fig. 2. The main architecture

The final system is a complex interplay between sensor-, agent- and network technology, machine learning and AI techniques brought together in a hybrid architecture. The hardware (CE radiation level requirements, water and shock proof, easy to mount and maintain) consists of:

- CAN bus architecture: Guaranteed delivery
- Odys Intelligent rudder control unit (IRCU): 20 kHz, max. 100 Amp (Extensive functions for self-diagnosis)
- Thetys Solid state digital motion sensor and compass
- Multifunction display
- Standard third party sensors with NMEA interface (e.g. B&G)

The software functionality involves:

- Agent based architecture
- Subsumption architecture
- Model builder: on line visual programming
- Real Time flow charting
- Relational database with third party datamining facility
- Web enabling
- Remote control and reporting

Machine Learning and AI techniques that are used:

- Watch man: Case Based Reasoning
- Helmsman: neural network on top of PID controller
- Advisor: nearest-neighbor search

- Agents and virtual sensors for symbol grounding
- Data-explorer with machine learning suite
- Waverider: 30 dimensional ARMA model
- Off line KDD effort: rule induction on the basis of fuzzy expert concepts

Several protypes of the Robsosail system have been tested over the years: the first version in the Single Handed Transatalantic in 2000, a second prototype was evaluated on board the Kingfisher during a trip from Brazil to the UK. A final evaluation was done on board of the Syllogic Sailing Lab during the Dual Round Britain and Ireland in 2002. In 2003 the first commercial version is available.

5   Lessons Learned

The Robosail application is a hybrid system that can be placed somewhere between pure rule-based systems and pure machine learning systems. The nature of these systems raises some interesting philosophical issues concerning the nature of rules and their linguistic representations. In the course of history people have *discovered* that certain systems can be built and controlled, without really understanding why this is the case. A sailing boat is such a system. It is what it is because of ill-understood hydro- and aerodynamical principles and has a certain form because the human body has to interact with it. It is thoroughly an anthropomorphic machine. Human beings can handle these systems, because they are the result of a long evolutionary process. Their senses are adapted to those regions of reality that are relatively stable and are sensitive to exactly those phase changes that give relevant information about the state of the systems. In a process of co-evolution the language to communicate about these concepts emerged. Specific concepts like 'wave', 'gust' and 'veering' exist because they mark relevant changes of the system. Their cognitive status however is complex, and it appears to be non-trivial to develop automated systems that discover these concepts on the basis of sensor data.

A deeper discussion of these issues would have to incorporate an analysis of the nature of rules that is beyond the scope of this paper. The rules of a game like chess exist independently of their verbal representation. We use the verbal representation to communicate with others about the game and to train young players. A useful distinction is the one between *constitutive rules* and *regulative rules*. The constitutive rules define the game. If they are broken the game stops. An example for chess would be: *You may not move a piece to a square already occupied by one of your own pieces.* Regulative rules define good strategies for the game. If you break them you diminish your chances of winning, but the game does not stop. An example of a regulative rule for chess would be: *When you are considering giving up some of your pieces for some of your opponent's, you should think about the values of the men, and not just how many each player possesses.* Regulative rules represent the experience of expert players. They have a certain fuzzyness and it is difficult to implement them in pure knowledge-based systems. The only way we can communicate about skills is in terms of regulative rules. The rule *If you sail clause hauled then luff in gust* is an example. Verbal reports of experts in terms of regulative rules can play an important role in the design of systems. From a formal point of view they reduce the complexity of the task. They tell us *where to look* in the state space of the system. From a cognitive point of view they play a similar role in teaching skills. They tell the student roughly what to do. The fine tuning of the skill is then a matter of training.

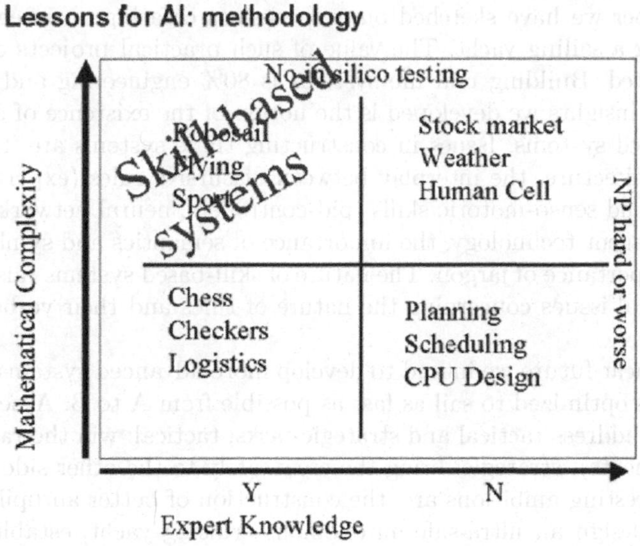

Fig. 3. A taxonomy of systems

This discussion suggests that we can classify tasks in two dimensions: 1) The expert dimension: Do human agents perform well on the task and can they report verbally on their actions and 2) The formal dimension: do we have adequate formal models of the task that allow us to perform tests in silico? For chess and a number of other tasks that were analyzed in the early stages of AI research the answer to both questions is yes. Operations research studies systems for which the first answer is no and the second answer is yes. For sailing the answer to the first question is positive, the answer to the second question negative. This is typical for skill-based systems. This situation has a number of interesting methodological consequences: we need to incorporate the knowledge of human experts into our system, but this knowledge in itself is fundamentally incomplete and needs to be embedded in an adaptive environment. Naturally this leads to issues concerning symbol grounding, modelling human judgements, hybrid architectures and many other fundamental questions relevant for the construction of ML applications in this domain.

A simple sketch of a methodology to develop skill-based systems would be:

– Select sensor type and range based on expert input
– Develop partial model based on expert terminology

- Create agents that emulate expert judgements
- Refine model using machine learning techniques
- Evaluate model with expert

6 Conclusion and Further Research

In this paper we have sketched our experiences creating an integrated system for steering a sailing yacht. The value of such practical projects can hardly be overestimated. Building real life systems is 80% engineering and 20% science. One of the insights we developed is the notion of the existence of a special class of skill-based systems. Issues in constructing these systems are: the need for a hybrid architecture, the interplay between discursive rules (expert system, rule induction)and senso-motoric skills (pid-controllers, neural networks), a learning approach, agent technology, the importance of semantics and symbol grounding and the importance of jargon. The nature of skill-based systems raises interesting philosophical issues concerning the nature of rules and their verbal representations.

In the near future we intend to develop more advanced systems. The current autopilot is optimized to sail as fast as possible from A to B. A next generation would also address tactical and strategic tasks, tactical: win the race (modelling your opponents), strategic: bring the crew safely to the other side of the ocean. Other interesting ambitions are: the construction of better autopilots for multi-hulls, the design an ultra-safe autonomous cruising yacht, establish an official speed record for autonomous sailing yachts and deploy the Robosail technology in other areas like the Automotive industry and aviation industry.

References

1. Frank Bethwaite. *High Performance Sailing*. Thomas Reed Publications, 1996.
2. Xia Chris Harris, Hong and Qiang Gan. *Adaptive Modelling, Estimation and Fusion from Data*. Springer, 2002.
3. Ross Garret. *The Symmetry of sailing*. Sheridan House, 1987.
4. Edwin Hutchins. *Cognition in the wild*. MIT Press, 1995.
5. N.K. Poulsen M. Nørgaard, O. Ravn and L.K. Hansen. *Neural Networks for Modelling and Control of Dynamic Systems*. Springer, 2000.
6. C.A. Marchaj. *Sail Performance, theory and practice*. Adlard Coles Nautical, 1996.
7. Tano Shun'Ichi Takeshi Furuhashi and Hans-Arno Jacobson. *Deep Fusion of Computational and Symbolic Processing*. Physica Verlag, 2001.
8. Martijn van Aartrijk, Claudio Tagliola, and Pieter Adriaans. Ai on the ocean: The robosail project. In Frank van Harmelen (Editor), editor, *Proceedings of the 15th European Conference on Artificial Intelligence*, pages 653–657. IOS Press, 2002.
9. Cornelis van Rietschoten and Barry Pickthall. *Blue Water Racing*. Dodd, Mead & Company, 1985.

Two-Eyed Algorithms and Problems

Leo Breiman

Department of Statistics, University of California, Berkeley
leo@stat.Berkeley.Edu

Two-eyed algorithms are complex prediction algorithms that give accurate predictions and also give important insights into the structure of the data the algorithm is processing. The main example I discuss is RF/tools, a collection of algorithms for classification, regression and multiple dependent outputs. The last algorithm is a preliminary version and further progress depends on solving some fascinating questions of the characterization of dependency between variables.

An important and intriguing aspect of the classification version of RF/tools is that it can be used to analyze unsupervised data–that is, data without class labels. This conversion leads to such by-products as clustering, outlier detection, and replacement of missing data for unsupervised data.

The talk will present numerous results on real data sets. The code (f77) and ample documentation for RFtools is available on the web site
www.stat.berkeley.edu/RFtools.

References

1. Leo Breiman. Random forests. *Machine Learning*, 45(1):5–32, 2001.

N. Lavrač et al. (Eds.): PKDD 2003, LNAI 2838, p. 9, 2003.
© Springer-Verlag Berlin Heidelberg 2003

Next Generation Data Mining Tools: Power Laws and Self-similarity for Graphs, Streams and Traditional Data

Christos Faloutsos

School of Computer Science, Carnegie Mellon University, Pittsburgh, PA
christos@cs.cmu.edu

Abstract. What patterns can we find in a bursty web traffic? On the web or internet graph itself? How about the distributions of galaxies in the sky, or the distribution of a company's customers in geographical space? How long should we expect a nearest-neighbor search to take, when there are 100 attributes per patient or customer record? The traditional assumptions (uniformity, independence, Poisson arrivals, Gaussian distributions), often fail miserably. Should we give up trying to find patterns in such settings?

Self-similarity, fractals and power laws are extremely successful in describing real datasets (coast-lines, rivers basins, stock-prices, brain-surfaces, communication-line noise, to name a few). We show some old and new successes, involving modeling of graph topologies (internet, web and social networks); modeling galaxy and video data; dimensionality reduction; and more.

Introduction – Problem Definition

The goal of data mining is to find patterns; we typically look for the Gaussian patterns that appear often in practice and on which we have all been trained so well. However, here we show that these time-honored concepts (Gaussian, Poisson, uniformity, independence), often fail to model real distributions well. Further more, we show how to fill the gap with the lesser-known, but even more powerful tools of self-similarity and power laws.

We focus on the following applications:

- Given a cloud of points, what patterns can we find in it?
- Given a time sequence, what patterns can we find? How to characterize and anticipate its bursts?
- Given a graph (e.g., social, or computer network), how does it look like? Which is the most important node? Which nodes should we immunize first, to guard against biological or computer viruses?

All three settings appear extremely often, with vital applications. Clouds of points appear in traditional relational databases, where records with k-attributes become points in k-d spaces; e.g. a relation with patient data (age, blood pressure, etc.); in geographical information systems (GIS), where points can be, e.g.,

N. Lavrač et al. (Eds.): PKDD 2003, LNAI 2838, pp. 10–15, 2003.

cities on a two-dimensional map; in medical image databases with, for example, three-dimensional brain scans, where we want to find patterns in the brain activation [ACF+93]; in multimedia databases, where objects can be represented as points in feature space [FRM94]. In all these settings, the distribution of k-d points is seldom (if ever) uniform [Chr84], [FK94]. Thus, it is important to characterize the deviation from uniformity in a succinct way (e.g. as a sum of Gaussians, or something even more suitable). Such a description is vital for data mining [AIS93],[AS94], for hypothesis testing and rule discovery. A succinct description of a k-d point-set could help reject quickly some false hypotheses, or could help provide hints about hidden rules.

A second, very popular class of applications is time sequences. Time sequences appear extremely often, with a huge literature on linear [BJR94], and non-linear forecasting [CE92], and the recent surge of interest on sensor data [OJW03] [PBF03] [GGR02]

Finally, graphs, networks and their surprising regularities/laws have been attracting significant interest recently. The applications are diverse, and the discoveries are striking. The World Wide Web is probably the most impressive graph, which motivated significant discoveries: the famous Kleinberg algorithm [Kle99] and its closely related PageRank algorithm of Google fame [BP98]; the fact that it obeys a "bow-tie" structure [BKM+00], while still having a surprising small diameter [AJB99]. Similar startling discoveries have been made in parallel for power laws in the Internet topology [FFF99], for Peer-to-Peer (gnutella/Kazaa) overlay graphs [RFI02], and for who-trusts-whom in the epinions.com network [RD02]. Finding patterns, laws and regularities in large real networks has numerous applications, exactly because graphs are so general and ubiquitous: Link analysis, for criminology and law enforcement [CSH+03]; analysis of virus propagation patterns, on both social/e-mail as well as physical-contact networks [WKE00]; networks of regulatory genes; networks of interacting proteins [Bar02]; food webs, to help us understand the importance of an endangered species.

We show that the theory of fractals provide powerful tools to solve the above problems.

Definitions

Intuitively, a set of points is a fractal if it exhibits self-similarity over all scales. This is illustrated by an example: Figure 1(a) shows the first few steps in constructing the so-called *Sierpinski triangle*. Figure 1(b) gives 5,000 points that belong to this triangle. Theoretically, the Sierpinski triangle is derived from an equilateral triangle ABC by excluding its middle (triangle A'B'C') and by recursively repeating this procedure for each of the resulting smaller triangles. The resulting set of points exhibits 'holes' in any scale; moreover, each smaller triangle is a *miniature replica* of the whole triangle. In general, the characteristic of fractals is this *self-similarity* property: parts of the fractal are similar (exactly or statistically) to the whole fractal. For our experiments we use 5,000 sample points from the Sierpinski triangle, using Barnsley's algorithm of Iterated Function Systems [BS88] to generated these points quickly.

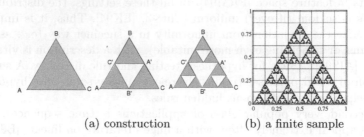

(a) construction   (b) a finite sample

Fig. 1. Theoretical fractals: the Sierpinski triangle (a) the first 3 steps of its recursive construction (b) a finite sample of it (5K points)

Notice that the resulting point set is neither a 1-dimensional Euclidean object (it has infinite length), nor 2-dimensional (it has zero area). The solution is to consider *fractional* dimensionalities, which are called *fractal dimensions*. Among the many definitions, we describe the *correlation* fractal dimension, D, because it is the easiest to describe and to use.

Let $nb(\epsilon)$ be the average number of neighbors of an arbitrary point, within distance ϵ or less. For a real, finite cloud of E-dimensional points, we follow [Sch91] and say that this data set is *self-similar in the range of scales* r_1, r_2 if

$$nb(\epsilon) \propto \epsilon^D \qquad r_1 \leq \epsilon \leq r_2 \qquad (1)$$

The *correlation integral* is defined as the plot of $nb(\epsilon)$ versus ϵ in log-log scales; for self-similar datasets, it is linear with slope D.

Notice that the above definition of fractal dimension D encompasses the traditional Euclidean objects: lines, line segments, circles, and all the standard curves have $D=1$; planes, disks and standard surfaces have $D=2$; Euclidean volumes in E-dimensional space have $D = E$.

Discussion – How Frequent Are Self-similar Datasets?

The reader might be wondering whether any real datasets behave like fractals, with linear correlation integrals. *Numerous* the real datasets give linear correlation integrals, including longitude-latitude coordinates of stars in the sky, population-versus-area of the countries of the world [FK94]; several geographic datasets [BF95] [FK94]; medical datasets [FG96]; automobile-part shape datasets [BBB+97,BBKK97].

There is overwhelming evidence from multiple disciplines that fractal datasets appear *surprisingly* often [Man77](p. 447),[Sch91]:

- coast lines and country borders ($D \approx 1.1$ - 1.3);
- the periphery of clouds and rainfall patches ($D \approx 1.35$)[Sch91](p.231);
- the distribution of galaxies in the universe ($D \approx 1.23$);
- stock prices and random walks (D=1.5)
- the brain surface of mammals ($D \approx 2.7$);
- the human vascular system ($D = 3$, because it has to reach every cell in the body!)
- even traditional Euclidean objects have linear box-counting plots, with integer slopes

Discussion – Power Laws

Self-similarity and power laws are closely related. A *power law* is a law of the form

$$y = f(x) = x^a \qquad (2)$$

Power laws are the only laws that have no characteristic scales, in the sense that they remain power laws, even if we change the scale: $f(c * x) = c^a * x^a$

Exactly for this reason, power laws and self-similarity appear often together: if a cloud of points is self similar, it has no characteristic scales; any law/pattern it obeys, should also have no characteristic scale, and it should thus be a power law.

Power laws also appear extremely often, in diverse settings: in text, with the famous Zipf law [Zip49]; in distributions of income (the Pareto law); in scientific citation analysis (Lotka law); in distribution of areas of lakes, islands and animal habitats (Korcak's law [Sch91,HS93,PF01]) in earthquake analysis (Gutenberg-Richter law [Bak96]; in LAN traffic [LTWW94]; in web click-streams [MF01]; and countless more settings.

Conclusions

Self-similarity and power laws can solve data mining problems that traditional methods can not. The two major tools that we cover in the talk are: (a) the "correlation integral" [Sch91] for a set of points and (b) the "rank-frequency" plot [Zip49] for categorical data. The former can estimate the intrinsic dimensionality of a cloud of points, and it can help with dimensionality reduction [TTWF00], axis scaling [WF02], and separability [TTPF01]. The rank-frequency plot can spot power laws, like the Zipf's law, and many more.

References

ACF' 93.   Manish Arya, William Cody, Christos Faloutsos, Joel Richardson, and Arthur Toga. QBISM: A prototype 3-D medical image database system. *IEEE Data Engineering Bulletin*, 16(1):38–42, March 1993.

AIS93.   Rakesh Agrawal, Tomasz Imielinski, and Arun Swami. Mining association rules between sets of items in large databases. In *Proc. ACM SIGMOD*, pages 207–216, Washington, DC, May 26-28 1993.

AJB99.   R. Albert, H. Jeong, and A.-L. Barabasi. Diameter of the world-wide web. *Nature*, 401:130–131, September 1999.

AS94.   Rakesh Agrawal and Ramakrishnan Srikant. Fast algorithms for mining association rules in large databases. In *Proc. of VLDB Conf.*, pages 487–499, Santiago, Chile, Sept. 12-15 1994.

Bak96.   Per Bak. How nature works : The science of self-organized criticality, September 1996.

Bar02.   Albert-Laszlo Barabasi. *Linked: The New Science of Networks*. Perseus Publishing, first edition, May 2002.

BBB'97.   Stefan Berchtold, Christian Boehm, Bernhard Braunmueller, Daniel A. Keim, and Hans-Peter Kriegel. Fast similarity search in multimedia databases. In *SIGMOD Conference*, pages 1–12, 1997.

BBKK97.   Stefan Berchtold, Christian Boehm, Daniel A. Keim, and Hans-Peter Kriegel. A cost model for nearest neighbor search in high-dimensional data space. *PODS*, pages 78–86, 1997.

BF95.   Alberto Belussi and Christos Faloutsos. Estimating the selectivity of spatial queries using the 'correlation' fractal dimension. In *Proc. of VLDB*, pages 299–310, Zurich, Switzerland, September 1995.

BJR94.   George E.P. Box, Gwilym M. Jenkins, and Gregory C. Reinsel. *Time Series Analysis: Forecasting and Control.* Prentice Hall, Englewood Cliffs, NJ, 3rd edition, 1994.

BKM'00.   Andrei Broder, Ravi Kumar, Farzin Maghoul1, Prabhakar Raghavan, Sridhar Rajagopalan, Raymie Stata, Andrew Tomkins, and Janet Wiener. Graph structure in the web: experiments and models. In *WWW Conf.*, 2000.

BP98.   Sergey Brin and Lawrence Page. The anatomy of a large-scale hypertextual (web) search engine. *Computer Networks and ISDN Systems*, 30(1–7):107–117, 1998.

BS88.   M.F. Barnsley and A.D. Sloan. A better way to compress images. *Byte*, pages 215–223, January 1988.

CE92.   M. Castagli and S. Eubank. *Nonlinear Modeling and Forecasting.* Addison Wesley, 1992. Proc. Vol. XII.

Chr84.   S. Christodoulakis. Implication of certain assumptions in data base performance evaluation. *ACM TODS*, June 1984.

CSH'03.   H. Chen, J. Schroeder, R. Hauck, L. Ridgeway, H. Atabaksh, H. Gupta, C. Boarman, K. Rasmussen, and A. Clements. Coplink connect: Information and knowledge management for law enforcement. *CACM*, 46(1):28–34, January 2003.

FFF99.   Michalis Faloutsos, Petros Faloutsos, and Christos Faloutsos. On power-law relationships of the internet topology. In *SIGCOMM*, pages 251–262, 1999.

FG96.   Christos Faloutsos and Volker Gaede. Analysis of the z-ordering method using the hausdorff fractal dimension. *VLDB*, September 1996.

FK94.   Christos Faloutsos and Ibrahim Kamel. Beyond uniformity and independence: Analysis of R-trees using the concept of fractal dimension. In *Proc. ACM SIGACT-SIGMOD-SIGART PODS*, pages 4–13, Minneapolis, MN, May 24-26 1994. Also available as CS-TR-3198, UMIACS-TR-93-130.

FRM94.   Christos Faloutsos, M. Ranganathan, and Yannis Manolopoulos. Fast subsequence matching in time-series databases. In *Proc. ACM SIGMOD*, pages 419–429, Minneapolis, MN, May 25-27 1994. 'Best Paper' award; also available as CS-TR-3190, UMIACS-TR-93-131, ISR TR-93-86.

GGR02.   Minos N. Garofalakis, Johannes Gehrke, and Rajeev Rastogi. Querying
           and mining data streams: You only get one look. *ACM SIGMOD*, page
           635, June 2002. (tutorial).

HS93.   Harold M. Hastings and George Sugihara. *Fractals: A User's Guide for the
           Natural Sciences*. Oxford University Press, 1993.

Kle99.   Jon M. Kleinberg. Authoritative sources in a hyperlinked environment.
           Journal of the ACM, 46(5):604–632, 1999.

LTWW94.   W.E. Leland, M.S. Taqqu, W. Willinger, and D.V. Wilson. On the self-
           similar nature of ethernet traffic. *IEEE Transactions on Networking*,
           2(1):1–15, February 1994. (earlier version in SIGCOMM '93, pp 183-193).

Man77.   B. Mandelbrot. *Fractal Geometry of Nature*. W.H. Freeman, New York,
           1977.

MF01.   Alan L. Montgomery and Christos Faloutsos. Identifying web browsing
           trends and patterns. *IEEE Computer*, 34(7):94–95, July 2001.

OJW03.   C. Olston, J. Jiang, and J. Widom. Adaptive filters for continuous queries
           over distributed data streams. *ACM SIGMOD*, 2003.

PBF03.   Spiros Papadimitriou, Anthony Brockwell, and Christos Faloutsos. Adap-
           tive, hands-off stream mining. *VLDB*, September 2003.

PF01.   Guido Proietti and Christos Faloutsos. Accurate modeling of region data.
           IEEE TKDE, 13(6):874–883, November 2001.

RD02.   M. Richardson and P. Domingos. Mining knowledge-sharing sites for viral
           marketing. In *SIGKDD*, pages 61–70, Edmonton, Canada, 2002.

RFI02.   M. Ripeanu, I. Foster, and A. Iamnitchi. Mapping the gnutella network:
           Properties of large-scale peer-to-peer systems and implications for system
           design. *IEEE Internet Computing Journal*, 6(1), 2002.

Sch91.   Manfred Schroeder. *Fractals, Chaos, Power Laws: Minutes from an Infinite
           Paradise*. W.H. Freeman and Company, New York, 1991.

TTPF01.   Agma Traina, Caetano Traina, Spiros Papadimitriou, and Christos Falout-
           sos. Tri-plots: Scalable tools for multidimensional data mining. *KDD*,
           August 2001.

TTWF00.   Caetano Traina, Agma Traina, Leejay Wu, and Christos Faloutsos. Fast
           feature selection using the fractal dimension,. In *XV Brazilian Symposium
           on Databases (SBBD)*, Paraiba, Brazil, October 2000.

WF02.   Leejay Wu and Christos Faloutsos. Making every bit count: Fast nonlinear
           axis scaling. *KDD*, July 2002.

WKE00.   Chenxi Wang, J. C. Knight, and M. C. Elder. On computer viral infection
           and the effect of immunization. In *ACSAC*, pages 246–256, 2000.

Zip49.   G.K. Zipf. *Human Behavior and Principle of Least Effort: An Introduction
           to Human Ecology*. Addison Wesley, Cambridge, Massachusetts, 1949.

Taking Causality Seriously:
Propensity Score Methodology Applied to
Estimate the Effects of Marketing Interventions

Donald B. Rubin

Department of Statistics, Harvard University, Cambridge MA
rubin@stat.harvard.edu

Propensity score methods were proposed by Rosenbaum and Rubin (1983, Biometrika) as central tools to help assess the causal effects of interventions. Since their introduction two decades ago, they have found wide application in a variety of areas, including medical research, economics, epidemiology, and education, especially in those situations where randomized experiments are either difficult to perform, or raise ethical questions, or would require extensive delays before answers could be obtained. Rubin (1997, Annals of Internal Medicine) provides an introduction to some of the essential ideas. In the past few years, the number of published applications using propensity score methods to evaluate medical and epidemiological interventions has increased dramatically. Rubin (2003, Erlbaum) provides a summary, which is already out of date.

Nevertheless, thus far, there have been few applications of propensity score methods to evaluate marketing interventions (e.g., advertising, promotions), where the tradition is to use generallly inappropriate techniques, which focus on the prediction of an outcome from an indicator for the intervention and background characteristics (such as least-squares regression, data mining, etc.). With these techniques, an estimated parameter in the model is used to estimate some global "causal" effect. This practice can generate grossly incorrect answers that can be self-perpetuating: polishing the Ferrari rather than the Jeeps "causes" them to continue to win more races than the Jeeps ¡=¿ visiting the high-prescribing doctors rather than the low-prescribing doctors "causes" them to continue to write more prescriptions.

This presentation will take "causality" seriously, not just as a casual concept implying some predictive association in a data set, and will show why propensity score methods are superior in practice to the standard predictive approaches for estimating causal effects. The results of our approach are estimates of individual-level causal effects, which can be used as building blocks for more complex components, such as response curves. We will also show how the standard predictive approaches can have important supplemental roles to play, both for refining estimates of individual-level causal effect estimates and for assessing how these causal effects might vary as a function of background information, both important uses for situations when targeting an audience and/or allocating resources are critical objectives.

The first step in a propensity score analysis is to estimate the individual scores, and there are various ways to do this in practice, the most common

N. Lavrač et al. (Eds.): PKDD 2003, LNAI 2838, pp. 16–22, 2003.
© Springer-Verlag Berlin Heidelberg 2003

being logisitic regression. However, other techniques, such as probit regression or discriminant analysis are also possible, as are the robust methods based on the t-family of long tailed distributions. Other possible methods include highly non-linear methods such as CART or neural nets. A critical feature of estimating propensity scores is that diagnosing the adequacy of the resulting fit is very straightforward, and in fact guides what the next steps in a full propensity score analysis should be. This diagnosing takes place without access to the outcome variables (e.g., sales, number of prescriptions) so that that objectivity of the analysis is maintained. In some cases, the conclusion of the diagnostic phase must be that inferring causality from the data set at hand is impossible without relying on heroic and implausible assumptions, and this can be very valuable information, information that is not directly available from traditional approaches.

Marketing applications from the practice of AnaBus, Inc. will also be presented. AnaBus currently has a Small Business Innovative Research Grant from the US NIH to implement essential software to allow the implementation of the full propensity score approach to estimating the effects of interventions. Other examples will also be presented if time permits, for instance, an application from the current litigation in the US on the effects of cigarette smoking (Rubin, 2002, Health Services Outcomes Research).

An extensive reference list from the author is included. These references are divided into five categories. First, general articles on inference for causal effects not having a focus on matching or propensity scores. Second, articles that focus on matching methods before the formulation of propensity score methods – some of these would now be characterized as examples of propensity score matching. Third, articles that address propensity score methods explicitly, either theoretically or through applications. Fourth, articles that document, by analysis and/or by simlulation, the superiority of propensity-based methods, especially when used in combination with model-based adjustments, over model-based methods alone. And fifth, introductions and reviews of propensity score methods. The easiest place for a reader to start is with the last collection of articles.

Such a reference list is obviously very idiosyncratic and is not meant to imply that only the author has done good work in this area. Paul Rosenbaum, for example, has been an extremely active and creative contributor for many years, and his text book "Observational Studies" is truly excellent. As another example, Rajeev Deheija and Sadek Wahba's 1999 article in the Journal of the American Statistical Association had been very influential, especially in economics.

References

General Causal Inference Papers

(1974). "Estimating Causal Effects of Treatments in Randomized and Nonrandomized Studies." Journal of Educational Psychology, 66, 5, pp. 688-701.

(1977). "Assignment to Treatment Group on the Basis of a Covariate." Journal of Educational Statistics, 2, 1, pp. 1-26. Printer's correction note 3, p. 384.

(1977). "Assignment to Treatment Group on the Basis of a Covariate." Journal of Educational Statistics, 2, 1, pp. 1-26. Printer's correction note 3, p. 384.

(1978). "Bayesian Inference for Causal Effects: The Role of Randomization." The Annals of Statistics, 7, 1, pp. 34-58.

(1983). "Assessing Sensitivity to an Unobserved Binary Covariate in an Observational Study with Binary Outcome." The Journal of the Royal Statistical Society, Series B, 45, 2, pp. 212-218. (With P.R. Rosenbaum).

(1983). "On Lord's Paradox." Principles of Modern Psychological Measurement: A Festschrift for Frederick Lord, Wainer and Messick (eds.). Erlbaum, pp. 3-25. (With P.W. Holland).

(1984). "Estimating the Effects Caused by Treatments." Discussion of "On the Nature and Discovery of Structure" by Pratt and Schlaifer. Journal of the American Statistical Association, 79, pp. 26-28. (With P.R. Rosenbaum).

(1984). "William G. Cochran's Contributions to the Design, Analysis, and Evaluation of Observational Studies." W.G. Cochran's Impact on Statistics, Rao and Sedransk (eds.). New York: Wiley, pp. 37-69.

(1986). "Which Ifs Have Causal Answers?" Discussion of Holland's "Statistics and Causal Inference." Journal of the American Statistical Association, 81, pp. 961-962.

(1988). "Causal Inference in Retrospective Studies." Evaluation Review, pp. 203-231. (With P.W. Holland).

(1990). "Formal Modes of Statistical Inference for Causal Effects." Journal of Statistical Planning and Inference, 25, pp. 279-292.

(1990). "Neyman (1923) and Causal Inference in Experiments and Observational Studies." Statistical Science, 5, 4, pp. 472-480.

(1991). "Dose-Response Estimands: A Comment on Efron and Feldman." Journal of the American Statistical Association, 86, 413, pp. 22-24.

(1994). "Intention-to-Treat Analysis and the Goals of Clinical Trials." Clinical Pharmacology and Therapeutics, 87, 1, pp. 6-15. (With L.B. Sheiner).

(1996). "Identification of Causal Effects Using Instrumental Variables." Journal of the American Statistical Association, 91, 434, as Applications Invited Discussion Article with discussion and rejoinder, pp. 444-472. (With J.D. Angrist and G.W. Imbens).

(1997). "Bayesian Inference for Causal Effects in Randomized Experiments with Noncompliance." The Annals of Statistics, 25, 1, pp. 305-327. (With G. Imbens).

(1997). "Estimating Outcome Distributions for Compliers in Instrumental Variables Models." Review of Economic Studies, 64, pp. 555-574. (With G. Imbens).

(1998). "More Powerful Randomization-Based p-values in Double-Blind Trials with Noncompliance." Statistics in Medicine, 17, pp. 371-385, with discussion by D.R. Cox, pp. 387-389.

(1999). "Addressing Complications of Intention-To-Treat Analysis in the Combined Presence of All-or-None Treatment-Noncompliance and Subsequent Missing Outcomes." Biometrika, 86, 2, pp. 366-379. (With C. Frangakis).

(1999). "Causal Inquiry in Longitudinal Observational Studies," Discussion of 'Estimation of the Causal Effect of a Time-varying Exposure on the Marginal Mean of a Repeated Binary Outcome' by J. Robins, S. Greenland and F-C. Hu. Journal of the American Statistical Association, 94, 447, pp. 702-703. (With C.E. Frangakis).

(1999). "Teaching Causal Inference in Experiments and Observational Studies." Proceedings of the Section on Statistical Education of the American Statistical Association, pp. 126-131.

(2000). "Statistical Issues in the Estimation of the Causal Effects of Smoking Due to the Conduct of the Tobacco Industry." Chapter 16 in Statistical Science in the Courtroom, J. Gastwirth (ed.). New York: Springer-Verlag, pp. 321-351.

(2000). "The Utility of Counterfactuals for Causal Inference." Comment on A.P. Dawid, "Causal Inference Without Counterfactuals". Journal of the American Statistical Association, 95, 450, pp. 435-438.

(2000). "Causal Inference in Clinical and Epidemiological Studies via Potential Outcomes: Concepts and Analytic Approaches." Annual Review of Public Health, 21, pp. 121-145. (With R.J.A. Little).

(2000). "Statistical Inference for Causal Effects in Epidemiological Studies Via Potential Outcomes." Proceedings of the XL Scientific Meeting of the Italian Statistical Society, Florence, Italy, April 26-28, 2000, pp. 419-430.

(2001). "Estimating The Causal Effects of Smoking." Statistics in Medicine, 20, pp. 1395-1414.

(2001). "Self-Experimentation for Causal Effects." Comment on 'Surprises From Self-Experimentation: Sleep, Mood, and Weight', by S. Roberts. Chance, 14, 2, pp. 16-17.

(2001). "Estimating the Effect of Unearned Income on Labor Supply, Earnings, Savings and Consumption: Evidence from a Survey of Lottery Players." American Economic Review, 19, pp. 778-794. (With G.W. Imbens and B. Sacerdote).

(2002). "Statistical Assumptions in the Estimation of the Causal Effects of Smoking Due to the Conduct of the Tobacco Industry." [CD-ROM] In Social Science Methodology in the New Millennium. Proceedings of the Fifth International Conference on Logic and Methodology (J. Blasius, J. Hox, E. de Leeuw and P. Schmidt, eds.), October 6, 2000, Cologne, Germany. Opladen, FRG: Leske + Budrich. P023003.

(2002). "School Choice in NY City: A Bayesian Analysis of an Imperfect Randomized Experiment", with discussion and rejoinder. Case Studies in Bayesian Statistics, Vol. V. New York: Springer-Verlag. C. Gatsonis, B. Carlin and A. Carriquiry (eds.), pp. 3-97. (With Barnard, J., Frangakis, C. and Hill, J.)

(2002). "Principal Stratification in Causal Inference." Biometrics, 58, 1. pp. 21-29. (With C. Frangakis).

(2002). "Clustered Encouragement Designs with Individual Noncompliance: Bayesian Inference with Randomization, and Application to Advance Directive Forms." With discussion and rejoinder, Biostatistics, 3, 2, pp. 147-177. (With C.E. Frangakis and X.-H. Zhou.)

(2002). "Discussion of 'Estimation of Intervention Effects with Noncompliance: Alternative Model Specification,' by Booil Jo". Journal of Educational and Behavioral Statistics, 27, 4, pp. 411-415. (With F. Mealli.)

(2003). "Assumptions Allowing the Estimation of Direct Causal Effects: Discussion of 'Healthy, Wealthy, and Wise? Tests for Direct Causal Paths Between Health and Socioeconomic Status' by Adams et al.'". Journal of Econometrics, 112, pp. 79-87. (With F. Mealli.)

(2003). "A Principal Stratification Approach to Broken Randomized Experiments: A Case Study of Vouchers in New York City." Journal of the American Statistical Association, 98, 462, with discussion and rejoinder. (With J. Barnard, C. Frangakis, and J. Hill.)

(2003). "Assumptions When Analyzing Randomized Experiments with Noncompliance and Missing Outcomes." To appear in Health Services Outcome Research Methodology. (With F. Mealli.)

(2003). "Hypothesis: A Single Clinical Trial Plus Causal Evidence of Effectiveness is Sufficient for Drug Approval." Clinical Pharmacology and Therapeutics, 73, pp. 481-490. (With C. Peck and L.B. Sheiner.)

(2003). "Teaching Statistical Inference for Causal Effects in Experiments and Observational Studies." To appear in The Journal of Educational and Behavioral Statistics.

Matching Methods, Pre-propensity Score Paper

(1973). "Matching to Remove Bias in Observational Studies." Biometrics, 29, 1, pp. 159-183. Printer's correction note 30, p. 728.

(1976). "Multivariate Matching Methods that are Equal Percent Bias Reducing, I: Some Examples." Biometrics, 32, 1, pp. 109-120. Printer's correction note p. 955.

(1976). "Multivariate Matching Methods that are Equal Percent Bias Reducing, II: Maximums on Bias Reduction for Fixed Sample Sizes." Biometrics, 32, 1, pp. 121-132. Printer's correction note p. 955.

(1980). "Bias Reduction Using Mahalanobis' Metric Matching." Biometrics, 36, 2, pp. 295-298. Printer's Correction p. 296 ((5,10) = 75%).

Propensity Score Techniques and Applications

(1983). "The Central Role of the Propensity Score in Observational Studies for Causal Effects." Biometrika, 70, pp. 41-55. (With P. Rosenbaum).

(1984). "Reducing Bias in Observational Studies Using Subclassification on the Propensity Score." Journal of the American Statistical Association, 79, pp. 516-524. (with P.R. Rosenbaum).

(1985). "The Use of Propensity Scores in Applied Bayesian Inference." Bayesian Statistics, 2, Bernardo, DeGroot, Lindley and Smith (eds.). North Holland, pp. 463-472.

(1985). "Constructing a Control Group Using Multivariate Matched Sampling Incorporating the Propensity Score." The American Statistician, 39, pp. 33-38. (With P.R. Rosenbaum).

(1985). "The Bias Due to Incomplete Matching." Biometrics, 41, pp. 103-116. (With P.R. Rosenbaum).

(1992). "Projecting from Advance Data Using Propensity Modelling". The Journal of Business and Economics Statistics, 10, 2, pp. 117-131. (With J.C. Czajka, S.M., Hirabayashi, and R.J.A. Little).

(1992). "Affinely Invariant Matching Methods with Ellipsoidal Distributions." The Annals of Statistics, 20, 2, pp. 1079-93. (With N. Thomas).

(1992). "Characterizing the Effect of Matching Using Linear Propensity Score Methods with Normal Covariates." Biometrika, 79, 4, pp. 797-809. (With N. Thomas).

(1995). "In Utero Exposure to Phenobarbital and Intelligence Deficits in Adult Men." The Journal of the American Medical Association, 274, 19, pp. 1518-1525. (With J. Reinisch, S. Sanders and E. Mortensen.)

(1996). "Matching Using Estimated Propensity Scores: Relating Theory to Practice." Biometrics, 52, pp. 249-264. (With N. Thomas).

(1999). "On Estimating the Causal Effects of Do Not Resuscitate Orders." Medical Care, 37, 8, pp. 722-726. (With M. McIntosh).

(1999). "The Design of the New York School Choice Scholarship Program Evaluation". Research Designs: Inspired by the Work of Donald Campbell, L. Bickman (ed.). Thousand Oaks, CA: Sage. Chapter 7, pp. 155-180. (With J. Hill and N. Thomas).

(2000). "Estimation and Use of Propensity Scores with Incomplete Data." Journal of the American Statistical Association, 95, 451, pp. 749-759. (With R. D'Agostino, Jr.).

(2002). "Using Propensity Scores to Help Design Observational Studies: Application to the Tobacco Litigation." Health Services & Outcomes Research Methodology, 2, pp. 169-188, 2001.

Matching & Regression Better than Regression Alone

(1973). "The Use of Matched Sampling and Regression Adjustment to Remove Bias in Observational Studies." Biometrics, 29, 1, pp. 184-203.

(1973). "Controlling Bias in Observational Studies: A Review." Sankhya - A, 35, 4, pp. 417-446. (With W.G. Cochran).

(1979). "Using Multivariate Matched Sampling and Regression Adjustment to Control Bias in Observational Studies." The Journal of the American Statistical Association, 74, 366, pp. 318-328.

(2000). "Combining Propensity Score Matching with Additional Adjustments for Prognostic Covariates." Journal of the American Statistical Association, 95, 450, pp. 573-585. (With N. Thomas.)

Propensity Score Reviews

(1997). "Estimating Causal Effects From Large Data Sets Using Propensity Scores." Annals of Internal Medicine, 127, 8(II), pp. 757-763.

(1998). "Estimation from Nonrandomized Treatment Comparisons Using Subclassification on Propensity Scores." Nonrandomized Comparative Clinical Studies, U. Abel and A. Koch (eds.) Dusseldorf: Symposion Publishing, pp. 85-100.

(2003). "Estimating Treatment Effects From Nonrandomized Studies Using Subclassification on Propensity Scores." To appear in Festschrift for Ralph Rosnow. Erlbaum publishers.

Efficient Statistical Pruning of Association Rules

Alan Ableson[1] and Janice Glasgow[2]

Department of Math and Stats
ableson@mast.queensu.ca
School of Computing
janice@cs.queensu.ca
Queen's University
Kingston, Ontario, Canada

Abstract. Association mining is the comprehensive identification of frequent patterns in discrete tabular data. The result of association mining can be a listing of hundreds to millions of patterns, of which few are likely of interest. In this paper we present a probabilistic metric to filter association rules that can help highlight the important structure in the data. The proposed filtering technique can be combined with maximal association mining algorithms or heuristic association mining algorithms to more efficiently search for interesting association rules with lower support.

1 Introduction

Association mining is the process of identifying frequent patterns in a tabular dataset, usually requiring some *minimum support*, or frequency of the pattern in the data [2]. The discovery of frequent patterns in the data is usually followed by the construction of *association rules*, which portray the patterns as predictive relationships between particular attribute values. Unfortunately, since association mining is an exhaustive approach, it is possible to generate many more patterns than a user can reasonably evaluate. Furthermore, many of these patterns may be redundant. Thus, is it important to develop informed and efficient pruning systems for association mining rules.

A number of methods to filter association mining results have been published, and they can be classified along several lines. First, the filtering can be *objective* or *subjective* [11]. Our goal is to design an objective, or purely computational, filter, both for inter-discipline generality and to avoid the bias introduced by subjective evaluation. In a separate categorization, rule filtering can be done on a rule-by-rule basis [7], or in an incremental manner where rules deemed interesting are gradually added to a rule list set [6] or probability model [12]. We focus our attention on the rule-by-rule approach. This approach is appropriate in application areas where dense data tables lead to the generation of millions of association rules, a set too large to use directly in incremental filtering algorithms. As an added advantage, filtering rules independently allows the straightforward use of batch parallelization of the filtering to produce linear speed-ups.

N. Lavrač et al. (Eds.): PKDD 2003, LNAI 2838, pp. 23–34, 2003.
© Springer-Verlag Berlin Heidelberg 2003

2 Background

It has long been understood in statistics that while not every significant feature of a dataset is interesting, an interesting feature *must* be statistically significant. Non-significant results, by definition, are those features that can be explained as a random effect, and therefore not worthy of further study. For example, an association rule predicting a customer action with 100% accuracy would not be significant if it only covers 2 customers in a large database, and so would not be interesting. This aspect of interestingness is sometimes referred to as the *reliability* of a rule [11].

In many association rule filtering approaches, the measure of significance or reliability has been largely ad-hoc, stemming from Boolean logical theory rather than statistical theory. Boolean approaches typically combine the *support* and *confidence* of a rule in a dual-ordering approach, trying to maximize both support and confidence, with an implied maximization of reliability [5]. Related logical mechanisms for removing redundant association rules use the concept of *closed* itemsets [8,21]. Closure-based methods are most effective in noise-free problems. In these problems, large sets of records contain exactly the same associations, allowing the pruning of redundant subsets of the items without loss of information. In noisier data, fewer rules can be considered redundant according to closure properties, limiting their effectiveness. For such noisy dataset, statistical techniques have been used by other authors, focusing either on pruning association rules [14,19] or identifying correlated attributes involved in association rules [12,17].

3 Problem Statement

In our approach to association rule filtering, we will follow the approach of Liu et al. [14] and use a statistical model to evaluate the reliability of association rules, with a focus on association rules with a single value/item in the consequent of the rule. Furthermore, we pose our problem as one of association rule mining over relational tables, rather than itemsets.

In mining over relational tables, the *input* consists of a table T with a set of N records, $R = \{r_1, \ldots, r_N\}$, and n attributes, $A = \{a_1, \ldots, a_n\}$; all of the attributes take on a discrete set of values, $Dom(a_i)$. We assume the table contains no missing values.

The *output* of an association mining exercise is a set of *patterns*, where a pattern X associates each attribute in a subset of A with a particular value,

$$X = \{< a_{x_1}, v_1 >, < a_{x_2}, v_2 >, ..., < a_{x_k}, v_k >\}$$

such that $v_i \in Dom(a_{x_i})$. A pattern that contains k attribute/value pairs is called a *k-th order* pattern. A record r_i of the table *matches* or *instantiates* a pattern X if, for each attribute a_j in X, r_i contains the value v_j for a_j.

Each of the patterns (also called *itemsets*) has a number of descriptive parameters. We define the *support* of a pattern, $Sup(X|T)$, as the cardinality of the set X_T, denoted $|X_T|$, such that $X_T = \{r_i \in T | r_i$ is an instance of pattern $X\}$. Wherever the table T is the original input table, we simply write $Sup(X)$.

Often patterns are combined to make a larger pattern, through the *composition* of two smaller patterns, X and Y, denoted $X{\circ}Y$. The composition of two patterns produces a longer pattern including the attribute/value pairs of both input patterns:

$$X{\circ}Y = \{< a_{x_1}, v_{x_1} >, ..., < a_{x_k}, v_{x_k} >, < a_{y_1}, v_{y_1} >, ..., < a_{y_m}, v_{y_m} >\}$$

To avoid degeneracies, we define composition only for patterns with non-overlapping attributes. Composition of patterns of order k_1 and k_2 results in a pattern of order $k_1 + k_2$.

An *association rule* is a pairing of two patterns, $X \to Z$, and is interpreted interpreted as a causal or correlational relationship. The *support* and *confidence* of an association rule, denoted $Sup(X \to Z)$ and $Conf(X \to Z)$ are defined in terms of support for the patterns and their compositions:

$$Sup(X \to Z) = Sup(X{\circ}Z), \quad Conf(X \to Z) = \frac{Sup(X{\circ}Z)}{Sup(X)}$$

4 Reliable Association Rules

In an association mining study, the set of patterns generated can be straightforwardly turned into a set of association rules [2]. The output of association rule generation will often be a long list of rules, $X_i \to Z_i$, with the number of rules determined by the table T, the minimum support required for any rule, *minsup*, and possibly a minimum confidence constraint, *minconf*. The two questions we want to address in our filtering, for any particular rule $X \to Z$, are

1. Is the rule $X \to Z$ reliable (statistically significant), given $Sup(X)$, $Sup(Z)$, and $Sup(X{\circ}Z)$?
2. Is the rule $X \to Z$ an unreliable extension of a lower-order rule, $Y \to Z$, where $X = Y{\circ}Q$, for some non-predictive pattern Q?

We approach both questions using a statistical sampling model. Imagine an urn filled with N balls, each ball representing one record of the input table T. For each record, its ball is colored red if it matches the consequent pattern Z, and black if does not. A sample is then taken from the urn of all those balls matching the predictive or antecendent pattern X. We then ask ourselves the question, "is the distribution of red and black balls matching the pattern X significantly different from what we would expect from a random scoop of $Sup(X)$ balls from the urn?" This question can be answered using the hypergeometric probability distribution, which exactly represents the probability of this sampling-without-replacement model. [16]

Specifically, if we have the four values $Sup(X)$, $Sup(Z)$, $Sup(\bar{Z})$, and $Sup(X{\circ}Z)$ (with shorthand $S_X = Sup(X)$, $S_{X{\circ}Z} = Sup(X{\circ}Z)$, etc. for conciseness), the hypergeometric probability distribution, H, is given by:

$$H(n = S_{X{\circ}Z} | S_Z, S_{\bar{Z}}, S_X) = \frac{C(S_Z, n)C(S_{\bar{Z}}, S_X - n)}{C(S_Z + S_{\bar{Z}}, S_X)} \tag{1}$$

where $C(N, x) = x!(N - x)!/N!$ is the number of ways to choose x records from a collection of N records.

The function H gives the probability of *exactly* $S_{X \circ Z}$ red balls being selected. To compute a significance value, we need to calculate the *cumulative probability*, or the probability of an outcome as extreme as $S_{X \circ Z}$, where either $P(n \geq S_{X \circ Z})$, or $P(n \leq S_{X \circ Z})$ through the following sums (where n is the number of red balls):

$$P(n \geq S_{X \circ Z} | S_Z, S_{\bar{Z}}, S_X) = \sum_{k=S_{X \circ Z}}^{S_X} H(k | S_Z, S_{\bar{Z}}, S_X) \tag{2}$$

$$P(n \leq S_{X \circ Z} | S_Z, S_{\bar{Z}}, S_X) = \sum_{k=0}^{S_{X \circ Z}} H(k | S_Z, S_{\bar{Z}}, S_X) \tag{3}$$

These formulae produce the *p-values* of the rule under the null hypothesis of the hypergeometric sampling model. A low *p*-value in Equation 2 indicates a higher-than-random confidence for the rule $X \to Z$, while a low *p*-value in Equation 3 indicates an interesting *low* confidence rule.

It should be noted that other authors have used *p*-value ranking approaches, using other statistical measures. The χ^2 measure has been the most popular, but studies have also used the gini index, correlation coefficient, and interest factor (see [19] for a comprehensive presentation of many of these measures). Unfortunately, all of these measures implicitly rely on large sample assumptions, and so their *p*-value estimates become less reliable as the patterns' support decreases. The hypergeometric distribution, because its explicit counting approach, is an *exact* method, applicable to patterns at all support values. We believe that the hypergeometric is a more appropriate null hypothesis distribution when dealing with association rules that include patterns with low support (less than 50 records). In the rest of the paper, we use the hypergeometric distribution exclusively, keeping in mind that other distributions could be used.

5 Filtering Algorithms

The hypergeometric probability distribution, or any other appropriate null hypothesis distribution such as χ^2, can be used to test the significance of association rules in two ways. The simplest way is to evaluate the reliability of every association rule discovered, $X \to Z$ by computing its *p*-value (using Equations 3 and 2) with the *p*-values being computed relative to the baseline distribution of the consequent pattern, Z. In particular, for any rule, $X_i \to Z_i$ found during association mining, we can define the two-tailed *p*-value of rule, p_i, as follows:

$$p_i = min \left(\begin{array}{c} P(n \geq S_{X_i \circ Z_i} | S_Z, S_{\bar{Z}}, S_X) \\ P(n \leq S_{X_i \circ Z_i} | S_Z, S_{\bar{Z}}, S_X) \end{array} \right)$$

If p_i is less than some user-defined *p*-value threshold p_{max}, we consider the rule $X_i \to Z_i$ to be interesting.

This p-value ranking is simple, and is computationally linear in the number of association rules found. It provides a straightforward statistical approach to rank and filter association rules. The disadvantages are that many similar rules will be evaluated, and, if a particular rule is significant, e.g. $X \to Z$, then adding spurious extensions to the antecedent is also likely to produce a significant rule e.g. $(X \circ Y) \to Z$, even if Y is a pattern which has no predictive relationship with Z.

To overcome the problems of spurious rule extensions, a statistical model may be used in a incremental way to gradually prune away non-interesting parts of an association rule, leaving only the interesting part. This process is laid out in Algorithm 1.

Algorithm 1 *Incremental Probabilistic Pruning* Input: *A table (T), a minimum support (minsup), and a p-value cutoff (p_{max}). Output: a set of association rules, R_{out}.*

1. *Create an empty set of association rules, $R_{out} = \phi$.*
2. *Construct a set of association rules, $I = \{X_i \to Z_i\}$ from the set of records in T, constraining the search with the minimum support, minsup.*
3. *For each association rule $X_i \to Z_i$, (called $X \to Z$ for clarity in this step),*
   (a) *Refer to T to find the supports related to the association rule, $S_{X \to Z}$,*
   (b) *Compute the two-tailed p-value of the association rule given every possible sub-rule $Y_j \to Z$, where $X = Y_j \circ Q_j$, where Q_j is a first-order pattern, and Y_j is one order less than the pattern X. Take the maximum of the p-values over the possible patterns Y_j:*

$$p_{ij} = min \left(\begin{array}{l} P(n \geq S_{X \circ Z} | S_{Y_j \circ Z}, S_{Y_j \circ \bar{Z}}, S_X) \\ P(n \leq S_{X \circ Z} | S_{Y_j \circ Z}, S_{Y_j \circ \bar{Z}}, S_X) \end{array} \right)$$

$$p_i = \overset{max}{Y_j} (p_{ij})$$

   using Equations 2 and 3 to compute the hypergeometric probabilities.
   (c) *If $p_i < p_{max}$, then none of the sub-rules $Y_{ij} \to Z_i$ can explain the rule $X_i \to Z_i$, so add the rule $X_i \to Z_i$ to the collection R_{out}.*
   (d) *If $p_i \geq p_{max}$, find the sub-rule $Y_{ij} \to Z_i$ for which the p_{ij} was maximal, and return to step 3, with $Y_{ij} \to Z_i$ instead of $X_i \to Z_i$.*

6 Implications of Pruning Algorithm for Association Mining

The idea of using subsuming patterns to define the relative interestingness of association rules was proposed earlier by Liu et. al. [14], using the χ^2 distribution rather than hypergeometric to measure rule significance. They performed their pruning following the incremental Apriori approach [1]. This involved finding and pruning all first-order rules, then second-order rules, etc. In dense data tables, as found by Liu et al., there are many association rules with relatively high support (above 20%), but few rules that pass by significance pruning as outlines in Algorithm 1. However, since the rules with high support are most likely to be those already known in the domain, it behooves us to search for interesting rules of lower support, which is hopelessly infeasible using a pure Apriori-based approach.

Fortunately, other association mining algorithms exist which do not require the explicit construction of every possible sub-pattern for an association rule. There are several algorithms that efficiently search for *maximal frequent patterns* [13,9]. A frequent pattern X is *maximal* if has no superset that is frequent: $Sup(X) \geq minsup$, and $Sup(X \circ Y) < minsup$ for any pattern Y. Two examples of algorithms that identify maximal itemsets are Max-Miner [13] and GenMax [9]. There are also heuristic association mining algorithms, such as SLAM [18] which do not use the recursive search strategy of the previous algorithms. Our hypothesis is that using the incremental pruning procedure in Algorithm 1, we should identify the roughly the same set of interesting rules whether we start from a relatively small set of maximal frequent patterns (e.g. GenMax), the complete set of frequent patterns (e.g. Apriori), or a heuristically-generated subset of patterns (e.g. SLAM).

7 Applications

There were two particular questions we wanted to answer in evaluating our association rule pruning. The first was what percentage of rules identified on different datasets could be safely be pruned away using Algorithm 1. The second question was how effective incomplete searches for frequent patterns followed by pruning would be compared to more exhaustive searches followed by pruning. Incomplete association mining algorithms can be orders of magnitude faster than exhaustive algorithms, and produce orders of magnitude fewer associations. If the pruning leads to roughly the same results using these different association mining tools, then there is a considerable computational advantage to be gained by pruning only the rules generated by the incomplete association mining algorithms.

In all the experiments, we computed p-values using the hypergeometric probability calculation described in [20]. We found the algorithm sufficiently fast for the number of records encountered, and it provided exact estimates of the probabilities for rules with small support. If faster computations are necessary, the hypergeometric distribution can be approximated by an appropriate binomial, normal or χ^2 distribution.

7.1 Description of Datasets

To evaluate the pruning technique, one simulated collection and two real-world datasets were considered.

The simulated datasets were designed to test the sensitivity and specificity of the pruning approach in Algorithm 1. Tables were created with 51 binary-valued attributes (values "0" and "1") and 1,000 records. The first 50 attributes were treated as the predictor attributes, and the 51st attribute was the *outcome attribute*, or the consequent attribute for association rules. The goal was to find patterns predictive of the positive value, "1", of this 51st attribute. The 50 predictor attributes were generated independently, using a uniform probability for each binary value. With this distribution, all predictive patterns of size n have an expected support of $1,000/2^n$.

The distribution for the outcome attribute was defined using two parameters, n_p and o_p, where n_p specified the size of the predictive pattern, and o_p the relative odds of a positive outcome if a record was an instance of the predictive pattern. For ease of identification, a pattern of size n_p was considered to be the pattern of n_p 1's on the first n_p attributes. For example, if we chose $n_p = 3$ and $o_p = 4$, then records being an instance of the pattern "1,1,1" as the values for the first three attributes would have a 4 times greater chance of having a 1 in the outcome than the other records. Records not matching the predictive pattern had a probability of $1/2$ of being an outcome of 1.

The rationale behind this simulation is that it generates dense datasets with large numbers of association rules, of which only a single rule is actually predictive. By controlling the size of the predictive rule by n_p, we are effectively setting the support to $1000/2^{n_p}$, and by setting the odds we can control the interestingness of the rule.

The real datasets we studied have been used previously in evaluations of algorithms for maximal frequent pattern identification. For clarity, we focused our attention on the *chess*, and *mushroom* datasets, (available from the UCI Machine Learning Repository [4]). The chess dataset contains 3,196 records with 23 attributes, while the mushroom dataset contains 8,124 records of 23 attributes.

7.2 Association Mining Algorithms

For each of the datasets, we performed searches for patterns associated with an outcome of interest. In the simulated datasets, we searched for patterns with a '1' in the fifty-first attribute; for the chess dataset, we searched for patterns that contained the outcome "won", and for the mushroom dataset, we searched for patterns that contained the outcome "edible".

We performed the search for these predictive patterns using four different association mining algorithms. Our goal was to decide whether the pruning technique required the explicit listing of all associations or whether more efficient but less exhaustive association mining searches could provide the same results after pruning. The four algorithms we used were Apriori [2], FPMine* [10], SLAM [18], and MAX.

Apriori was used in its normal fashion to find a complete set of patterns, given a particular minimum support constraint. We modified the FPMine algorithm of [10] slightly so it would not output every possible subset of patterns found when it had uncovered a path-tree in its recursion step; we called our variation FPMine*. With this change, FPMine* outputs fewer associations at a given minimum support than Apriori, but the set is still generally large and will include the maximal frequent patterns. SLAM is a heuristic association

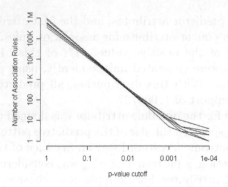

Fig. 1. Log-log plot of the number of association rules reported at different p-value cutoffs for Algorithm 1, applied to the simulated data. Different lines indicate different parameter settings, n_p and o_p, of the simulator for data generation.

mining algorithm, which tends to generate fewer associations than Apriori and FPMine*, but can use a lower *minsup*. MAX was written as a simple recursive search for maximal patterns (see [13] and [9] for more efficient algorithms for mining maximal patterns).

8   Results

We present the results for the filtering of association rules below, discussing first the results using simulated datasets and then using the real datasets.

8.1   Simulated Datasets

For identifying rules on the simulated datasets, we only used the Apriori algorithm, with a minimum support cutoff of *minsup* = 40 records out of 1,000. With this support, Apriori generated between 500,000 and 1,000,000 association rules, of which only a single rule was actually predictive. Since Algorithm 1 takes a p-value cutoff as a parameter, we first studied the number of unique association rules left using different p-value cutoffs. The graph in Figure 1 shows the steep decline in the number of interesting rules as the p-value threshold is lowered[1].

From the graph in Figure 1, it is clear that the pruning dramatically reduced the number of patterns reported, so few non-predictive patterns are reported when the p-value threshold is set high enough. Our next concern was the sensitivity of the test: how frequently was the real pattern reported after pruning? We considered pattern sizes (n_p) from {1,2,3,4}, and predictive odds (o_p) from

---

* It should be noted that the pruning rule is performing multiple statistical tests for each of the many association rules, and so a "typical" p-value cutoff like 0.01 is not appropriate: lower (more significant) values should be used to obtain more realistic significance estimates. A discussion of multiple testing is presented in [3].

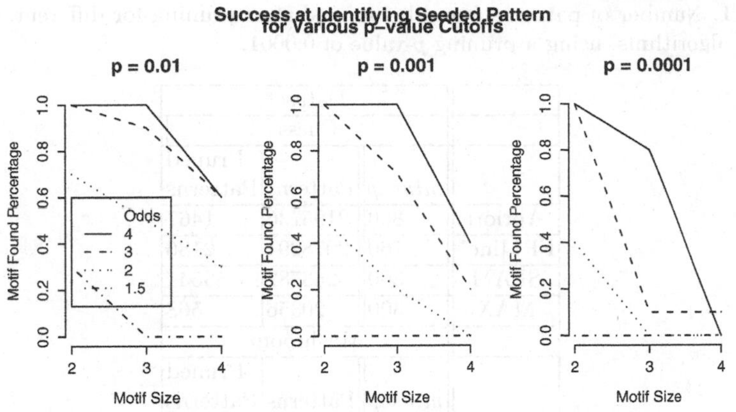

Success at Identifying Seeded Pattern for Various p-value Cutoffs

Fig. 2. For the simulated data, using different p-values, the seeded motif is reported less often if it is relatively infrequent (larger motif sizes) or less predictive (lower odds).

{1.5, 2, 3, 4}. For each possible pair (n_p, o_p), we generated ten tables using the simulator distribution, performed a search for frequent patterns (*minsup* of 40), and pruned the resulting set using Algorithm 1.

In Figure 2, we see the percentage of the runs for which the correct motif was found after pruning, for different p values of the pruning algorithm. As expected, when the motifs were shorter and had higher odds, they were more significant and more consistently reported. Motifs with odds of 1.5 were only reported occasionally when the motif size was short (size 2), and never otherwise, even for least restrictive p-value of 0.01. This is unsurprising, given the low odds and relatively low prevalence of the rule; the occasional significant finding is the result of sampling variation.

Motifs of size 3 or 4 with odds of 1.5 have too low a support to ever be significant even at $p = 0.01$ levels.

The results of the association mining on simulated data indicate that the approach is effective at removing the vast majority of statistically non-interesting patterns, while still identifying real motifs if they achieve the specified statistical confidence level. We now consider the application of the pruning algorithm in the context of real data.

8.2   Real Datasets

For the real datasets, chess and mushroom, we used all four association mining algorithms, Apriori, FPMine*, SLAM and MAX. For Apriori and FPMine*, we experimented to find minimum support values that produced roughly 2 million patterns associated with the outcome of interest (winning the chess game, edible mushrooms). Since FPMine* does not report all patterns above the minimum support threshold, it could use a lower minimum support, and identify less frequent patterns. We ran MAX at various levels of support for which run times were reasonable (less than an hour). SLAM, being an iterative procedure, can

Table 1. Number of patterns found before and after pruning for different association mining algorithms, using a pruning p-value of 0.0001.

	Dataset		
	Chess		
	$minsup$	Patterns	Pruned Patterns
Apriori	860	2195736	1467
FPMine*	760	2078291	2186
SLAM	300	243782	3541
MAX	300	20356	568
	Mushroom		
	$minsup$	Patterns	Pruned Patterns
Apriori	280	2334446	3182
FPMine*	70	2337408	5122
SLAM	40	26950	1563
MAX	40	4859	231

be run for any number of iterations, regardless of the minimum support. We selected a run-time of 10 minutes on each dataset to search for patterns (similar to Apriori and FPMine*). For ease of comparison, SLAM used a minimum support equal to that used with MAX.

After each association mining algorithm was used on both datasets, the patterns found were pruned using Algorithm 1. The number of patterns before and after the pruning are given in Table 1.

In Table 1, we see the clear monotonicity in the minimum supports used for the different algorithms. Furthermore, the incomplete searches (SLAM and MAX) generate orders of magnitude fewer associations than the more complete searches (Apriori and FPMine*). When different p-value cutoffs were used in the pruning, the relative number of pruned motifs for each algorithm stayed roughly constant (not shown).

We also studied the relative overlap of the patterns reported by each method. We used the patterns generated by Apriori as the baseline for comparison. In general, using the FPMine* algorithm to find patterns with lower support resulted in a duplication of the patterns found by Apriori, with the addition of new significant patterns with lower support. For the chess dataset, all of the pruned patterns from Apriori were found in the pruned FPMine* patterns, with FPMine* having an additional 719 novel patterns. For the mushroom dataset, there were some patterns found in the pruned Apriori results (169 patterns), while FPMine* reported an additional 2109 patterns beyond the set found by Apriori. This indicated that overall, if more significant patterns were desired, then using an approach like FPMine*, with a relatively low minimum support, would be more rewarding than performing a complete search with higher support using Apriori.

The patterns found while pruning the SLAM and MAX results were quite different from those found by the more complete search methods. In the chess dataset, of MAX's pruned patterns, roughly two-thirds were not present in the patterns from Apriori, but 1123 patterns from Apriori were not found by MAX. In this situation, there is argument for using several search algorithms to maximize the likelihood of finding a comprehensive set of interesting patterns.

As we hoped, the heuristic association mining algorithm SLAM provided a complementary approach between the extremes of discovering all patterns and only the maximal patterns. It generated far fewer patterns than the complete methods, of which a much higher proportion pruned down to unique significant patterns.

9   Discussion

We have shown that the proposed filtering algorithm can be used to reliably detect truly predictive patterns in data tables with many spurious associations. We have further shown that the algorithm can be used to complement a variety of association mining algorithms, from complete searches to maximal assocation mining. Regardless of the search algorithm, the number of significant associations is always far smaller than the raw set returned, indicating the usefulness of statistical pruning as a post-processing step for any association mining. Furthermore, by selecting different p-value cutoffs for the pruning, the size of the final set of patterns can be easily controlled.

There are several desirable features about this filtering algorithm. Firstly, the filtering can be applied to every predictive pattern independently of the others, meaning that batch parallelism could be used to simultaneously prune a large number of patterns. The parallel overhead would the re-combination of the relatively small number of pruned patterns into a single set of unique patterns.

Also, the proposed pruning method can be used to evaluate the statistical reliability of rules against a more general probability model. In our examples, we evaluated the support of various patterns by counting records in the dataset. However, a more general probability model could be used to compute the expected support levels, such as the maximum entropy model described in [15].

10   Conclusion

We have presented a statistical approach for filtering and pruning predictive patterns identified through association mining. We have further shown that this filtering can be used with a variety of association mining algorithms, allowing a progressive filtering of a large collection of predictive patterns down to a relatively small set of significant patterns. This pruning can be performed in isolation, or as a pre-processing step for more computationally expensive pattern filtering algorithms.

Acknowledgments

We would like to thank Molecular Mining Corporation for permission to use SLAM algorithm, and the Natural Science and Engineering Research Council of Canada and the Institute for Robotics and Intelligent System for financial support for the research. We would also like to thank the three anonymous referees for their helpful comments and for identifying important previous work.

References

1. R. Agrawal and R. Srikant. Fast algorithms for mining association rules. In *Proc. 20th Int. Conf. Very Large Data Bases, VLDB*, pages 487–499, 12–15 1994.
2. R. Agrawal and T. Imielinski A. Swami. Mining association rules between sets of items in large databases. In *ACM SIGMOD Intl. Conference on Management of Data*, pages 207–216, 26–28 1993.
3. Y. Benjamini and Y. Hochberg. Controlling the false discovery rate: a practical and powerful approach to multiple testing. *Journal of the Royal Statistical Society, Series B*, 57(1):389–300, 1995.
4. C.L. Blake and C.J. Merz. UCI repository of machine learning databases, 1998.
5. D. Shah et al. Interestingness and pruning of mined patterns. In *ACM SIGMOD Workshop on Research Issues in Data Mining and Knowledge Discovery*, 1999.
6. H. Toivonen et al. Pruning and grouping of discovered association rules, 1995.
7. M. Klemettinen et al. Finding interesting rules from large sets of discovered association rules. In *CIKM 1994)*, pages 401–407, 1994.
8. Y Bastide et al. Mining minimal non-redundant association rules using frequent closed itemsets. *Lecture Notes in Computer Science*, 1861:972, 2000.
9. K. Gouda and M.J. Zaki. Efficiently mining maximal frequent itemsets. In *ICDM*, pages 163–170, 2001.
10. J. Han, J. Pei, and Y. Yin. Mining frequent patterns without candidate generation. In *ACM SIGMOD Intl. Conference on Management of Data*, pages 1–12, 05 2000.
11. F. Hussain, H. Liu, and H. Lu. Relative measure for mining interesting rules. In *The Fourth European Conference on Principles and Practice of Knowledge Discovery in Databases*, 2000.
12. S. Jaroszewicz and D. A. Simovici. Pruning redundant association rules using maximum entropy principle. In *Advances in Knowledge Discovery and Data Mining, PAKDD*, pages 135–147, 2002.
13. R. J. Bayardo Jr. Efficiently mining long patterns from databases. In *ACM SIGMOD Intl. Conference on Management of Data*, pages 85–93, 1998.
14. Bing Liu, Wynne Hsu, and Yiming Ma. Pruning and summarizing the discovered associations. In *Knowledge Discovery and Data Mining*, pages 125–134, 1999.
15. Dmitry Pavlov, Heikki Mannila, and Padhraic Smyth. Beyond independence: Probabilistic models for query approximation on binary transaction data. Technical Report UCI-ICS TR-01-09, UC Irvine, 2001.
16. S. M. Ross. *Introduction to Probability Models*. Academic Press, 1972.
17. C. Silverstein S. Brin, R. Motwani. Beyond market baskets: Generalizing association rules to correlations. In *SIGMOD Conference*, pages 265–276, 1997.
18. E. Steeg, D. A. Robinson, and E. Willis. Coincidence detection: A fast method for discovering higher-order correlations in multidimensional data. In *KDD 1998*, pages 112–120, 1998.
19. P. Tan, V. Kumar, and J. Srivastava. Selecting the right interestingness measure for association patterns. In *ACM SIGKDD*, 2002.
20. T. Wu. An accurate computation of the hypergeometric distribution function. *ACM Transactions on Mathematical Software (TOMS)*, 19(1):33–43, 1993.
21. M. Zaki. Generating non-redundant association rules. In *KDD 2000*, pages 34–43, 2000.

Majority Classification
by Means of Association Rules

Elena Baralis and Paolo Garza

Politecnico di Torino
Corso Duca degli Abruzzi 24, 10129 Torino, Italy
{baralis,garza}@polito.it

Abstract. Associative classification is a well-known technique for structured data classification. Most previous work on associative classification based the assignment of the class label on a single classification rule. In this work we propose the assignment of the class label based on simple majority voting among a group of rules matching the test case.
We propose a new algorithm,L_M^3, which is based on previously proposed algorithm L^3. L^3 performed a reduced amount of pruning, coupled with a two step classification process. L_M^3 combines this approach with the use of multiple rules for data classification. The use of multiple rules, both during database coverage and classification, yields an improved accuracy.

1   Introduction

Association rules [1] describe the co-occurrence among data items in a large amount of collected data. Recently, association rules have been also considered a valuable tool for classification purposes. Classification rule mining is the discovery of a rule set in the training database to form a model of the data, the classifier. The classifier is then used to classify appropriately new data for which the class label is unknown [12]. Differently from decision trees, association rules consider the simultaneous correspondence of values of different attributes, hence allowing to achieve better accuracy [2,4,8,9,14].

Most recent approaches to associative classification (e.g., CAEP [4], CBA [9], ADT [14], and L^3 [2]) use a single classification rule to assign the class label to new data whose label is unknown. A different approach, based on the use of multiple association rules to perform classification of new data has been proposed in CMAR [8], where it has been shown that this technique yields an increase in the accuracy of the classifier. We believe that this technique can be applied orthogonally to almost any type of classifier. Hence, in this paper we propose L_M^3, a new algorithm which incorporates multiple rule classification into L^3, a levelwise classifier previously proposed in [2].

L^3 was based on the observation that most previous approaches, when performing pruning to reduce the size of the rule base obtained from association rule mining, may go too far and discard also useful knowledge. We extend this

N. Lavrač et al. (Eds.): PKDD 2003, LNAI 2838, pp. 35–46, 2003.
© Springer-Verlag Berlin Heidelberg 2003

idea to considering multiple rules to perform classification of new data. In this paper, we propose L_M^3, a new classification algorithm that combines the lazy pruning approach of L^3, which has been shown to yield accurate classification results, with a rule assignment technique that selects the class label basing its decision on a group of eligible rules which are drawn either from the first or the second level of the classifier.

The paper is organized as follows. Section 2 introduces the problem of associative classification. In Section 3 we present the classification algorithm L_M^3, by describing both the generation of the two levels of the classifier and the classification of test data by means of majority voting applied to its two levels. Section 4 provides experimental results which validate the L_M^3 approach. Finally, in Section 5 we discuss the main differences between our approach and previous work on associative classification, and Section 6 draws conclusions.

2 Associative Classification

The database is represented as a relation R, whose schema is given by k distinct attributes $A_1 \ldots A_k$ and a class attribute C. Each tuple in R can be described as a collection of pairs *(attribute, integer value)*, plus a class label (a value belonging to the domain of class attribute C). Each pair *(attribute, integer value)* will be called *item* in the reminder of the paper. A training case is a tuple in relation R, where the class label is known, while a test case is a tuple in R where the class label is unknown.

The attributes may have either a categorical or a continuous domain. For categorical attributes, all values in the domain are mapped to consecutive positive integers. In the case of continuous attributes, the value range is discretized into intervals, and the intervals are also mapped into consecutive positive integers[1]. In this way, all attributes are treated uniformly.

A classifier is a function from $A_1, \ldots, A_n$ to C, that allows the assignment of a class label to a test case. Given a collection of training cases, the classification task is the generation of a classifier able to predict the class label for test cases with high accuracy.

Association rules [1] are rules in the form $X \rightarrow Y$. When using them for classification purposes, X is a set of items, while Y is a class label. A case d is said to match a collection of items X when $X \subseteq d$. The quality of an association rule is measured by two parameters, its support, given by the number of cases matching $X \cup Y$ over the number of cases in the database, and its confidence given by the the number of cases matching $X \cup Y$ over the number of cases matching X. Hence, the classification task can be reduced to the generation of the most appropriate set of association rules for the classifier. Our approach to such task is described in the next section.

---

[1] The problem of discretization has been widely dealt with in the machine learning community (see, e.g., [5]) and will not be discussed further in this paper.

3   Majority Classification

In this paper, we introduce the use of multiple association rules to perform classification of structured data, in the levelwise classifier L^3 [2]. In L^3, a lazy pruning technique is proposed, which only discards "harmful" rules, i.e., rules that only misclassify training cases. Lazy pruning is coupled with a two levels classification approach. Rules that would be discarded by currently used pruning techniques are included in the second level of the classifier and used only when first level rules are not able to classify a test case.

Majority selection of the class label requires (a) selecting a group of good quality rules matching the case to be classified, and (b) assigning the appropriate class with simple majority voting among selected rules. To obtain a good quality rule set in step (a), a wide selection of rules from which to extract matching rules should be available. In Section 3.1 we describe how association rules are extracted, while in section 3.2 we discuss how the rules that form the model of the classifier are selected. Finally, in Section 3.3 the majority classification technique is presented.

3.1   Association Rule Extraction

Analogously to L^3, in L_M^3 abundance of classification rules allows a wider choice of rules both for rule selection when the classifier is generated, and for new case classification. Hence, during the rule extraction phase, the support threshold should be set to zero. Only the confidence threshold should be used to select good quality rules. Unfortunately, no rule mining algorithm extracting rules only with a confidence threshold is currently available[2].

In L_M^3, the extraction of classification rules is performed by means of an adaptation of the well-known FP-growth algorithm [7], which only extracts association rules with a class label in the head. Analogously to [8], we also perform pruning based on χ^2 (see below) during the rule extraction process.

3.2   Pruning Techniques and Classifier Generation

In L_M^3 two pruning techniques are applied: χ^2 pruning and lazy pruning. χ^2 is a statistical test widely used to analyze the dependence between two variables. The use of χ^2 as a quality index for association rules is proposed for the first time in [11] and is also used in [8] for pruning purposes. This type of pruning was not performed in L^3. However, we performed a large number of experiments, which have shown that rules which do not match the χ^2 threshold are usually useless for classification purpose. Since the use of χ^2 test heavily reduces the size of the rule set, it may significantly increase the efficiency of the following steps without deteriorating the informative content (quality) of the rule set after pruning. We perform χ^2 pruning during the classification rule extraction step.

---

[2] Some attempt in this direction has been proposed in [13], but its scalability is unclear.

Even with χ^2 pruning, if a low minimim support threshold is used, a huge rule set may be generated during the extraction phase. However, most of these rules may be useful [2]. The second pruning technique used in L_M^3 is the lazy pruning technique proposed in L^3.

Before performing lazy pruning, a global order is imposed on the rule base. Rules are first sorted on descending confidence, next on descending support, then on descending length (number of items in the body of the rule), and finally lexicographically on items. The only significant difference with respect to most previous work ([8], [9]) is rule sorting on descending length. Most previous approaches prefer short rules over long rules. The reason for our choice is to give a higher rank in the ordering to more specific rules (rules with a larger number of items in the body) over generic rules, which may lead to misclassification. Note that, since shorter rules are not pruned, they can be considered anyway.

The idea behind lazy pruning [2] is to discard from the classifier only the rules that do not correctly classify any training case, i.e., the rules that only negatively contribute to the classification of training cases. To this end, after rule sorting, we cover the training cases to detect "harmful" rules (see Figure 1), using a database coverage technique. However, to allow a wider selection of rules for majority classification, a different approach is taken in the generation of the classifier levels. In L_M^3 a training document is removed from the data set when it is covered by δ rules, while in L^3 each training case is removed as soon as is covered by one rule. Hence, by setting $\delta = 1$ the lazy pruning performed by L_M^3 degenerates in that of L^3.

Lines 1-26 of the pseudocode in Figure 1 show our approach. The first rule r in the sort order is used to classify each case d still in *data* (lines 3-11). Each case d covered by r is included in the set $r.dataClassified$, and the counter $d.covered$ is increased. When d is covered by δ rules ($d.covered = \delta$), d is removed from *data* (line 9). The appropriate counter of r is increased (lines 6-7), depending on the correctness of the label.

After all cases in *data* have been considered, r is checked. If rule r only classified training cases wrongly (lines 12-18), then r is discarded, and the counter of each case classified by r is decreased by one. Cases included in $d.covered$ and removed before (line 9), because covered by δ rules, are included again in *data* (line 15).

The loop (lines 2-20) is repeated for the next rule in the order, considering the cases still in *data*. The loop ends when either the data set or the rule set are empty. The remaining rules are divided in two groups (lines 21-26), which will form the two levels of the classifier:

Level I which includes rules that have already correctly classified at least one training case,
Level II which includes rules that have not been used during the training phase, but may become useful later.

Rules in each level are ordered following the global order described above.

Rules in level I provide a high level model of each class. Rules in level II, instead, allow us to increase the accuracy of the classifier by capturing "special"

Procedure generateClassifier($rules$,$data$,δ)
1. r = first rule of $rules$;
2. while ($data$ not empty) and (r not NULL) {
3.   for each d in $data$ {
4.   if r matches d {
5. r.dataClassified = r.dataClassified $\cup$ d;
6.   if (d.class==r.class) r.right++;
7.   else r.wrong++;
8. d.matched++;
9.   if (d.matched==δ) delete d from $data$;
10.   }
11.   }
12.   if r.wrong>0 and r.right==0 {
13.   delete r from $rules$;
14.   for each d in r.dataClassified {
15.   if (d.matched==δ) $data$=$data$ $\cup$ d;
16. d.matched- -;
17.   }
18.   }
19.   r=next rule from $rules$;
20.}
21. for each r in $rules$ {
22.   if r.right>0
23. $levelI$ = $levelI$ $\cup$ r;
24.   else
25. $levelII$ = $levelII$ $\cup$ r;
26. }

Fig. 1. L_M^3 classifier generation

cases which are not covered by rules in the first level. Even if the levels are used similarly to L^3, their size may be significantly different. In particular, the use of δ generally increases the size of the first level, compared to the first level of L^3. Hence, L_M^3 is characterized by a first level which is "more fat" that that of L^3. In Section 4 it is shown that this technique may provide a higher accuracy than L^3, but the model includes more rules and is hence somewhat less readable as a high level description of the classifier. However, we note that the readability of the classifier generated by L_M^3 is still better than that of non-associative classifiers (e.g., Naive-Bayes [6]).

3.3   Classification

Majority classification is performed by considering multiple classification rules to assign the class label to a test case. The first step is the selection of a group of rules matching the given test case. When rules in the group yield different class labels, a simple majority voting technique is used to assign the class label. The size of the rule group (i.e., the maximum number of rules used to classify new cases) may vary, depending on the number of rules matching the new case. It is limited by an upper bound, defined by the parameter max_rules.

This technique is combined with the two levels of the L^3 classifier. To build a rule group, rules in level I of the classifier are first considered. If no rule in this level matches the test case, then rules in level II are considered. Hence, rules in a rule group are never selected from both levels.

When a new case is to be classified, the first level is considered. The algorithm selects (at most) the first max_rules rules in the first level matching the case. When at least one rule matches the case, the matching process stops either at the end of the first level, or when the upper limit max_rules is reached.

Selected rules are divided in sets, one for each class label. Then, simple majority voting takes place. The rule set with the largest cardinality assigns the class label to the new case. A different approach, based on the evaluation of a weight for each rule set using χ^2 (denoted as $\chi^2 - max$), is proposed in CMAR [8]. We performed a wide set of experiments which showed that the average accuracy obtained by using this method is slightly lower than that given by the simple majority technique described above, differently from what is reported in [8]. The difference between our results and those reported in [8] may be due to the elimination of redundand rules applied in CMAR and not in L^3_M.

If no rule in the first level matches the test case, then rules in level II are considered. Both matching process and label assignment are repeated analogously for this level.

We observe that the use of $\delta > 1$ during the lazy pruning phase is necessary when using the simple majority technique described before to classify new cases. Indeed, if δ is set to one, only few rules are included in the first level, which becomes very thin. In this case, just a couple of rules may be available in the first level for matching and majority voting, and the selection of the appropriate class label may degenerate to the case of single rule classification.

L^3_M usually contains a large number of rules. In particular, the first level of the classifier contains a limited number of rules, which during the training phase covered some training cases. The cardinality of the first level is comparable to the size of the rule set in CMAR, while most previous approaches, including L^3 first level, were characterized by a smaller rule set. As shown in Section 4, this level performs the "heavy duty" classification of most test cases and provides a general model of each class. By contrast, level II of the classifier usually contains a large number of rules which are seldom used. These rules allow the classification of some more cases, which cannot be covered by rules in the first level.

Since level I usually contains about 10^2-10^3 rules, it can easily fit in main memory. Thus, the main classification task can be performed efficiently. Level II, in our experiments, included around 10^5-10^6 rules. Rules were organized in a compact list, sorted as described in Section 3.2. Level II of L^3 could generally be loaded in main memory as well. Of course, if the number of rules in the second level further increases (e.g., because the support threshold is further lowered to capture more rules with high confidence), efficient access may become difficult.

4   Experimental Results

In this section we describe the experiments to measure accuracy and classification efficiency for L_M^3. We compared L_M^3 with the classification algorithms CBA [9], CMAR [8], C4.5 [12], and with its previous version with single rule classification L^3 [2]. The differences between our approach and the above algorithms is further discussed in Section 5. A large set of experiments has been performed, using 26 data sets downloaded from UCI Machine Learning Repository [3]. The experiments show that L_M^3 achieves a larger average accuracy (+0.47% over the best previous, i.e., L^3), and has best accuracy on 10 data sets over 26.

For classification rule extraction the mininum support threshold has been set to 1%, a standard value used by previous associative classifiers. For 5 data sets (auto,hypo,iono,sick,sonar) the mininum support threshold has been set to 5%, to limit the number of generated rules. The confidence constraint has not been enforced, i.e., $minconf=0$. We have adopted the same technique used by CBA to discretize continuous attributes. A 10 fold cross validation test has been used to compute the accuracy of the classifier. All the experiments have been performed on a 1000Mhz Pentium III PC with 1.5G main memory, running RedHat Linux 7.2.

Recall from Section 3 that the performance of L_M^3 depends on the values of two parameters: δ and max_rules. δ is used during the training phase and sets the maximal number of rules that can match a document. max_rules is used during the classification phase and sets an upper bound on the size of the selected rule group before voting. A huge amount of experiments has been performed, using different values for the parameters δ and max_rules. Unfortunately, it has not been possible to find overall optimal values for the parameters. However, we have devised values, denoted as default values, that yield a good average result, and are sufficiently appropriate for every data set considered in the experiments. The default values are $\delta = 9$, $max\_rules = 9$. These values may be used for "normal" classification usage, for any data set.

We report in Figure 2 the variation of accuracy with varying δ for different values of max_rules. For many data distributions, of which data set TicTac is representative, accuracy tends to be stable after a given threshold for parameter values. Hence, default values and optimal values tend to be very close. A different behavior is shown by data set Cleve, for which high accuracy is associated with very specific values of the parameters (e.g., optimal values are $\delta = 2$, $max\_rules = 9$). For these exceptional cases, optimal values can only be computed by running a vast number of experiments in which values of the parameters are varied and average accuracy is evaluated on a ten fold. The value pair that yields the best accuracy is finally selected. This technique requires fine tuning for a specific data set and should be used only when very high accuracy is needed.

Table 1 compares the accuracy of L_M^3 with the accuracy of L^3, C4.5, CBA and CMAR, obtained using standard values for all the parameters. In particular, the columns of Table 1 are: (1) name of data set, (2) number of attributes, (3) number of classes, (4) number of cases (records), (5) accuracy of C4.5, (6) accuracy of

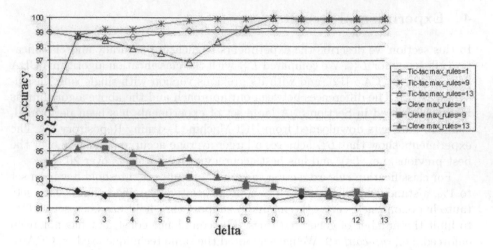

Fig. 2. Variation of accuracy with varying δ

CBA, (7) accuracy of CMAR, (8) accuracy of L^3, (9) accuracy of L^3_M with default values ($\delta = 9$, $max\_rules = 9$, identical for all data sets), (10) accuracy obtained using only the first level of L^3_M (always with default parameters), (11) improvement in accuracy given by the second level of L^3_M, and (12) accuracy of L^3_M with optimal values (different values of parameters for each data set).

L^3_M, with default values, has best average accuracy ($+0.47\%$ with respect to L^3) and best accuracy on 10 of the 26 UCI data sets. Only for 7 data sets the accuracy achieved by L^3_M is lower than that achieved by L^3, while for 15 data sets the accuracy is larger. Hence, the use of majority voting can improve the approach proposed by L^3.

We ran experiments to separate the contribution in accuracy improvement due to the use of multiple rules during the classification phase, and to the second level. In particular, we compared the accuracy obtained by only using rules in level I of L^3_M with the accuracy obtained by using both levels. The results of the experiments are reported in Table 1. The related columns of Table 1 are: (10) accuracy of L^3_M using only rules in the first level, (11) difference between L^3_M with both levels (column (9)) and L^3_M with only first level (column (10)). By considering only rules in the first level, L^3_M achieves best accuracy on 9 of the UCI data sets, and has average accuracy higher than L^3 ($+0.25\%$). This result shows the significant effect due to multiple rule usage in the first level.

The effect of the second level in L^3_M is definitely less relevant than in L^3. We observe an increase in accuracy given by the second level only in 8 data sets, and the average increase in accuracy is $+0.22\%$. The increase given by the second level of L^3 is more relevant [2]. In particular, for 20 data sets the second level is useful, and an average accuracy increase of $+1.67\%$ is given by the use of the second level. These results highlight that the second level is very useful when δ is set to 1 (L^3) and the first level is very thin. However, its contribution is less

Table 1. Comparison of L_M^3 accuracy with respect to previous algorithms

Name	A	C	R	C4.5	CBA	CMAR	L^3	L_M^3 default values	Only I level	Δ_{acc}	L_M^3 optimal values
Anneal	38	6	898	94.8	**97.9**	97.3	96.2	96.4	96.4	0.00	96.4
Austral	14	2	690	84.7	84.9	**86.1**	85.7	**86.1**	86.1	0.00	**86.4**
$Auto^{(*)}$	25	7	205	**80.1**	78.3	78.1	81.5	78.5	76.6	1.90	81.5
Breast	10	2	699	95.0	96.3	96.4	95.9	**96.6**	96.6	0.00	**96.7**
Cleve	13	2	303	78.2	82.8	**82.2**	82.5	82.5	82.5	0.00	**86.4**
Crx	15	2	690	84.9	84.7	84.9	84.4	**85.5**	85.1	0.40	**85.9**
Diabetes	8	2	768	74.2	76.7	74.5	75.8	**78.6**	78.6	0.00	**79.0**
German	20	2	1000	72.3	73.4	**74.9**	73.8	74.5	74.5	0.00	74.7
Glass	9	7	214	68.7	73.9	70.1	**76.6**	75.7	75.7	0.00	**76.6**
Heart	13	2	270	80.8	81.9	82.2	**84.4**	83.3	83.3	0.00	**84.4**
Hepatic	19	2	155	80.6	81.8	80.5	**81.9**	**81.9**	81.3	0.60	**83.2**
Horse	22	2	368	82.6	82.1	82.6	**82.9**	82.1	81.8	0.30	**83.2**
$Hypo^{(*)}$	25	2	3163	**99.2**	98.9	98.4	95.2	97.5	97.5	0.00	97.5
$Iono^{(*)}$	34	2	351	90.0	92.3	91.5	**93.2**	92.8	92.0	0.80	**93.2**
Iris	4	3	150	**95.3**	94.7	94.0	93.3	93.3	93.3	0.00	94.0
Labor	16	2	57	79.3	86.3	89.7	91.2	**96.5**	96.5	0.00	**96.5**
Led7	7	10	3200	**73.5**	71.9	72.5	72.0	72.4	72.4	0.00	72.8
Lymph	18	4	148	73.5	77.8	83.1	**85.1**	84.5	83.8	0.70	**85.1**
Pima	8	2	768	75.5	72.9	75.1	**78.4**	78.0	78.0	0.00	**79.2**
$Sick^{(*)}$	29	2	2800	**98.5**	97.0	97.5	94.7	94.7	94.7	0.00	94.7
$Sonar^{(*)}$	60	2	208	70.2	77.5	79.4	78.9	**81.7**	81.7	0.00	**81.7**
Tic-tac	9	2	958	99.4	99.6	99.2	98.4	**100.0**	100.0	0.00	**100.0**
Vehicle	18	4	846	72.6	68.7	68.8	73.1	**73.2**	73.0	0.20	**73.2**
Waveform	21	3	5000	78.1	80.0	**83.2**	82.1	82.8	82.8	0.00	82.8
Wine	13	3	178	92.7	95.0	95.0	98.3	**98.9**	98.3	0.60	**98.9**
Zoo	16	7	101	92.2	96.8	**97.1**	95.1	97.0	97.0	0.00	97.0
Average				83.34	84.69	85.22	85.88	**86.35**	86.13	0.22	**86.84**

$(*)Minimum\ support\ threshold\ 5\%$

relevant for $\delta > 1$, when the first level is already rich enough to allow a good coverage of test cases. The second level remains always useful to capture special cases and allows a further increase in accuracy.

In column (12) of Table 1 are reported the accuracy results obtained by using optimal values of the δ and max_rules parameters for each data set. The effect of fine tuning the parameters values is significant, since it yields an increase of about $+1\%$ compared to L^3, and 0.49% with respect to L_M^3 with default values. Furthermore, the classifier shows best accuracy on 17 data sets.

Table 2 allows us to compare the structure and usage of the two levels for L^3 and L_M^3(with default values for the parameters). In order to analyze only the effect of multiple rule selection on the size of the two levels, L^3 has been modified to incorporate χ^2 pruning. This allows us to observe the difference in level size between L_M^3 and L^3 due exclusively to the level assignment technique based on multiple rule selection.

In Table 2 the comparison of the number of rules in the first level of L^3 (column (3)) and of L_M^3 (column (4)) shows that the first level of L_M^3 is about an order of magnitude larger. We performed other experiments, not reported here, using different values for δ, which showed that the first level of L_M^3 is approximately δ times larger than the first level of L^3. The only exception to this rule is data set Wine, where the size of the first level in L_M^3 and L^3 is comparable. In this case, the second level becomes more useful, since rules in the first level are not enough to cover all test cases. We can conclude that the first level of L_M^3 trades a reduced readability in favor of an increased accuracy and the value of δ allows to fine tune the tradeoff between these two features.

In Table 2 is also reported the number of rules in the second level for L^3 (column (5)) and L_M^3 (column (6)). We observe that the second level of L_M^3 is usually slightly smaller than that of L^3. This may be due to two different effects. (1) The first level of L_M^3 is larger and contains some rules that would have been assigned to the second level of L^3. This effect is particularly evident in the case of data set Iris, where the total number of rules is rather small. In this case, most rules migrate from the second level to the first level, leaving an almost empty second level. (2) The multiple matching technique used for generating the classifier causes L_M^3 to analyze more rules, which L^3 did not consider at all. If these rules make only mistakes, they are pruned by L_M^3, but not by L^3 (L^3 considers them unused and assigns them directly to the second level).

We also analyzed the performance of L_M^3 during the classification of test data. The classification time is not affected by the use of the second level, because it is used rarely (see column(8) of Table 2). The average time for classifying a new case is about 1ms, and is comparable to that reported for L^3. With respect to memory usage, since the size of both levels is not dramatically different for L_M^3 and L^3, the same considerations already reported in [2] hold also for L_M^3.

5   Previous Related Work

CMAR [8] is the first associative classification algorithm where multiple rules are used to classify new cases. CMAR proposes a suite of different pruning techniques: pruning of specialistic rules, use of the χ^2 coefficient, and database coverage. In L_M^3 pruning based on the χ^2 coefficient is adopted, but specialistic rules are not pruned. Our database coverage technique is more tolerant, since it allows more rules to cover the same training case. This effect depends on the value of the δ parameter, discussed in Section 4. A similar parameter is available in CMAR (denoted as δ), but its suggested value allows a lower number of rules during the selection step. Hence, in CMAR useful rules may be pruned, thus reducing the overall accuracy of the classifier. This problem has been denoted as overpruning in [2]. Furthermore, we use simple majority voting to assign the final class label to a test case, while in CMAR a more complex weighting technique

Table 2. Usage of the two levels

Name	R	Rules I level L^3	Rules I level L_M^3	Rules II level L^3	Rules II level L_M^3	Use of I level L_M^3	Use of II level L_M^3
Anneal	898	38	358	169802	168851	99.44	0.56
Austral	690	152	1458	171638	159165	100.00	0.00
Breast	699	51	516	6241	5407	100.00	0.00
Cleve	303	74	724	16481	14676	100.00	0.00
Crx	690	159	1422	341382	322675	99.57	0.43
Diabetes	768	65	360	466	180	100.00	0.00
German	1000	291	2420	62359	57786	100.00	0.00
Glass	214	30	274	1385	1047	100.00	0.00
Heart	270	56	506	3449	2725	100.00	0.00
Hepatic	155	31	313	185453	184757	98.71	1.29
Horse	368	97	888	179345	177803	99.73	0.27
Iris	150	8	82	88	13	100.00	0.00
Labor	57	13	85	209	119	100.00	0.00
Led7	3200	75	318	1159	980	100.00	0.00
Lymph	148	40	302	1442098	1441055	95.95	4.05
Pima	768	64	362	472	174	100.00	0.00
Tic-tac	958	28	599	3258	2566	100.00	0.00
Vehicle	846	180	1433	2408341	2406231	99.53	0.47
Wine	178	8	11	122249	122116	99.40	0.60
Zoo	101	10	72	1515389	1515288	100.00	0.00
Average						99.63	0.37

based on χ^2 is proposed. Experiments show that our technique is both simpler and more effective.

The L_M^3 algorithm derives its two level approach from the L^3 algorithm proposed in [2] and enhances L^3 with the introduction of classification based on multiple rules. However, introducing majority voting requires a larger first level, which may reduce the readability of the model with respect to L^3. Hence, the selection of an appropriate value for the δ parameter allows the fine tuning of the richness of the first level. We observe that L^3 can be seen as a degenerate case of L_M^3, when both δ and *max_rules* parameters are set to 1.

Associative classification has been first proposed in CBA [9]. CBA, based on the Apriori algorithm, extracts only a limited number of association rules (max 80000). Furthermore, it applies a database coverage pruning technique that significantly reduces the number of rules in the classifier, thus losing relevant knowledge. A new version of the algorithm has been presented [10], in which the use of multiple supports is proposed, together with a combination of C4.5 and Naive-Bayes classifiers. Unfortunately, none of these techniques addresses the overpruning problem described in [2].

ADT [14] is a different classification algorithm based on association rules, combined with decision tree pruning techniques. All rules with a confidence greater or equal to a given threshold are extracted and more specific rules are pruned. A decision tree is created based on the remaining association rules, on which classical decision tree pruning techniques are applied. Analogously to other algorithms, the classifier is composed by a small number of rules and prone to the overpruning problem.

6   Conclusions

In this paper we have described L_M^3, an associative classifier which combines levelwise classification with majority voting. This approach is a natural extension of the concept of exploiting rule abundance for associative classification, initially proposed in [2]. In [2] rule abundance was only pursued when selecting rules to form the classifier by performing lazy pruning. With L_M^3 we extend the same concept to the classification phase, by considering multiple rules for label assignment. Experiments show that the adopted approach allows a good increase in accuracy with respect to previous approaches. The main disadvantage of this approach is the (slightly) reduced readability of the first level of the classifier, which should provide a general model of classes.

References

1. R. Agrawal, T. Imilienski, and A. Swami. Mining association rules between sets of items in large databases. *In SIGMOD'93 , Washington DC*, May 1993.
2. E. Baralis and P. Garza. A lazy approach to pruning classification rules. *In ICDM'02, Maebashi, Japan*, December 2002.
3. C. Blake and C. Merz. UCI repository of machine learning databases, 1998.
4. G. Dong, X. Zhang, L. Wong, and J. Li. CAEP: Classification by aggregating emerging patterns. *In Int. Conf. on Discovery Science, Tokyo, Japan*, Dec. 1999.
5. U. Fayyad and K. Irani. Multi-interval discretization of continuos-valued attributes for classification learning. *In IJCAI'93*, 1993.
6. N. Friedman, D. Geiger, and M. Goldszmidt. Bayesian network classifiers. *Machine Learning, 29:131-163*, 1997.
7. J. Han, J. Pei, and Y. Yin. Mining frequent patterns without candidate generation. *In SIGMOD'00, Dallas, TX*, May 2000.
8. W. Li, J. Han, and J. Pei. CMAR: Accurate and efficient classification based on multiple class-association rules. *In ICDM'01, San Jose, CA*, November 2001.
9. B. Liu, W. Hsu, and Y. Ma. Integrating classification and association rule mining. *In KDD'98, New York, NY*, August 1998.
10. B. Liu, Y. Ma, and K. Wong. Improving an association rule based classifier. *In PKDD'00, Lyon, France*, Sept. 2000.
11. R. Motwani, S. Brin, and C. Silverstein. Beyond market baskets: Generalizing association rules to correlation. *ACM SIGMOD'97 , Tucson, Arizona*, May 1997.
12. J. Quinlan. *C4.5: program for classification learning*. Morgan Kaufmann, 1992.
13. K. Wang, Y. He, D. W. Cheung, and F. Y. L. Chin. Mining confident rules without support requirement. *In CIKM'01, Atlanta, GA*, November 2001.
14. K. Wang, S. Zhou, and Y. He. Growing decision trees on support-less association rules. *In KDD'00, Boston, MA*, August 2000.

Adaptive Constraint Pushing
in Frequent Pattern Mining

Francesco Bonchi[1,2,3], Fosca Giannotti[1,2],
Alessio Mazzanti[1,3], and Dino Pedreschi[1,3]

[1] Pisa KDD Laboratory
http://www-kdd.cnuce.cnr.it
[2] ISTI - CNR Area della Ricerca di Pisa, Via Giuseppe Moruzzi, 1 - 56124 Pisa, Italy
Giannotti@cnuce.cnr.it
[3] Department of Computer Science, University of Pisa
Via F. Buonarroti 2, 56127 Pisa, Italy
{bonchi,mazzanti,pedre}@di.unipi.it

Abstract. Pushing monotone constraints in frequent pattern mining can help pruning the search space, but at the same time it can also reduce the effectiveness of anti-monotone pruning. There is a clear tradeoff. Is it better to exploit more monotone pruning at the cost of less anti-monotone pruning, or viceversa? The answer depends on characteristics of the dataset and the selectivity constraints. In this paper, we deeply characterize this trade-off and its related computational problem. As a result of this characterization, we introduce an adaptive strategy, named *ACP* (Adaptive Constraint Pushing) which exploits any conjunction of monotone and anti-monotone constraints to prune the search space, and level by level adapts the pruning to the input dataset and constraints, in order to maximize efficiency.

1 Introduction

Constrained itemsets mining is a hot research theme in data mining [3,4,5,6]. The most studied constraint is the frequency constraint, whose anti-monotonicity is used to reduce the exponential search space of the problem. Exploiting the anti-monotonicity of the frequency constraint is also known as the *apriori trick* [1]: this is a valuable heuristic that drastically reduces the search space making the computation feasible in many cases. Frequency is not only computationally effective, it is also semantically important since frequency provides "support" to any discovered knowledge. For these reasons frequency is the base constraints and in general we talk about *frequent itemsets mining*. However, many other constraints can facilitate user focussed exploration and control as well as reduce the computation. For instance, a user could be interested in mining all frequently purchased itemsets having a total price greater than a given threshold and containing at least two products of a given brand. Classes of constraints sharing nice properties have been individuated. The class of anti-monotone constraints is the most effective and easy to use in order to prune the search space. Since

N. Lavrač et al. (Eds.): PKDD 2003, LNAI 2838, pp. 47–58, 2003.

any conjunction of anti-monotone constraints is an anti-monotone constraint, we can use all the constraints in a conjunction to make the *apriori trick* more selective.

The dual class, monotone constraints, has been considered more complicated to exploit and less effective in pruning the search space. As highlighted by Boulicaut and Jeudy in [3], pushing monotone constraints can lead to a reduction of anti-monotone pruning. Therefore, when dealing with a conjunction of monotone and anti-monotone constraints we face a tradeoff between anti-monotone and monotone pruning.

In [2] we have shown that the above consideration holds only if we focus completely on the search space of all itemsets, which is the approach followed so far. With the novel algorithm ExAnte we have shown that an effective way of attacking the problem is to reason on both the itemsets search space and the transactions input database *together*. In this way, pushing monotone constraints does not reduce anti-monotone pruning opportunities, on the contrary, such opportunities are boosted. Dually, pushing anti-monotone constraints boosts monotone pruning opportunities: the two components strengthen each other recursively. ExAnte is a pre-processing data reduction algorithm which reduces dramatically both the search space and the input dataset. It can be coupled with any constrained patterns mining algorithm, and it is always profitable to start any constrained patterns computation with an ExAnte preprocess. Anyway, after the ExAnte preprocessing, when computing frequent patterns we face again the tradeoff between anti-monotone and monotone pruning.

The Tradeoff

Suppose that an itemset has been removed from the search space because it does not satisfy a monotone constraint. This pruning avoids checking support for this itemset, but however if we check its support and find it smaller than the threshold, we may prune away all the supersets of this itemset. In other words, by monotone pruning we risk to loose anti-monotone pruning opportunities given by the removed itemset. The tradeoff is clear [3]: pushing monotone constraint can save tests on anti-monotone constraints, however the results of these tests could have lead to more effective pruning.

On one hand we can exploit all the anti-monotone pruning with an *apriori* computation, checking the monotone constraint at the end, and thus not performing any monotone constraint pushing. We call this strategy *g&t* (generate and test). On the other hand, we can exploit completely any monotone pruning opportunity, but the price to pay is less anti-monotone pruning. We call this strategy *mcp* (monotone constraint pushing).

No one of the two extremes outperforms the other on every input dataset and conjunction of constraints. The best strategy depends of the characteristics of the input and the optimum is usually in between the two extremes.

In this paper, we introduce a general strategy, *ACP*, that balances the two extremes adaptively. Both monotone and anti-monotone pruning are exploited in a level-wise computation. Level by level, while acquiring new knowledge about the dataset and selectivity of constraints, ACP adapts its behavior giving more power to one pruning over the other in order to maximize efficiency.

Problem Definition

Let $Items = \{x_1, ..., x_n\}$ be a set of distinct literals, usually called **items**. An **itemset** X is a non-empty subset of $Items$. If $k = |X|$ then X is called a **k-itemset**. A **transaction** is a couple $\langle tID, X \rangle$ where tID is the transaction identifier and X is the content of the transaction (an itemset). A **transaction database** TDB is a set of transactions. An itemset X is **contained** in a transaction $\langle tID, Y \rangle$ if $X \subseteq Y$. Given a transaction database TDB the subset of transaction which contain an itemset X is named $TDB[X]$. The **support** of an itemset X, written $supp_{TDB}(X)$ is the cardinality of $TDB[X]$. Given a user-defined **minimum support** δ, an itemset X is called **frequent** in TDB if $supp_{TDB}(X) \geq \delta$. This the definition of the frequency constraint $\mathcal{C}_{freq}[TDB]$: if X is frequent we write $\mathcal{C}_{freq}[TDB](X)$ or simply $\mathcal{C}_{freq}(X)$ when the dataset is clear from the context. Let $Th(\mathcal{C}) = \{X | \mathcal{C}(X)\}$ denotes the set all itemsets X that satisfy constraint $\mathcal{C}$. The *frequent itemset mining problem* requires to compute the set of all frequent itemsets $Th(\mathcal{C}_{freq})$. In general given a conjunction of constraints $\mathcal{C}$ the *constrained itemset mining problem* requires to compute $Th(\mathcal{C})$; the *constrained frequent itemsets mining problem* requires to compute $Th(\mathcal{C}_{freq}) \cap Th(\mathcal{C})$.

Definition 1. Given an itemset X, a constraint $\mathcal{C}_{AM}$ is anti-monotone if:

$$\forall Y \subseteq X : \mathcal{C}_{AM}(X) \Rightarrow \mathcal{C}_{AM}(Y)$$

Definition 2. Given an itemset X, a constraint $\mathcal{C}_M$ is monotone if:

$$\forall Y \supseteq X : \mathcal{C}_M(X) \Rightarrow \mathcal{C}_M(Y)$$

independently from the given input transaction database.

Observe that the independence from the input transaction database is necessary since we want to distinguish between simple monotone constraints and global constraints such as the "*infrequency constraint*": $(supp_{TDB}(X) \leq \delta)$. This constraint is still monotone but it is dataset dependent and it requires dataset scans in order to be computed.

Since any conjunction of monotone constraints is a monotone constraint, in this paper we consider the problem:

$$Th(\mathcal{C}_{freq}) \cap Th(\mathcal{C}_M).$$

The concept of border is useful to characterize the solution space of the problem.

Definition 3. Given an anti-monotone constraint $\mathcal{C}_{AM}$ and a monotone constraint $\mathcal{C}_M$ we define their borders as:

$$B(\mathcal{C}_{AM}) = \{X | \forall Y \subset X : \mathcal{C}_{AM}(Y) \wedge \forall Z \supset X : \neg \mathcal{C}_{AM}(Z)\}$$

tID	itemset
1	b,d
2	b,d,e
3	b,c,d,e
4	c,e
5	c,d,e
6	c,d
7	a,d,e

item	price
a	4
b	3
c	5
d	6
e	15

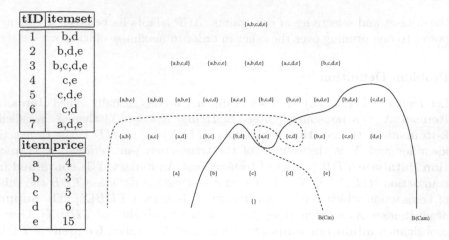

Fig. 1. The borders $B(\mathcal{C}_M)$ and $B(\mathcal{C}_{freq})$ for the transaction database and the price table on the left with $\mathcal{C}_M \equiv sum(X.prices) \geq 12$ and $\mathcal{C}_{AM} \equiv supp_{TDB}(X) \geq 2$.

$$B(\mathcal{C}_M) = \{X | \forall Y \supset X : \mathcal{C}_M(Y) \wedge \forall Z \subset X : \neg \mathcal{C}_M(Z)\}$$

Moreover, we distinguish between positive and negative borders. Given a general constraint $\mathcal{C}$ we define:

$$B^+(\mathcal{C}) = B(\mathcal{C}) \cap Th(\mathcal{C}) \quad B^-(\mathcal{C}) = B(\mathcal{C}) \cap Th(\neg \mathcal{C})$$

In Figure 1 we show the borders of two constraints: the anti-monotone constraint $supp(X) \geq 2$, and the monotone one $sum(X.prices) \geq 12$. In the given situation the borders are:

$$B^+(\mathcal{C}_M) = \{e, abc, abd, acd, bcd\} \quad B^+(\mathcal{C}_{freq}) = \{bde, cde\}$$
$$B^-(\mathcal{C}_M) = \{ab, ac, ad, bc, bd, cd\} \quad B^-(\mathcal{C}_{freq}) = \{a, bc\}$$

The solutions to our problem are the itemsets that lie under the anti-monotone border and over the monotone one: $R = \{e, be, ce, de, bde, cde\}$.

Our Contributions:

In the next section we provide a through characterization of the addressed computational problem, and we compare the two opposite extreme strategies. Then we introduce a general adaptive strategy, named *ACP*, which manages the two extremes in order to adapt its behavior to the given instance of the problem. The proposed strategy has the following interesting features:

- It exploits both monotone and anti-monotone constraints in order to prune the search space.
- It is able to adapt its behavior to the given input in order to maximize efficiency. This is the very first adaptive algorithm in literature on the problem.
- It computes the support of every solution itemset, which is necessary when we want to compute association rules.
- Being a level-wise solution it can be implemented exploiting the many optimization techniques and smart data structure studied for the apriori algorithm.

2   Level-Wise Solutions

Strategy $g \& t$ performs an *apriori* computation and then tests among frequent itemsets which ones satisfy also the monotone constraint. Strategy *mcp*, introduced by Boulicaut and Jeudy [3] works the opposite. The border $B^+(\mathcal{C}_M)$ is considered already computed and is given in input. Only itemsets over that border, and hence in $Th(\mathcal{C}_M)$, are generated by a special generation procedure **generate.** . Therefore we just need to check frequency for these candidates. The procedure **generate.** takes in input the set of the solutions at the last iteration R_k, the border $B^+(\mathcal{C}_M)$, and the maximal cardinality of an element in $B^-(\mathcal{C}_M)$, which we denote *maxb*. In the rest of this paper we use $Items_k$ to denote the set of all k-itemsets.

Procedure: generate$_m$($k, R_k, B^+(\mathcal{C}_M)$)

1. **if** $k = 0$ **then return** $B^+(\mathcal{C}_M) \cap Items$
2. **else if** $k \leq maxb$ **then**
3. **return** $generate_1(R_k, Items) \cup (B^+(\mathcal{C}_M) \cap Items_{k+1})$
4. **else if** $k > maxb$ **then**
5. **return** $generate_{apriori}(R_k)$

Where:

- $generate_1(R_k, X) = \{A \cup B | A \in R_k \land B \in X\}$
- $generate_{apriori}(R_k) = \{X | X \in Items_{k+1} \land \forall Y \in Items_k : Y \subseteq X . Y \in R_k\}$

The procedure **generate.** creates as candidates only supersets of itemsets which are solution at the last iteration. Thus these candidates only need to be checked against $\mathcal{C}_{freq}$ since they surely satisfy $\mathcal{C}_M$. These candidates are generated adding to a solution a 1-itemset. Unluckily we can not use the apriori trick completely with this strategy. In fact a candidate itemset can not be pruned away simply because all its subsets are not solution, since some of them could have not been considered at all. What we can do is prune whenever we know that at least one subset of the candidate itemset is not a solution because it does not satisfy $\mathcal{C}_{freq}$. This pruning is performed on the set of candidates C_k by the following procedure.

Strategy: *mcp*

1. $C_1 := generate_m(0, \emptyset, B^+(\mathcal{C}_M)); R_0 := \emptyset; \ k := 1$
2. **while** $C_k \neq \emptyset$ **or** $k \leq maxb$ **do**
3. $C_k := prune_m(R_{k-1}, C_k)$
4. **test** $\mathcal{C}_{freq}(\mathbf{C_k}); \ R_k := Th(\mathcal{C}_{freq}) \cap (C_k)$
5. $C_{k+1} := generate_m(k, R_k, B^+(\mathcal{C}_M)); \ k := k + 1$
6. **end while**
7. **return** $\bigcup_{i=1}^{k-1} R_i$

Procedure: prune$_m$(R_{k-1}, C_k)

1. $C' := C_k$
2. **for all** $S \in C_k$ **do for all** $S' \subseteq S : S' \in Items_{k-1}$
3. **do if** $S' \notin R_{k-1} \land \mathcal{C}_m(S')$ **then** remove S from C'
4. **return** C'

Example 4. Consider the executions of strategy *mcp* and strategy *g&t* on the dataset and the constraints in Figure 1, focussing on the numbers of checking of $\mathcal{C}_{freq}$. At the first iteration strategy *mcp* produces a unique candidate $C_1 = \{e\}$ which is the only 1-itemset in $B^+(\mathcal{C}_M)$. This candidate is checked for the anti-monotone constraint and it results to be a solution $R_1 = \{e\}$. At the second iteration 4 candidates are produced $C_2 = \{ae, be, ce, de\}$. Only *ae* does not satisfy $\mathcal{C}_{AM}$, hence $R_2 = \{be, ce, de\}$. At the third iteration 7 candidates are produced $C_3 = \{abc, abd, acd, bcd, bce, bde, cde\}$. Only two of these pass the anti-monotone checking: $R_3 = \{bde, cde\}$. Finally $C_4 = \emptyset$. Therefore, with the given dataset and constraints, strategy *mcp* performs $1 + 4 + 7 = 12$ checking of $\mathcal{C}_{freq}$. Strategy *g&t* uses a normal *apriori* computation in order to find $Th(\mathcal{C}_{freq})$ and then check the satisfaction of $\mathcal{C}_M$. It performs 13 checking of $\mathcal{C}_{freq}$.

Example 5. This example is borrowed by [3]. Suppose we want to compute frequent itemsets $\{X | supp(X) \geq 100 \wedge |X| \geq 10\}$. This is a conjunction of an anti-monotone constraint (frequency) with a monotone one (cardinality of the itemset ≥ 10). Strategy *mcp* generates no candidate of size lower than 10. Every itemset of size 10 is generated as candidate and tested for frequency in one database scan. This leads to at least $\binom{n}{10}$ where $n = |Items|$ candidates and, as soon as n is large this turns to be intractable. On the other hand strategy *g&t* generates candidates that will never be solutions, but this strategy remains tractable ever for large n.

The two examples show that no one of the two strategies outperforms the other on every input dataset and conjunction of constraints.

2.1   Strategies Analysis

We formally analyze the search space explored by the two extreme level-wise strategies. To this purpose we focus on the number of frequency tests, since the monotone constraint is cheaper to test.

Definition 6. Given a strategy $\mathcal{S}$ the number of frequency test performed by $\mathcal{S}$ is indicated with $|\mathcal{C}_{freq}|_{\mathcal{S}}$.

Generally, a strategy $\mathcal{S}$ checks for frequency a portion of $Th(\mathcal{C}_M)$ (which can produce solutions) and a portion of $Th(\neg\mathcal{C}_M)$ (which can not produce solutions):

$$|\mathcal{C}_{freq}|_{\mathcal{S}} = \gamma|Th(\neg\mathcal{C}_M)| + \beta|Th(\mathcal{C}_M)| \qquad \gamma, \beta \in [0, 1] \qquad (1)$$

The *mcp* strategy has $\gamma = 0$, but evidently it has a β much larger than strategy *g&t*, since it can not benefit from the pruning of infrequent itemsets in $Th(\neg\mathcal{C}_M)$, as we formalize later. Let us further characterize the portion of $Th(\neg\mathcal{C}_M)$ explored by strategy *g&t* as:

$$\gamma|Th(\neg\mathcal{C}_M)| = \gamma_1|Th(\neg\mathcal{C}_M) \cap Th(\mathcal{C}_{freq})| + \gamma_2|Th(\neg\mathcal{C}_M) \cap B^-(\mathcal{C}_{freq})| \qquad (2)$$

For *g&t* strategy $\gamma_1 = \gamma_2 = 1$: it explores all frequent itemsets $(Th(\mathcal{C}_{freq}))$ and candidate frequent itemsets that results to be infrequent $(B^-(\mathcal{C}_{freq}))$ even if they are in $Th(\neg\mathcal{C}_M)$ and thus can not produce solutions. Let us examine what happens over the monotone border. We can further characterize the explored portion of $Th(\mathcal{C}_M)$ as:

$$\beta|Th(\mathcal{C}_M)| = \beta_1|B^+(\mathcal{C}_M) \cap (Th(\neg\mathcal{C}_{freq}) \setminus B^-(\mathcal{C}_{freq}))| + \beta_2|R| + \qquad (3)$$
$$\beta_3|B^-(\mathcal{C}_{freq}) \cap Th(\mathcal{C}_M)|$$

Trivially $\beta_2 = 1$ for any strategy which computes all the solutions of the problem. Moreover, also β_3 is always 1 since we can not prune these border itemsets in any way. The only interesting variable is β_1, which depends from γ_2. Since the only infrequent itemsets checked by strategy $g\&t$ are itemsets in $B^-(\mathcal{C}_{freq})$, it follows that for this strategy $\beta_1 = 0$. On the other hand, strategy mcp generates as candidates all itemsets in $B^+(\mathcal{C}_M)$ (see line 3 of $generate_m$ procedure), thus for this strategy $\beta_1 = 1$. the following proposition summarizes the number of frequency tests computed by the two strategies.

Proposition 7.
$$|\mathcal{C}_{freq}|_{g\&t} = |Th(\neg\mathcal{C}_M) \cap Th(\mathcal{C}_{freq})|$$
$$+|Th(\neg\mathcal{C}_M) \cap B^-(\mathcal{C}_{freq})| + |R| + |B^-(\mathcal{C}_{freq}) \cap Th(\mathcal{C}_M)|$$
$$|\mathcal{C}_{freq}|_{mcp} = |B^+(\mathcal{C}_M) \cap (Th(\neg\mathcal{C}_{freq}) \setminus B^-(\mathcal{C}_{freq}))| + |R| + |B^-(\mathcal{C}_{freq}) \cap Th(\mathcal{C}_M)|$$

In the next section we introduce an adaptive algorithm which manages this tradeoff, balancing anti-monotone and monotone pruning w.r.t. the given input.

3   Adaptive Constraint Pushing

The main drawback of strategy $g\&t$ is that it explores portions of search space in $Th(\neg\mathcal{C}_M)$ which will never produce solutions ($\gamma_1 = 1$); while the main drawback of strategy mcp is that it can generate candidate itemsets in $Th(\neg\,\mathcal{C}_{freq}) \setminus B^-(\mathcal{C}_{freq})$ that would have been already pruned by a simple *apriori* computation ($\beta_1 = 1$). This is due to the fact that strategy mcp starts computation bottom-up from the monotone border and has no knowledge about the portion of search space below such a border($Th(\neg\,\mathcal{C}_M)$). However, some knowledge about small itemsets which do not satisfy $\mathcal{C}_M$ could be useful to have a smaller number of candidates over the border. But on the other hand, we need some additional computation below the monotone border in order to have some knowledge. Once again we face the tradeoff. The basic idea of a general adaptive pushing strategy (ACP) is to explore only a *portion* of $Th(\neg\,\mathcal{C}_M)$: this computation will never create solutions, but if well chosen it can prune heavily the computation in $Th(\mathcal{C}_M)$. In other terms, it tries to balance γ and β_1. To better understand we must further characterize the search space $Th(\neg\mathcal{C}_M)$:

$$\gamma|Th(\neg\mathcal{C}_M)| = \gamma_1|Th(\neg\mathcal{C}_M) \cap Th(\mathcal{C}_{freq})| + \gamma_2|Th(\neg\mathcal{C}_M) \cap B^-(\mathcal{C}_{freq})| + \qquad (4)$$
$$+\gamma_3|Th(\neg\mathcal{C}_M) \cap (Th(\neg\mathcal{C}_{freq}) \setminus B^-(\mathcal{C}_{freq}))|$$

Strategy ACP tries to reduce γ_1 but the price to pay is a possible reduction of γ_2. Since the portion of search space $Th(\neg\mathcal{C}_M) \cap B^-(\mathcal{C}_{freq})$ is helpful to prune, a reduction of γ_2 yields a reduction of pruning opportunities. This can lead to the exploration of a portion of search space ($\gamma_3 > 0$) that would have not been explored by a $g\&t$ strategy. This phenomenon can be seen as a virtual raising of the frequency border: suppose we loose an itemset of the frequency border, we will later explore some of its supersets and obviously find them infrequent. By the point of view of the strategy these are frequency border itemsets, even if they are not really in $B^-(\mathcal{C}_{freq})$. The optimal *ideal* strategy would have $\gamma_1 = \gamma_3 = 0$ and $\gamma_2 = 1$ since we are under the monotone border and we are just looking for infrequent itemsets in order to have pruning in $Th(\mathcal{C}_M)$. Therefore, a general *ACP* strategy should explore a portion of $Th(\neg \mathcal{C}_M)$ in order to find infrequent itemsets, trying not to loose pieces of $Th(\neg \mathcal{C}_M) \cap B^-(\mathcal{C}_{freq})$.

Proposition 8.

$$|\mathcal{C}_{freq}|_{ideal} = |Th(\neg \mathcal{C}_M) \cap B^-(\mathcal{C}_{freq})| + |R| + |B^-(\mathcal{C}_{freq}) \cap Th(\mathcal{C}_M)|$$

$$|\mathcal{C}_{freq}|_{acp} = \gamma|Th(\neg \mathcal{C}_M)| + \beta|Th(\mathcal{C}_M)| \ \textit{(as defined in equations (3) and (4))}.$$

Two questions arise:

1. *What is a "good" portion of candidates?*
2. *How large this set of candidates should be?*

The answer to the first question is simply: *"itemsets which have higher probabilities to be found infrequent"*. The answer to the second question is what the adaptivity of *ACP* is about. We define a parameter $\alpha \in [0, 1]$ which represents the fraction of candidates to be chosen among all possible candidates. This parameter is initialized after the first scan of the dataset using all information available, and it is updated level by level with the newly collected knowledge.

Let us now introduce some notation useful for the description of the algorithm.

- $L_k \subseteq \{I | I \in (Items_k \cap Th(\mathcal{C}_{freq}) \cap Th(\neg \mathcal{C}_M))\}$
- $N_k \subseteq \{I | I \in (Items_k \cap Th(\neg \mathcal{C}_{freq}) \cap Th(\neg \mathcal{C}_M))\}$
- $R_k = \{I | I \in (Items_k \cap Th(\mathcal{C}_{freq}) \cap Th(\mathcal{C}_M))\}$
- $P_k = \{I | I \in Items_k \wedge \forall n, m < k.(\nexists L \subset I.L \in B_m) \wedge (\nexists J \subset I.J \in N_n)\}$
- $B_k = \{I | I \in (B^+(\mathcal{C}_M) \cap P_k)\}$
- $E_k = \{I | I \in (Th(\neg \mathcal{C}_M) \cap P_k)\}$

L_k is the set of frequent k-itemsets which are under the monotone border: these have been checked for frequency even if they do not satisfy the monotone constraint hoping to find them infrequent; N_k is the set of infrequent k-itemsets under the monotone border: these are itemsets used to prune over the monotone border; R_k is the set of solutions k-itemsets; P_k is the set of itemsets potentially frequent (none of their subsets have been found infrequent) and potentially in $B^+(\mathcal{C}_M)$ (none of their subsets have been found satisfying $\mathcal{C}_M$). B_k is the subset of elements in P_k which satisfy $\mathcal{C}_M$ and hence are in $B^+(\mathcal{C}_M)$ since all their subsets are in $Th(\neg \mathcal{C}_M)$. E_k is the subset of elements in P_k which still do not satisfy $\mathcal{C}_M$. From this set is chosen an α-portion of elements to be checked against frequency constraint named $C_k^{\mathcal{U}}$ (candidates $\mathcal{U}$nder). This selection is indicated as $\alpha \otimes E_k$. Finally we have the set of candidates in which we can find solutions $C_k^{\mathcal{O}}$ (candidates $\mathcal{O}$ver) which is the set of candidates over the monotone border. The frequency test for these two candidates sets is performed with a unique database scan and data structure. Itemsets in $C_k^{\mathcal{O}}$ which satisfy $\mathcal{C}_{freq}$ will be solutions; itemsets in $C_k^{\mathcal{U}}$ which do not satisfy $\mathcal{C}_{freq}$ go in N_k and will prune itemsets in $C_j^{\mathcal{O}}$ for some $j > k$.

We now introduce the pseudo-code for the generic adaptive strategy. In the following with the sub-routine **generate....**, we mean **generate.** followed by the pruning of itemsets which are superset of itemsets in N (we call it **prune..**), followed by **prune.** (decribed in Section 2).

Strategy: generic *ACP*

---

1. $R_0, N := \emptyset; \quad C_1 := Items$
2. **test** $\mathcal{C}_{\mathbf{freq}}(\mathbf{C_1}) \Rightarrow C_1 := Th(\mathcal{C}_{freq}) \cap (C_1)$
3. **test** $\mathcal{C}_{\mathbf{M}}(\mathbf{C_1}) \Rightarrow R_1 := Th(\mathcal{C}_M) \cap (C_1); \; L_1 := Th(\neg\,\mathcal{C}_M) \cap (C_1)$
4. $P_2 := generate_{apriori}(L_1)$
5. $k := 2$
6. **while** $P_k \neq \emptyset$ **do**
7.   **test** $\mathcal{C}_{\mathbf{M}}(\mathbf{P_k}) \Rightarrow B_k := Th(\mathcal{C}_M) \cap (P_k); E_k := Th(\neg\mathcal{C}_M) \cap (P_k)$
8. $C_k^{\mathcal{O}} := generate_{over}(R_{k-1}, C_1, N) \cup B_k$
9. $initialize/update(\alpha)$
10. $C_k^{\mathcal{U}} := \alpha \otimes E_k$
11.   **test** $\mathcal{C}_{\mathbf{freq}}(\mathbf{C_k^{\mathcal{U}}} \cup \mathbf{C_k^{\mathcal{O}}}) \Rightarrow$
      $R_k := Th(\mathcal{C}_{freq}) \cap C_k^{\mathcal{O}}; \; L_k := Th(\mathcal{C}_{freq}) \cap C_k^{\mathcal{U}}; \; N_k := Th(\neg\mathcal{C}_{freq}) \cap C_k^{\mathcal{U}}$
12. $N := N \cup N_k$
13. $P_{k+1} := generate_{apriori}(E_k \setminus N_k)$
14. $k := k+1$
15. **end while**
16. $C_k := generate_{over}(R_{k-1}, C_1, N)$
17. **while** $C_k \neq \emptyset$ **do**
18.   **test** $\mathcal{C}_{\mathbf{freq}}(\mathbf{C_k}); R_k := Th(\mathcal{C}_{freq}) \cap (C_k)$
19. $C_{k+1} := generate_{apriori}(R_k)$
20. $k := k + 1$
21. **end while**
22. **return** $\bigcup_{i=1}^{k-1} R_i$

It is worthwhile to highlight that the pseudo-code given in Section 2 for strategy *mcp*, which is a theoretical strategy, does not perform the complete first anti-monotone test, while strategies *ACP* and *g&t* perform it. This results to be a reasonable choice on our toy-example, but it turns to be a suicide choice on every reasonably large dataset. Anyway, we can imagine that any practical implementation of *mcp* would perform at least this first anti-monotone test. We call this practical implementation strategy *mcp**. Moreover, strategy *mcp** does not take the monotone border in input but it discovers it level-wise as *ACP* does. In our experiments we will use *mcp** instead of *mcp*.

Note that:

- if $\alpha = 0$ constantly, then $ACP \equiv$ strategy *mcp**;
- if $\alpha = 1$ constantly, then $ACP \equiv$ strategy *g&t*;

Our adaptivity parameter can be seen as a setting knob which ranges from 0 to 1, from an extreme to the other. To better understand how *ACP* works, we show the execution given the input in Figure 1.

3.1 Run-through Example

Strategy ACP starts with $C_1 = \{a, b, c, d, e\}$, tests the frequency constraint, tests the monotone constraint and finds the first solution, $R_1 = \{e\}$ and $L_1 = \{b, c, d\}$. Now (Line 4) we generate the set of 2-itemsets potentially frequent and potentially in $B^+(\mathcal{C}_M)$: $P_2 = \{bc, bd, cd\}$. At this point we enter in the loop from line 6 to 15. The set P_2 is checked for $\mathcal{C}_M$, and it turns out that no element in P_2 satisfies the monotone constraint: thus $B_2 = \emptyset$, $E_2 = P_2$. At line 8 ACP generates candidates for the computation over the monotone border $C_2^O = \{be, ce, de\}$ and performs the two pruning procedure that in this case have no effects. At this point ACP initializes our adaptivity parameter $\alpha \in [0, 1]$. The procedure $initialize(\alpha)$ can exploit all the information collected so far, such as number of transactions, total number of 1-itemsets, support threshold, number of frequent 1-itemsets and their support, number of solutions at the first iteration. For this example suppose that α is initialized to 0.33. Line 10 assigns to $C_k^{\mathcal{U}}$, a portion equals to α of E_k. Intuitively, since we want to find infrequent itemsets in order to prune over the monotone border, the best third is the 2-itemset which has the subsets with the lowest support, therefore $C_2^{\mathcal{U}} = \{bc\}$. Line 11 performs the frequency test for both set of candidates sharing a unique dataset scan. The count of support gives back four solutions $R_2 = \{be, ce, de\}$, moreover we have $L_2 = \emptyset$ and $N_2 = \{bc\}$. Then ACP generates $P_3 = \emptyset$ (line 13) and exits the loop (line 6). At line 16 we generate $C_3 = \{bde, cde\}$, we check their support (line 18) and obtain that $R_3 = C_3$; finally we obtain $C_4 = \emptyset$ and we exit the second loop. Algorithm ACP performs $5 + 4 + 2 = 11$ tests of frequency.

4   Adaptivity Strategies and Optimization Issues

In the previous section we have introduced a generic strategy for adaptive constraint pushing. This can not really be considered an algorithm since we have left not instantiated the *initialize/update* function for the adaptivity parameter α (line 9), as well as the α-selection (line 10). In this section we propose a very simple adaptivity strategy for α and our first experimental results. We believe that many other different adaptivity strategies can be defined and compared.

Since we want to select itemsets which are most likely infrequent, the simplest idea is to estimate on the fly, using all information available at the moment, a support measure for all candidates itemsets below the monotone border. Then the α-selection (line 10) will simply choose among all itemsets in E_k the α-portion with lowest estimated support.

In our first set of experiments, we have chosen to estimate the support for an itemset using only the real support value of items belonging to the given itemset, and balancing two extreme conditions of the correlation measure among the items: complete independence and maximal correlation. In the former the estimated itemset support is obtained as the product, and in the latter as the minimum, of relative support of the items belonging to the itemset. Also for the α-adaptivity we have chosen a very simple strategy. The parameter α is initialized w.r.t. the number of items which satisfy frequency and monotone constraint at the first iteration. Then at every new iteration it adapts its value

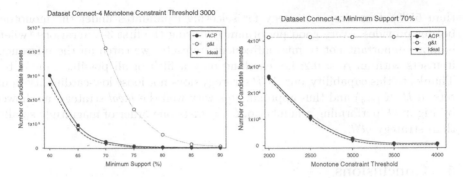

Fig. 2. Number of candidate itemsets tested against C_{freq}.

according to the results of the α-selection at the previous iteration. Let us define the α-focus as the ratio of itemsets found infrequent among α-selected itemsets. An α-focus very close to 1 (greater than 0.98) suggests that we have selected and counted too few candidates and thus we raise the α parameter for the next iteration accordingly. An α-focus less then 0.95 suggests that we are selecting and counting too much candidates and thus produces a shrink of α.

These two proposed strategies for estimating candidates support and for the adaptivity of α do not exploit all available information, but they allow an efficient implementation, and they experimentally exhibit very good candidates-selection capabilities.

Experimental Results

Since ACP balances the tradeoff between frequency and a monotone constraint, it gives the best performance when the two components are equally strong, i.e. no constraint is much more selective than the other. On sparse datasets frequency is always very selective even at very low support levels: joining it with an equally strong monotone constraint would result in an empty set of solutions. Therefore, ACP is particularly interesting in applications involving dense datasets.

In Figure 2, we show a comparison of the 4 strategies $g\&t$, mcp^*, $ideal$ and ACP, based on the portion of search space explored, i.e. the number of C_{freq} tests performed, on the well-known dense dataset $connect$-4 [1], for different support thresholds and monotone constraints. In order to create a monotone constraint we have attached to each item a value v selected using a normal distribution. Then we have chosen as monotone constraint the sum of values v in an itemset to be greater than a given threshold.

Strategy mcp^* always performs very poorly and its results could not be reported in the graph in Figure 2. Strategy $g\&t$ explores a portion of search space that obviously does not depend by the monotone constraint. On this dense dataset it performs poorly and becomes hard to compute for low supports (less

---

[1] http://www.ics.uci.edu/~mlearn/MLRepository.html

than 55%). Our simple strategy for selecting candidates under the monotone border provides a very good performance: during the first 3-4 iterations (where is more important not to miss infrequent itemsets) we catch all the infrequent itemsets with an $\alpha \approx 0.2$; i.e. checking only a fifth of all possible candidates. Thanks to this capability, our ACP strategy does not loose low-cardinality itemsets in $B^-(\mathcal{C}_{freq})$ and thus approximates very well the *ideal* strategy, as showed by Figure 2, performing a number of $\mathcal{C}_{freq}$ tests one order of magnitude smaller than strategy $g\&t$.

5 Conclusions

In this paper, we have deeply characterized the problem of the computation of a conjunction of monotone and anti-monotone constraints. As a result of this characterization, we introduce a generic adaptive strategy, named ACP (Adaptive Constraint Pushing) which exploits any conjunction of monotone and anti-monotone constraints to prune the search space. We have introduce an adaptivity parameter, called α which can be seen as a setting knob which ranges from 0 (favorite monotone pruning) to 1 (favorite anti-monotone pruning); and level by level adapts itself to the input dataset and constraints, giving more power to one pruning over the other in order to maximize efficiency. The generic algorithmic architecture presented can be instantiated with different adaptivity strategy for α. In this paper we have presented a very simple strategy which does not do not exploit all available information, but it still provides very good selection capability and it allows an efficient implementation.

References

1. R. Agrawal and R. Srikant. Fast Algorithms for Mining Association Rules in Large Databases. In *Proceedings of the Twentieth International Conference on Very Large Databases*, pages 487–499, Santiago, Chile, 1994.
2. F. Bonchi, F. Giannotti, A. Mazzanti, and D. Pedreschi. ExAnte: Anticipated data reduction in constrained pattern mining. In *Proceedings of the 7th European Conference on Principles and Practice of Knowledge Discovery in Databases (PKDD03)*, 2003.
3. J.-F. Boulicaut and B. Jeudy. Using constraints during set mining: Should we prune or not? In *Actes des Seizième Journées Bases de Donnúes Avancúes BDA'00, Blois (F)*, pages 221–237, 2000.
4. G. Grahne, L. Lakshmanan, and X. Wang. Efficient mining of constrained correlated sets. In *16th International Conference on Data Engineering (ICDE' 00)*, pages 512–524. IEEE, 2000.
5. R. T. Ng, L. V. S. Lakshmanan, J. Han, and A. Pang. Exploratory mining and pruning optimizations of constrained associations rules. In *Proceedings of the ACM SIGMOD International Conference on Management of Data (SIGMOD-98)*, volume 27,2 of *ACM SIGMOD Record*, pages 13–24, New York, June 1–4 1998. ACM Press.
6. J. Pei, J. Han, and L. V. S. Lakshmanan. Mining frequent item sets with convertible constraints. In *ICDE'01*, pages 433–442, 2001.

ExAnte: Anticipated Data Reduction in Constrained Pattern Mining

Francesco Bonchi[1,2,3], Fosca Giannotti[1,2],
Alessio Mazzanti[1,3], and Dino Pedreschi[1,3]

[·] Pisa KDD Laboratory*
http://www-kdd.cnuce.cnr.it
[·] ISTI - CNR Area della Ricerca di Pisa, Via Giuseppe Moruzzi, 1 - 56124 Pisa, Italy
Giannotti@cnuce.cnr.it
[·] Department of Computer Science, University of Pisa
Via F. Buonarroti 2, 56127 Pisa, Italy
{bonchi,mazzanti,pedre}@di.unipi.it

Abstract. Constraint pushing techniques have been proven to be effective in reducing the search space in the frequent pattern mining task, and thus in improving efficiency. But while pushing anti-monotone constraints in a level-wise computation of frequent itemsets has been recognized to be always profitable, the case is different for monotone constraints. In fact, monotone constraints have been considered harder to push in the computation and less effective in pruning the search space. In this paper, we show that this prejudice is ill founded and introduce ExAnte, a pre-processing data reduction algorithm which reduces dramatically both the search space and the input dataset in constrained frequent pattern mining. Experimental results show a reduction of orders of magnitude, thus enabling a much easier mining task. ExAnte can be used as a pre-processor with any constrained pattern mining algorithm.

1 Introduction

Constrained itemset mining i.e., finding all itemsets included in a transaction database that satisfy a given set of constraints, is an active research theme in data mining [3,6,7,8,9,10,11,12]. The most studied constraint is the frequency constraint, whose anti-monotonicity is used to reduce the exponential search space of the problem. Exploiting the anti-monotonicity of the frequency constraint is known as *apriori trick* [1,2]: it dramatically reduces the search space making the computation feasible. Frequency is not only computationally effective, it is also semantically important since frequency provides "support" to any discovered knowledge. For these reasons frequency is the base constraint of what is generally referred to as *frequent itemset mining*. However, many other constraints can facilitate user-focussed exploration and control, as well as reduce

---

* The present research is founded by "Fondazione Cassa di Risparmio di Pisa" under the "WebDigger Project".

N. Lavrač et al. (Eds.): PKDD 2003, LNAI 2838, pp. 59–70, 2003.

the computation. For instance, a user could be interested in mining all frequently purchased itemsets having a total price greater than a given threshold and containing at least two products of a given brand. Among these constraints, classes have been individuated which exhibit nice properties. The class of anti-monotone constraints is the most effective and easy to use in order to prune the search space. Since any conjunction of anti-monotone constraints is in turn anti-monotone, we can use the *apriori trick* to exploit completely the pruning power of the conjunction: the more anti-monotone constraints, the more selective the *apriori trick* will be.

The dual class, monotone constraints, has been considered more complicated to exploit and less effective in pruning the search space. As highlighted by Boulicaut and Jeudy in [3], pushing monotone constraints can lead to a reduction of anti-monotone pruning. Therefore, when dealing with a conjunction of monotone and anti-monotone constraints we face a tradeoff between anti-monotone and monotone pruning. Our observation is that the above consideration holds only if we focus completely on the search space of all itemsets, which is the approach followed by the work done so far.

In this paper we show that the most effective way of attacking the problem is to reason on both the itemsets search space and the transactions input database *together*. In this way, pushing monotone constraints does not reduce anti-monotone pruning opportunities, on the contrary, such opportunities are boosted. Dually, pushing anti-monotone constraints boosts monotone pruning opportunities: the two components strengthen each other recursively. We prove our previous statement by introducing ExAnte, a pre-processing data reduction algorithm which reduces dramatically both the search space and the input dataset in constrained frequent pattern mining.

ExAnte can exploit any constraint which has a monotone component, therefore also succinct monotone constraints [9] and convertible monotone constraints [10,11] can be used to reduce the mining computation. Being a preprocessing algorithm, ExAnte can be coupled with any constrained pattern mining algorithm, and it is always profitable to start any constrained patterns computation with an ExAnte preprocess. The correctness of ExAnte is formally proven in this paper, by showing that the reduction of items and transaction database does not affect the set of constrained frequent patterns, which are solutions to the given problem, as well as their support. We discuss a thorough experimentation of the algorithm, which points out how effective the reduction is, and which potential benefits it offers to subsequent frequent pattern computation.

Our Contributions:

Summarizing, the data reduction algorithm proposed in this paper is characterized by the following:

- ExAnte uses, for the first time, the real synergy of monotone and anti-monotone constraints to prune the search space and the input dataset: the total benefit is greater than the sum of the two individual benefits.

- ExAnte can be used with any constraint which has a monotone component: therefore also succinct monotone constraints and convertible monotone constraints can be exploited.
- ExAnte maintains the exact support of each solution itemset: a necessary condition if we want to compute Association Rules.
- ExAnte can be used to make feasible the discovery of particular patterns which can be discovered only at very low support level, for which the computation is unfeasible for traditional algorithms.
- Being a pre-processing algorithm, ExAnte can be coupled with any constrained pattern mining algorithm, and it is always profitable to start any constrained pattern computation with an ExAnte preprocess.
- ExAnte is efficient and effective: even a very large input dataset can be reduced of an order of magnitude in a small computation.
- A thorough experimental study has been performed with different monotone constraints on various datasets (both real world and synthetic datasets), and the results are described in details.

2   Problem Definition

Let $Items = \{x_1, ..., x_n\}$ be a set of distinct literals, usually called **items**. An **itemset** X is a subset of $Items$. If $|X| = k$ then X is called a **k-itemset**. A **transaction** is a couple $\langle tID, X \rangle$ where tID is the unique transaction identifier and X is the content of the transaction (an itemset). A **transaction database** TDB is a finite set of transactions. An itemset X is contained in a transaction $\langle tID, Y \rangle$ if $X \subseteq Y$. Given a transaction database TDB the subset of transactions which contain an itemset X is denoted $TDB[X]$. The **support** of an itemset X, written $supp_{TDB}(X)$ is the cardinality of $TDB[X]$. Given a user-defined **minimum support** δ, an itemset X is called **frequent** in TDB if $supp_{TDB}(X) \geq \delta$. This defines the frequency constraint $C_{freq}[TDB]$: if X is frequent we write $C_{freq}[TDB](X)$ or simply $C_{freq}(X)$ when the dataset is clear from the context.

Let $Th(C) = \{X | C(X)\}$ denotes the set all itemsets X that satisfy constraint C. The *frequent itemset mining problem* requires to compute the set of all frequent itemsets $Th(C_{freq})$. In general given a conjunction of constraints C the *constrained itemset mining problem* requires to compute $Th(C)$; the *constrained frequent itemsets mining problem* requires to compute $Th(C_{freq}) \cap Th(C)$.

We now formally define the notion of anti-monotone and monotone constraints.

Definition 1. Given an itemset X, a constraint C_{AM} is anti-monotone if

$$\forall Y \subseteq X : C_{AM}(X) \Rightarrow C_{AM}(Y)$$

If C_{AM} holds for X then it holds for any subset of X.   □

The frequency constraint is clearly anti-monotone. This property is used by the APRIORI algorithm with the following heuristic: if an itemset X does not satisfy C_{freq}, then no superset of X can satisfy C_{freq}, and hence they can be pruned.

This pruning can affect a large part of the search space, since itemsets form a lattice. Therefore the APRIORI algorithm operates in a level-wise fashion moving bottom-up on the itemset lattice, and each time it finds an infrequent itemset it prunes away all its supersets.

Definition 2. Given an itemset X, a constraint $\mathcal{C}_M$ is monotone if:

$$\forall Y \supseteq X : \mathcal{C}_M(X) \Rightarrow \mathcal{C}_M(Y)$$

independently from the given input transaction database. If $\mathcal{C}_M$ holds for X then it holds for any superset of X.   □

Note that in the last definition we have required a monotone constraint to be satisfied independently from the given input transaction database. This is necessary since we want to distinguish between simple monotone constraints and global constraints such as the *"infrequency constraint"*:

$$supp_{TDB}(X) \leq \delta.$$

This constraint is still monotone but has different properties since it is dataset dependent and it requires dataset scans in order to be computed. Obviously, since our pre-processing algorithm reduces the transaction dataset, we want to exclude the infrequency constraint from our study. Thus, our study focuses on "local" monotone constraints, in the sense that they depend exclusively on the properties of the itemset (as those ones in Table 1), and not on the underlying transaction database.

The general problem that we consider in this paper is the mining of itemsets which satisfy a conjunction of monotone and anti-monotone constraints:

$$Th(\mathcal{C}_{AM}) \cap Th(\mathcal{C}_M).$$

Since any conjunction of anti-monotone constraints is an anti-monotone constraint, and any conjunction of monotone constraints is a monotone constraint, in this paper we focus on the problem given by the conjunction of frequency anti-monotone constraint $(\mathcal{C}_{AM} \equiv supp_{TDB}(X) \geq \delta)$, with various simple monotone constraints (see Table 1).

$$Th(\mathcal{C}_{freq}) \cap Th(\mathcal{C}_M).$$

However, our algorithm can work with any conjunction of anti-monotone constraints, provided that the frequency constraint is included in the conjunction: the more anti-monotone constraints, the larger the data reduction will be.

3 Search Space and Input Data Reduction

As already stated, if we focus only on the itemsets lattice, pushing monotone constraint can lead to a less effective anti-monotone pruning. Suppose that an itemset has been removed from the search space because it does not satisfy some

Table 1. Monotone constraints considered in our analysis.

Monotone constraint	$\mathcal{C}_M \equiv$
cardinality	$|X| \geq n$
sum of prices	$sum(X.prices) \geq n$
maximum price	$max(X.prices) \geq n$
minimum price	$min(X.prices) \leq n$
range of prices	$range(X.prices) \geq n$

monotone constraints $\mathcal{C}_M$. This pruning avoids checking support for it, but it may be that if we check support, the itemset could result to be infrequent, and thus all its supersets could be pruned away. By monotone pruning an itemset we risk to lose anti-monotone pruning opportunities given from the itemset itself. The tradeoff is clear [3]: pushing monotone constraint can save tests on anti-monotone constraints, however the results of these tests could have lead to more effective pruning. In order to obtain a real amalgam of the two opposite pruning strategies we have to consider the constrained frequent patterns problem in its whole: not focussing only on the itemsets lattice but considering it together with the input database of transactions. In fact, as proved by the theorems in the following section, monotone constraints can prune away transactions from the input dataset *without losing solutions*. This monotone pruning of transactions has got another positive effect: while reducing the number of transactions in input it reduces the support of items too, hence the total number of frequent 1-itemsets. In other words, the monotone pruning of transactions strengthens the anti-monotone pruning. Moreover, infrequent items can be deleted by the computation and hence pruned away from the transactions in the input dataset. This anti-monotone pruning has got another positive effect: reducing the size of a transaction which satisfies a monotone constraint can make the transaction violates the monotone constraint. Therefore a growing number of transactions which do not satisfy the monotone constraint can be found. We are clearly inside a loop where two different kinds of pruning cooperates to reduce the search space and the input dataset, strengthening each other step by step until no more pruning is possible (a fix-point has been reached). This is precisely the idea underlying ExAnte.

3.1 ExAnte Properties

In this section we formalize the basic ideas of ExAnte. First we define the two kinds of reduction, then we prove the completeness of the method. In the next section we provide the pseudo-code of the algorithm.

Definition 3 (μ-reduction). Given a transaction database TDB and a monotone constraint $\mathcal{C}_M$, we define the μ-reduction of TDB as the dataset resulting from pruning the transactions that do not satisfy $\mathcal{C}_M$.

$$\mu[TDB]_{\mathcal{C}_M} = \{\langle tID, X \rangle \mid \langle tID, X \rangle \in TDB \wedge X \in Th(\mathcal{C}_M)\}$$

□

Definition 4 (α-reduction). Given a transaction database TDB, a transaction $\langle tID, X \rangle$ and a frequency constraint $\mathcal{C}_{freq}[TDB]$, we define the α-reduction of $\langle tID, X \rangle$ as the transaction resulting from pruning the items in X that do not satisfy $\mathcal{C}_{freq}[TDB]$.

$$\alpha[\langle tID, X \rangle]_{\mathcal{C}_{freq}[TDB]} = \langle tID, (F_1 \cap X) \rangle$$

Where: $F_1 = \{i \in Items \,|\, \{i\} \in Th(\mathcal{C}_{freq}[TDB])\}$. We define the α-reduction of TDB as the dataset resulting from the α-reduction of all transactions in TDB. □

The following two key theorems state that we can always μ-reduce and α-reduce a dataset without reducing the support of solution itemsets. Moreover, since satisfaction of $\mathcal{C}_M$ is independent from the transaction dataset, all solution itemsets will still satisfy it. Therefore, we can always μ-reduce and α-reduce a dataset without losing solutions.

Theorem 5 (μ-reduction correctness). *Given a transaction database TDB, a monotone constraint $\mathcal{C}_M$, and a frequency constraint $\mathcal{C}_{freq}$, we have that:*

$$\forall X \in Th(\mathcal{C}_{freq}[TDB]) \cap Th(\mathcal{C}_M) : supp_{TDB}(X) = supp_{\mu[TDB]_{\mathcal{C}_M}}(X).$$

Proof. Since $X \in Th(\mathcal{C}_M)$, all transactions containing X will also satisfy $\mathcal{C}_M$ for the monotonicity property. Therefore no transaction containing X will be μ-pruned (in other words: $TDB[X] \subseteq \mu[TDB]_{\mathcal{C}_M}$). This, together with the definition of support, implies the thesis. □

Theorem 6 (α-reduction correctness).
Given a transaction database TDB, a monotone constraint $\mathcal{C}_M$, and a frequency constraint $\mathcal{C}_{freq}$, we have that:

$$\forall X \in Th(\mathcal{C}_{freq}[TDB]) \cap Th(\mathcal{C}_M) : supp_{TDB}(X) = supp_{\alpha[TDB]_{\mathcal{C}_{freq}}}(X).$$

Proof. Since $X \in Th(\mathcal{C}_{freq})$, all subsets of X will be frequent (by the anti-monotonicity of frequency). Therefore no 1-itemsets in X will be α-pruned (in other words: $TDB[X] \subseteq \alpha[TDB]_{\mathcal{C}_{freq}}$) . This, together with the definition of support, implies the thesis. □

3.2   ExAnte Algorithm

The two theorems above suggest a fix-point computation. ExAnte starts the first iteration as any frequent pattern mining algorithm: counting the support of singleton items. Items that are not frequent are thrown away once and for all. But during this first count only transactions that satisfy $\mathcal{C}_M$ are considered. The other transactions are signed to be pruned from the dataset (μ-reduction). Doing so we reduce the number of interesting 1-itemsets. Even a small reduction of this number represents a huge pruning of the search space. At this point

ExAnte deletes from alive transactions all infrequent items (α-reduction). This pruning can reduce the monotone value (for instance, the total sum of prices) of some alive transactions, possibly resulting in a violation of the monotone constraints. Therefore we have another opportunity of μ-reducing the dataset. But μ-reducing the dataset we create new opportunities for α-reduction, which can turn in new opportunities for μ-reduction, and so on, until a fix-point is reached. The pseudo-code of ExAnte algorithm follows:

Procedure: **ExAnte**($TDB, \mathcal{C}_M, min\_supp$)

---

1. $I = \emptyset$;
2. **forall** transactions t in TDB **do**
3.   **if** $\mathcal{C}_M(t)$ **then forall** items i in t **do**
4. $i.count++$; **if** $i.count == min\_supp$ **then** $I = I \cup \{i\}$;
5. $old\_number\_interesting\_items = |Items|$;
6. **while** $|I| < old\_number\_interesting\_items$ **do**
7. $TDB = \alpha[TDB]_{\mathcal{C}_{freq}}$;
8. $TDB = \mu[TDB]_{\mathcal{C}_M}$;
9. $old\_number\_interesting\_items = |I|$;
10. $I = \emptyset$;
11.   **forall** transactions t in TDB **do**
12.   **forall** items i in t **do**
13. $i.count + +$;
14.   **if** $i.count == min\_supp$ **then** $I = I \cup \{i\}$;
15. **end while**

Fig. 1. The ExAnte algorithm pseudo-code.

Clearly, a fix-point is eventually reached after a finite number of iterations, as at each step the number of alive items strictly decreases.

3.3   Run–through Example

Suppose that the transaction and price datasets in Table 2 are given. Suppose that we want to compute frequent itemsets ($min\_supp = 4$) with a sum of prices ≥ 45. During the first iteration the total price of each transaction is checked to avoid using transactions which do not satisfy the monotone constraint. All transaction with a sum of prices ≥ 45 are used to count the support for the singleton items. Only the fourth transaction is discarded. At the end of the count we find items a, e, f and h to be infrequent. Note that, if the fourth transaction had not been discarded, items a and e would have been counted as frequent. At this point we perform an α-reduction of the dataset: this means removing a, e, f and h from all transactions in the dataset. After the α-reduction we have more opportunities to μ-reduce the dataset. In fact transaction 2, which at the beginning has a total price of 63, now has its total price reduced to 38 due to the pruning of a and e. This transaction can now be pruned away. The same

Table 2. Run-through Example: price table (a) and transaction database (b), items and their supports iteration by iteration (c).

item	price
a	5
b	8
c	14
d	30
e	20
f	15
g	6
h	12

(a)

tID	Itemset	Total price
1	b,c,d,g	58
2	a,b,d,e	63
3	b,c,d,g,h	70
4	a,e,g	31
5	c,d,f,g	65
6	a,b,c,d,e	77
7	a,b,d,f,g,h	76
8	b,c,d	52
9	b,e,f,g	49

(b)

Supports			
Items	1_{st}	2_{nd}	3_{rd}
a	3	†	†
b	7	4	4
c	5	5	4
d	7	5	4
e	3	†	†
f	3	†	†
g	5	3	†
h	2	†	†

(c)

reasoning holds for transactions number 7 and 9. At this point ExAnte counts once again the support of alive items with the reduced dataset. The item g which initially has got a support of 5 now has become infrequent (see Table 2 (c) for items support iteration by iteration). We can α-reduce again the dataset, and then μ-reduce. After the two reductions transaction number 5 does not satisfy anymore the monotone constraint and it is pruned away. ExAnte counts again the support of items on the reduced datasets but no more items are found to have turned infrequent. The fix-point has been reached at the third iteration: the dataset has been reduced from 9 transactions to 4 transactions (number 1,3,6 and 8), and interesting itemsets have shrunk from 8 to 3 (b, c and d). At this point any constrained frequent pattern mining algorithm would find very easily the unique solution to problem which is the 3-itemset $\{b, c, d\}$.

4   Experimental Results

In this section we deeply describe the experimental study that we have conducted with different monotone constraints on various datasets. In particular, the monotone constraints used in the experimentation are in Table 1. In addition, we have experimented a harder to exploit constraint: $avg(X.prices) \geq n$. This constraint is clearly neither monotone nor anti-monotone, but can exhibit a monotone (or anti-monotone) behavior if items are ordered by ascending (or descending) price, and frequent patterns are computed following a prefix-tree approach. This class of constraints, named *convertible*, has been introduced in [10]. In our experiments the constraint $avg(X.prices) \geq n$ is treated by inducing a weaker monotone constraint: $max(X.prices) \geq n$. Note that in every reported experiment we have chosen monotone constraints thresholds that are not very selective: there are always solutions to the given problem. In the experiments reported in this paper we have used two datasets. "IBM" is a synthetic dataset obtained with the most commonly adopted dataset generator, available from IBM Almaden[1]. We have generate a very large dataset since we have not been

. http://www.almaden.ibm.com/software/quest/Resources/datasets/syndata.html#assocSynData

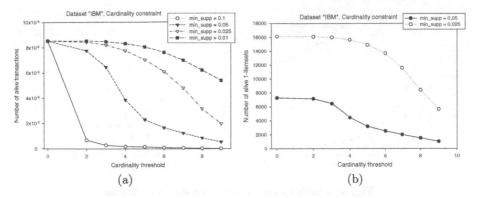

Fig. 2. Transactions reduction (a), and interesting 1-itemsets reduction (b), on dataset "IBM".

Table 3. Characteristics of the datasets used in the experiments.

Dataset	Transactions	Items	Max Trans Size	Avg Trans Size
IBM	8,533,534	100,000	37	11.21
Italian	186,824	4800	31	10.42

Dataset	Min Price	Max Price	Avg Price
Italian	100	900,000	6454.87

able to find a real-world dataset over one million transactions. "Italian" is a real-world dataset obtained from an Italian supermarket chain within a market-basket analysis project conducted by our research lab, few years ago (note that the prices are in the obsolete currency Italian Lira).

For a more detailed report of our experiments see [5]. In Figure 2 (a) the reduction of the number of transactions w.r.t the cardinality threshold is shown for four different support thresholds on the synthetic dataset. When the cardinality threshold is equal to zero the number of transactions equals the total number of transactions in the database, since there is no monotone pruning. Already for a low support threshold as 0.1% with a cardinality threshold equals to 2 the number of transactions decreases dramatically. Figure 2 (b) describes the reduction of number of interesting 1-itemsets on the same dataset.

As already stated, even a small reduction in the number of relevant 1-itemsets represents a very large pruning of the search space. In our experiments, as a measure of the search space explored, we have considered the number of candidate itemsets generated by a level-wise algorithm such as Apriori. In Figure 3 is reported a comparison of the number of candidate itemsets generated by Apriori and by *ExAnteApriori* (ExAnte pre-processing followed by Apriori) on the "Italian" dataset with various constraints. The dramatic search space reduction is evident, and it will be confirmed by computation time reported in the next

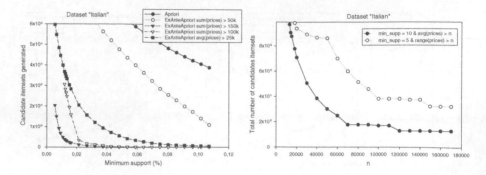

Fig. 3. Search space reduction on dataset "Italian".

section. How the number of candidate itemsets shrinks by increasing strength of the monotone constraint is also reported in Figure 3. This figure also highlights another interesting feature of ExAnte: even at very low support level (min_supp = 5 on a dataset of 186,824 transactions) the frequent patterns computation is feasible if coupled with a monotone constraint. Therefore, ExAnte can be used to make feasible the discovery of particular patterns which can be discovered only at very low support level, for instance:

- extreme purchasing behaviors (such as patterns with a very high average of prices);
- very long patterns (using the cardinality constraint coupled with a very low support threshold).

We report run-time comparison between Apriori and *ExAnteApriori*. We have chosen Apriori as the "standard" frequent pattern mining algorithm. Recall that every frequent pattern mining algorithm can be coupled with ExAnte pre-processing obtaining similar benefits. Execution time is reported in Figure 4. The large search space pruning reported in the previous section is here confirmed by the execution time.

5   Related Work

Being a pre-processing algorithm, ExAnte can not be directly compared with any previously proposed algorithm for constrained frequent pattern mining. However, it would be interesting to couple ExAnte data reduction with those algorithms and to measure the improve in efficiency. Among constrained frequent pattern mining algorithms, we would like to mention $\mathcal{FIC}^{\mathcal{M}}$ [11] and the recently proposed DualMiner [4].

6   Conclusions and Future Work

In this paper we have introduced ExAnte, a pre-processing data reduction algorithm which reduces dramatically the search space the input dataset, and hence

Iteration	Transactions	1-itemsets
0	17306	2010
1	13167	1512
2	11295	1205
3	10173	1025
4	9454	901
5	9005	835
6	8730	785
7	8549	754
8	8431	741
9	8397	736
10	8385	734
11	8343	729
12	8316	726
13	8312	724
14	8307	722
15	8304	722
Execution time: 1.5 sec		

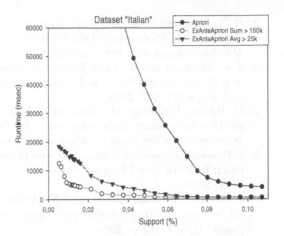

Fig. 4. A typical execution of ExAnte: Dataset "Italian" with $min\_sup = \%40$ and sum of prices ≥ 100000 (on the left); and a runtime comparison between Apriori and ExAnteApriori with two different constraints (on the right).

the execution time, in constrained frequent pattern mining. We have proved experimentally the effectiveness of our method, using different constraints on various datasets. Due to its capacity in focussing on any particular instance of the problem, ExAnte exhibits very good performance also when one of the two constraints (the anti-monotone or the monotone) is not very selective. This feature makes ExAnte useful to discover particular patterns which can be discovered only at very low support level, for which the computation is unfeasible for traditional algorithms.

We are actually developing a new algorithm for constrained frequent pattern mining, which will take full advantage of ExAnte pre-processing. We are also interested in studying in which other mining tasks ExAnte can be useful. We will investigate its applicability to the constrained mining of closed itemsets, sequential patterns, graphs structure, and other complex kinds of data and patterns. ExAnte executable can be downloaded by our web site:
http://www-kdd.cnuce.cnr.it/

Acknowledgements

We are indebted with Laks V.S. Lakshmanan, who first suggested the problem to the first author.

References

1. R. Agrawal, T. Imielinski, and A. N. Swami. Mining association rules between sets of items in large databases. In P. Buneman and S. Jajodia, editors, *Proceedings of the 1993 ACM SIGMOD International Conference on Management of Data*, pages 207–216, Washington, D.C., 26–28 May 1993.

2. R. Agrawal and R. Srikant. Fast Algorithms for Mining Association Rules in Large Databases. In *Proceedings of the Twentieth International Conference on Very Large Databases*, pages 487–499, Santiago, Chile, 1994.

3. J.-F. Boulicaut and B. Jeudy. Using constraints during set mining: Should we prune or not? In *Actes des Seizième Journúes Bases de Donnúes Avancúes BDA'00, Blois (F)*, pages 221–237, 2000.

4. C. Bucila, J. Gehrke, D. Kifer, and W. White. Dualminer: A dual-pruning algorithm for itemsets with constraints. In *Proceedings of the 8th ACM SIGKDD International Conference on Knowledge Discovery and Data Mining*, 2002.

5. F.Bonchi, F.Giannotti, A.Mazzanti, and D.Pedreschi. Exante: a preprocessing algorithm for constrained frequent pattern mining. Technical Report ISTI-B4-2003-07, ISTI, 2003.

6. G. Grahne, L. Lakshmanan, and X. Wang. Efficient mining of constrained correlated sets. In *16th International Conference on Data Engineering (ICDE' 00)*, pages 512–524. IEEE, 2000.

7. J. Han, L. V. S. Lakshmanan, and R. T. Ng. Constraint-based, multidimensional data mining. *Computer*, 32(8):46–50, 1999.

8. L. V. S. Lakshmanan, R. T. Ng, J. Han, and A. Pang. Optimization of constrained frequent set queries with 2-variable constraints. *SIGMOD Record (ACM Special Interest Group on Management of Data)*, 28(2), 1999.

9. R. T. Ng, L. V. S. Lakshmanan, J. Han, and A. Pang. Exploratory mining and pruning optimizations of constrained associations rules. In *Proceedings of the ACM SIGMOD International Conference on Management of Data (SIGMOD-98)*, volume 27,2 of *ACM SIGMOD Record*, pages 13–24, New York, June 1–4 1998. ACM Press.

10. J. Pei and J. Han. Can we push more constraints into frequent pattern mining? In R. Ramakrishnan, S. Stolfo, R. Bayardo, and I. Parsa, editors, *Proceedinmgs of the 6th ACM SIGKDD International Conference on Knowledge Discovery and Data Mining (KDD'00)*, pages 350–354, N. Y., Aug. 20–23 2000. ACM Press.

11. J. Pei, J. Han, and L. V. S. Lakshmanan. Mining frequent item sets with convertible constraints. In *(ICDE'01)*, pages 433–442, 2001.

12. R. Srikant, Q. Vu, and R. Agrawal. Mining association rules with item constraints. In D. Heckerman, H. Mannila, D. Pregibon, and R. Uthurusamy, editors, *Proc. 3rd Int. Conf. Knowledge Discovery and Data Mining, KDD*, pages 67–73. AAAI Press, 14–17 Aug. 1997.

Minimal k-Free Representations of Frequent Sets

Toon Calders[1] and Bart Goethals[2]

. University of Antwerp, Belgium
. Helsinki Institute for Information Technology, Finland

Abstract. Due to the potentially immense amount of frequent sets that can be generated from transactional databases, recent studies have demonstrated the need for concise representations of all frequent sets. These studies resulted in several successful algorithms that only generate a lossless subset of the frequent sets. In this paper, we present a unifying framework encapsulating most known concise representations. Because of the deeper understanding of the different proposals thus obtained, we are able to provide new, provably more concise, representations. These theoretical results are supported by several experiments showing the practical applicability.

1 Introduction

The frequent itemset mining problem is by now well known [1]. We are given a set of items $\mathcal{I}$ and a database $\mathcal{D}$ of subsets of $\mathcal{I}$. The elements of $\mathcal{D}$ are called transactions. An *itemset* $I \subseteq \mathcal{I}$ is some set of items; its *support* in $\mathcal{D}$, denoted $support(I, \mathcal{D})$, is defined as the number of transactions in $\mathcal{D}$ that contain all items of I. An itemset is called *s-frequent* in $\mathcal{D}$ if its support in $\mathcal{D}$ exceeds s. The database $\mathcal{D}$ and the minimal support s are omitted when they are clear from the context. The goal is now, given a minimal support threshold and a database, to find all frequent itemsets. The set of all frequent itemsets is denoted $\mathcal{F}(\mathcal{D}, s)$, the set of infrequent sets is denoted $\overline{\mathcal{F}}(\mathcal{D}, s)$.

Recent studies on frequent itemset mining algorithms resulted in significant performance improvements. However, if the minimal support threshold is set too low, or the data is highly correlated, the number of frequent itemsets itself can be extremely large. To overcome this problem, recently several proposals have been made to construct a concise representation [13] of the frequent itemsets, instead of mining all frequent itemsets: *Closed sets* [2,4,14,15,16], *Free sets* [5], *Disjunction-Free Sets* [6,10], *Generalized Disjunction-Free Generators* [12,11], and *Non-Derivable Itemsets* [8].

A *Concise Representation of frequent sets* is a subset of all frequent sets with their supports that contains enough information to construct all frequent sets with their support. Therefore, based on the representation, for each itemset I, we must be able to (a) decide whether I is frequent, and (b) if I is frequent, produce its support.

Mannila et al. [13] introduced the notion of a concise representation in a more general context. Our definition resembles theirs, but for reasons of simplicity we only concentrate on representations that are exact, and for frequent itemsets.

N. Lavrač et al. (Eds.): PKDD 2003, LNAI 2838, pp. 71–82, 2003.
© Springer-Verlag Berlin Heidelberg 2003

For representations the term *concise* will refer to their space-efficiency; that is, $\mathcal{R}$ is called *more concise* than $\mathcal{R}'$ if for every database $\mathcal{D}$ and support threshold s, $\mathcal{R}(\mathcal{D}, s)$ is smaller than or equal to $\mathcal{R}'(\mathcal{D}, s)$.

We introduce new representations based on the deduction rules for support presented in [8]. Many of the proposals in the literature, such as the *free sets* [5], the *disjunction-free sets* [6,10], the *generalized disjunction-free sets* [12,11], the *disjunction-free generators* [10], the *generalized disjunction-free generators* [11,12], and the *non-derivable itemsets* [8] representations, will be shown to be manifestations of this method. As such, the proposed method serves as a unifying framework for these representations.

The organization of the paper is as follows. In Section 2 we briefly describe different concise representations in the literature. Section 3 revisits the deduction rules introduced in [8]. In Section 4, a unifying framework for different concise representations is given, based on the deduction rules. Also new, minimal, representations are introduced. In Section 5 we present the results of experiments concerning the size of the different representations.

2   Related Work

Closed Sets. The first successful concise representation was the closed set representation introduced by Pasquier et al. [14]. In short, a *closed set* is an itemset such that its frequency does not equal the frequency of any of its supersets. The collection of the frequent closed sets together with their supports is a concise representation. This representation will be denoted *ClosedRep*.

Generalized Disjunction-Free Sets.   [11,12] Let X, Y be two disjunct itemsets. The *disjunctive rule* $X \to \bigvee Y$ is said to *hold in the database* $\mathcal{D}$, if every transaction in $\mathcal{D}$ that contains X, also contains at least one item of Y. A set I is called *generalized disjunction-free* if there do not exist disjunct subsets X, Y of I such that $X \to \bigvee Y$ holds. The set of all generalized disjunction free sets is denoted *GDFree*.

In [12], a representation based on the frequent generalized disjunction-free sets is introduced. On the one hand, based on the supports of all subsets of a set I (including I), it can be decided whether I is generalized disjunction-free or not. On the other hand, if a disjunctive rule $X \to \bigvee Y$ holds, the support of every superset I of $X \cup Y$ can be constructed from the supports of its subsets. For example, $a \to b \vee c$ holds if and only if for every superset X of abc,

$$supp(X) = supp(X - b) + supp(X - c) - supp(X - bc) .$$

Hence, if we know that a rule $X \to \bigvee Y$ holds, there is no need to store supersets of $X \cup Y$ in the representation.

However, the set of frequent generalized disjunction-free sets *FGDFree* is not a representation. We illustrate this with an example. Suppose that *FGDFree* completed with the supports is $\{(\emptyset, 10), (a, 5), (b, 4), (c, 3), (ab, 3)\}$. What conclusion should be taken for the set ac? There can be two reasons for ac to be left out of the representation: (a) because ac is infrequent, or (b) because ac is not generalized disjunction-free. Furthermore, suppose that ac was left out because it is not generalized disjunction-free. Since we have no clue which disjunctive rule holds for ac, we cannot produce its support. Hence, *FGDFree* completed with the supports of the sets clearly is not a representation. This problem is resolved in [12] by adding a part of the *border* of the set *FGDFree* to the representation.

Definition 1. *Let S be a set of itemsets.* $\mathcal{B}(S) = \{J \mid J \notin S, \forall J' \subset J : J' \in S\}$.

Suppose that we also store the sets in $\mathcal{B}(FGDFree)$ in the representation. Let I be a set not in $FGDFree \cup \mathcal{B}(FGDFree)$. There exists a set $J \subset I$ in $\mathcal{B}(FGDFree)$. The set J is either infrequent, or not generalized disjunction-free. If J is infrequent, then I is as well. If J is not generalized disjunction-free, then the supports of all subsets of J (including the support of J) allow for determining the rule $X \rightarrow \bigvee Y$ that holds for J. Hence, we know a rule $X \rightarrow \bigvee Y$ that holds for I $(X, Y \subseteq J \subset I)$. Therefore, from the supports of all strict subsets of I, we can derive the support of I using this rule. Using induction on the cardinality of I, it can easily be proven that $FGDFree \cup \mathcal{B}(FGDFree)$ completed with the supports is a representation. For the details, we refer to [11,12].

It is also remarked in [12] that it is not necessary to store the complete border $\mathcal{B}(FGDFree)$. For example, we could decide to leave out the infrequent sets. When reconstructing the complete set of frequent itemsets, we will be able to recognize these infrequent sets in the border because they are the only sets that have all their strict subsets in *FGDFree*, but that are not in the representation themselves. Other alternatives are the *generalized disjunction-free generators representation* (*GDFreeGenRep*) [12] and the representations in Section 4.

Free and Disjunction-Free Sets. [5,6,10] Free and disjunction-free sets are special cases of generalized disjunction-free sets. For free sets, the righthand side of the rules $X \rightarrow \bigvee Y$ is restricted to singletons, for disjunction free sets to singletons and pairs. Hence, a set I is free if and only if there does not exist a rule $X \rightarrow a$ that holds with $X \cup \{a\} \subseteq I$, and I is disjunction-free if there does not exists a rule $X \rightarrow a \vee b$ that holds with $X \cup \{a, b\} \subseteq I$. The free and disjunction-free sets are denoted respectively by *Free* and *DFree*, the frequent free and frequent disjunction-free sets by *FFree* and *FDFree*.

Again, neither *FFree* nor *FDFree* completed with the supports form a concise representation. The reasons are the same as explained for the generalized disjunction-free sets above. Hence, for the representations based on the free sets and the disjunction-free sets, (parts of) the border must be stored as well. Which parts of the border are stored can have a significant influence on the size of the representations, since the border is often very large, sometimes even larger than the total number of frequent itemsets.

However, the parts of the border that are stored in the representations presented in [5,6,10,11,12] are often far from optimal. In this paper we describe a unifying framework for these disjunctive-rule based representations. This framework is based on the deduction rules for support presented in [8] and revisited in Section 3. The framework allows a neat description of the different strategies used in the free, disjunction-free and generalized disjunction-free based representations. Due to the deeper understanding of the problem resulting from the unifying framework, we are able to find new and more concise representations that drastically reduce the number of sets to be stored.

3   Deduction Rules

In this section we review the deduction rules introduced in [8]. These rules derive bounds on the support of an itemset I if the supports of all strict subsets of I are known. In [7], it is shown that these rules are sound and complete; that is, they compute the best possible bounds.

Let a *generalized itemset* be a conjunction of items and negations of items. For example, $G = \{a, b, \overline{c}, d\}$ is a generalized itemset. A transaction T *contains a general itemset* $G = X \cup \overline{Y}$ if $X \subseteq T$ and $T \cap Y = \emptyset$. The *support of a generalized itemset* G *in a database* $\mathcal{D}$ is the number of transactions of $\mathcal{D}$ that contain G.

We say that a general itemset $G = X \cup \overline{Y}$ is *based* on itemset I if $I = X \cup Y$. From the well known inclusion-exclusion principle [9], we know that for a given general itemset $G = X \cup \overline{Y}$ based on I,

$$support(G) = \sum_{X \subseteq J \subseteq I} (-1)^{|J \setminus X|} support(J) \ .$$

Since $supp(G)$ is always larger than or equal to 0, we derive

$$\sum_{X \subseteq J \subseteq I} (-1)^{|J \setminus X|} support(J) \geq 0$$

If we isolate $supp(I)$ in this inequality, we obtain the following bound on the support of I:

$$supp(I) \leq \sum_{X \subseteq J \subset I} (-1)^{|I \setminus J|+1} supp(J) \qquad \text{If } |I \setminus J| \text{ odd}$$

$$supp(I) \geq \sum_{X \subseteq J \subset I} (-1)^{|I \setminus J|+1} supp(J) \qquad \text{If } |I \setminus J| \text{ even}$$

This rule will be denoted $\mathcal{R}_I(X)$. Depending of the sign of the coefficient of $supp(I)$, the bound is a lower or an upper bound. If $|I \setminus X|$ is odd, $\mathcal{R}_I(X)$ is an upper bound, otherwise it is a lower bound. Thus, given the supports of all subsets of an itemset I, we can derive lower and upper bounds on the support of I with the rules $\mathcal{R}_I(X)$ for all $G = X \cup \overline{Y}$ based on I.

We denote the greatest lower bound on I by $LB(I)$ and the least upper bound by $UB(I)$. The complexity of the rules $\mathcal{R}_I(X)$ increases exponentially with the cardinality of $I \setminus X$. The number $|I \setminus X|$ is called the *depth* of rule $\mathcal{R}_I(X)$. Since calculating all rules is often tedious, we sometimes restrict ourselves to only rules of limited depth. More specifically, we denote the greatest lower and least upper bounds on the support of I resulting from evaluation of rules up to depth k by $LB_k(I)$ and $UB_k(I)$. Hence, the interval $[LB_k(I), UB_k(I)]$ are the bounds calculated by the rules $\{\mathcal{R}_I(X) \mid X \subseteq I, |I \setminus X| \leq k\}$.

Example 1. Consider the following database:

TID	Items
1	a
2	b
3	c
4	a, b
5	a, c
6	b, c
7	a, b, c

$$supp(abc) \geq 0$$
$$\leq s_{ab} = 2$$
$$\leq s_{ac} = 2$$
$$\leq s_{bc} = 2$$
$$\geq s_{ab} + s_{ac} - s_a = 0$$
$$\geq s_{ab} + s_{bc} - s_b = 0$$
$$\geq s_{ac} + s_{bc} - s_c = 0$$
$$\leq s_{ab} + s_{ac} + s_{bc} - s_a - s_b - s_c + s_\emptyset = 1$$

The rules above are the rules $\mathcal{R}_{abc}(X)$ for X respectively $abc, ab, ac, bc, a, b, c, \emptyset$. The first rule has depth 0, the following three rules depth 1, the next three rules depth 2, and the last rule has depth 3. Hence, $LB_0(abc) = 0$, $LB_2(abc) = 0$, $UB_1(abc) = 2$, $UB_3(abc) = 1$. □

Links Between $\mathcal{R}_I(X)$, the support of $X \cup \overline{Y}$, and $X \to \bigvee Y$. Let I be an itemset, and $G = X \cup \overline{Y}$ a generalized itemset based on I. From the derivation of the rule $\mathcal{R}_I(X)$, it can be seen that the difference between the bound calculated by it, and the actual support of I equals the support of $X \cup \overline{Y}$. Hence, the bound calculated by $\mathcal{R}_I(X)$ equals $supp(I)$ if and only if $supp(X \cup \overline{Y}) = 0$. It is also true that the disjunctive rule $X \to \bigvee Y$ holds if and only if $supp(X \cup \overline{Y}) = 0$. Indeed, if $supp(X \cup \overline{Y})$ is 0, then there are no transactions that contain X but do not contain any of the items in Y. Therefore, we obtain the following theorem.

Theorem 1. *Let I be an itemset, and $G = X \cup \overline{Y}$ a generalized itemset based on I. The following are equivalent:*
(a) The bound calculated by $\mathcal{R}_I(X)$ equals the support of I,
(b) $supp(G) = 0$, and
(c) The disjunctive rule $X \to \bigvee Y$ holds. □

Example 2. We continue Example 1. Since the bound 1 calculated by $\mathcal{R}_{abc}(\emptyset)$ equals $supp(abc)$, $supp(\overline{abc})$ must be 0. Indeed, there is no transaction that contains none of $a, b,$ or c. Hence, the disjunctive rule $\emptyset \to a \vee b \vee c$ holds. On the other hand, the difference between the bound calculated by $\mathcal{R}_{abc}(a)$ and the actual support of abc is 1. Hence, $supp(a \cup \overline{bc}) = 1$. □

4   Unifying Framework

In [8] we introduced the NDI representation based on the deduction rules which we repeated in Section 3. The NDI-representation was defined as follows:

$$NDIRep(\mathcal{D}, s) =_{def} \{(I, supp(I, \mathcal{D})) \mid supp(I, \mathcal{D}) \geq s, LB(I) \neq UB(I)\}$$

Hence, if a set I is not in the representation, then either $LB(I) = UB(I)$, and hence the support of I is determined uniquely by the deduction rules, or I is infrequent. A set I with $LB(I) = UB(I)$ is called a *derivable itemset* (DI), otherwise it is called a *non-derivable itemset* (NDI). Derivability is anti-monotone, which allows an Apriori-like algorithm [8].

NDIRep is the only representation that is based on logical implication. For every set I not in the representation, I is either infrequent in every database consistent with the supports in *NDIRep*, or every such database gives the same support to I. All other representations are based on additional assumptions. For example, in the disjunction-free generators representation there is an explicit assumption that all sets in the border of *FGDFree* that are not in the representation, are not free. Such assumptions make it possible to reduce the size of the representations.

In this section, we add similar assumptions to the NDI-based representations. In order to do this, we identify different groups of itemsets: itemsets that are frequent versus those that are infrequent, sets that have support equal to the lower bound, equal to the upper bound, etc. Based on these groups a similar strategy as for the free, the disjunction-free, and the generalized disjunction-free representations will be followed. We identify minimal sets of groups that need to be stored in order to obtain a representation.

4.1 k-Free Sets

The k-free sets will be a key tool in the unified framework.

Definition 2.
A set I is said to be k-free, if $supp(I) \neq LB_k(I)$ and $supp(I) \neq UB_k(I)$.
A set I is said to be ∞-free, if $supp(I) \neq LB(I)$, and $supp(I) \neq UB(I)$.
The set of all k-free (∞-free) sets is denoted $Free_k$ ($Free_\infty$).   □

As the next lemma states, these definitions cover freeness, disjunction-freeness, and generalized disjunction-freeness. The proof is based on Theorem 1, but is omitted because of space restrictions.

Lemma 1. *Let I be an itemset.*

- *I is free if and only if I is 1-free*
- *I is disjunction free if and only if I is 2-free.*
- *I is generalized disjunction-free if and only if I is ∞-free.*

k-freeness is anti-monotone; if a set I is k-free, then all its subsets are k-free as well. Moreover, if $supp(J) = LB_k(J)$ ($supp(J) = UB_k(J)$), then also $supp(I) = LB_k(I)$ ($supp(I) = UB_k(I)$), for all $J \subseteq I$.

4.2   Groups in the Border

Let now $FFree_k$ be the frequent k-free sets. As we argued in Section 2 for the generalized disjunction-free representations, $FFree_k$ is not a representation. Indeed, if a set I is not in the representation, there is no way to know whether I was left out of the representation because I is infrequent, or because $supp(I) = LB_k(I)$,

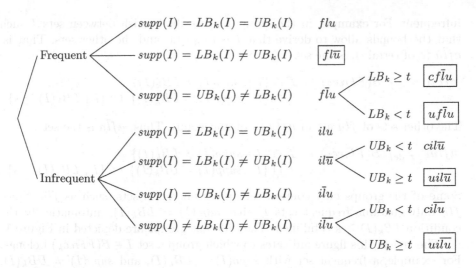

Fig. 1. This tree classifies every set in $\mathcal{B}(FFree_k)$ in the right group. Only the groups that are in a rectangle need to be stored in a representation.

or because $supp(I) = UB_k(I)$. To resolve this problem, parts of the border $\mathcal{B}(FFree_k)$ have to be stored as well. If we can restore the border exactly, then also the other frequent sets can be determined. This can be seen as follows: if a set I is not in $\mathcal{B}(FFree_k)$, and not in $FFree_k$, then it has a subset J in the border. If this set J is infrequent, then so is I. If $supp(J) = LB_k(J)$, then also $supp(I) = LB_k(I)$, and, if $supp(J) = UB_k(J)$, then also $supp(I) = UB_k(I)$ (Lemma 1). Hence, if we can restore the complete border, then we can restore all necessary information.

The sets in $\mathcal{B}(FFree_k)$ can be divided in different groups, depending on whether they are frequent or not, have frequency equal to the lower bound or not, and have frequency equal to the upper bound or not. In order to make the discussion easier, we introduce a 3-letter notation to denote the different groups in the border. The first letter denotes whether the sets in the group are frequent: f is frequent, i is infrequent. The second letter is l if the sets I in the group have $supp(I) = LB_k(I)$, otherwise it is $\bar{l}$. The third letter is u for groups with $supp(I) = UB_k(I)$, and $\bar{u}$ otherwise. The rule depth k is indicated as a subscript to the notation. For example, $f\bar{l}u_k$ denotes the group

$$f\bar{l}u_k =_{def} \mathcal{B}(FFree_k) \cap \mathcal{F} \cap \{I \mid supp(I) \neq LB_k(I)\} \\ \cap \{I \mid supp(I) = UB_k(I)\} ,$$

and $i\bar{l}\bar{u}_k$ denotes the group

$$i\bar{l}\bar{u}_k =_{def} \mathcal{B}(FFree_k) \cap \overline{\mathcal{F}} \cap \{I \mid supp(I) = LB_k(I)\} \\ \cap \{I \mid supp(I) \neq UB_k(I)\} .$$

We split some of the groups even further, based on whether or not the bounds $LB_k(I)$, and $UB_k(I)$ allow to conclude that a set is certainly frequent or certainly infrequent. For example, in the group $f\bar{l}u$, we distinguish between sets I such that the bounds allow to derive that I is frequent, and the other sets. That is, $cf\bar{l}u$ (c of certain), is the set

$$cf\bar{l}u_k =_{def} \mathcal{B}(FFree_k) \cap \mathcal{F} \cap \{I \mid supp(I) \neq LB_k(I)\}$$
$$\cap \{I \mid supp(I) = UB_k(I)\} \cap \{I \mid LB_k(I) \geq s\}.$$

The other sets of $f\bar{l}u$ are in $uf\bar{l}u$ (u of uncertain). Thus, $uf\bar{l}u$ is the set

$$uf\bar{l}u_k =_{def} \mathcal{B}(FFree_k) \cap \mathcal{F} \cap \{I \mid supp(I) \neq LB_k(I)\}$$
$$\cap \{I \mid supp(I) = UB_k(I)\} \cap \{I \mid LB_k(I) < s\}.$$

Some of the groups only contain certain or uncertain sets, such as $f\bar{l}\bar{u}$. Since $f\bar{l}\bar{u}$ only contains frequent sets I with $supp(I) = LB_k(I)$, automatically the condition $LB_k(I) \geq s$ is fulfilled. The different groups are depicted in Figure 1.

The tree in this figure indicates to which group a set $I \in \mathcal{B}(FFree_k)$ belongs. For example, a frequent set with $supp(I) = LB_k(I)$, and $supp(I) \neq UB_k(I)$, takes the upper branch at the first split, since it is frequent, and the second branch in the second split. Notice that there are no groups with code $f\bar{l}u$, because sets that are frequent and have a frequency that equals neither the lower, nor the upper bound, must be in $FFree_k$ and hence cannot be in $\mathcal{B}(FFree_k)$. To make notations more concise, we will sometimes leave out some of the letters. For example, fl_k denotes the union $flu_k \cup f\bar{l}u_k$, and $i\bar{l}_k$ denotes $i\bar{l}u_k \cup ci\bar{l}u_k \cup ui\bar{l}u_k$.

4.3   Representations Expressed with $FFree_k$ and the Groups

We can express many of the existing representations in function of $FFree_k$ for a certain k, and a list of groups in the border of $FFree_k$. Table 1 describes different existing representations in this way. The correctness of this table is proven in [7]. The first line of the table for example, states that the free sets representation actually is

$$(\{(I, supp(I)) \mid I \in FFree_1\}, f\bar{l}\bar{u}_1, ci\bar{l}\bar{u}_1, ui\bar{l}\bar{u}_1, \overline{ci\bar{l}u}_1, ui\bar{l}\bar{u}_1) .$$

We do not differentiate between a representation that stores the different groups separately, or in one set; that is, storing the one set $flu \cup f\bar{l}u$ is considered the same as storing the pair of sets $(flu, f\bar{l}u)$. The reason for this is that for space usage the difference between the two is not significant.

The notation $f\bar{u}_{\infty,1}$ and $i\bar{u}_{\infty,1}$ for the generalized disjunction-free generators representation indicates that in this representation, $FFree_\infty$ is used as basis, but for pruning the border $\mathcal{B}(FFree_\infty)$, only rules up to depth 1 are used. In the experiments however, we will use the other rules for pruning the border as well, and hence we report a slightly better size for this representation.

4.4   Minimal Representations

We can not distinguish between two itemsets within the same group if we only use comparisons between their lower and upper bound, their support, and the

Table 1. Representations in function of $FFree_k$ and the groups in $\mathcal{B}(FFree_k)$. $DFreeGenRep$ denotes the disjunction-free generators representation, $GDFreeGenRep$ the generalized disjunction-free generators representation.

Representation	Base	with frequency	without frequency
$FreeRep$	$FFree.$		$\overline{u}.$
$DFreeRep$	$FFree.$	complete border	
$DFreeGenRep$	$FFree.$	$fl\overline{u}.$	$i\overline{u}.$
$GDFreeRep$	$FFree_\infty$	complete border	
$GDFreeGenRep$	$FFree_\infty$	$f\overline{u}_{\infty,.}$	$i\overline{u}_{\infty,.}$
$NDIRep$	$FFree_\infty$	$flu_\infty, fl\overline{u}_\infty$	

minimal support threshold. Hence, we can think of the different groups as being equivalence classes. We will now concentrate on which of these classes have to be stored to get a minimal representation.

Instead of storing the complete border in a representation, we can restrict ourselves to only some of the groups. It is, for example, not necessary to store the groups flu and ilu, because every set I in these two groups has $supp(I) = LB_k(I) = UB_k(I)$, and thus, its support is derivable. Furthermore, it is not necessary to store the sets in $i\overline{l}u$, $cil\overline{u}$, and $ci\overline{l}u$, because these sets have $UB_k(I) < s$ and thus are certainly infrequent. In Figure 1, the groups which cannot be excluded directly are indicated with boxes. The other groups can always be reconstructed, based on $FFree_k$.

Notice that for all these groups, there is no need to store the supports of the sets in it. For example, for $fl\overline{u}_k$ all sets I in $fl\overline{u}_k$ have $supp(I) = LB_k(I)$. Hence, we can derive the support of a set I if we know that I is in $fl\overline{u}_k$. Similar observations hold for the other groups as well. In the proposed representations, each group is stored separately.

We can reduce the number of groups even more. For some subsets $\mathcal{G} = \{g_1, \ldots, g_n\}$ of the remaining groups $\{fl\overline{u}_k, cf\overline{l}u_k, uf\overline{l}u_k, uil\overline{u}_k, ui\overline{l}u_k\}$, the structure

$$(\{(I, supp(I)) \mid I \in FFree_k\}, g_1, \ldots, g_n)$$

will be a representation, and for some groups $\mathcal{G}$ will not. We denote the structure associated with $\mathcal{G}$ and rules up to depth k with $\mathcal{S}_k(\mathcal{G})$.

The structure $\mathcal{S}_k(\{fl\overline{u}_k, cf\overline{l}u_k\})$ is a representation for every k, but neither $\mathcal{S}_k(\{fl\overline{u}_k\})$, nor $\mathcal{S}_k(\{cf\overline{l}u_k\})$ are. Hence, $\mathcal{S}_k(\{fl\overline{u}_k, cf\overline{l}u_k\})$ is a minimal representation among the representations $\mathcal{S}_k(\mathcal{G})$. The only minimal sets of groups $\mathcal{G}$ such that the associated structures are representations are:

$$\mathcal{G}_1 = \{fl\overline{u}, uf\overline{l}u\} \ , \qquad \mathcal{G}_2 = \{cf\overline{l}u, uf\overline{l}u\} \ ,$$
$$\mathcal{G}_3 = \{fl\overline{u}, uil\overline{u}, ui\overline{l}u\} \ , \text{ and} \qquad \mathcal{G}_4 = \{cf\overline{l}u, uil\overline{u}, ui\overline{l}u\} \ .$$

Theorem 2. [7] *Let* $\mathcal{G} \subseteq \{fl\overline{u}, cf\overline{l}u, uf\overline{l}u, uil\overline{u}, ui\overline{l}u\}$. $\mathcal{S}_k(\mathcal{G})$ *is a representation if and only if either* $\mathcal{G}_1 \subseteq \mathcal{G}$, *or* $\mathcal{G}_2 \subseteq \mathcal{G}$, *or* $\mathcal{G}_3 \subseteq \mathcal{G}$, *or* $\mathcal{G}_4 \subseteq \mathcal{G}$.

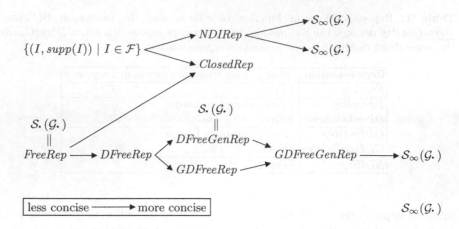

Fig. 2. Relation between the different representations.

For the proof we refer to [7]. The theorem implies that representations $\mathcal{S}_\infty(G_1)$, $\mathcal{S}_\infty(G_2)$, $\mathcal{S}_\infty(G_3)$, and $\mathcal{S}_\infty(G_4)$ are minimal. Thus, all representations in Table 1, have at least one $\mathcal{S}_k(\mathcal{G})$ that is more concise. The relations between the different representations are given in Figure 2. For proofs of the relations see [7].

5   Experiments

To empirically evaluate the newly proposed concise representations, we experimented with several database benchmarks used in [16]. Due to space limitations, we only report results for the BMS-Webview-1 dataset, containing 59 602 transactions, created from click-stream data from a small dot-com company which no longer exists [17], and the pumsb* dataset, containing 100 000 transactions from census data from which items that occur more frequently than 80% are removed [3]. Each experiment finished within minutes (mostly seconds) on a 1GHz Pentium IV PC with 1GB of main memory.

Figure 3 shows the total number of itemsets that is stored for each of the four new representations, together with the previously known minimal representations, i.e., the non-derivable itemsets, the closed itemsets, and the generalized disjunction-free generators.

In both experiments, the representations $\mathcal{S}_\infty(\mathcal{G}_1)$ and $\mathcal{S}_\infty(\mathcal{G}_2)$ have more or less the same size. This is not very surprising, since the parts of the border these two representations store have a big overlap. Also the representations $\mathcal{S}_\infty(\mathcal{G}_3)$ and $\mathcal{S}_\infty(\mathcal{G}_4)$ are almost equal in size. Again we see that $\mathcal{G}_3$ and $\mathcal{G}_4$ are almost equal.

Notice also that for BMS-Webview-1 the representations *GDFreeGenRep* and $\mathcal{S}_\infty(\mathcal{G}_3)$ have the same size. The reason for this can be found in Figure 2. In this figure we see that the size of *GDFreeGenRep* is between the sizes of $\mathcal{S}_2(\mathcal{G}_3)$ and $\mathcal{S}_\infty(\mathcal{G}_3)$. Therefore, the fewer rules of depth more than 2 that need to evaluated in order to get optimal bounds, the closer *GDFreeGenRep* will be to $\mathcal{S}_\infty(\mathcal{G}_3)$. In

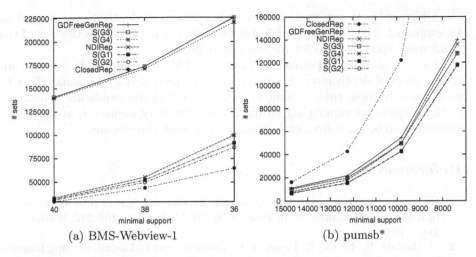

Fig. 3. Number of sets in concise representations for varying minimal support.

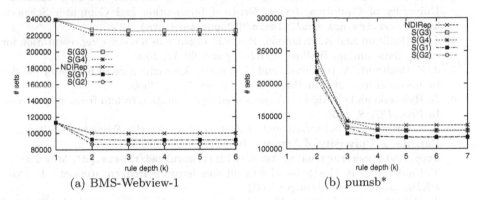

Fig. 4. Number of sets in concise representations of BMS-Webview-1 for varying rule depth.

Figure 4, the effect of varying rule depth is given. The plot shows the sizes of the representations $\mathcal{S}_k(\mathcal{G}_i)$ for different values of k. For the BMS-Webview-1 dataset, evaluating rules of depth greater than 2 does not give any additional gain. In the pumsb* dataset, some gain is still achieved with rules of depth 3. Hence, in the BMS-Webview-1 dataset, *GDFreeGenRep* and $\mathcal{S}_\infty(\mathcal{G}_3)$ have similar size, and in the pumsb*-dataset, there is a slight difference in the part of the border that is stored. In the BMS-Webview-1 dataset, the total number of sets in representations $\mathcal{S}_\infty(\mathcal{G}_1)$ and $\mathcal{S}_\infty(\mathcal{G}_2)$ is smaller than all other representations, except for the closed sets. However, in the pumsb* dataset, the closed set representation is much larger than all others. As can be seen, $\mathcal{S}_\infty(\mathcal{G}_3)$ and $\mathcal{S}_\infty(\mathcal{G}_4)$ sometimes contain more sets, which was expected since these representations also include infrequent sets.

Additionally, to get these results, only rules up to depth 3 were needed to be evaluated. This is illustrated in Figure 4, in which we plotted the size of the condensed representation for varying rule depth.

Also, for all other experiments almost no additional gain resulted from evaluating rules of depth larger than 3. As a consequence, the additional effort to evaluate only these rules is almost negligible during the candidate generation of the frequent set mining algorithm. Indeed, for every itemset I, at most $\binom{|I|}{3}$ rules need to be evaluated, each containing at most three terms.

References

1. R. Agrawal, T. Imilienski, and A. Swami. Mining association rules between sets of items in large databases. In *Proc. ACM SIGMOD*, pages 207–216, Washington, D.C., 1993.
2. Y. Bastide, R. Taouil, N. Pasquier, G. Stumme, and L. Lakhal. Mining frequent patterns with counting inference. *SIGKDD Explorations*, 2(2):66–75, 2000.
3. C.L. Blake and C.J. Merz. *UCI Repository of machine learning databases*. University of California, Irvine, Dept. of Information and Computer Sciences, http://www.ics.uci.edu/~mlearn/MLRepository.html, 1998.
4. J.-F. Boulicaut and A. Bykowski. Frequent closures as a concise representation for binary data mining. In *Proc. PaKDD*, pages 62–73, 2000.
5. J.-F. Boulicaut, A. Bykowski, and C. Rigotti. Approximation of frequency queries by means of free-sets. In *Proc. PKDD*, pages 75–85, 2000.
6. A. Bykowski and C. Rigotti. A condensed representation to find frequent patterns. In *Proc. PODS*, 2001.
7. T. Calders. *Axiomatization and Deduction Rules for the Frequency of Itemsets*. PhD thesis, University of Antwerp, Belgium, http://win-www.ruca.ua.ac.be/u/calders/download/thesis.pdf, May 2003.
8. T. Calders and B. Goethals. Mining all non-derivable frequent itemsets. In *Proc. PKDD*, pages 74–85. Springer, 2002.
9. J. Galambos and I. Simonelli. *Bonferroni-type Inequalities with Applications*. Springer, 1996.
10. M. Kryszkiewicz. Concise representation of frequent patterns based on disjunction-free generators. In *Proc. ICDM*, pages 305–312, 2001.
11. M. Kryszkiewicz and M. Gajek. Concise representation of frequent patterns based on generalized disjunction-free generators. In *Proc. PaKDD*, pages 159–171, 2002.
12. M. Kryszkiewicz and M. Gajek. Why to apply generalized disjunction-free generators representation of frequent patterns? In *Proc. ISMIS*, pages 382–392, 2002.
13. H. Mannila and H. Toivonen. Multiple uses of frequent sets and condensed representations. In *Proc. KDD*, 1996.
14. N. Pasquier, Y. Bastide, R. Taouil, and L. Lakhal. Discovering frequent closed itemsets for association rules. In *Proc. ICDT*, pages 398–416, 1999.
15. J. Pei, J. Han, and R. Mao. Closet: An efficient algorithm for mining frequent closed itemsets. In *ACM SIGMOD Workshop DMKD*, Dallas, TX, 2000.
16. M.J. Zaki and C. Hsiao. ChARM: An efficient algorithm for closed association rule mining. In *TR 99-10, Computer Science, Rensselaer Polytechnic Institute*, 1999.
17. Z. Zheng, R. Kohavi, and L. Mason. Real world performance of association rule algorithms. In *Proc. KDD*, pages 401–406, 2001.

Discovering Unbounded Episodes
in Sequential Data*

Gemma Casas-Garriga

Departament de Llenguatges i Sistemes Informàtics
Universitat Politècnica de Catalunya
Jordi Girona Salgado 1-3, Barcelona
gcasas@lsi.upc.es

Abstract. One basic goal in the analysis of time-series data is to find frequent interesting episodes, i.e, collections of events occurring frequently together in the input sequence. Most widely-known work decide the interestingness of an episode from a fixed user-specified window width or interval, that bounds the length of the subsequent sequential association rules. We present in this paper, a more intuitive definition that allows, in turn, interesting episodes to grow during the mining without any user-specified help. A convenient algorithm to efficiently discover the proposed unbounded episodes is also implemented. Experimental results confirm that our approach results useful and advantageous.

1 Introduction

A well-defined problem in Knowledge Discovery in Databases arises from the analysis of sequences of data, where the main goal is the identification of frequently-arising patterns or subsequences of events. There are at least two related but somewhat different models of the sequential pattern mining. In one of them each piece of data is a sequence (such as the aminoacids of a protein, the banking operations of a client, or the occurences of recurrent illnesses), and one desires to find patterns common to several pieces of data (proteins with similar biological functions, clients of a similar profile, or plausible consequences of medical decisions). See [2] or [7] for an introduction to this model of a sequential database. The second model of sequential pattern matching is the slightly different approach proposed in [6], where data come in a single, extremely long stream, e.g. a sequence of alarms in a telecommunication network, in which some recurring patterns, called *episodes*, are to be found.

Both problems seem similar enough, but we concentrate here on the second one of finding episodes in a single sequence. Abstractly, such ordered data can be viewed as a string of events, where each event has an associated time of occurrence. An example of an event sequence is represented in Figure 1. Here A, B and C are the event types, such as the diferent types of user actions marked on a time line.

---

* This work is supported in part by EU ESPRIT IST-1999-14186 (ALCOM-FT), and MCYT TIC 2002-04019-C03-01 (MOISES)

N. Lavrač et al. (Eds.): PKDD 2003, LNAI 2838, pp. 83–94, 2003.

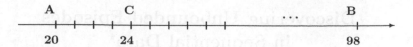

Fig. 1. A sequence of events

We briefly describe the current approaches to interesting episodes, point out some disadvantages, and then propose, as our main contribution, an alternative approach for defining a new kind of serial episodes, i.e *unbounded episodes*. We finally explain how previous algorithms for finding frequent sets can be applied to our approach, and suggest an interpretation of parallel episodes as summaries of serial episodes, with the corresponding algorithmic consequences. Finally, we describe the results of a number of preliminary experiments with our proposals.

2   Framework Formalization

To formalize the framework of the time-series data we follow the terminology, notation, and setting of [6]. The input of the problem is a *sequence* of events. Given a set E of event types, an *event* is a pair (A, t) where $A \in E$ is an event type and t is its occurrence time.

An event sequence is a triple (s, T_s, T_e), where T_s is called the starting time of the sequence, T_e is the ending time, and s has the form: $s = \langle (A_1, t_1!), .., (!A_n, t_n) \rangle$ where A_i is an event type, and t_i is the associated occurrence time, with $T_s \leq t_i < t_{i+1} \leq T_e$ for all $i = 1, \ldots, n - 1$. The time t_i can be measured in any time unit, since this is actually irrelevant for our algorithms and proposals.

2.1   Episodes

Our desired output for each input sequence is a set of frequent episodes. An *episode* is a partially ordered collection of events occurring together in the given sequence. Episodes can be described as directed acyclic graphs. Consider, for instance episodes α, β and γ in Figure 2. Episode $\alpha = B \rightarrow C$ is a *serial episode*: event type B occurs before event type C in the sequence. Of course, there can be other events occurring between these two in the sequence. Episode $\beta = \{A, B\}$ is a *parallel episode*: events A and B occur frequently close in the sequence, but there are no constraints about the order of their appearences. Finally, episode γ is an example of *hybrid episode*: it occurs in a sequence if there are occurences of A and B and these precede an occurrence of C, possibly, again, with other intervening events.

More formally, an episode can be defined as a triple $(V, \leq, g)$ where: V is a set of nodes, $\leq$ is a partial order relation on V, and $g : V \rightarrow E$ is a mapping associating each node with an event type. We also define the size of an episode as the number of events it contains, i.e, $|V|$. The interpretation of an episode is that events in $g(V)$ must occur in the order described by $\leq$. In this paper we will only deal with serial and parallel episodes.

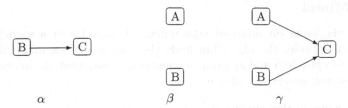

Fig. 2. Types of episodes

Definition 1. *An episode* $\beta = (V', \leq', g')$ *is a* **subepisode** *of* $\alpha = (V, \leq, g)$, *noted by* $\beta \subseteq \alpha$, *if there exists an injective mapping* $f : V' \rightarrow V$ *such that* $g'(v) = g(f(v))$ *for all* $v \in V'$, *and for all* $v, w \in V'$ *with* $v \leq' w$ *also* $f(v) \leq f(w)$.

3 Classical Approaches to Define Interesting Episodes

In the analysis of sequences we are interested in finding all *frequent* episodes from a class of episodes which can be *interesting* to the user. In this section we will mainly take the classical widely-used work of [6] as a reference, i.e, we state that to be considered interesting, the events of an episode must occur close enough in time.

3.1 Winepi

In the first approach of [6], the user defines how close the events of an interesting episode should be by giving the width of the *time window* within which the episode must occur. The number of possible windows of a certain width win in the sequence (s, T_s, T_e) is exactly: $T_e - T_s + win - 1$, and we denote by $W(s, win)$ the set of all these windows of size win. Thereby, the frequency of an episode α in s is defined to be:

$$fr(\alpha, s, win) = \frac{|\{w \in W(s, win)|\alpha \text{ occurs in } w\}|}{|W(s, win)|}$$

So, an episode is frequent according to the number of windows where that episode has occured, or to its ratio to the total number of possible such windows in the sequence. To be frequent, the ratio $fr(\alpha, s, win)$ of an episode must be over a minimum user-specified real value. The Winepi approach applies the Apriori algorithm to find the frequency of all the candidate episodes in the sequence as though each sliding window were a transaction with ordered events.

Once the frequent interesting episodes are discovered from the sequence, the second goal of the approach is to create the **episode association rules** that hold over a certain minimum confidence. For all episodes $\beta \subseteq \alpha$, an **episodal rule** $\beta \Rightarrow \alpha$ **holds with confidence:**

$$conf(\beta \Rightarrow \alpha) = \frac{fr(\alpha, s, win)}{fr(\beta, s, win)}$$

3.2   Minepi

Minepi is based on *minimal occurrences* of episodes in a sequence. For each frequent episode, the algorithm finds the location of its minimal occurrences. Given an episode α and an event sequence s, we say that the interval $w = [t_s, t_e)$ is a *minimal occurrence* of α in s, if:

(1) α occurs in the window w.
(2) α does not occur in any proper subwindow on w.

Basically, the applied algorithm is Apriori: it locates, for every episode going from the smaller ones to larger ones, its minimal occurrences. In the candidate generation phase, the location of minimal occurrences of a candidate α is computed as a temporal join of the minimal occurrences of two subepisodes of α.

This approach differs from Winepi in the fact that it does not use a frequency ratio to decide when an episode is frequent. Instead, an episode will be considered frequent when its number of minimal occurrences is over an *integer* value given by the user. This is a consequence of the fact that the lengths of the minimal occurrences vary, so that a uniform ratio could be misleading. One advantage of this approach is that allows the user to find final rules with two windows widths, one for the left-hand side and one for the whole rule, such as "if A and B occur within 15 seconds, then C follows withing 30 seconds". So, in this approach an **episode association rule** is an expression $\beta[win_1] \Rightarrow \alpha[win_2]$, where β and α are episodes such that $\beta \subseteq \alpha$, and win_1 and win_2 are integers specifying interval widths. The informal interpretation of the rule is that if episode β has a minimal occurrence at interval $[t_s, t_e)$ with $t_e - t_s \leq win_1$, then episode α occurs at interval $[t_s, t_e')$ for some t_e' such that $t_e' - ts \leq win_2$.

The **confidence of an episode association rule** $\beta[win_1] \Rightarrow \alpha[win_2]$ with $\beta \subseteq \alpha$ and two user-specified interval widths win_1 and win_2 is the following:

$$conf(\beta[win_1] \Rightarrow \alpha[win_2]) =$$
$$\frac{|\{[t_s, t_e) \in mo(\beta) \ s.t \ t_e - t_s \leq win_1 \ and \ [t_s, t_s + win_2) \in mo(\alpha)\}|}{|\{[t_s, t_e) \in mo(\beta) \ and \ t_s - t_e \leq win_1\}|}$$

where $mo(\alpha)$ are the set of minimal occurrences of the episode α in the original input sequence. So, even if there is no fixed window size (as occurred in Winepi approach) and apparently minimal occurrences are not restricted in length, now the user needs to specify the time bounds win_1 and win_2 for the generation of the subsequent episode rules and their confidences. These values force minimal occurrences to be bounded in a fixed interval size of at most win_2 time units during the mining process.

3.3   Some Disadvantages of These Previous Approaches

We summarize below some of the observed disadvantages in Winepi and Minepi.

– In Winepi the window width is fixed by the user and it remains fixed throughout the mining. Consequently, the size of the discovered episodes is limited.

Winepi just reduces the problem of mining the long event sequence to a sequential database (such as in [2]), where now each transaction is a fixed window.

- In Minepi the user specifies two time bounds for the creation of the subsequent episode association rules. These intervals make the final minimal occurrences to be bounded in size, since just those occurrences contained within the bounds are counted.

- Both Minepi and Winepi require the end user to fix one parameter with not much guidance on how to do it. Intervals or windows too wide can lead to misleading episodes where the events are widely separated among them; so, the subsequent rules turn out to be uninformative. On the other hand, interval or windows set too tight give rise to overlapping episodes: if there exists an interesting episode, α, whose size is larger than the fixed window width, then that episode will never fit in any window and, consequently, α will be discovered just partially.

- Minepi does not use a frequency ratio to decide whether an episode is frequent. This makes difficult the application of sampling in the algorithms of finding frequent episodes.

- In case the user decides to find the episode association rules for a different time bound (a different window size in Winepi or a different interval length for Minepi), then the algorithm that finds the source of frequent episodes has to be run again, incurring in a inconvenient overhead.

- Both approaches do not seem truly compatible for those problems where the adjancency of the events in the discovered episodes is a must (such as protein function identification). Neither Winepi or Minepi allow to set this kind of restriction between the events of an interesting episode.

4   Unbounded Episodes

In order to avoid all these drawbacks and be able to enlarge the window width automatically throughout the mining process, we propose the following approach. We will consider a serial or parallel episode interesting if it fulfills the following two properties:

(1) Its *correlative events* have a gap of at most *tus* time units (see figure 3).
(2) It is frequent.

$$\overset{tus \quad tus}{A \to B \to C} \qquad \overset{tus \quad tus}{\{A, \quad B, \quad C\}}$$

$$\alpha \qquad\qquad \beta$$

Fig. 3. Example of serial and parallel unbounded episodes

So, in our proposal, the measure of interestingness is based on *tus*, the **time-unit separation** between correlative events in the episode. This number of time units must be specified by the user. The above two episodes α and β are examples of the interpretation of our approach. In the serial episode α, the distance between A and B is tus, and the distance between B and C is also tus time units. Besides, despite not specifying the distance between events A and C, it can be clearly seen they are at most $2 \times tus$ time units away. In the parallel episode β, distance between correlative events A, B and C, regardless of the order of their appearences in the sequence, must be of at most tus time units. More generally, an episode of size e may span up to $(e-1) \times tus$ time units.

Now, every episode that is candidate to be frequent, will be searched in windows whose width will be delimited by the episode size: an episode with e events will be searched in all windows in the sequence of width $(e-1) \times tus$ time units. Thus, the window width is not bounded, nor is the size of the episode, and both will grow automatically, if necessary, during the mining. This explains the name chosen: we are mining *unbounded episodes*.

At this point, it is worth mentioning the work of [8] (contributing with the algorithm cSPADE) and [7] (the algorithm GSP). These two papers integrate inside the mining process the possibility to define a max-gap constraint between the elements of the frequent sequences found in a sequential database. However, this max-gap constraint in [8] or [7] does not lead to an unbounded class of patterns as we present here. The reason is that they work on the sequential database problem, and so, the frequent mined patterns turn out to be naturally bounded by the lenght of the transactions in the database.

With our approach the window width is allowed to grow automatically without any predetermined limits. The frequency of an episode can be defined in the following way: let us denote by $W_k(s, win)$ the total set of windows in a sequence (s, T_i, T_f) of a fixed width $win = k \times tus$ time units (the number of such windows in the sequence is $T_e - T_f + win - 1$). Then:

Definition 2. *The* frequency *of an episode α of size $k+1$ in a sequence(s, T_i, T_f) is:*

$$fr(\alpha, s, tus) = \frac{|\{w \in W_k(s, win) | \alpha \ occurs \ in \ w\}|}{|W_k(s, win)|}$$

where $win = (|\alpha| - 1) \times tus = k \times tus$.

Note that the dependence on win, for fixed α, is here simply a more natural way to reflect the dependence on the user-supplied parameter tus, but both correspond to the same fact since win and tus are linearly correlated.

To sum up, every episode α will be frequent if its frequency is over a minimum user-specified frequency, that is, according to the number of windows in which it occurs; however, the width of that window depends on the number of events in α. So, the effect in the algorithm is that, as an episode size becomes bigger and the number of its events increases, the proper window in which that episode is searched also increases its width; and simultaneously the ratio that has to be compared with the user-specified desired frequency is appropriately adjusted.

4.1   Episode Association Rule with Unbounded Episodes

The approach of mining unbounded episodes will be flexible enough to allow the generation of association rules according to two interval widths (one for the left hand side, and one for the whole rule as occurred with Minepi).

An **unbounded episode rule** will be an expresion $\beta[n_l] \Rightarrow \alpha[n_r]$, where β and α are unbounded episodes such that $\beta \subseteq \alpha$, and n_l and n_r are integers such that $n_l = |\beta|$ and $n_r = |\alpha| - |\beta|$. The informal interpretation of these two new variables n_l and n_r is the number of events occurring in the left hand side (n_l) and new events implied in the right hand side (n_r) of the rule respectively.

So, we can rewrite any unbounded episode rule $\beta[n_l] \Rightarrow \alpha[n_r]$ in terms of a rule with two window widths $\beta[w_1] \Rightarrow \alpha[w_2]$ by considering $w_1 = (n_l - 1) \times tus$ and $w_2 = n_r \times tus$. This transformation will lead to an easy and informative interpretation of the rule: "if events in β occur within w_1 time units, then, the rest of the events in α will follow within w_2 time units".

One of the advantages of this proposed approach is that focusing our episode search on the time-unit separation between events, will allow to generate the best unbounded episode rule $\beta[n_l] \Rightarrow \alpha[n_r]$ (and so, the best rule $\beta[w_1] \Rightarrow \alpha[w_2]$) without fixing any other extra parameter: neither n_l or n_r will be user-specified for any rule, since these values will be chosen from the best antecedent and consequent maximizing the value of confidence for that rule (or in other words, n_l and n_r will be uniquely determined by the size of the episode being the antecedent and the size of the episode being the consequent in the best rule according to confidence ratio).

Since in our approach we have a ratio of frequency support, we can define the **confidence of a rule** $\beta \Rightarrow \alpha$ for $\beta \subseteq \alpha$ as:

$$conf(\beta \Rightarrow \alpha) = \frac{fr(\alpha, s, tus)}{fr(\beta, s, tus)}$$

where the value of $fr(\alpha, s, tus)$ for a fixed α, depends on the occurrences of α in all windows of lenght $(|\alpha| - 1) \times tus$ in the sequence. Note that since β is a subepisode of α, the rule right-hand side α contains information about the relative location of each event in it, so the "new" events in the rule right-hand can actually be required to be positioned between events in the left-hand side. The rules defined here are also rules that point forward in time (rules that point backwards can be defined in a similar way).

As we see, the values n_l and n_r of a rule do not affect the confidence, and they can be determined after having chosen the best rule by following the procedure:

for each maximal episode α,
$$\beta[|\beta|] \Rightarrow \alpha[|\alpha| - |\beta|] = arg.max\{conf(\beta \Rightarrow \alpha) \ s.t \ \beta \subseteq \alpha\}$$

So, the final windows widths $(w_1 = |\beta| \times tus$ and $w_2 = (|\alpha| - |\beta|) \times tus)$ are determined by the best rule in terms of confidence, and this can vary from one rule to the other, adapting always to the best combination. Note that instead of confidence, any other well-defined metric for episodes could be used to select the best rule in this procedure, and so, different unbounded rules would be taken.

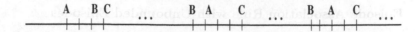

Fig. 4. Example of an event sequence

Example in figure 4 will serve to illustrate the advantatges of our unbounded episode approach. The sequence of this figure shows that we could consider frequent the episodes: $\beta = \{A, B\}$ and $\gamma = \{A, B\} \to C$ (as they are represented as a graph in figure 2). The best association rule we can find in this example is the following: $\{A, B\} \Rightarrow C$, that should have a confidence of 1 for this presented piece of sequence.

For Winepi, at least a fixed window of 5 time units of width should be specified to find both β and γ. But this parameter depends on the user and it is not intuitive enough to chose the right value. In this example, if the user decides a window width of 3 time units, then the episode γ would never be fully discovered and the rule will never be generated.

With Minepi, the problem comes when specifying the two windows widths for the episode association rule. In case the user decides $win_1 = 3$ and $win_2 = 4$, the generated rule would be $\beta[3] \Rightarrow \gamma[4]$, that has a confidence of just 1/3 in this example. It is not the best association rule, and it is due to the value of $win_2 = 4$, that it is set too tight. Besides, if the user wants to specify a wider win_2, the algorithm finding frequent minimal occurrences has to be run again.

For the unbounded approach however, one would find both β and γ by just specifing a big enough value for tus. This is an intuitive parameter, and the best subsequent episode rule in terms of confidence would be $\beta[2] \Rightarrow \gamma[1]$, with a confidence of 1. This rule can be transformed in terms of two window widths and interpret the following "if A and B occur within tus time units, then C will follow in next tus time units".

4.2   Advantages of Our Approach

We shortly summarize some advantages of our unbounded proposal.

- Since the window increases its width along with the episode size, the final frequent episodes do not overlap unnecesarily, and their size is not limited.
- The unbounded episode rules have better quality in terms of confidence without any previous user help.
- Unbounded episodes generalizes Minepi and Winepi in that episodes found with a window width of x time units can be found with our approach using a distance of $x - 1$ time units between correlative events.
- The application of sampling techniques are allowed.
- Once the frequent unbounded episodes are mined, finding the episode rules with two windows widths is straight. What is more, the user can try different windows widths for the rules, and chose the best width according to some statistical metric. This does not affect the previous mining and the discove-

red unbounded episodes, and they are always the same once we are in the generating rule phase.

- Our proposal can be adapted to the sequential-database style by imposing a wide gap between the different pieces of data.

On the whole, we can say that unbounded episodes are more general and intuitive than Minepi or Winepi approaches. In particular, these unbounded episodes can be very useful in contexts such as the classification of documents or the intrusion detection systems. As argued in [5], a drawback of subsequence patterns is that they are not suitable for classifying long strings over small alphabet, since a short subsequence pattern matches with almost all long strings. So, the larger the episodes found in a text the better for the future predictions.

5   Algorithms to Mine Unbounded Episodes

Our proposed definition of unbounded episodes is flexible enough to still allow the use of previous algorithms. Besides, to prove the flexibility of the proposal, we also adapt here our strategy from [3], Best-First strategy, which is a non-trivial evolution of Dynamic Itemset Counting (DIC, [4]) and provides better performance than both Apriori and DIC. For better understanding, we give a brief account of how our Best-First strategy works.

Similarly to DIC, our strategy keeps cycling through the data as many times as necessary, counting the support of a number of candidate itemsets. Whenever one of them reaches the threshold that declares it frequent, it immediately "notifies" this fact to all itemsets one unit larger than it. In this way, potential future candidates keep being informed of whether each of their immediate predecessors is frequent. When all of them are, the potential candidate is promoted to candidate and its support starts to be counted. DIC follows a similar pattern but only tries to generate new candidates every M processed transactions: running it with $M = 1$ would be similar to Best-First strategy, but would incur overheads that our algorithm avoids thanks to the previous online information of which subsets of the potential candidates are frequent at each moment.

To follow the same structure, our new algorithm for mining episodes, called Episodal Best-First (**EpiBF**), will distiguish two sets of episodes: 1/ **candidate episodes** whose frequency is being counted, and 2/ **potential candidates** that will be incorcorated as candidates as soon as the monotonicity property of frequency is fulfilled. Hovever, given that now we are using our unbounded approach of interestingness, we must relax the monotonicity property frequency for pruning unwanted candidates. Other Breadth-First algorithms, like Apriori or DIC, can be also easily applied by taking into account that at each new scan of the database for candidates of size k, the window width must be incremented conveniently (i.e, $(k - 1) \times tus$). Apart from that, we also have to relax here the monotonicity property of frequency used in the candidate generation phase as we will see in short. We discuss separately the case of serial episodes first.

5.1   Discovering Serial Episodes

In case of using our approach of unbounded episodes, the well-known mono-tonicity property of frequency (stating that any frequent episode has all its subepisodes also frequent) does not hold: that is, a frequent unbounded episode could have some subepisode not frequent. For instance, let us consider the unbounded serial episode $A \rightarrow B \rightarrow C$, and its subepisode $A \rightarrow C$. They refer to two different classes of unbounded interestingness: while in $A \rightarrow B \rightarrow C$, events A and C are separated for at most $2 \times tus$ time units, in its subepisode $A \rightarrow C$ the events are separated at most tus time units. So, it might well be that since the gap between events is different in both episodes, $A \rightarrow C$ is not frequent while $A \rightarrow B \rightarrow C$ is frequent. We cannot use this property to prune unwanted candidates.

But we will relax this notion here and we will just consider those subepisodes whose events follow an adjacency of tus time-unit separation. For instance, to consider the episode $A \rightarrow B \rightarrow C$ a good candidate that deserves to be counted in the data, one has to find frequent just the subepisodes $A \rightarrow B$ and $B \rightarrow C$ (i.e, the overlapping parts of an unbounded episode). Then, it is true that any frequent unbounded episode has all its overlapping parts frequent.

Now, the algorithm EpiBF for serial episodes goes in the following way. It starts by initializing the set of candidate episodes with all episodes of size 2, and the set of potential candidates with all episodes of size 3. Then, it goes on counting the frequency of all the candidate episodes until this set becomes empty. When one of these candidate episodes of size k achieves the state of frequent, it increments counters corresponding to all the potential candidates of size $k + 1$ that we can obtain by adding one more event before it or after it. This growth leads to unbounded episodes. On the other hand, when a potential candidate of size $k + 1$ finds that both subepisodes of size k, obtained by chopping off either end, have been declared frequent, then it will be incorporated in the set of candidate episodes.

It is important to highlight that, in this algorithm, the set of candidate episodes can be made up of episodes of diferent sizes, and each episode α of size k must be searched and counted in all windows of width $(k-1) \times tus$ time units. This fact forces EpiBF to handle windows of different sizes at the same time by simply taking, at every step, the largest window for the longest episode in the set of candidate episodes. The rest of episodes in the set of candidate episodes will be searched in the proper subwindows.

5.2   Discovering Serial and Parallel Episodes Simultaneously

In case of mining parallel episodes the problem can be reduced efficiently to mining serial episodes in the following way. Every parallel episode of size k lumps together up to $k!$ serial episodes. For instance, the parallel episode $\{A, B\}$ gathers the following two serial episodes: $A \rightarrow B$ and $B \rightarrow A$. In this case, a serial episode will be called *participant* of a parallel episode. Clearly, any serial episode is participant of one, and only one, parallel episode.

Let us discuss what could be the meaning of parallel episode mining. Clearly, if a frequent parallel episode has some (but not all) participants already frequent, the desired output is the list of such frequent serial episodes: the parallel one, given alone, provides less information. In such cases we should not move from the serial episodes to the parallel one, unless actually *all* of them are frequent: in this last case, the parallel episode is an effective way of representing this fact. Thus, according to our proposal, in order to be considered interesting, a parallel episode α must fullfil one of the two following conditions: either

1. by adding up the frequency of the serial episodes that are participants of α, we reach the user-specified minimal frequency, *but* no serial episode participant of α is frequent alone; or
2. every serial episode participant of α is frequent.

From the point of view of the algorithm, any used strategy will mine serial episodes, but these serial episodes can refer to parallel ones too. Thereby, the set of candidate episodes will be made of serial ones, while the set of potential candidates will be composed of parallel episodes. This means that the algorithm will be counting the support of serial candidates, as in the previous case; however, when declaring one of these serial episodes, α, frequent or non-frequent, the notification must go to that parallel episode which α is participant of.

6 Experiments

In this section we present the results of running (a probabilistic version of the) EpiBF algorithm on a variety of different data collections.

First, we experimented, as in [6], with protein sequences. We used data in the PROSITE database of the ExPASy WWW molecular biology server of the Geneva University Hospital and University of Geneva [10]. The purpose of this experiment is to identify specific patterns in sequences so as to determine to which family of protein they belong. The sequences in the family we selected ("DNA mismatch repair proteins I", PROSITE entry PS00058, the same one used in [6] for comparison), are known to contain the string GFRGEAL. This string represents a serial episode of seven consecutive symbols separated by 1 unit of time among them. Parameter tus was set to 1, and the support threshold was set to 15, for the 15 individual sequences in the original data. Note that no previous knowledge of the pattern to be found is involved in this parameter setting.

As expected, we found in the database the pattern GFRGEAL along with 3,755 more serial episodes (whether maximal or not), most of them much shorter. When comparing our approach against previous ones, we see that both Winepi and Minepi need to know in advance the length of the expected pattern in the protein sequence, in order to fix the window width. However, it is usual that we do not know which pattern is to be found in a sequence; so, one must try the experiment with different window widths.

In order to see the flexibility of the unbounded episodes, we also run experiments with text data collections. In particular, we used a part of a text extracted from "Animal Farm" by Orwell [9]. Once again, setting *tus* close to 1 and considering each letter a new event, we are able to find frequent prepositions, articles, suffixes of words, and concatenations of words (such as "to", "in", "at", "ofthe", "was", "her", "ing"...). This experiment could have been done considering an event to be every new word in the text; this will lead to unbounded episodes as a tool to classify other new texts.

When it comes to the general performance of the method, we found that, naturally, the larger the value of the parameter *tus*, the more discovered episodes. Besides, discovering our serial and parallel episodes simultaneously, allows the algorithm to discover parallel patterns when hardly serial patterns are found in the database. For example, fed with the first 40,000 digits of the Champernowne sequence (012345678910111213141516...), with a high frequency threshold of 50% and digits far apart at most 15 positions in the episodes (*tus* = 15), only 3 serial episodes were found but we discovered 15 other parallel episodes.

7   Conclusions

We present in this paper a more intuitive approach for interesting episodes. This proposal overcomes the disadvantages of the widely-used previous approaches (Minepi and Winepi), and it turns out to be an adaptive approach for categorical time-series data. The algorithmic consequences of the unbounded episodes are also discused and implemented. Finally, we have also introduced a new way of considering parallel episodes as a set of participant serial episodes. First experiments prove to be promising, but more experimentation on the different values of *tus* and their consequences in the subsequent rules is on the way.

References

1. R.Agrawal, H.Mannila, R.Srikant, H.Toivonen and I.Verkamo. Fast Discovery of Association Rules. *Advances in Knowledge Discovery and Data Mining.* 1996.
2. R.Agrawal and R.Srikant. Mining Sequential Patterns. *Proc. of the Int. Conf. on Data Engineering.* 1995.
3. J.Baixeries, G.Casas-Garriga, and J.L.Balcázar. A Best First Strategy for Finding Frequent Sets. *Extraction et gestion des connaissances* (EGC'2002), 100–106. 2002.
4. S.Brin, R.Motwani, J.Ullman and S.Tsur. Dynamic Itemset Counting and Implication Rules for Market Basket Data. *Int. Conf. Management of Data.* 1997.
5. M.Hiaro, S.Inenaga, A.Shinohara, M.Takeda and S.Arikawa. A Practical Algorithm to Find the Best Episode Patterns. *Int. Conf. on Discovery Science*, 235–440. 2001.
6. H.Mannila, H.Toivonen and I.Verkamo. Discovery of frequent episodes in event sequences. *Proc. Int. Conf. on Knowledge Discovery and Data Mining.* 1995.
7. R.Srikant and R.Agrawal. Mining Sequential Patterns: Generalizations And Performance Improvements.*Proc. 5th Int. Conf. Extending Database Technology.* 1996.
8. M.J.Zaki. Sequence Mining in Categorical Domains: Incorporating Constrains. *Proc. Int. Conf. on Information and knowledge management* ,422–429. 2000.
9. Data Analysis Challenge, http://centria.di.fct.unl.pt/ida01/
10. Geneva University Hospital and University of Geneva, Switzerland. ExPASy Molecular Biology Server. http://www.expasy.ch/

Mr-SBC: A Multi-relational Naïve Bayes Classifier

Michelangelo Ceci, Annalisa Appice, and Donato Malerba

Dipartimento di Informatica, Università degli Studi
via Orabona, 4 - 70126 Bari - Italy
{ceci,appice,malerba}@di.uniba.it

Abstract. In this paper we propose an extension of the naïve Bayes classification method to the multi-relational setting. In this setting, training data are stored in several tables related by foreign key constraints and each example is represented by a set of related tuples rather than a single row as in the classical data mining setting. This work is characterized by three aspects. First, an integrated approach in the computation of the posterior probabilities for each class that make use of first order classification rules. Second, the applicability to both discrete and continuous attributes by means a supervised discretization. Third, the consideration of knowledge on the data model embedded in the database schema during the generation of classification rules. The proposed method has been implemented in the new system Mr-SBC, which is tightly integrated with a relational DBMS. Testing has been performed on two datasets and four benchmark tasks. Results on predictive accuracy and efficiency are in favour of Mr-SBC for the most complex tasks.

1 Introduction

Many inductive learning algorithms assume that the training set can be represented as a single table, where each row corresponds to an example and each column to a predictor variable or to the *target* variable Y. This assumption, also known as *single-table assumption* [23], seems quite restrictive in some data mining applications, where data are stored in a database and are organized into several tables for reasons of efficient storage and access. In this context, both predictor variables and the target variable are represented as attributes of distinct tables (relations).

Although in principle it is possible to consider a single relation reconstructed by performing a relational join operation on the tables, this approach is fraught with many difficulties in practice [2,11]. It produces an extremely large, and impractical to handle, table with lots of data being repeated. A different approach is the construction of a single central relation that summarizes and/or aggregates information which can be found in other tables. Also this approach has some drawbacks, since information about how data were originally structured is lost. Consequently, the (multi-)relational data mining approach has been receiving considerable attention in the literature, especially for the classification task [1,10,15,20,7].

In the traditional classification setting [18], data are generated independently and with an identical distribution from an unknown distribution P on some domain **X** and are labelled according to an unknown function g. The domain of g is spanned by m independent (or predictor) random variables X_i (both numerical and categorical), that

N. Lavrač et al. (Eds.): PKDD 2003, LNAI 2838, pp. 95–106, 2003.
© Springer-Verlag Berlin Heidelberg 2003

is $\mathbf{X}=X_1\times X_2\times...\times X_m$, while the range of g is a finite set $Y=\{C_1, C_2, ..., C_L\}$, where each C_i is a distinct class. An inductive learning algorithm takes a training sample $S=\{(\mathbf{x}, y) \in \mathbf{X} \times Y \mid y=g(\mathbf{x}) \}$ as input and returns a function f which is hopefully close to g on the domain $\mathbf{X}$. A well-known solution is represented by the Naïve Bayesian Classifiers [3], which aim to classify any $x \in \mathbf{X}$ is the class maximizing the *posterior probability* $P(C_i|x)$ that the observation x is of class C_i, that is:

$$f(x)= arg\ max_i\ P(C_i|x)$$

By applying the Bayes theorem, $P(C_i|x)$ can be reformulated as follows:

$$P(C_i|x) = \frac{P(C_i)P(x|C_i)}{P(x)}$$

where the term $P(x|C_i)$ is in turn estimated by means of the *naïve Bayes assumption*:

$$P(x|C_i)=P(x_1,x_2,...,x_m|C_i)=P(x_1|C_i) \times P(x_2|C_i) \times...\times P(x_m|C_i)$$

This assumption is clearly false if the predictor variables are statistically dependent. However, even in this case, the naïve Bayesian classifier can give good results [3].

In this paper we present a new approach to the problem of learning classifiers from relational data. In particular, we intend to extend the naïve Bayes classification to the case of relational data. Our proposal is based on the induction of a set of first-order classification rules in the context of naive Bayesian classification.

Studies on first-order naïve Bayes classifiers have already been reported in the literature. In particular, Pompe and Kononenko [20] proposed a method based on a two-step process. The first step uses the ILP-R system [21] to learn a hypothesis in the form of a set of first-order rules and then, in the second step, the rules are probabilistically analyzed. During the classification phase, the conditional probability distributions of individual rules are combined naïvely according to the naïve Bayesian formula.

Flach and Lachiche proposed a similar two-step method, however, unlike the previous one, there is no learning of first-order rules in the first step. Alternatively, a set of patterns (first-order conditions) is generated that are used afterwards as attributes in a classical attribute-value naive Bayesian classifier [7]. 1BC, the system implementing this method, views individuals as structured objects and distinguishes between *structural* predicates referring to parts of individuals (e.g. atoms within molecules), and *properties* applying to the individual or one or several of its parts (e.g. a bond between two atoms). An *elementary first-order feature* consists of zero or more structural predicates and one property.

An evolution of 1BC is represented by the system 1BC2 [16], where no preliminary generation of first-order conditions is present. Predicates whose probabilities have to be estimated are dynamically defined on the basis of the individual to classify. Therefore, this is a form of *lazy* learning, which defers processing of its inputs (i.e., the estimation of the posterior probability according to the Bayesian statistical framework) until it receives requests for information (the class of the individual). Computed probabilities are discarded at the end of the classification process. Probability estimates are recursively computed and problems of non-termination in the computation may also occur.

An important aspect of the first two (*eager*) approaches is that they keep the phases of first-order rules/conditions generation and of probability estimation separate. In particular, Pompe and Kononenko use ILP-R to induce first-order rules [21], while

1BC uses TERTIUS [8] to generate first order features. Then, the probabilities are computed for each first-order rule or feature. In the classification phase, the two approaches are similar to a multiple classifier because they combine the results of two algorithms. However, most first-order features or rules share some literals and this approach takes into account the related probabilities more than once. To overcome this problem it is necessary to rely on an integrated approach, so that the computation of probabilities on shared literals can be separated from the computation of probabilities on the remaining literals.

Systems implementing one of the three above approaches work on a set of main-memory Prolog facts. In real-world applications, where facts correspond to tuples stored on relational databases, some pre-processing is required in order to transform tuples into facts. However, this has some disadvantages. First, only part of the original hypothesis space implicitly defined by foreign key constraints can be represented after some pre-processing. Second, much of the pre-processing may be unnecessary, since a part of the hypothesis described by Prolog facts space may never be explored, perhaps because of early pruning. Third, in applications where data can frequently change, pre-processing has to be frequently repeated. Finally, database schemas provide the learning system free of charge with useful knowledge of data model that can help to guide the search process. This is an alternative to asking the users to specify a language bias, such as in 1BC or 1BC2.

A different approach has been proposed by Getoor [13] where the Statistical Relational Models (SRM) are learnt taking advance from the tightly integration with a database. SRMs are models very similar to Bayesian Networks. The main difference is that the input of a SRM learner is the relational schema of the database and the tuples of the tables in the relational schema.

In this paper the system Mr-SBC (Multi-Relational Structural Bayesian Classifier) is presented. It implements a new learning algorithm based on an integrated approach of first-order classification rules with naive Bayesian classification, in order to separate the computation of probabilities of shared literals from the computation of probabilities for the remaining literals. Moreover, Mr-SBC is tightly integrated with a relational database as in the work by Getoor, and handles categorical as well as numerical data through a discretization method.

The paper is organized as follows. In the next section the problem is introduced and defined. The induction of first-order classification rules is presented in Section 3, the discretization method is explained in Section 4 and the classification model is illustrated in Section 5. Finally, experimental results are reported in Section 6 and some conclusions are drawn.

2   Problem Statement

In traditional classification systems that operate on a single relational table, an observation (or individual) is represented as a tuple of the relational table. Conversely, in Mr-SBC, which induces first-order classifiers from data stored in a set $S=\{T_0, T_1, \ldots, T_h\}$ of tables of a relational database, an individual is a tuple t of a *target* relation T joined with all the tuples in S which are related to t following a foreign key path. Formally, a foreign key path is defined as follows:

Def 1. *A foreign key path is an ordered sequence of tables* $\vartheta=(T_{i_1}, T_{i_2},..., T_{i_s})$, where

- $\forall j=1, ...,s, T_{i_j} \in S$
- $\forall j=1..s-1, T_{i_{j+1}}$ has a foreign key to the table T_{i_j}

In Fig.1 an example of foreign key paths is reported. In this case, S={MOLECULE, ATOM, BOND} and the foreign keys are: A_M_FK, B_M_FK, A_A_FK1, A_A_FK1. If the target relation T is MOLECULE then five foreign key paths exists. They are: (MOLECULE), (MOLECULE,ATOM), (MOLECULE, BOND), (MOLECULE, ATOM, BOND) and (MOLECULE, ATOM, BOND). The last two are equal because the bond table has two foreign keys referencing the table atom.

A formal definition of the learning problem solved by MR-SBC is the following problem:

Given:

- A *training set* represented by means of h relational tables S={$T_0,T_1,...,T_h$} of a relational database D.
- A set of *primary key constraints* on tables in S.
- A set of *foreign key constraints* on tables in S.
- A *target relation* $T(x_1,..., x_n)\in S$
- a *target* discrete *attribute* y in T, different from the primary key of T.

Find:

A naive Bayesian classifier which predicts the value of y for some individual represented as a tuple in T (with possibly UNKNOWN value for y) and related tuples in S according to foreign key paths.

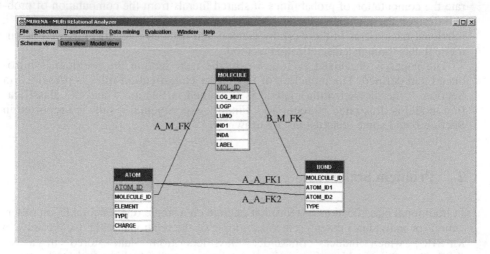

Fig. 1. An example of a relational representation of training data of the Mutagenesis database.

3 Generation of First-Order Rules

Let R' be a set of first-order classification rules for the classes $\{C_1, C_2,..., C_L\}$, and I an individual to be classified and defined as above. The individual can be logically represented as a set of ground facts, the only exception being the fact associated to the target relation T, where the argument corresponding to the target attribute y is a variable Y. A rule $R_j \in R'$ *covers* I, if a substitution θ exists, such that $R_j\theta \subseteq I\theta$. The application of the substitution to I is required to ground the only variable Y in I to the same constant as that reported in R_j for the target attribute. Let R be the subset of rules in R' that cover I, that is $R=\{R_j \in R' \mid R_j \text{ covers } I \}$. The first-order naïve Bayes classifier for the individual I, $f(I)$, is defined as follows:

$$f(I)= arg\ max_i\ P(C_i|R) = arg\ max_i\ \frac{P(C_i)P(R|C_i)}{P(R)}$$

The value $P(C_i)$ is the prior probability of the class C_i. Since $P(R)$ is independent of the class C_i, it does not affect $f(I)$, that is,

$$f(I)= arg\ max_i\ P(C_i)P(R|C_i) \tag{1}$$

The computation of $P(R|C_i)$ depends on the structure of R. Therefore, it is important to clarify how first-order rules are built in order to associate them with a probability measure. As already pointed out, Pompe and Kononenko use the first-order learning system ILP-R to induce the set of rules R'. This approach is very expensive and does not take into account the bias automatically determined by the constraints in the database. On the other hand, Flach and Lachiche use Tertius to determine the structure of *first-order features* on the basis of the structure of the individuals. The system Tertius deals with learning first-order logic rules from data lacking an explicit classification predicate. Consequently, the learned rules are not restricted to predicate definitions as in supervised inductive logic programming. Our solution is similar to that proposed by Flach since the structure of classification rules is determined on the basis of the structure of the individuals. The main difference is that the classification predicate is considered during the generation of the rules.

All predicates in classification rules generated by Mr-SBC are binary and can be of two different types.

Def 2. A binary predicate p is a *structural* predicate associated to a table $T_i \in S$ if a foreign key FK in T_i exists that references a table $T_{ij} \in S$. The first argument of p represents the primary key of T_{ij} and the second argument represents the primary key of T_i.

Def 3. A binary predicate p is a *property* predicate associated to a table $T_i \in S$, if the first argument of p represents the primary key of T_i and the second argument represents another attribute in T_i which is neither the primary key of T_i nor a foreign key in T_i.

Def 4. A first order classification rule associated to the *foreign key path* ϑ is a clause in the form:

$$p_0(A_1,y):- p_1(A_1,A_2), p_2(A_2,A_3), ..., p_{s-1}(A_{s-1},A_s), p_s(A_s,c).$$

where

1. p_0 is a property predicate associated to the target table T and to the target attribute y.

2. $\vartheta = (T_{i_1}, T_{i_2}, \ldots, T_{i_s})$ is a *foreign key path* such that for each $k=1, \ldots, s\text{-}1$: p_k is a structural predicate associated to the table T_{i_k}

3. p_s is a property predicate associated to the table T_{i_s}.

An example of a first-order rule is the following:

   molecule_Label(A, active) :- molecule_Atom(A,B), atom_Type(B,'[22..27]').

Mr-SBC searches all possible classification rules by means of a breadth-first strategy and iterates over some refining steps. A refining step is biased by the possible foreign key paths and consists of the addition of a new literal, the unification of two variables and, in the case of a property predicate, in the instantiation of a variable. The search strategy is biased by the structure of the database because each refining step is made only if the generated first-order classification rule can be associated to a foreign key path. However, the number of refinement steps is upper bounded by a user-defined constant MAX_LEN_PATH.

4   Discretization

In Mr-SBC continuous attributes are handled through supervised discretization. Supervised discretization methods utilize the information on the class labels of individuals to partition a numerical interval into bins. The proposed algorithm sorts the observed values of a continuous feature and attempts to greedily divide the domain of the continuous variable into bins, such that each bin contains only instances of one class. Since such a scheme could possibly lead to one bin for each observed real value, the algorithm is constrained to merge bins in a second step. Merging of two contiguous bins is performed when the increase of entropy is lower than a user-defined threshold (MAX_GAIN). This method is a variant of the one-step method 1RD by Holte [14] for the induction of one-level decision trees, that proved to work well with the Naïve Bayes Classifier [4]. It is also different from the one-step method by Fayyad and Irani [6] that recursively splits the initial interval according to the class information entropy measure until a stopping criterion based on the Minimum Description Length (MDL) principle is verified.

5   The Computation of Probabilities

According to the naïve Bayes assumption, the attributes are considered independent. However, this assumption is clearly false for the attributes that are primary keys or foreign keys. This means that the computation of $P(R|C_i)$ in equation (1) depends on the structures of rules in R. For instance, if R_1 and R_2 are two rules of class C_i, that share the same structure and differ only for the property predicates in their bodies

   R_1: $\beta_{1,0} : -\beta_{1,1}, \ldots, \beta_{1,K_1-1}, \beta_{1,K_1}$

   R_2: $\beta_{2,0} : -\beta_{2,1}, \ldots, \beta_{2,K_2-1}, \beta_{2,K_2}$

where

$$K_j = K_2 \text{ and } \beta_{1,1} = \beta_{2,1}, \beta_{1,2} = \beta_{2,2},...,\beta_{1,K_1-1} = \beta_{2,K_2-1}$$

then $P(\{R_1, R_2\}|C_i) = P(\beta_{1,0} \cap (\beta_{1,1},...,\beta_{1,K_1-1}) \cap \beta_{1,K_1} \cap \beta_{2,K_2} |C_i) =$

$$P(\beta_{1,0} \cap (\beta_{1,1},...,\beta_{1,K_1-1})|C_i) \cdot P(\beta_{1,K_1} \cap \beta_{2,K_2} | \beta_{1,0} \cap (\beta_{1,1},...,\beta_{1,K_1-1}) \cap C_i)$$

The first term takes into account the structure common to both rules while the second term refers to the conditional probability of satisfying the property predicates in the rules given the common structure.

The latter probability can be factorized under the naïve Bayes assumption, that is:

$$P(\beta_{1,K_1} \cap \beta_{2,K_2} | \beta_{1,0} \cap (\beta_{1,1},...,\beta_{1,K_1-1}) \cap C_i) =$$

$$P(\beta_{1,K_1} | \beta_{1,0} \cap (\beta_{1,1},...,\beta_{1,K_1-1}) \cap C_i) \cdot P(\beta_{2,K_2} | \beta_{1,0} \cap (\beta_{1,1},...,\beta_{1,K_1-1}) \cap C_i)$$

According to this approach the conditional probability of the structure is computed only once. This approach differs from that proposed in the works of Pompe and Kononenko [20] and Flach [7] where the factorization would multiply the structure probability twice.

By generalizing to a set of classification rules we have:

$$P(C_i)P(R|C_i) = P(C_i)P(structure)\prod_j P(R_j|structure) \qquad (2)$$

where the term *structure* takes into account the class C_i and the structural parts of the rules in R.

If the classification rule $R_j \in R$ is in the form $\beta_{j,0} : -\beta_{j,1},...,\beta_{j,K_j-1},\beta_{j,K_j}$ where $\beta_{j,0}$ and β_{j,K_j} are property predicates and $\beta_{j,1},\beta_{j,2},...,\beta_{j,K_j-1}$ are structural predicates, then:

$$P(R_j|structure) = P(\beta_{j,K_j} | \beta_{j,0},\beta_{j,1},...,\beta_{j,K_j-1}) = P(\beta_{j,K_j} | C_i,\beta_{j,1},...,\beta_{j,K_j-1})$$

where C_i is the value of the target attribute in the head of the clause ($\beta_{j,0}$). To compute this probability, we use the Laplace estimation:

$$P(\beta_{j,K_j} | C_i,\beta_{j,1},...,\beta_{i,K_j-1}) = \frac{\#(\beta_{j,K_j},C_i,\beta_{j,1},...,\beta_{j,K_j-1})+1}{\#(C_i,\beta_{j,1},...,\beta_{j,K_j-1})+F}$$

where F is the number of possible values of the attribute to which the β_{j,K_j} property predicate is associated. Laplace's estimate is used in order to avoid null probabilities in the equation (2). In practice, the value at the nominator is the number of individuals which satisfy that conjunction $\beta_{j,K_j},C_i,\beta_{j,1},...,\beta_{j,K_j-1}$, in other words, the number of individuals covered by the rule $\beta_{j,0} : -\beta_{j,1},...,\beta_{j,K_j-1},\beta_{j,K_j}$. It is determined by a "select count (*)" SQL instruction. The value of the denominator is the number of individuals covered by the rule $\beta_{j,0} : -\beta_{j,1},...,\beta_{j,K_j-1}$.

The term $P(structure)$ in the equation (2) is computed as follows: Let $B=\{(\beta_{j,1},\beta_{j,2},...,\beta_{j,t})| j=1..s \text{ and } t=1, ..., K_j-1\}$ the set of all distinct sequences of structural predicates in the rules of R. Then

$$P(structure)= \prod_{seq \in B} P(seq) \qquad (3)$$

To compute *P(seq)* it is necessary to introduce the definition of the probability JP that a join query is satisfied, for this purpose, the formulation provided in [11] can be useful. Let $\vartheta=(T_{i_1}, T_{i_2},\ldots, T_{i_s})$ be a *Foreign Key Path*, then:

$$JP(\vartheta)=JP(T_{i_1},\ldots,T_{i_s})=\frac{|\bowtie(T_{i_1}\times\ldots\times T_{i_s})|}{|T_{i_1}|\times\ldots\times|T_{i_s}|}$$

where $\bowtie(T_{i_1}\times\ldots\times T_{i_s})$ is the result of the join between the tables $T_{i_1},\ldots,T_{i_s}$.

We must remember that each sequence *seq* is associated to a foreign key path ϑ. If $seq=(\beta_{j,1},\beta_{j,2},\ldots,\beta_{j,t})$ there are two possibilities: either a prefix of *seq* is in *B* or not. By denoting as T_{j_h} the table related to $\beta_{j,h}$, $h=1,\ldots,t$, the probability P(seq) can be recursively defined as follows:

$$P(seq)=\begin{cases}JP(T_{j_1},\ldots,T_{j_t}) & \text{if } seq \text{ has no prefix in B}\\[2mm]\dfrac{JP(T_{j_1},\ldots,T_{j_t})}{P(seq')} & \text{if } seq' \text{ is the longest prefix of } seq \text{ in B}\end{cases}$$

This formulation is necessary in order to compute the formula (3) considering both dependent and independent events. Since *P(structure)* takes into account the class, *P(seq)* is computed separately for each class.

6   Experimental Results

MR-SBC has been implemented as a module of the system MURENA and has been empirically evaluated on the Mutagenesis datasets and on Biodegradability datasets.

6.1   Results on Mutagenesis

These datasets, taken from the MLNET repository, concern the problem of identifying the mutagenic compounds [19] and have been extensively used to test both inductive logic programming (ILP) systems and (multi-)relational mining systems. We considered, analogously to related experiments in the literature, the "regression friendly" dataset of 188 elements.

A recent study on this database [22] recognizes five levels of background knowledge for mutagenesis which can provide richer descriptions of the examples. In this study we used only the first three levels of background knowledge in order to compare the performance of Mr-SBC with other methods for which experimental results are available in the literature. Table 1 shows the first three sets of background knowledge used in our experiments, where $BK_i \subseteq BK_{i+1}$ for i=0, ..., 2. The greater the BK, the more complex the learning problem.

The dataset is analyzed by means of a 10-fold cross-validation, that is, the target table is first divided into ten blocks of near-equal size and distribution of class values, and then, for every block, a subset of tuples in *S* related to the tuples in the target table block are extracted. In this way, ten databases are created. Mr-SBC is trained on nine databases and tested on the hold-out database. Mr-SBC has been executed with the following parameters: MAX_LEN_PATH=4 and MAX_GAIN= 0.5.

Table 1. Background knowledge for Mutagenesis database.

Background	Description
BK_0	Consists of those data obtained with the molecular modelling package QUANTA. For each compound it obtains the atoms, bonds, bond types, atom types, and partial charges on atoms.
BK_1	Consists of Definitions in *B0* plus indicators *ind1*, and *inda* in molecule table.
BK_2	Variables (attributes) *logp*, and *lumo* are added to definitions in BK_1.

Table 2. Accuracy comparison on the set of 188 regression friendly elements of Mutagenesis. Results for Progol2, Foil, Tilde are taken from [1]. Results for Progol_1 are taken from [22]. The results for 1BC are taken from [9]. Results for 1BC2 are taken from [16]. Results for MRDTL are taken from [17]. The values are the results of 10-fold cross-validation.

System	Accuracy(%)		
	BK_0	BK_1	BK_2
Progol_1	79	86	86
Progol_2	76	81	86
Foil	61	61	83
Tilde	75	79	85
MRDTL	67	**87**	88
1BC2	72.9	---	72.9
1BC	**80.3**	---	87.2
Mr-SBC	76.5	81	**89.9**

Experimental results on predictive accuracy are reported in Table 2 for increasing complexity of the models. A comparison to other results reported in the literature is also made. Mr-SBC has the best performance for the most complex task (BK_2) with an accuracy of almost 90%, while it ranks third for the simplest task. Interestingly, the predictive accuracy increases with the complexity of the background knowledge, which means that the variables added in BK_1 and BK_2 are meaningful and Mr-SBC takes advantages of that.

As regards execution time (see Table 3). The time required by Mr-SBC increases with the complexity of the background knowledge. Mr-SBC is generally considerably faster than competing systems, such as Progol, Foil, Tilde and 1BC, that do not operate on data stored in a database. Moreover, except for the task BK_0, Mr-SBC performs better that MRDTL which works on a database. It is noteworthy that the trade-off between accuracy and complexity is in favour of Mr-SBC.

The average number of extracted rules for each fold is quite high (55.9 for BK_0, 59.9 for BK_1, and 64.8 for BK_2). Some rules are either redundant or cover very few individuals. Therefore, some additional stopping criteria are required to avoid the generation of these rules and to reduce further the cost complexity of the algorithm.

6.2   Results on Biodegradability

The Biodegradability dataset has already been used in the literature for both regression and classification tasks [5]. It consists of 328 structural chemical molecules described in terms of atom and bond. The target variable for machine learning systems is the natural logarithm of the arithmetic mean of the low and high estimate of the HTL (Half-Life Time) for acqueous biodegradation in aerobic conditions, measured in hours. We use a discretized version in order to apply classification systems to the problem. As in [5], four classes have been defined: chemicals degrade *fast, moderately, slowly* or are *resistant*.

Table 3. Time comparison of the set of 188 regression friendly elements of Mutagenesis. Results for Progol2, Foil, Tilde are taken from [1]. Results for Progol_1 are taken from [22]. Results for MRDTL are taken from [17]. The results of MR-SBC have been taken on a PIII WIN2k platform.

System	Time (Secs)		
	BK_0	BK_1	BK_2
Progol_1	8695	4627	4974
Progol_2	117000	64000	42000
Foil	4950	9138	0.5
Tilde	41	170	142
MRDTL	0.85	170	142
1BC2	--	--	--
1BC	--	--	--
MR-SBC	36	42	48

Table 4. Accuracy comparison on the set of 328 chemical molecules of Biodegradability. Results for Mr-SBC and Tilde are reported.

Fold	Mr-SBC	Tilde Pruned
0	**0.90909**	0.69697
1	**0.87878**	0.81818
2	0.84848	**0.90909**
3	0.87878	0.87879
4	**0.78788**	0.69697
5	0.84848	**0.90909**
6	0.90625	0.90625
7	**0.87879**	0.81818
8	0.87500	**0.93750**
9	**0.93939**	0.72727
Average	*0.87509*	*0.82983*

The dataset is analyzed by means of a 10-fold cross-validation. For each database Mr-SBC and Tilde are trained on nine databases and tested on the hold-out database. Mr-SBC has been executed with the following parameters: MAX_LEN_PATH=4 and MAX_GAIN= 0.5. Experimental results on predictive accuracy are reported in Table 4. They are in favour of Mr-SBC on the average of accuracy varying the fold.

7   Conclusions

In the paper, a multi-relational data mining system with a tight integration to a relational DBMS is described. It is based on the induction of a set of first-order classification rules in the context of naive Bayesian classification. It presents several differences with respect to related works. First, it is based on an integrated approach, so that the contribution of literals shared by several rules to the posterior probability is computed only once. Second, it works both on discrete and continuous attributes. Third, the generation of rules is based on the knowledge of a data model embedded in the database schema. The proposed method has been implemented in the new system Mr-SBC and tested on four benchmark tasks. Results on predictive accuracy are in favour of our system for the most complex tasks. Mr-SBC also proved to be efficient.

As future work, we plan to extend the comparison of Mr-SBC to other multi-relational data mining systems on a larger set of benchmark datasets. Moreover, we intend to frame the proposed method in a transduction inference setting, where both labelled and unlabelled data are available for training. Finally we intend to integrate Mr-SBC in a document processing system that makes extensive use of machine learning tools to reach a high adaptivity to different tasks.

Acknowledgments

This work has been supported by the annual Scientific Research Project "Scoperta di conoscenza in basi di dati: metodi e tecniche efficienti e robuste per dati complessi" Year 2002 funded by the University of Bari. The authors thank Hendrik Blockeel for providing mutagenesis and biodegradability datasets.

References

1. Blockeel, H. Top-down induction of first order logical decision trees. PhD dissertation, Department of Computer Science, Katholieke Universiteit Leuven, 1998.
2. De Raedt, L. Attribute-value learning versus Inductive Logic Programming: the Missing Links (Extended Abstract). In *Proceedings of the 8 th International Conference on Inductive Logic Programming*, volume 1446 of *Lecture Notes in Artificial Intelligence*, Springer-Verlag, 1998.
3. Domingos, P. & Pazzani, M.. On the optimality of the simple bayesian classifier under zero-one loss. *Machine Learning*, 29(2-3), pp. 103-130, 1997.
4. Dougherty, J., Kohavi, R., Sahami, M.: *Supervised and unsupervised discretization of continuous features*. In: Machine Learning: Proc of 12th International Conference. Morgan Kaufmann, pp.194-202. 1995.
5. Dzeroski S., Blockeel H., Kramer S., Kompare B., Pfahringer B., and Van Laer W.. Experiments in predicting biodegradability. *Proceedings of the Ninth International Workshop on Inductive Logic Programming* (S. Dzeroski and P. Flach, eds.), LNAI, vol. 1634, Springer, pp. 80-91, 1999.
6. Fayyad U.M., Irani K.B., Multi-interval discretization of continuous-valued attributes for classification learning. In Proc. Of the 13th International Joint Conference on Artificial Intelligence. pp.1022—1027, 1994.
7. Flach P.A. and Lachiche N.. Decomposing probability distributions on structured individuals. In Paula Brito, Joaquim Costa, and Donato Malerba, editors, *Proceedings of the ECML2000 workshop on Dealing with Structured Data in Machine Learning and Statistics*, pages 33--43, Barcelona, Spain, May 2000.
8. Flach P.A. and Lachiche N.. *Confirmation-guided discovery of first-order rules with Tertius*. Machine Learning, 2000.

9. Flach P. and Lachiche N.. First-order Bayesian Classification with 1BC. Submitted. Downloadable from http://hydria.u-strasbg.fr/~lachiche/1BC.ps.gz
10. Friedman, N., Getoor, L., Koller, D., and Pfeffer, A. Learning probabilistic relational models. In *Proceedings of the 6 th International Joint Conference on Artificial Intelligence*, Morgan Kaufman, 1999.
11. Getoor, L. Multi-relational data mining using probabilistic relational models: research summary. In: A. J. Knobbe, and D. M. G. van der Wallen, editors. *Proceedings of the First Workshop in Multi-relational Data Mining*, 2001.
12. Getoor L., Koller D., Taskar B. Statistical models for relational data. In Proceedings of the KDD-2002 Workshop on Multi-Relational Data Mining, pages 36-55, Edmonton, CA, 2002.
13. Getoor L.. Learning Statistical Models from Relational Data, Ph.D. Thesis, Stanford University, December, 2001.
14. Holte, R.C. Very simple classification rules perform well on most commonly used datasets, Machine Learning 11, pp. 63-90, 1993.
15. Krogel, M., and Wrobel, S. Transformation-Based Learning Using Multirelational Aggregation. In Céline Rouveirol and Michèle Sebag, editors, *Proceedings of the 11 th International Conference on Inductive Logic Programming*, vol. 2157 of *Lecture Notes in Artificial Intelligence*, Springer-Verlag, 2001.
16. Lachiche N. and Flach P.. 1BC2: a true first-order Bayesian classifier. In Claude Sammut and Stan Matwin, ed., *Proceedings of the Thirteenth International Workshop on Inductive Logic Programming (ILP'02)*, Sydney, Australia. LNAI 2583, Springer. pp. 133-148. 2003.
17. Leiva H.A.:MRDTL: *A multi-relational decision tree learning algorithm.* Master thesis, University of Iowa, USA, 2002.
18. Mitchell, T. Machine Learning. McGraw Hill, 1997.
19. Muggleton S. H., Bain M., Hayes-Michie J., Michie D.. An experimental comparison of human and machine learning formalisms. In Proc. Sixth International Workshop on Machine Learning, Morgan Kaufmann, San Mateo, CA, pp. 113--118, 1989.
20. Pompe U. and Kononenko I.. Naive Bayesian classifier within ILP-R. In L. De Raedt, editor, *Proc. of the 5th Int. Workshop on Inductive Logic Programming*, pages 417--436. Dept. of Computer Science, Katholieke Universiteit Leuven, 1995.
21. Pompe U., Kononenko I.. Linear space induction in first order logic with relief. In R. Kruse, R. Viertl. & G. Della Riccia (Eds.), *CISM Lecture Notes.* Udine Italy, 1994.
22. Srinivasan, A., King, R. D., and Muggleton, S. The role of background knowledge: using a problem from chemistry to examine the performance of an ILP program. Technical Report PRG-TR-08-99, Oxford University ComputingLaboratory, Oxford, 1999.
23. Wrobel, S. Inductive logic programming for knowledge discovery in databases. In: D eroski, S., N. Lavrač(eds.): Relational Data Mining, Springer: Berlin, pp. 74-101. 2001.

SMOTEBoost: Improving Prediction
of the Minority Class in Boosting

Nitesh V. Chawla[1], Aleksandar Lazarevic[2],
Lawrence O. Hall[3], and Kevin W. Bowyer[4]

[1] Business Analytic Solutions, Canadian Imperial Bank of Commerce (CIBC)
BCE Place, 161 Bay Street, 11th Floor, Toronto, ON M5J 2S8, Canada
nitesh.chawla@cibc.ca
[2] Department of Computer Science, University of Minnesota
200 Union Street SE, Minneapolis, MN 55455, USA
aleks@cs.umn.edu
[3] Department of Computer Science and Engineering, University of South Florida
ENB 118, 4202 E. Fowler Avenue, Tampa, FL 33620, USA
hall@csee.usf.edu
[4] Department of Computer Science and Engineering
384 Fitzpatrick Hall, University of Notre Dame, IN 46556, USA
kwb@cse.nd.edu

Abstract. Many real world data mining applications involve learning from imbalanced data sets. Learning from data sets that contain very few instances of
the minority (or interesting) class usually produces biased classifiers that have a
higher predictive accuracy over the majority class(es), but poorer predictive accuracy over the minority class. SMOTE (Synthetic Minority Over-sampling
TEchnique) is specifically designed for learning from imbalanced data sets.
This paper presents a novel approach for learning from imbalanced data sets,
based on a combination of the SMOTE algorithm and the boosting procedure.
Unlike standard boosting where all misclassified examples are given equal
weights, SMOTEBoost creates synthetic examples from the rare or minority
class, thus indirectly changing the updating weights and compensating for
skewed distributions. SMOTEBoost applied to several highly and moderately
imbalanced data sets shows improvement in prediction performance on the minority class and overall improved *F-values*.

1   Motivation and Introduction

Rare events are events that occur very infrequently, i.e. whose frequency ranges from
say 5% to less than 0.1%, depending on the application. Classification of rare events
is a common problem in many domains, such as detecting fraudulent transactions,
network intrusion detection, Web mining, direct marketing, and medical diagnostics.
For example, in the network intrusion detection domain, the number of intrusions on
the network is typically a very small fraction of the total network traffic. In medical
databases, when classifying the pixels in mammogram images as cancerous or not

N. Lavrač et al. (Eds.): PKDD 2003, LNAI 2838, pp. 107–119, 2003.
© Springer-Verlag Berlin Heidelberg 2003

[1], abnormal (cancerous) pixels represent only a very small fraction of the entire image. The nature of the application requires a fairly high detection rate of the minority class and allows for a small error rate in the majority class since the cost of misclassifying a cancerous patient as non-cancerous can be very high.

In all these scenarios when the majority class typically represents 98-99% of the entire population, a trivial classifier that labels everything with the majority class can achieve high accuracy. It is apparent that for domains with imbalanced and/or skewed distributions, classification accuracy is not sufficient as a standard performance measure. ROC analysis [2] and metrics such as *precision, recall* and *F-value* [3, 4] have been used to understand the performance of the learning algorithm on the minority class. The prevalence of class imbalance in various scenarios has caused a surge in research dealing with the minority classes. Several approaches for dealing with imbalanced data sets were recently introduced [1, 2, 4, 9-15].

A confusion matrix as shown in Table 1 is typically used to evaluate performance of a machine learning algorithm for rare class problems. In classification problems, assuming class "C" as the minority class of the interest, and "NC" as a conjunction of all the other classes, there are four possible outcomes when detecting class "C".

Table 1. Confusion matrix defines four possible scenarios when classifying class "C"

	Predicted Class "C"	**Predicted Class "NC"**
Actual class "C"	True Positives (TP)	False Negatives (FN)
Actual class "NC"	False Positives (FP)	True Negatives (TN)

From Table 1, *recall, precision* and *F-value* may be defined as follows:

$$Precision \quad = \quad TP / (TP + FP)$$

$$Recall \quad = \quad TP / (TP + FN)$$

$$F\text{-}value \quad = \quad \frac{(1+\beta^2) \cdot Re\,call \cdot Pr\,ecision}{\beta^2 \cdot Re\,call + Pr\,ecision},$$

where β corresponds to relative importance of *precision* vs. *recall* and it is usually set to 1. The main focus of all learning algorithms is to improve the *recall*, without sacrificing the *precision*. However, the *recall* and *precision* goals are often conflicting and attacking them simultaneously may not work well, especially when one class is rare. The *F-value* incorporates both *precision* and *recall*, and the "goodness" of a learning algorithm for the minority class can be measured by the *F-value*. While ROC curves represent the trade-off between values of TP and FP, the F-value basically incorporates the relative effects/costs of *recall* and *precision* into a single number.

It is well known in machine learning that a combination of classifiers can be an effective technique for improving prediction accuracy. As one of the most popular combining techniques, boosting [5] uses adaptive sampling of instances to generate a highly accurate ensemble of classifiers whose individual global accuracy is only moderate. There has been significant interest in the recent literature for embedding cost-sensitivity in the boosting algorithm. CSB [6] and AdaCost boosting algorithms [7] update the weights of examples according to the misclassification costs. Karakoulas and Shawe-Taylor's ThetaBoost adjusts the margins in the presence of unequal loss functions [8]. Alternatively, Rare-Boost [4, 9] updates the weights of the examples differently for all four entries shown in Table 1.

In this paper we propose a novel approach for learning from imbalanced data sets, SMOTEBoost, that embeds SMOTE [1], a technique for countering imbalance in a dataset, in the boosting procedure. We apply SMOTE during each boosting iteration in order to create new synthetic examples from the minority class. SMOTEBoost constructs focuses on the minority class examples sampled for each boosting iteration, and constructs new examples. Experiments performed on data sets from several domains have shown that SMOTEBoost is able to achieve a higher *F-value* than SMOTE applied to a classifier, standard boosting algorithm, AdaCost [7] and first smote then boosting for each of the datasets. We also provide a *precision-recall* analysis of the approaches.

2   Synthetic Minority Oversampling Technique - SMOTE

SMOTE (Synthetic Minority Oversampling Technique) was proposed to counter the effect of having few instances of the minority class in a data set [1]. SMOTE creates synthetic instances of the minority class by operating in the "feature space" rather than the "data space". By synthetically generating more instances of the minority class, the inductive learners, such as decision trees (e.g. C4.5 [16]) or rule-learners (e.g. RIPPER [17]), are able to broaden their decision regions for the minority class. We deal with nominal (or discrete) and continuous attributes differently in SMOTE. In the nearest neighbor computations for the minority classes we use Euclidean distance for the continuous features and the Value Distance Metric (with the Euclidean assumption) for the nominal features [1, 18, 19]. The new synthetic minority samples are created as follows:

- For the continuous features
  - o   Take the difference between a feature vector (minority class sample) and one of its k nearest neighbors (minority class samples).
  - o   Multiply this difference by a random number between 0 and 1.
  - o   Add this difference to the feature value of the original feature vector, thus creating a new feature vector
- For the nominal features
  - o   Take majority vote between the feature vector under consideration and its k nearest neighbors for the nominal feature value. In the case of a tie, choose at random.
  - o   Assign that value to the new synthetic minority class sample.

Using this technique, a new minority class sample is created in the neighborhood of the minority class sample under consideration. The neighbors are proportionately utilized depending upon the amount of SMOTE. Hence, using SMOTE, more general regions are learned for the minority class, allowing the classifiers to better predict unseen examples belonging to the minority class. A combination of SMOTE and under-sampling creates potentially optimal classifiers as a majority of points from the SMOTE and under-sampling combination lie on the convex hull of the family of ROC curves [1, 2].

3   SMOTEBoost Algorithm

In this paper, we propose a SMOTEBoost algorithm that combines the Synthetic Minority Oversampling Technique (SMOTE) and the standard boosting procedure. We want to utilize SMOTE for improving the prediction of the minority classes, and we want to utilize boosting to not sacrifice accuracy over the entire data set. Our goal is to better model the minority class in the data set, by providing the learner not only with the minority class instances that were misclassified in previous boosting iterations, but also with a broader representation of those instances. We want to improve the overall accuracy of the ensemble by focusing on the difficult minority (positive) class cases, as we want to model this class better, with minimal accuracy degradation for the majority class The goal is to improve our True Positives (TP).

The standard boosting procedure gives equal weights to all misclassified examples. Since boosting algorithm samples from a pool of data that predominantly consists of the majority class, subsequent samplings of the training set may still be skewed towards the majority class. Although boosting reduces the variance and the bias in the final ensemble, it might not be as effective for data sets with skewed class distributions.. Boosting algorithm (Adaboost) treats both kinds of errors (FP and FN) in a similar fashion, and therefore sampling distributions in subsequent boosting iterations could have a larger composition of majority class cases.

Our goal is to reduce the bias inherent in the learning procedure due to the class imbalance. Introducing SMOTE in each round of boosting will enable each learner to learn from more of the minority class cases, thus learning broader decision regions for the minority class. We only SMOTE for the minority class examples in the distribution D_t at the iteration t. This has an implicit effect of increasing the sampling weights of minority class cases, as new examples are created in D_t. The synthetically created minority class cases are discarded after learning a classifier at iteration t. That is, they are not added to the original training set, and new examples are constructed in each iteration t, by sampling from D_t. The error-estimate after each boosting iteration is on the original training set. Thus, we try to maximize the margin for the skewed class dataset, by adding new minority class cases before learning a classifier in a boosting iteration. We also conjecture that introducing the SMOTE procedure also increases the diversity amongst the classifiers in the ensemble, as in each iteration we produce a different set of synthetic examples, and therefore different classifiers. The amount of SMOTE is a parameter that can vary for each data set. It will be useful to know a priori the amount of SMOTE to be introduced for each data set. We believe that utilizing a validation set to set the amount of SMOTE before the boosting iterations can be useful.

The combination of SMOTE and the boosting procedure that we present here is a variant of the AdaBoost.M2 procedure [5]. The proposed SMOTEBoost algorithm, shown in Fig. 1, proceeds in a series of T rounds. In every round a weak learning algorithm is called and presented with a different distribution D_t altered by emphasizing particular training examples. The distribution is updated to give wrong classifications higher weights than correct classifications. Unlike standard boosting, where the distribution D_t is updated uniformly for examples from both the majority and minority classes, in the SMOTEBoost technique the distribution D_t is updated such that the examples from the minority class are oversampled by creating synthetic minority class examples (See Line 1, Fig. 1). The entire weighted training set is given to the weak learner to compute the weak hypothesis h_t. At the end, the different hypotheses are combined into a final hypothesis h_{fn}.

- Given: Set S $\{(x_1, y_1), \ldots , (x_m, y_m)\}$ $x_i \in X$, with labels $y_i \in Y = \{1, \ldots, C\}$, where C_p, $(C_p < C)$ corresponds to a minority (positive) class.
- Let B = $\{(i, y): i = 1,\ldots,m, y \neq y_i\}$
- Initialize the distribution D_1 over the examples, such that $D_1(i) = 1/m$.
- For $t = 1, 2, 3, 4, \ldots T$
  1. Modify distribution D_t by creating N synthetic examples from minority class C_p using the SMOTE algorithm
  2. Train a weak learner using distribution D_t
  3. Compute weak hypothesis $h_t: X \times Y \to [0, 1]$
  4. Compute the pseudo-loss of hypothesis h_t:
     $$\varepsilon_t = \sum_{(i,y)\in B} D_t(i, y)(1 - h_t(x_i, y_i) + h_t(x_i, y))$$
  5. Set $\beta_t = \varepsilon_t / (1 - \varepsilon_t)$ and $w_t = (1/2)\cdot(1 - h_t(x_i, y) + h_t(x_i, y_i))$
  6. Update D_t: $D_{t+1}(i, y) = (D_t(i, y)/Z_t)\cdot\beta_t^{w_t}$
     where Z_t is a normalization constant chosen such that D_{t+1} is a distribution.
- Output the final hypothesis: $h_{fn} = \arg\max_{y\in Y}\sum_{t=1}^{T}(\log\frac{1}{\beta_t})\cdot h_t(x, y)$

Fig. 1. The SMOTEBoost algorithm

We used RIPPER [17], a learning algorithm that builds a set of rules for identifying the classes while minimizing the amount of error, as the classifier in our SMOTEBoost experiments. RIPPER is a rule-learning algorithm based on the separate-and-conquer strategy. We applied SMOTE with different values for the parameter N that specifies the amount of synthetically generated examples.

4   Experiments

Our experiments were performed on the four data sets summarized in Table 2. For all data sets, except for the KDD Cup-99 intrusion detection data set [20, 21], the reported (averaged) values for *recall, precision* and *F-value* were obtained by performing 10-fold cross-validation. For the KDDCup-99 data set however, the separate intrusion detection test set was used to evaluate the performance of proposed algorithms. Since the original training and test data sets have totally different distributions due to novel intrusions introduced in the test data, for the purposes of this paper, we modified the data sets in order to have similar distributions in the training and test data. Therefore, we first merged the original training and test data sets and then sampled 69,980 network connections from this merged data set in order to reduce the size of the data set. The sampling was performed only from majority classes (normal background traffic and the DoS attack category), while other classes (Probe, U2R, R2L) remained intact. Finally, the new train and test data sets used in our experiments were obtained by randomly splitting the sampled data set into equal size subsets. The distribution of network connections in the new test data set is given in Table 2. Unlike the KDDCup-99 intrusion data set that has a mixture of both nominal and continuous features, the remaining data sets (mammography [1], satimage [22], phoneme [23]) have all continuous features. For the satimage data set we chose the smallest class as the minority class and collapsed the remaining classes into one class as was done in [24]. This procedure gave us a skewed 2-class dataset, with 5809 majority class examples and 626 minority class examples.

Table 2. Summary of data sets used in experiments

Data set	Number of majority class instances			Number of minority class instances		Number of classes
KDDCup-99 Intrusion	DoS	Probe	Normal	U2R	R2L	5
	13027	2445	17400	136	1982	
Mammography	10923			260		2
Satimage	5809			626		2
Phoneme	3818			1586		2

When experimenting with SMOTE and the SMOTEBoost algorithm, different values for the SMOTE parameter N, ranging between 100 and 500, were used for the minority classes. Since the KDD Cup'99 data set has two minority classes U2R and R2L that are not equally represented in the data set, different combinations of SMOTE parameters were investigated for these two minority classes (values 100, 300, and 500 were used for the U2R class while the value 100 was used for the R2L class). The values of the SMOTE parameters for U2R class were higher than the SMOTE parameter values for R2L class, since the U2R class is rarer than the R2L class in KDD-Cup 1999 data set (R2L has a larger number of examples). Our experimental results showed that the higher values of SMOTE parameters for the R2L class could lead to over-fitting and decreasing the prediction performance on that class (since SMOTEBoost achieved only minor improvements for the R2L class, these results are not reported here due to space limitations).

The experimental results for all four data sets are presented in Tables 3 to 6 and in Figures 2 to 4. It is important to note that these tables report only the prediction performance for the minority classes from four data sets, since prediction of the majority class was not of interest in this study. Moreover, precision captures the FP's introduced in the classification. So, F-value includes the estimate for majority class examples wrongly classified. Due to space limitations, the figures with *precision* and *recall* trends over the boosting iterations, along with the *F-value* trends for the representative SMOTE parameter were not shown for the R2L class from KDDCup'99 data as well as for the satimage data set. In addition, the left and the right parts of the reported Figures do not have the same scale due to the fact that the range of changes in *recall* and *precision* shown in the same graph is much larger than the change of the *F-value*.

Table 3. Final values for *recall*, *precision* and *F-value* for minority U2R class when proposed methods are applied on KDDCup-99 intrusion data set. (N_{u2r} corresponds to the SMOTE parameter for U2R class, while N_{r2l} corresponds to the SMOTE parameter for R2L class)

Method			Recall	Precision	F-value	Method			Recall	Precision	F-value
Standard RIPPER			57.35	84.78	68.42	Standard Boosting			80.15	90.083	84.83
	N_{u2r}	N_{r2l}	Recall	Precision	F-value		N_{u2r}	N_{r2l}	Recall	Precision	F-value
SMOTE	100	100	80.15	88.62	84.17	SMOTE -Boost	100	100	84.2	93.9	**88.8**
	300	100	74.26	92.66	82.58		300	100	87.5	88.8	88.15
	500	100	68.38	86.11	71.32		500	100	84.6	92.0	88.1
First SMOTE then Boost	N_{u2r}	N_{r2l}	Recall	Precision	F-value	Ada- Cost	Cost factor		Recall	Precision	F-value
	100	100	81.6	90.92	**86.01**		c = 2		83.1	96.6	**89.3**
	300	100	82.5	89.30	85.77		c = 5		83.45	95.29	88.98
	500	100	82.9	89.12	85.90						

Table 4. Final values for *recall*, *precision* and *F-value* for minority class when proposed methods are applied on *mammography* data set

Method		Recall	Precision	F-value	Method		Recall	Precision	F-value
Standard RIPPER		48.12	74.68	58.11	Standard Boosting		59.09	77.05	66.89
SMOTE	N = 100	58.04	64.96	61.31	SMOTE -Boost	N = 100	61.73	76.59	**68.36**
	N = 200	62.16	60.53	60.45		N = 200	62.63	74.54	68.07
	N = 300	62.55	56.57	58.41		N = 300	64.16	69.92	66.92
	N = 500	64.51	53.81	58.68		N = 500	61.37	70.41	65.58
First SMOTE then Boost	N = 100	60.22	76.16	**67.25**	Ada- Cost	Cost factor	Recall	Precision	F-value
	N = 200	62.61	72.10	67.02		2	59.83	69.07	**63.01**
	N = 300	63.92	70.26	66.94		5	68.45	55.12	59.36
	N = 500	64.14	69.80	66.85					

Table 5. Final values for *recall*, *precision* and *F-value* for minority class when proposed methods are applied on *Satimage* data set

Method		Recall	Precision	F-value	Method		Recall	Precision	F-value
Standard RIPPER		47.43	67.92	55.50	Standard Boosting		58.74	80.12	67.78
SMOTE	$N = 100$	65.17	55.88	59.97	SMOTE-Boost	$N = 100$	63.88	77.71	70.12
	$N = 200$	74.89	48.08	58.26		$N = 200$	65.35	73.17	69.04
	$N = 300$	76.32	47.17	57.72		$N = 300$	67.87	72.68	**70.19**
	$N = 500$	77.96	44.51	56.54		$N = 500$	67.73	69.5	68.6
First SMOTE then Boost	$N = 100$	64.69	72.53	**68.65**	Ada-Cost	*Cost factor*	*Recall*	*Precision*	*F-value*
	$N = 200$	69.23	67.10	68.15		2	64.85	54.58	**58.2**
	$N = 300$	67.25	69.92	68.56		5	60.85	56.01	57.6
	$N = 500$	67.84	68.02	67.93					

Table 6. Final values for *recall*, *precision* and *F-value* for minority class when proposed methods are applied on *phoneme* data set

Method		Recall	Precision	F-value	Method		Recall	Precision	F-value
Standard RIPPER		62.28	69.13	65.15	Standard Boosting		76.1	77.07	76.55
SMOTE	$N = 100$	82.18	59.91	68.89	SMOTE-Boost	$N = 100$	81.86	73.66	**77.37**
	$N = 200$	85.88	58.51	69.59		$N = 200$	84.86	76.47	76.47
	$N = 300$	89.79	56.15	69.04		$N = 300$	86	66.76	75.16
	$N = 500$	94.2	50.22	65.49		$N = 500$	88.46	65.16	75.04
First SMOTE then Boost	$N = 100$	82.05	72.34	**76.89**	Ada-Cost	*Cost factor*	*Recall*	*Precision*	*F-value*
	$N = 200$	85.25	68.97	76.25		2	76.83	75.71	**75.99**
	$N = 300$	87.37	66.38	75.44		5	85.05	68.71	75.9
	$N = 500$	89.21	64.73	75.03					

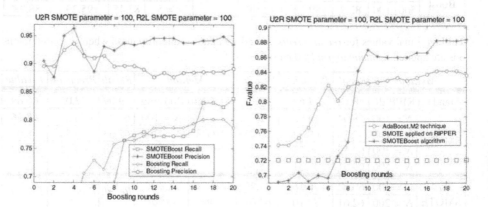

Fig. 2. Precision, Recall, and F-values for the minority U2R class when the SMOTEBoost algorithm is applied on the KDDCup 1999 data set

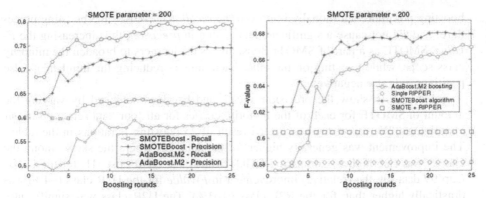

Fig. 3. *Precision*, *Recall*, and *F-values* for the minority class when the SMOTEBoost algorithm is applied on the Mammography data set

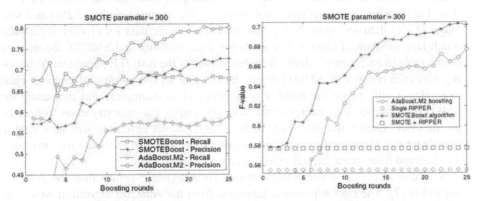

Fig. 4. Precision, Recall, and F-values for the minority class when the SMOTEBoost algorithm is applied on the Satimage data set

Analyzing Figures 2 to 4 and Tables 3 to 6, it is apparent that SMOTEBoost achieved higher *F-values* than the other presented methods including standard boosting, AdaCost, SMOTE with the RIPPER classifier and the standard RIPPER classifier, although the improvement varied with different data sets. We have also compared SMOTEBoost to the procedure "First SMOTE, then Boost" when we first apply SMOTE and then perform boosting in two separate steps. It is SMOTEBoost's apparent improvement in *recall*, while not causing a significant degradation in *precision* that improves the over-all *F-value*. Tables 3 to 6 include the *precision*, *recall*, and *F-value* for the various methods at different amounts of SMOTE (best values are given in bold). These reported values indicate that SMOTE applied with the RIPPER classifier has the effect of improving the *recall* of the minority class due to improved coverage of the minority class examples, while at the same time SMOTE causes a decrease in *precision* due to an increased number of false positive examples. Thus, SMOTE is more targeted to the minority class than standard boosting or RIPPER. On the other hand, standard boosting is able to improve both the *recall* and *precision* of a single classifier, since it gives all errors equal weights. SMOTE embedded within the boosting procedure additionally improved the *recall* achieved by the boosting procedure, and did not cause a significant degradation in *precision*, thus increasing the *F-value*. SMOTE as a part of SMOTEBoost allows the learners to broaden the minority class scope, while boosting on the other hand aims at reducing the number of false positives and false negatives.

Tables 3 to 6 show the precision, recall, and F-values achieved by varying the amount of SMOTE for each of the minority classes for all four data sets used in our experiments. We report the aggregated result of 25 boosting iterations in the tables. The improvement was generally higher for the data sets where the skew among the classes was also higher. Comparing SMOTEBoost and AdaBoost.M1, for the KDD-Cup'99 data set, the (relative) improvement in *F-value* for the U2R class (~4%) was drastically higher than for the R2L class (0.61%). The U2R class was significantly less represented in the data set than the R2L class (the number of U2R examples was around 15 times smaller than the number of examples from the R2L class). In addition, the (relative) improvements in *F-value* for the mammography (2.2%) and satimage (3.4%) data sets were better than for the phoneme data set (1.4%), which had much less imbalanced classes. For phoneme data, boosting and SMOTE Boost were comparable to each other, while for higher values of the SMOTE parameter N, boosting was even better than SMOTEBoost. In this data set the number of majority class examples is only twice the number of minority class examples, and increasing the SMOTE parameter N to values larger than 200 causes the minority class to become the majority. Hence, the classifiers in the SMOTEBoost ensemble will now tend to over-learn the minority class, causing a higher degradation in *precision* for the minority class and therefore a reduction in *F-value*.

We have also shown that SMOTEBoost gives higher *F-values* than the AdaCost algorithm [7]. The cost-adjustment functions from the AdaCost algorithm were chosen as follows: $\beta_- = 0.5*c + 0.5$ and $\beta_+ = -0.5*c + 0.5$, where β_- and β_+ are the functions for mislabeled and correctly labeled examples, respectively. AdaCost causes a greater sampling from the minority class examples due to the β function in the boosting distribution. This implicitly has an effect of over-sampling with replication. SMOTEBoost on the other hand constructs new examples at each round of boosting, thus avoiding overfitting and achieving higher minority class classification performances than AdaCost. Although AdaCost improves the *recall* over AdaBoost, it significantly reduces the precision thus causing a reduction in *F-value*. It is also interesting to note that SMOTEBoost achieves better *F-values* than the procedure "First SMOTE, then Boost" since in every boosting iteration new examples from minority class are generated, and thus, more diverse classifiers are created in the boosting ensemble. Finally, SMOTEBoost particularly focuses on the examples selected in the Dt, which are potentially misclassified or are on the classification boundaries.

5   Conclusions

A novel approach for learning from imbalanced data sets is presented. The proposed SMOTEBoost algorithm is based on the integration of the SMOTE algorithm within the standard boosting procedure. Experimental results from several imbalanced data sets indicate that the proposed SMOTEBoost algorithm can result in better prediction of minority classes than AdaBoost, AdaCost, "First SMOTE then Boost" procedure and a single classifier. Data sets used in our experiments contained different degrees of imbalance and different sizes, thus providing a diverse test bed.

The SMOTEBoost algorithm successfully utilizes the benefits from both boosting and the SMOTE algorithm. While boosting improves the predictive accuracy of classifiers by focusing on difficult examples that belong to all the classes, the SMOTE algorithm improves the performance of a classifier only on the minority class examples. Therefore, the embedded SMOTE algorithm forces the boosting algorithm to focus more on difficult examples that belong to the minority class than to the majority class. SMOTEBoost implicitly increases the weights of the misclassified minority class instances (false negatives) in the distribution D_t by increasing the number of minority class instances using the SMOTE algorithm. Therefore, in the subsequent boosting iterations SMOTEBoost is able to create broader decision regions for the minority class compared to the standard boosting. We conclude that SMOTEBoost can construct an ensemble of diverse classifiers and reduce the bias of the classifiers. SMOTEBoost combines the power of SMOTE in vastly improving the *recall* with the power of boosting in improving the *precision*. The overall effect is a better *F-value*.

Our experiments have also shown that SMOTEBoost is able to achieve higher F-values than AdaCost, due to SMOTE's ability to improve the coverage of the minority class when compared to the indirect effect of oversampling with replication in AdaCost.

Although the experiments have provided evidence that the proposed method can be successful for learning from imbalanced data sets, future work is needed to address its possible drawbacks. First, automatic determination of the amount of SMOTE will not only be useful when deploying SMOTE as an independent approach, but also for combining SMOTE and boosting. Second, our future work will also focus on investigating the effect of mislabeling noise on the performance of SMOTEBoost, since it is known that boosting does not perform well in the presence of noise.

Acknowledgments

This research was partially supported by the US Department of Energy through the San-dia National Labs ASCI VIEWS Data Discovery Program contract number DE-AC04-76DO00789 and by Army High Performance Computing Research Center contract number DAAD19-01-2-0014. The content of the work does not necessarily reflect the position or policy of the government and no official endorsement should be inferred. Access to computing facilities was provided by AHPCRC and the Minnesota Supercomputing Institute. We also thank Philip Kegelmeyer for his helpful feedback. We would also like to thank anonymous reviewers for their useful comments on the paper.

References

1. N. V. Chawla, K.W. Bowyer, L. O. Hall, W. P. Kegelmeyer, SMOTE: Synthetic Minority Over-Sampling Technique, *Journal of Artificial Intelligence Research*, vol. 16, 321-357, 2002.
2. F. Provost, T. Fawcett, Robust Classification for Imprecise Environments, *Machine Learning*, vol. 42/3, pp. 203-231, 2001.
3. M. Buckland, F. Gey, The Relationship Between Recall and Precision, *Journal of the American Society for Information Science*, 45(1):12--19, 1994.
4. M. Joshi, V. Kumar, R. Agarwal, Evaluating Boosting Algorithms to Classify Rare Classes: Comparison and Improvements, *First IEEE International Conference on Data Mining*, San Jose, CA, 2001.
5. Y. Freund, R. Schapire, Experiments with a New Boosting Algorithm, *Proceed-ings of the 13th International Conference on Machine Learning*, 325-332, 1996.
6. K. Ting, A Comparative Study of Cost-Sensitive Boosting Algorithms, *Proceedings of 17th International Conference on Machine Learning*, 983-990, Stanford, CA, 2000.
7. W. Fan, S. Stolfo, J. Zhang, P. Chan, AdaCost: Misclassification Cost-Sensitive Boosting, *Proc. of 16th International Conference on Machine Learning*, Slovenia, 1999.
8. G. Karakoulas, J. Shawe-Taylor, Optimizing Classifiers for Imbalanced Training Sets. In Kearns, M., Solla, S., and Cohn, D., editors. *Advances in Neural Information Processing Systems* 11, MIT Press, 1999.
9. M.Joshi, R. Agarwal, V. Kumar, Predicting Rare Classes: Can Boosting Make Any Weak Learner Strong?, *Proceedings of Eighth ACM Conference ACM SIGKDD Conference on Knowledge Discovery and Data Mining*, Edmonton, Canada, 2002.
10. M.Joshi, R. Agarwal, PNrule: A New Framework for Learning Classifier Models in Data Mining (A Case-study in Network Intrusion Detection), *First SIAM Conference on Data Mining*, Chicago, IL, 2001.
11. P. Chan, S. Stolfo, Towards Scalable Learning with Non-uniform Class and Cost Distributions: A Case Study in Credit Card Fraud Detection, *Proceedings of Fourth ACM Conference ACM SIGKDD International Conference on Knowledge Discovery and Data Mining*, 164-168, New York, NY, 1998.
12. M. Kubat, R. Holte, and S. Matwin, Machine Learning for the Detection of Oil Spills in Satellite Radar Images, *Machine Learning*, vol. 30, pp. 195-215, 1998.
13. N. Japkowicz, The Class Imbalance Problem: Significance and Strategies, *Proceedings of the 2000 International Conference on Artificial Intelligence (IC-AI'2000): Special Track on Inductive Learning*, Las Vegas, Nevada, 2000.
14. D. Lewis and J. Catlett, Heterogeneous Uncertainty Sampling for Supervised Learning, *Proceedings of the Eleventh International Conference of Machine Learning*, San Francisco, CA, 148-156, 1994.
15. C. Ling and C. Li, Data Mining for Direct Marketing Problems and Solutions, *Proceedings of the Fourth International Conference on Knowledge Discovery and Data Mining*, New York, NY, 1998.
16. J. Quinlan, *C4.5: Programs for Machine Learning*. San Mateo, CA: Morgan Kaufman, 1992.
17. W. Cohen, Fast Effective Rule Induction, *Proceedings of the 12th International Conference on Machine Learning*, Lake Tahoe, CA, 115-123, 1995.

18. C. Stanfill, D. Waltz, Toward Memory-based Reasoning, *Communications of the ACM*, vol. 29, no. 12, pp. 1213-1228, 1986.
19. S. Cost, S. Salzberg, A Weighted Nearest Neighbor Algorithm for Learning with Symbolic Features, *Machine Learning*, vol. 10, no. 1, pp. 57-78, 1993.
20. KDD-Cup 1999 Task Description, http://kdd.ics.uci.edu/databases/kddcup99/task.html
21. R. Lippmann, D. Fried, I. Graf, J. Haines, K. Kendall, D. McClung, D. Weber, S. Webster, D. Wyschogrod, R. Cunningham, M. Zissman, Evaluating Intrusion Detection Systems: The 1998 DARPA Off-line Intrusion Detection Evaluation, *Proceedings DARPA Information Survivability Conference and Exposition (DISCEX) 2000*, Vol 2, pp. 12-26, IEEE Computer Society Press, Los Alamitos, CA, 2000.
22. C. Blake and C. Merz, UCI Repository of Machine Learning Databases http://www.ics.uci.edu/~mlearn/~MLRepository.html, Department of Information and Computer Sciences, University of California, Irvine, 1998.
23. F. Provost, T. Fawcett, R. Kohavi, The Case Against Accuracy Estimation for Comparing Induction Algorithms, *Proceedings of 15th International Conference on Machine Learning*, 445-453, Madison, WI, 1998.
24. ELENA project, ftp.dice.ucl.ac.be in directory pub/neural-nets/ELENA/databases

Using Belief Networks and Fisher Kernels for Structured Document Classification

Ludovic Denoyer and Patrick Gallinari

Laboratoire d'Informatique de Paris VI
LIP6, France
{ludovic.denoyer,patrick.gallinari}@lip6.fr

Abstract. We consider the classification of structured (e.g. XML) textual documents. We first propose a generative model based on Belief Networks which allows us to simultaneously take into account structure and content information. We then show how this model can be extended into a more efficient classifier using the Fisher kernel method. In both cases model parameters are learned from a labelled training set of representative documents. We present experiments on two collections of structured documents: WebKB which has become a reference corpus for HTML page classification and the new INEX corpus which has been developed for the evaluation of XML information retrieval systems.

Keywords: textual document classification, structured document, XML corpus, Belief Networks, Fisher Kernel.

1 Introduction

The development of large electronic document collections and Web resources has been paralleled by the emergence of structured document format proposals. They are aimed at encoding content information in a suitable form, for a variety of information needs. These document formats allow us to enrich the document content with additional information (document logical structure, meta-data, comments, etc) and to store and access the documents in a more efficient way. Some proposals have already gained somea popularity and description languages like XML are already widely used by different communities. For text documents, these representations encode both structural and content information.

With the development of structured collections, there is a need to develop information access methods which may take all the benefit of these richer representations and also allow to answer new information access challenges and new user needs. Current Information Retrieval (IR) methods have mainly been developed for handling flat document representations and cannot be easily adapted to deal with structured representations.

In this paper, we focus on the particular task of structured document categorization. We propose methods for exploiting both the content and the structure information for this task. Our core model is a generative categorization model based on belief networks (BN). This work offers a natural framework for encoding structured representations and allows us to perform inference both on

N. Lavrač et al. (Eds.): PKDD 2003, LNAI 2838, pp. 120–131, 2003.

whole documents and on document subparts. We then show how to turn this generative model into a discriminant classification model using the Fisher kernel trick. Paper is organized as follows: we make in 2 a brief review of previous work on structured document classification, we describe in 3 the type of structured document we are working on, we then introduce in 4 our generative model and the discriminant model in 5. Section 6 presents a series of experiments performed on two textual collections, the WebKB [20] and the INEX Corpus [7].

2   Previous Works

Text categorization is a classical information retrieval task which has motivated a large amount of work over the last fews years. Most categorization models have been designed for handling bag of words representations and do not consider word ordering or document structure. Generally speaking, classifiers fall into two categories: generative models which estimate class conditional densities $P(document/Class)$ and discriminant models which directly estimate the posterior probabilities $P(Class/document)$. The naive Bayes model [12] for example is a popular generative categorization model whereas among discriminative techniques support vector machines [10] have been widely used over the last few years. [17] makes a complete review of flat document categorization methods. More recently, models which take into account sequence information have been proposed [3]. Classifying structured document is a new challenge both from IR and machine learning perspectives. For the former, flat text classifiers do not lead to natural extensions for structured documents, however there has been recently some interest in the classification of HTML pages. For the latter, the classification of structured data is an open problem since most classifiers have been designed for vector or sequence representations, and only a few formal frameworks allow to consider simultaneously content and structure informations. We briefly review below recent work in these different areas.

The expansion of the Web has motivated a series of works on Web page categorization - viz. the last two Trec competitions [19]. Web pages are built from different type of information (title, links, text, etc) which play different roles. There has been several attempts to combine these information sources in order to increase page categorization scores ([5],[21]). Chakrabarti ([2]) proposes to use the information contained in neightboring documents of an HTML pages. All these approaches which deal only with HTML, propose simple schemes either for encoding the page structure or for exploiting the different types of information by combining basic classifiers. These models exploit a priori knowledge about the particular semantics of HTML tags, and as such cannot be extended to more complex languages like XML where tags may be defined by the user. We will see that our model does not exploit this type of semantics and is able to learn from data the importance of tag information.

Some authors have proposed more principled approaches to deal with the general problem of structured document categorization. These models are not specific to HTML even when they are tested on HTML databases due to the lack of a reference XML corpus. [4] for example propose the Hidden Tree Markov Model (HTMM) which is an extension of HMMs to a structured representation. They consider tree structured documents where in each node (structural element), terms are generated by a node specific HMM. [16] have proposed a Bayesian network for classifying structured documents. This is a discriminative model which directly computes the posterior probability corresponding to the document relevance for each class. [22] present an extension of the Naive Bayes model to semi-structured documents where essentially global word frequencies estimators are replaced with local estimators computed for each path element. [18] propose to use Probabilistic Relationnal Models to classify structured document and more precisely Web pages.

For the ad-hoc IR task, Bayesian networks (BN) have been used for information retrieval for some time. Inquery [1] retrieval engine operates on flat text while more recent proposal handle structured documents,e.g. [14], [15]. Outside the field of information retrieval, some models have been proposed to handle structured data. The hierarchical HMM (HHMM) [6] is a generalization of HMMs to structured data, it has been tested on handwriting recognition and on the analysis of English sentences, similar HMM extensions have been used for multi-agent modeling [13]. However, inference and learning algorithms in these models are too computationally demanding for handling large IR tasks. The inference complexity for HHMM is $O(NT^3)$ where N is the number of states in their HMM and T the length of the text in words, for comparison our model is more like $O(N + T)$ as will be seen later.

The core model we propose is a generative model which has been developed for the categorization of any tree like document structure (typically XML documents). This model bears some similarities with the one in [4], however, their model is adapted to the semantic of HTML documents and considers only the inclusion relation between two document parts. Ours is generic and can be used for any type of structured document. Even when tags do not convey semantic information, it allows considering different types of relations between structured elements: inclusion, depth in the hierarchical document, etc. This model could be considered as a special case of the HHMM [6] since it is simpler and since HHMM can be represented as particular BNs [13]. It is computationally much less demanding and has been designed for handling large document collections. This generative model is then extended into a discriminant one using the method of the Fisher Kernel. For that, we extend to the case of structured data the ideas initially proposed by [9] for sequences.

Our main contributions are a new generative model for the categorization of large collections of structured documents and its extension via the use of Fisher kernels into a discriminant model. We also describe for the first time to our knowledge experiments on a large corpus of structured XML documents (INEX) developed in 2002 for ad-hoc retrieval tasks.

3   Document Structure

We represent a structured document d as a Directed Acyclic Graph (DAG). Each node of the graph represents a structural entity of the document, and each edge represents a hierarchical relation between two entities (for example, a paragraph is included in a section, two paragraphs are on the same level of the hierarchy, etc). For keeping inference complexity to a reasonable level, we do not consider circular relations which might appear in some documents (e.g. Web sites), this restriction is not too severe since this definition already encompasses many different types of structured documents.

Each node of the DAG is composed of **a label** (for example, labels can be *section, paragraph, title* and represent the structural semantic of a document) and **a textual information** (which is the textual content associated to this node if any).

A structured document then contains three types of information: **the logical structure information** represented by the edges of the DAG (the position of the tag in an XML document), **the label information** (the name of the tag in an XML document) and **the textual information**. Figure 1 gives a simple example of structured document.

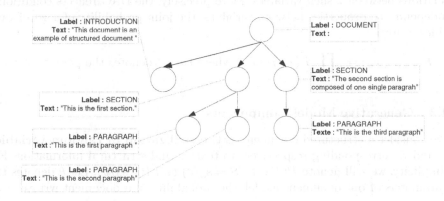

Fig. 1. An example of structured document represented as a Direct Acyclic Graph. This document is composed of an introduction and two sections. Each part of the document is represented by a node with a **label** and a **textual information.**

4   A Generative Model for Structured Documents

We now present a generative model for structured documents. It is based on BNs and allows to handle these 3 types of information. This model can be used with any XML document without using *a priori* informations about the semantic of the structure. We first briefly introduce BNs and then describe the different elements of the model.

4.1   Notations

We will use the following notations, let:

- d: be a structured document
- s_d: be the structure of document d. $s_d = (\{s_d^i\}, pa(s_d^i))$ where $\{s_d^i\}_{i \in [1..|s_d|]}$ is the set of node labels ($|s_d|$ is the number of structured nodes for d), $s_d^i \in \Lambda$ with Λ the set of possible labels. $pa(s_d^i)$ are the parents of node $\{s_d^i\}$ in the structured document and describe the **logical structure information**.
- t_d: be the textual information in d. t_d is a set $\{t_d^i\}_{i \in [1..|s_d|]}$ of textual elements for each node i of the structured document.
- $\{w_{d,k}^i\}_{k \in [1..|t_d^i|]}$ is the set of words of part t_d^i in document d ($|t_d^i|$ is the number of words of part t_d^i) and $w_{d,k}^i$ is the kth word in t_d^i. $w_{d,k}^i \in V$ where V is the set of indexing terms in the corpus.

4.2   Base Model Using Belief Networks

Belief networks [11] are stochastic models for computing the joint probability distribution over a set of random variables. A BN is a DAG whose nodes are the random variables and whose edges correspond to probabilistic dependence relations between 2 such variables. More precisely, the DAG reflects conditional independence properties between variables, the joint probability of a set of variables writes:

$$P(x_1, ..., x_n) = \prod_{i=1..n} P(x_i/pa(x_i)) \text{ where } pa(x_i) \text{ denotes the parents of } x_i.$$

4.3   Generative Model Components

We consider a structured document as **the realization of random variables** T and S corresponding respectively to textual and structural information. For simplicity, we will denote $P(T = t_d, S = s_d/\theta)$ as $P(t_d, s_d/\theta)$. Let θ denotes the parameters of our document model, the probability of a document writes:

$$P(d|\theta) = P((t_d, s_d)|\theta) = P(s_d|\theta)P(t_d|s_d, \theta) \tag{1}$$

$P(s_d|\theta)$ is the **structural probability** of d and $P(t_d/\theta)$ is the **textual probability** of d given its structure s. Each document will be modeled via a BN, whose nodes correspond either to tag or textual information and whose directed edges encode the relations between the document elements. The whole corpus will then be represented as a series of BN models, 1 per document. The BN model of a document can be thought of as a model of the structured document generation, where the generation process goes as follows: someone who wants to create a document about a specific topic will sequentially and recursively create the do cument organization and then fill the corresponding nodes with text. For example he first creates sections after what, for each section, he creates subsections etc... recursively. At the end, in each "terminal" node, he will create the textual information of this part as a succession of words. This is a typical generative approach which extends to structured information the classical HMM approach for modeling sequences. The two components-structure and content-of the model are detailed below.

Structural Probability
The structural information of a document is encoded into the edges of the BN. Under the conditional independence assumption of the BN document model, the structural density of document d writes:

$$P(s_d|\theta) = \prod_{i=1}^{|s_d|} P(s_d^i|pa(s_d^i))$$ (2)

The BN structural parameters are then the $\{P(s_d^i|pa(s_d^i))\}$ which are the probabilities to observe s_d^i given its parents $pa(s_d^i)$ in the BN.

In order to have a robust estimation of the BN parameters, we will share sets of parameters among all the document models. We will make the hypothesis that the $\{P(s_d^i|pa(s_d^i))\}$ only depend on the labels of nodes s_d^i and $pa(s_d^i)$, i.e. two nodes in two different document models which share the same label and whose parents also share the same labels will have the same conditional probability $P(s_d^i|pa(s_d^i))$.

Within this framework, several BN models may be associated to a document d. Figure 2 illustrates two of the models we have been working with. The DAG structure of Model 2 is copied from the tree structure of the document and reflects only the inclusion relation. The same type of relation is used in [4]. Model 1 contains both inclusion information (vertical edges) and sequence information (horizontal edges). Both models are an overly simplified representation of the real dependencies between document parts. This allows to keep the complexity of learning and inference algorithms low. Statistical models that work best are often very simple compared to the underlying phenomenon (e.g. naive Bayes in text classification or Hidden Markov Models in speech recognition), practioners of BNs have experienced the same phenomenon. Note that other instances of our generic model could have also been used here.

Textual Probability
For modeling the textual content of a structured document, we make the following hypothesis:

- the probability of a word only depends on the label of the node that contains this word (first order dependency assumption).
- in a node, words are independent (Naive Bayes assumption)

The naive Bayes hypothesis is not mandatory here and any other term generative model (e.g. HMM) could be used instead, however this hypothesis allows for a robust density estimation and in our experiments more sophisticated models did not led to any performance improvement.

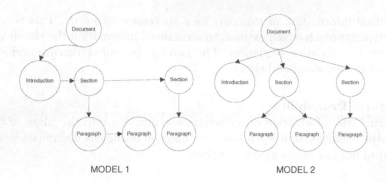

Fig. 2. Two possible structural belief networks constructed for the document presented in figure 1.

For a particular part t_d^i of document d, we then have:

$$P(t_d^i/s_d, \theta) = P(t_d^i/s_d^i, \theta) = \prod_{k=1}^{|t_d^i|} P(w_{d,k}^i|s_d^i, \theta) \qquad (3)$$

And for the entire document, we have:

$$P(t_d|s_d, \theta) = \prod_{i=1}^{i=|s_d|} \prod_{k=1}^{|t_d^i|} P(w_{d,k}^i|s_d^i, \theta) \qquad (4)$$

Final Belief Network

Combining equations 2 and 4, we get a generative structured document model

$$P(d|\theta) = \prod_{i=1}^{|s_d|} \left(P(s_d^i|pa(s_d^i), \theta) \prod_{k=1}^{|t_d^i|} P(w_{d,k}^i|s_d^i, \theta) \right) \qquad (5)$$

Equation 5 describes the contribution of structural and textual information in our model.

4.4   Learning

This model is completely defined by two sets of parameters, *transition* and *emission* probabilities respectively denoted by $P(s_i|s_j)$ and $P(w_i|s_j)$:

$$\theta = \{P(s_i|s_j)\}_{s_i, s_j \in \Lambda} \bigcup \{P(w_i|s_j)\}_{w_i \in V, s_j \in \Lambda}$$

In order to learn the θ, we use the Expectation Maximization (EM) algorithm for optimizing the maximum likelihood of the data. Since evidence is available for any variable in the BN model of a document, this simply amounts to a count for each possible value of the random variables.

Using equation 5, the log-likehood for all documents in the corpus D is:

$$L = \sum_{d \in D} \sum_{i=1}^{|s_d|} \left(\log P(s_d^i | pa(s_d^i), \theta) + \sum_{k=1}^{|t_d^i|} \log P(w_{d,k}^i | s_d^i), \theta) \right) \qquad (6)$$

Let us denote the model parameters $P(s_d^i | pa(s_d^i))$ and $P(w_{d,k}^i | s_d^i)$ by $\theta_{s_d^i, pa(s_d^i)}$ and $\theta_{w_{d,k}^i, s_d^i}$. Equation 6 then writes:

$$L = \sum_{d \in D} \left(\left(\sum_{i=1}^{|s_d|} \log \theta_{s_d^i, pa(s_d^i)} \right) + \left(\sum_{i=1}^{|s_d|} \sum_{k=1}^{|t_d^i|} \log \theta_{w_{d,k}^i, s_d^i} \right) \right) \qquad (7)$$

The learning algorithm then solves $\frac{\partial L}{\partial \theta_{n,m}} = 0$ with the constraint $\sum_{n} \theta_{n,m} = 1$.

Let $N_{n,m}^d$ be the number of times a part with label n has his parent with label m in document d or respectively the number of times a word with value n is in a part with label m for document d, the solution of the learning problem is:

$$\theta_{n,m} = \frac{\sum\limits_{d \in D} N_{n,m}^d}{\sum\limits_{i} \sum\limits_{d \in D} N_{i,m}^d} \qquad (8)$$

The complexity of the algorithm is $O(\sum_{d \in D} |s_d| + |t_d|)$. In a classical structured document, the number of node of the structural network is smaller than the number of words of the document. So the complexity is equivalent to $O(\sum_{d \in D} |t_d|)$ which is the classical learning complexity of the Naive Bayes algorithm. Note that, in the case of flat documents, our model is strictly equivalent to the classical Naive Bayes model.

5   Discriminant Approach

The above model could be used for different tasks, e.g. document classification or clustering or even for performing more sophisticated inferences on document parts, e.g. deciding which part of a document is relevant for a specific topic. For classification of whole documents which is the focus of this paper, discriminant approaches are most often preferred to generative ones since they usually score higher. We then propose below to derive from the generative model of structured document a discriminant model. For that, we follow the line of [9] who proposed to build a discriminant model from a generative sequence model. We show how this idea could be extended to our generative structured document model.

5.1   Fisher Kernel

Given a generative model with parameters θ for sequences, [9] propose to compute for each sequence x the Fisher score $U_d = \nabla_\theta log P(x/\theta)$ of the model for the sequence, i.e. the gradient of the log likelihood of x for model θ. For each sequence sample, this score is a vector of fixed dimensionality which explains how each parameter of the generative model contributes to generate the sequence.

Using this score, they then define a distance between two examples x and y as a kernel function:

$$K(x,y) = U_x^T M^{-1} U_y \text{ with } M = E_X[U_X^T U_X] \tag{9}$$

This kernel can then be used with any kernel classifier, (e.g. SVM) in order to classify the examples. The key idea here is to map the sequence information onto a vector of scores. This allows to make use of any classical vector discriminant classifier on this new representation and therefore to use well known and efficient vector classifiers for sequence classification. We show below that this idea may be naturally adapted to our structured generative model.

5.2   Fisher Kernel for the Structured Document Model

For our model, the Fisher Kernel can be easily computed. Using 7 we get:

$$\frac{\partial P(d/\theta)}{\partial \theta_{n,m}} = \frac{N_{n,m}^d}{\theta_{n,m}} \tag{10}$$

The Fisher kernel idea initially proposed for HMMs, naturally carries over to our structured data model. However, in practice, using the Fisher Kernel method is not straightforward. In order to make the method work, one must make different approximations, especially when the number of parameters of the generative model is high which is the case here. In our implementation, we make the following approximations:

- we first approximate the M matrix using the identity matrix like in [9]
- we then compute the gradient of the log likelihood wrt $2\sqrt{\theta_{n,m}}$ like in [8].

Let $\rho_{n,m} = 2\sqrt{\theta_{n,m}}$, we have: $\frac{\partial P(d/\theta)}{\partial \rho_{n,m}} = 2\frac{N_{n,m}^d}{\rho_{n,m}} = \frac{N_{n,m}^d}{\sqrt{\theta_{n,m}}}$

We use this last formula to compute the vector corresponding to each structured document d.

6   Experiments

6.1   Corpora

We use two corpora in our experiments.

WebKB corpus [20] is composed of 8282 HTML documents from computer science departments web sites. This is a reference corpus in the machine learning community for classifying HTML pages. It is composed of 7 topics (classes): *student, faculty, course, project, department, staff, other. Other* is a trash topic, and has been ignored here as it is usually done. We are then left with 4520 documents. We used Porter Stemming and pruned all words that appear in less than 5 documents. The size of the vocabulary V is 8038 terms. We only keep the tags with the higher frequency (*H1, H2,H3, TITLE,B,I,A*). We made a 5-fold cross-validation (80% on train and 20% on test).

INEX corpus [7] is the new reference corpus for Information Retrieval with XML documents. It was designed for ad-hoc retrieval. It is made up of articles from journals and proceedings of the IEEE Computer Society. All articles are XML documents. The collection contains approximately 15 000 articles from over 20 different journals or proceedings. We used Porter Stemming and pruned the words which appear in less than 50 documents. The final size of the vocabulary is about 50 000 terms and the number of tags is about 100. We made a random split using 50% for training and 50% for testing. The task was to classify articles into the right journal or proceedings (20 classes).

6.2   Results

We have used a Naive Bayes classifier as a baseline generative classifier and SVM ([10]) as a baseline discriminant model. Results appear in figures 4 and 3. *Macro-average* is obtained by averaging the percentage of correct classification for every class considered. *Micro-average* is obtained by weighting the average by the relative size of each class.

Let us consider the micro-average. On WebKB, the BN model achieves a mean 3 % improvement with regard to Naive Bayes. This is encouraging and superior to already published results on this dataset [4]. The Fisher model still

	course	department	staff	faculty	student	project	Macro	Micro
Naive Bayes	0.96	0.93	0.07	0.67	0.91	0.65	0.70	0.81
BN Model	0.96	0.82	0.03	0.72	0.93	0.76	0.70	0.83
SVM	0.90	0.79	0.17	0.85	0.91	0.77	0.73	0.85
Naive Bayes Fisher	0.95	0.77	0.17	0.82	0.91	0.71	0.72	0.85
BN Model Fisher	0.95	0.83	0.14	0.84	0.94	0.72	0.73	0.87

Fig. 3. Performance of 5 classifiers on WebKB corpus.

	Macro	Micro
Naive Bayes	0.61	0.64
BN Model	0.67	0.66
SVM	0.71	0.70
Naive Bayes Fisher	0.69	0.69
BN Model Fisher	0.72	0.71

Fig. 4. Performance of 5 classifiers on INEX corpus.

rises this score by 4%. This corresponds to 2% more than the baseline discriminant SVM. The structured generative document model is clearly superior to the flat Naive Bayes classifier, and the Fisher Kernel operating on the structured generative models compares well to the baseline SVM.

On the much larger INEX database, our generative model achieves about 2% micro-average improvement with regard to Naive Bayes and the Fisher Kernel method increases the BN score by about 6%, but only 1% compared to the baseline SVM. This confirms the good results obtained on WebKB. Note that, to our knowledge, these are the first classification results obtained on a real world large XML corpus.

These experiments show that it is important to take simultaneously into account structure and content information in HTML or XML documents. The proposed methods allow to model and combine the two types of information. Both the generative and discriminant models for structured documents offer a complexity similar to that of the baseline flat classification models while increasing the performances.

7   Conclusion and Perspectives

We have proposed a new generative model for structured textual documents representation. This model offers a general framework for handling different tasks like classification, clustering or more specialized structured document access problems. We focused here on the classification of whole documents and described how to extend this generative model into a more efficient discriminant classifier using the Fisher kernel idea. Experiments performed on two databases show that the proposed methods are indeed able to take simultaneously into account the structure and content informations and offer good performances compared to baseline classifiers.

References

1. Jamie P. Callan, W. Bruce Croft, and Stephen M. Harding. The INQUERY Retrieval System. In A. Min Tjoa and Isidro Ramos, editors, *Database and Expert Systems Applications*, pages 78–83, Valencia, Spain, 1992. Springer-Verlag.
2. Soumen Chakrabarti, Byron E. Dom, and Piotr Indyk. Enhanced hypertext categorization using hyperlinks. In Laura M. Haas and Ashutosh Tiwary, editors, *Proceedings of SIGMOD-98, ACM International Conference on Management of Data*, pages 307–318, Seattle, US, 1998. ACM Press, New York, US.
3. Ludovic Denoyer, Hugo Zaragoza, and Patrick Gallinari. HMM-based passage models for document classification and ranking. In *Proceedings of ECIR-01*, pages 126–135, Darmstadt, DE, 2001.
4. M. Diligenti, M. Gori, M. Maggini, and F. Scarselli. Classification of html documents by hidden tree-markov models. In *Proceedings of ICDAR*, pages 849–853, Seatle, 2001. WA (USA).
5. Susan T. Dumais and Hao Chen. Hierarchical classification of Web content. In Nicholas J. Belkin, Peter Ingwersen, and Mun-Kew Leong, editors, *Proceedings of SIGIR-00*, pages 256–263. ACM Press, 2000.

6. Shai Fine, Yoram Singer, and Naftali Tishby. The hierarchical hidden markov model: Analysis and applications. *Machine Learning*, 32(1):41–62, 1998.
7. Norbert Fuhr, Norbert Govert, Gabriella Kazai, and Mounia Lalmas. INEX: Initiative for the Evaluation of XML Retrieval. In *Proceedings ACM SIGIR 2002 Workshop on XML and Information Retrieval*, 2002.
8. Thomas Hofmann. Learning the similarity of documents: An information-geometric approach to document retrieval and categorization. In *Research and Development in Information Retrieval*, pages 369–371, 2000.
9. Tommi S. Jaakkola, Mark Diekhans, and David Haussler. Using the Fisher kernel method to detect remote protein homologies. In *Intelligent Systems for Molecular Biology Conference (ISMB'99)*, Heidelberg, Germany, August 1999. AAAI.
10. Thorsten Joachims. Text categorization with support vector machines: learning with many relevant features. In *Proceedings of ECML-98*, pages 137–142, Chemnitz, DE, 1998. Springer Verlag, Heidelberg, DE.
11. Jin H. Kim and Judea Pearl. A Computational Model for Causal and Diagnostic Reasoning in Inference Systems. In Alan Bundy, editor, *Proceedings of the 8th IJCAI*, Karlsruhe, Germany, August 1983. William Kaufmann.
12. David D. Lewis. Naive (Bayes) at forty: The independence assumption in information retrieval. In *Proceedings of ECML-98,*, pages 4–15, Chemnitz, DE, 1998. Springer Verlag, Heidelberg, DE.
13. K. Murphy and M. Paskin. Linear time inference in hierarchical hmms, 2001.
14. Sung Hyon Myaeng, Dong-Hyun Jang, Mun-Seok Kim, and Zong-Cheol Zhoo. A Flexible Model for Retrieval of SGML documents. In *Proceedings of the 21st Annual International ACM SIGIR*, pages 138–140, Melbourne, Australia, August 1998. ACM Press, New York.
15. Benjamin Piwowarki and Patrick Gallinari. A Bayesian Network Model for Page Retrieval in a Hierarchically Structured Collection. In *XML Workshop of the 25th ACM SIGIR Conference*, Tampere, Finland, 2002.
16. B. Piwowarski, L. Denoyer, and P. Gallinari. Un modele pour la recherche d'informations sur les documents structures. In *Proceedings of the 6emes journees Internationales d'Analyse Statistique des Donnees Textuelles (JADT2002)*.
17. Fabrizio Sebastiani. Machine learning in automated text categorization. *ACM Computing Surveys*, 34(1):1–47, 2002.
18. Ben Taskar, Peter Abbeel, and Daphne Koller. Discriminative probabilistic models for relationnal data. In *Eighteenh Conference on Uncertainty in Artificail Intelligence (UAI02)*, Edmonton, Canada, 2002.
19. Trec. Text REtrieval Conference (trec 2001), National Institute of Standards and Technology (NIST).
20. webKB. http://www-2.cs.cmu.edu/afs/cs.cmu.edu/project/theo-20/www/data/, 1999.
21. Yiming Yang, Seán Slattery, and Rayid Ghani. A study of approaches to hypertext categorization. *Journal of Intelligent Information Systems*, 18(2/3):219–241, 2002.
22. Jeonghee Yi and Neel Sundaresan. A classifier for semi-structured documents. In *Proceedings of the sixth ACM SIGKDD international conference on Knowledge discovery and data mining*, pages 340–344. ACM Press, 2000.

A Skeleton-Based Approach
to Learning Bayesian Networks from Data

Steven van Dijk, Linda C. van der Gaag, and Dirk Thierens

Universiteit Utrecht, Institute of Information and Computing Sciences
Decision Support Systems, PO Box 80.089, 3508 TB Utrecht, The Netherlands
{steven,linda,dirk}@cs.uu.nl

Abstract. Various different algorithms for learning Bayesian networks from data have been proposed to date. In this paper, we adopt a novel approach that combines the main advantages of these algorithms yet avoids their difficulties. In our approach, first an undirected graph, termed the *skeleton*, is constructed from the data, using zero- and first-order dependence tests. Then, a search algorithm is employed that builds upon a quality measure to find the best network from the search space that is defined by the skeleton. To corroborate the feasibility of our approach, we present the experimental results that we obtained on various different datasets generated from real-world networks. Within the experimental setting, we further study the reduction of the search space that is achieved by the skeleton.

1 Introduction

The framework of Bayesian networks has proven to be a useful tool for capturing and reasoning with uncertainty. A Bayesian network consists of a graphical structure, encoding a domain's variables and the probabilistic relationships between them, and a numerical part, encoding probabilities over these variables (Cowell et al., 1999). Building the graphical structure of a Bayesian network and assessing the required probabilities by hand is quite labour-intensive. With the advance of information technology, however, more and more datasets are becoming available that can be exploited for constructing a network automatically. *Learning* a Bayesian network from data then amounts to finding a graphical structure that, supplemented with maximum-likelihood estimates for its probabilities, most accurately describes the observed probability distribution.

Most state-of-the-art algorithms for learning Bayesian networks from data take one of two approaches: the use of *(in)dependence tests* (Rebane and Pearl, 1987; Spirtes and Glymour, 1991; de Campos and Huete, 2000) and the use of a *quality measure* (Cooper and Herskovits, 1992; Buntine, 1991; Heckerman et al., 1995; Lam and Bacchus, 1993). Although with both approaches encouraging results have been reported, they both suffer from some difficulties. With the first approach, a statistical test such as χ^2 is employed for examining whether or not two variables are dependent given some conditioning set of variables; the *order* of the test is the size of the conditioning set used. By starting with zero-order tests and selectively growing the conditioning set, in theory all (in)dependence statements can be recovered from the data and the network that generated the data can be reconstructed. In practice, however, the statistical test employed quickly becomes unreliable for higher orders, because the number of data

N. Lavrač et al. (Eds.): PKDD 2003, LNAI 2838, pp. 132–143, 2003.

required increases exponentially with the order. If the test would then return incorrect (in)dependence statements, errors could arise in the graphical structure. With the second approach, a quality measure such as MDL is used for assessing the quality of candidate graphs. The graphical structure yielding the highest score then is taken to be the one that best explains the observed data. This approach suffers from the size of the search space. To efficiently traverse the huge space of graphical structures, often a greedy search algorithm is used. Other algorithms explicitly constrain the space by assuming a topological ordering on the nodes of candidate structures. Both types of algorithm may inadvertently prune high-quality networks from the search space of structures.

In this paper, we adopt a novel approach to learning Bayesian networks from data that combines the main advantages of the two approaches outlined above. In our approach, first an undirected graph is constructed from the data using just zero- and first-order dependence tests. The resulting graph, termed the *skeleton* of the network under construction, is used to explicitly restrict the search space of graphical structures. In the second phase of our approach, a search algorithm is employed to traverse the restricted space. This algorithm orients or removes each edge from the skeleton to produce a directed graphical structure. To arrive at a fully specified Bayesian network, this structure is supplemented with maximum-likelihood estimates computed from the data. We experimented with two instances of our approach, building upon a simple hill-climber and upon a genetic algorithm, respectively, for the search algorithm. The results that we obtained compare favourably against various state-of-the-art learning algorithms.

The paper is organised as follows. In Section 2, we present the details of our approach. In doing so, we focus on the construction of the skeleton; an in-depth discussion of the design of a competent search algorithm for the second phase of our approach is presented elsewhere (van Dijk et al., 2003). The experimental results obtained with our approach are reported in Section 3. We analyse various properties of the skeleton in Section 4. We end the paper with a discussion of our approach in Section 5.

2 Skeleton-Based Learning

Our approach divides the task of learning a Bayesian network from data into two phases. In the first phase, a skeleton is constructed. This skeleton is taken as a template that describes all graphical structures that can be obtained by orienting or deleting its edges. In the second phase, the search space that is defined by the skeleton is traversed by means of a search algorithm. Focusing on the first phase, we discuss the construction of the skeleton in Section 2.1; in Section 2.2, we briefly review related work.

2.1 Constructing the Skeleton

We consider learning a Bayesian network from a given dataset. For ease of exposition, we assume that this dataset has been generated by sampling from a network whose graphical structure perfectly captures the dependences and independences of the represented distribution. The undirected graph underlying the structure of this network will be referred to as the *true skeleton*. In the first phase of our approach, we construct a skeleton from the available data to restrict the search space for the second phase. In

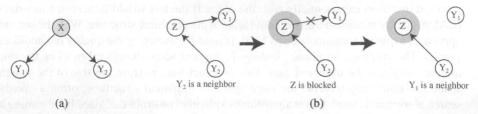

Y_2 is a neighbor   Z is blocked Y_1 is a neighbor

(a)   (b)

Fig. 1. (a) Two non-neighbouring variables Y_1 and Y_2 that are dependent yet become independent given X. (b) The separation graph $G(X)$ of X, with $L(X) = \{Z, Y_1, Y_2\}$. The variable Y_2 is identified as a neighbour of X since it has no incoming arcs. The arc from Y_2 to Z indicates Z to be a non-neighbour. Removal of the arc from Z to Y_1 reveals Y_1 to be a neighbour of X.

doing so, we aim to find a skeleton that is already close to the true skeleton. On the one hand, we try to avoid missing edges, because these could prune the best network from the search space. On the other hand, we try to minimise the number of additional edges, since these would unnecessarily increase the size of the space to be traversed. To construct an appropriate skeleton, we analyse the dependences and independences that are embedded in the dataset for the various different variables. To this end, a statistical test is employed. Well-known examples of such tests are the χ^2 statistic and the mutual information criterion. In the sequel, we will write $DT(X, Y \mid Z)$ if, for a given threshold value, the test indicates that the variables X and Y are dependent given the (possibly empty) conditioning set of variables Z; otherwise, we write $\neg DT(X, Y \mid Z)$.

When constructing the skeleton, we try to identify the true neighbours of each variable X. To this end, we begin by identifying all variables that have a zero-order dependence on X. If for a specific variable Y, the test employed fails to report a result, for example due to a lack of data, we assume that Y is independent of X. We now observe that, while neighbouring variables in the true skeleton are always dependent, the reverse does not hold: two dependent variables may be separated by one or more intervening variables, for example as in Figure 1(a). The list $L(X) = \{Y \mid DT(X, Y)\}$ obtained therefore includes neighbours as well as non-neighbours of X from the true skeleton. Since a non-neighbour Y of X is separated from X in the true skeleton by a set of true neighbours of X, we expect that $\neg DT(X, Y \mid Z)$ for some set $Z \subseteq L(X) \setminus \{Y\}$. We now use first-order tests to remove, from among the list $L(X)$, any non-neighbours of X. The skeleton then in essence is found by adding an edge between X and $Y \in L(X)$ if and only if $DT(X, Y \mid \{Z\})$ holds for all $Z \in L(X) \setminus \{Y\}$.

We note that using just first-order tests as outlined above, does not suffice for identifying *all* non-neighbours of X from among the list $L(X)$. In fact, a higher-order test may be required to establish a variable Y as a non-neighbour of X; for example, if $DT(X, Y \mid \{Z_1\})$, $DT(X, Y \mid \{Z_2\})$, and $\neg DT(X, Y \mid \{Z_1, Z_2\})$, a second-order test is needed for this purpose. By using higher-order tests, therefore, additional non-neighbours could be identified and a sparser skeleton could result. As we have argued before, however, the test employed quickly becomes unreliable for larger conditioning sets, thereby possibly giving rise to errors in the skeleton. Since the purpose of the skeleton is to *safely* reduce the search space, we restrict the tests employed to just zero- and first-order tests, and let the search algorithm remove the spurious edges.

We further note that the method described above, when applied straightforwardly, could erroneously remove some true neighbours of a variable X from the list $L(X)$. As an example we consider the neighbours Y_1 and Y_2 of X in the true skeleton, where Y_2 has Z for a second neighbour. Now, if the true neighbour Y_1 would exhibit a weak dependence on X and the non-neighbour Z would exert a very strong influence on X, then values for Z could hide the dependence of Y_1 on X. The dependence test then returns $\neg DT(X, Y_1 \mid \{Z\})$ and Y_1 would be identified as a non-neighbour of X. To support identification of the true neighbours of X, therefore, we construct an auxiliary directed graph termed the *separation graph* $G(X)$ of X. The variables from the list $L(X)$ are the nodes of the graph. There is an arc from a variable Z to a variable Y if $\neg DT(X, Y \mid \{Z\})$ for some $Z \in L(X) \setminus \{Y\}$. If the first-order test fails to establish dependence or independence, we assume dependence. In the separation graph $G(X)$, all variables without any incoming arcs are true neighbours of X, since these variables remain dependent on X regardless of the conditioning set used. The thus identified neighbours are used to find non-neighbours of X, by following their outgoing arcs. The outgoing arcs of these non-neighbours are removed, which may cause other variables to reveal themselves as neighbours. Figure 1(b) illustrates the basic idea of the separation graph. The process of identifying neighbours and non-neighbours is repeated until no more neighbours of X can be identified. Variables that are part of a cycle in the remaining separation graph are all marked as neighbours of X: these variables correspond to ambiguous situations, which are thus resolved safely, that is, without discarding possible neighbours. We note that the process of identifying neighbours requires at most $|L(X)|$ iterations. The skeleton is now built by finding the neighbours of every variable and connecting these.

2.2 Related Work

de Campos and Huete (2000) use a skeleton within a test-only approach. The skeleton is built by connecting variables for which no zero- or first-order test indicates independence. From the thus constructed skeleton, a directed graphical structure is derived without employing a search algorithm. The authors do suggest the use of such an algorithm, however. Cheng et al. (2002) also present a test-only approach that builds upon a skeleton constructed from lower-order dependence tests. Steck and Tresp (1999) deal with the construction of a usable skeleton when unreliable tests offer conflicting dependence statements. The *Hybrid Evolutionary Programming* (HEP) algorithm by Wong et al. (2002) takes an approach that is closely related to ours. Although the algorithm does not explicitly construct a skeleton, it does use zero- and first-order dependence tests to restrict the search space of graphical structures that is subsequently traversed by an MDL-based search algorithm. The HEP algorithm has shown high-quality performance on datasets generated from the well-known Alarm network.

3 Experiments

Our approach to learning Bayesian networks allows for various different instances. For the first phase, different dependence tests can be employed and for the second phase, different quality measures and different search algorithms can be used. We present two such instances in Section 3.1 and report on their performance in Section 3.2.

3.1 The Instances

To arrive at an instance of our approach, we have to specify a dependence test to be used in the construction of a skeleton as outlined in the previous section. In our experiments, we build upon the χ^2 test, using independence for its null-hypothesis. The χ^2 test calculates a statistic from the contingency table of the variables X and Y concerned:

$$stat(X,Y) = \sum_{i,j} \frac{(O_{ij} - E_{ij})^2}{E_{ij}},$$

where $O_{ij} = \hat{p}(x_i,y_j) \cdot N$ is the observed frequency of the combination of values (x_i,y_j) and $E_{ij} = \hat{p}(x_i) \cdot \hat{p}(y_j) \cdot N$ is the expected frequency of (x_i,y_j) if X and Y were independent; N denotes the size of the available dataset and $\hat{p}(x_i)$ denotes the proportion of x_i. The computed statistic is compared against a critical value s with $\int_s^\infty \chi_{df}^2(x) = \varepsilon$, for a given threshold ε and the χ_{df}^2 distribution with df degrees of freedom. If the statistic is higher than s, the null-hypothesis is rejected, that is, X and Y are established as being dependent. The test for dependence of X and Y given Z is defined analogously, taking independence of X and Y for every possible value of Z for the null-hypothesis. In our experiments, we use the threshold values $\varepsilon_0 = 0.005$ for the zero-order dependence test and $\varepsilon_1 = 0.05$ for the second-order dependence test. Choosing the χ^2 test allows for a direct comparison of our approach against the HEP algorithm mentioned above.

With the specification of a dependence test and its associated threshold values, the first phase of our approach has been detailed. To arrive at a fully specified instance, we now have to detail the search algorithm to be used for traversing the space of graphical structures and the quality measure it employs for comparing candidate structures. In our experiments, we use the well-known MDL quality measure (Lam and Bacchus, 1993). This measure originates from information theory and computes the *description length* of a Bayesian network and a given dataset; the description length equals the sum of the size of the network and the size of the dataset after it has been compressed given the network. While a more complex network can better describe the data and hence compress it to a smaller size than a simpler network, it requires a larger encoding to specify its arcs and associated probabilities. The best network for a given dataset now is the network that best balances its complexity and its ability to describe the data.

In our experiments, we further use two different search algorithms: a simple hill-climber and a genetic algorithm. The hillclimber sets out with the empty graph. In each step, it considers all pairs of neighbouring nodes from the skeleton and all possible changes to the graph under construction, that is, remove, insert, or reverse the considered arc; it then selects the change that improves the MDL score the most. This process is repeated until the score cannot be further improved. The genetic algorithm builds upon an encoding of graphical structures by strings of genes. Each gene corresponds with an edge in the skeleton and can be set in one of three states, matching absence and either orientation of the edge. A special-purpose recombination operator is used to guarantee good mixing and preservation of building blocks (van Dijk et al., 2003).

Table 1. Results of the experiments. The top part shows in the first column the MDL score of the original network, averaged over five datasets. The second column shows the results from the GA with the true skeleton, averaged over five datasets and five runs per dataset. The third column gives the results from the hillclimber on the true skeleton, averaged over five datasets. The bottom part lists the results obtained with the GA, with the HEP algorithm, and with the hillclimber.

	Original ($\overline{score} \pm sd$)	GA+true ($\overline{score} \pm \overline{sd}$)	HC+true ($\overline{score}$)
Alarm-250	7462.14±197.20	5495.05±0.02	5645.56
Alarm-500	10862.49±196.59	9272.49±0	9519.90
Alarm-2000	31036.54±236.00	30214.72±0	31677.65
Alarm-10000	138914.66±764.09	138774.32±0	145564.45
Oesoca-250	10409.23±88.20	7439.62±0	7535.68
Oesoca-500	15692.60±154.25	13034.87±0.19	13274.75
Oesoca-2000	46834.31±244.07	45115.91±2.77	45542.13
Oesoca-10000	213253.55±387.12	212364.09±1.23	213204.67

	GA ($\overline{score} \pm \overline{sd}$)	HEP ($\overline{score} \pm sd$)	HC ($\overline{score}$)
Alarm-250	5566.71±0.20	5523.50±11.74	5654.52
Alarm-500	9458.03±1.62	9260.87±35.97	9619.94
Alarm-2000	30563.03±2.29	30397.20±161.86	31727.86
Alarm-10000	138955.54±72.36	139499.51±530.41	144677.31
Oesoca-250	7703.43±0.82	9619.25±0.10	7743.33
Oesoca-500	13282.38±0.07	15388.89±16.75	13474.23
Oesoca-2000	45335.16±2.12	46504.41±40.18	45861.42
Oesoca-10000	212544.21±27.47	212446.04±276.97	214942.04

3.2 Experimental Results

We studied the two instances of our approach outlined above and compared their performance against that of the HEP algorithm. We used datasets that were generated by means of logic sampling from two real-world Bayesian networks. The well-known *Alarm* network was built to help anesthetists monitor their patients and is quite commonly used for evaluating the performance of algorithms for learning Bayesian networks. The *Oesoca* network was developed at Utrecht University, in close collaboration with experts from the Netherlands Cancer Institute; it was built to aid gastroenterologists in assessing the stage of oesophageal cancer and in predicting treatment outcome. Table 1 summarises the results obtained for datasets of four different sizes for each network. The results for the genetic algorithm and for the HEP algorithm are averaged over five different datasets and five runs of the algorithm per dataset; the results for the hillclimber are averaged over the five datasets. Depending on the size of the data set, running times ranged from two to 80 minutes for the GA, up to 30 minutes for the hillclimber, and up to seven minutes for the HEP algorithm. Calculation of the skeleton could take up to 50% of the total time of a run with the GA.

The bottom part of Table 1 shows that all three algorithms under study perform quite well. The table in fact reveals that the algorithms often yield a network that has a lower score than the original network, whose score is shown in the top part of the table. The fact that the original network may not be the one of highest quality can be attributed to the datasets being finite samples. Since the datasets are subject to sampling error, they may not accurately reflect all the (in)dependences from the original network. The distribution observed in the data may then differ from the distribution captured by the original network. From the bottom part of the table we further observe that the genetic algorithm and the HEP algorithm perform comparably. The small standard deviation revealed by the genetic algorithm indicates that it is likely to always give results of similar quality. Since the HEP algorithm reveals much more variation, the genetic algorithm may be considered the more reliable of the two algorithms.

The top part of Table 1 summarises the results obtained with the GA and with the hillclimber when given the true skeleton rather than the skeleton constructed from the data. We note that only a slight improvement in quality results from using the true skeleton. From this observation, we may conclude that, for practical purposes, the constructed skeleton is of high quality. The good performance of the hillclimber, moreover, is an indication of how much the learning task benefits from the use of the skeleton.

4 Analysis of Our Approach

We recall from Section 2 that our approach divides the task of learning a Bayesian network from data into two phases. In the first phase, a skeleton is constructed that is taken as a specification of part of the search space of graphical structures. In the second phase, the specified subspace is traversed by a search algorithm. The feasibility of our approach depends to a large extent on the properties of the computed skeleton. First of all, to avoid pruning optimal solutions from the search space, there should be no edges of the true skeleton missing from the computed skeleton. Secondly, there should be few additional edges: the more densely connected the computed skeleton is, the less feasible it is to traverse the specified subspace of graphical structures. Since it is very hard to prove theoretical results about the computed skeleton, we opt for an experimental investigation of its properties. In the subsequent sections, we compare the computed skeleton against the true skeleton in increasingly realistic situations.

4.1 Use of a Perfect Oracle

To investigate by how much a computed skeleton can deviate from the true skeleton, we performed an experiment in which we precluded the effects of sampling error and of inaccuracy from dependence tests. To this end, we constructed an *oracle* that reads the (in)dependences tested for from the structure of the original network. For the Alarm and Oesoca networks, we thus computed two skeletons each. For the first skeleton, we used zero-order dependence tests only: we connected each variable X to all variables having an unconditional dependence on X. For the construction of the second skeleton, we used zero- and first-order tests as outlined in Section 2. Table 2 reports the numbers of additional edges found in the computed skeletons compared against the true skeleton; the table further includes the results for a skeleton consisting of the complete graph.

Since we used a perfect oracle to establish dependence or independence, the computed skeletons include all edges from the true skeletons: there are no edges missing. Table 2 therefore gives insight in the reduction of the search space for the second phase

Table 2. Numbers of *additional* edges found in the skeletons constructed using an oracle. Results are shown for the complete skeleton, for the skeleton built by using just zero-order tests, and for the skeleton computed by the proposed approach. The true skeleton of the Alarm network includes 46 edges; the true skeleton of the Oesoca network includes 59 edges.

	Alarm	Oesoca
Complete skeleton	620	802
Zero-order skeleton	255	728
Computed skeleton	58	116

of our approach, that is achieved under the assumption that the *strengths* of the dependences do not affect the results from the dependence test. The table reveals that, under this assumption, sizable reductions are found. Note that further reductions could have been achieved by using higher-order dependence tests. Such tests, however, would have increased the computational demands for the construction of the skeleton. Moreover, in practice such tests would have quickly become highly unreliable.

4.2   Use of the χ^2 Test on a Perfect Dataset

In the experiments described in Section 3, we used the χ^2 test for studying dependence. We recall that, for two variables X and Y, the χ^2 test calculates a statistic $stat(X,Y)$. This statistic is compared against a critical value s to decide upon acceptance or rejection of the null-hypothesis of independence of X and Y. The critical value depends upon the threshold ε and upon the degrees of freedom df of the χ^2 distribution used. Writing $f(\varepsilon, df)$ for the function that yields the critical value s, we have that the test reports dependence if $stat(X,Y) > f(\varepsilon, df)$. Hence,

$$N \cdot c > f(\varepsilon, df),$$

where c is a constant that depends upon the marginal and joint probability distributions over X and Y. The choice of the threshold ε now directly influences the topology of the computed skeleton. If the threshold is set too low with respect to the size of the available dataset, weak dependences will escape identification and the skeleton will have edges missing. On the other hand, if the threshold is too high, coincidental correlations in the data will be mistaken for dependences and the skeleton will include spurious edges.

To study the impact of the thresholds used with the χ^2 test, we performed some experiments from which we precluded the effects of sampling error. For this purpose, we constructed *virtual* datasets that perfectly capture the probability distribution to be recovered: for the proportions $\hat{p}(C)$ reflected in these datasets, we thus have that $\hat{p}(C) = p(C)$, where $p(C)$ is the true distribution over the variables C. For the Alarm and Oesoca networks, we constructed various skeletons from virtual datasets of different sizes, using different thresholds. In our first experiment, we focused on skeletons that were constructed from zero-order dependence tests only. Figures 2(a) and 2(b) show the numbers of edges from the true skeletons of the Alarm network and of the Oesoca network, respectively, that are missing from these zero-order skeletons. Figure 2(a) reveals that the Alarm network consists of quite strong dependences that are effectively

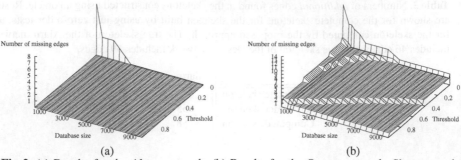

(a)   (b)

Fig. 2. (a) Results for the Alarm network. (b) Results for the Oesoca network. Shown are the numbers of edges missing from the skeleton built from just zero-order dependences.

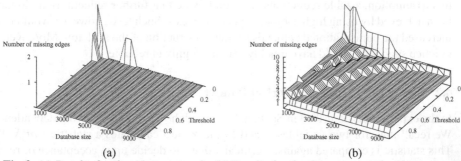

(a)   (b)

Fig. 3. (a) Results for the Alarm network. (b) Results for the Oesoca network. Shown are the numbers of edges missing from the skeleton built using zero- and first-order dependence tests.

found, even with small thresholds and small datasets. Figure 2(b) shows that a similar observation does not apply to the Oesoca network. This network models a very weak dependence that is only found with a threshold equal to zero, even for a dataset of size 10000. The other relatively weak dependences modelled by the network, are recovered by a tradeoff between the threshold used and the size of the dataset under study.

We further constructed skeletons using zero- and first-order tests, as described in Section 2. Once again, we used virtual datasets of different sizes and employed different thresholds for the first-order dependence test. For the zero-order test, we used, for the Alarm network, the highest threshold with which all edges from the true skeleton were recovered; for the Oesoca network, we used the highest threshold with which all dependences except the two weakest ones were found. Figures 3(a) and 3(b) show the numbers of edges from the true skeletons of the Alarm network and of the Oesoca network, respectively, that are missing from the thus computed skeletons. Figure 3(a) shows that the true skeleton of the Alarm network is effectively recovered, with almost all thresholds and dataset sizes. Figure 3(b) shows, once again, that the weaker dependences modelled by the Oesoca network are only found by a tradeoff between the threshold and the size of the dataset used.

Where Figure 3 shows the numbers of edges missing from the skeletons constructed using zero- and first-order tests, Figure 4 shows the numbers of edges in these skeletons that are absent from the true skeletons. Figures 4(a) and 4(b) thus show the numbers

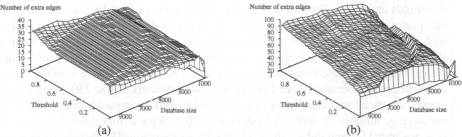

Fig. 4. (a) Results for the Alarm network. (b) Results for the Oesoca network. Shown are the numbers of spurious edges in the skeleton built using zero- and first-order dependence tests.

of spurious edges in the skeletons constructed for the Alarm and Oesoca networks, respectively. Figure 4(b) especially illustrates the tradeoff between recovering (almost) all edges of the true skeleton and excluding spurious ones. We further observe that the landscapes of the Figures 3(b) and 4(b) are not monotonically increasing or decreasing: Figure 3(b) reveals a valley and Figure 4(b) shows a ridge. These non-monotonicities are caused by a very weak dependency in the Oesoca network. With higher thresholds, the weakness of the dependency forestalls its identification, thereby hiding a true neighbour. Upon lowering the threshold, however, the neighbour is identified and thereby effectively changes the separation graph, which causes the observed ridge and valley.

Since we used virtual datasets, we precluded from our experiments the effects of sampling error. The Figures 3 and 4, therefore, give insight in the ability of our approach to construct a skeleton that is already close to the true skeleton, under the assumption that the dataset used perfectly captures the probability distribution to be recovered. The figures reveal that, under this assumption, most dependences are recovered with small threshold values, giving rise to good skeletons. We found, however, that carefully handcrafted, real-world networks may embed very weak dependences that would require high thresholds for their recovery from data.

4.3   Use of χ^2 with Sampled Datasets

The last, and most realistic situation that we address, involves datasets that were generated by means of logic sampling from a network under study. We recall that we used such datasets in our main experiments described in Section 3. We observe that sampled datasets differ from virtual datasets in two important aspects. Firstly, generated datasets show the effects of sampling errors, that is, the distribution observed in the dataset may differ slightly from the original distribution. Secondly, as generated datasets are finite, the dependence test used can fail to reliably establish dependence or independence. In our experiments, we adopted the common rule of thumb that the result of the χ^2 test can be considered reliable only if all cells in the contingency tables have expected frequencies larger than five. As before, we compared the skeletons computed from the sampled datasets against the true skeletons of the Alarm and Oesoca networks. The numbers of missing and additional edges are summarised only briefly due to space restrictions.

The differences between the true skeletons and the skeletons constructed from the sampled datasets, strongly depended upon the sizes of the dataset used and upon the thresholds employed. For low thresholds, the differences found were relatively modest. For example, with datasets of size 10 000 sampled from the Oesoca network and with thresholds equal to 0.05, the average number of edges missing from the computed skeleton was 3; the average number of additional edges was 81. With virtual datasets of the same size, these numbers were 2 and 59, respectively. With smaller datasets, these differences became larger. For example, with datasets of size 250, using the same thresholds, the average number of edges missing from the computed skeletons was 23.6; the average number of additional edges was 86.2. With virtual datasets of the same size, we found these numbers to be 11 and 22, respectively. The thresholds used were found to have a much stronger impact on the numbers of missing and additional edges. With datasets of size 2000, for example, raising the thresholds from $\varepsilon_0 = \varepsilon_1 = 0.05$ to $\varepsilon_0 = 0.1, \varepsilon_1 = 0.4$ served to double the size of the resulting skeleton.

From the above observations, we conclude that in realistic situations the thresholds used with the dependence test should be set of a relatively low value since using more liberal thresholds would result in an unfavourable tradeoff between the size of the resulting skeleton and its number of missing edges. We further conclude that sampling error can cause substantial deviations of the computed skeleton from the true one.

5 Discussion

Most state-of-the-art algorithms for learning Bayesian networks from data build upon either the use of (in)dependence tests or the use of a quality measure and search algorithm. While important progress has been made with both approaches, we have argued that there are some obstacles to their practicability. Within the first approach, for example, the statistical test employed quickly becomes unreliable for larger conditioning sets. The second approach suffers from the huge space of graphical structures to be traversed. We have proposed a novel approach that combines the main advantages of these earlier algorithms yet avoids their difficulties. In the first phase of our approach, we use zero- and first-order dependence tests to build an undirected skeleton for the network under construction. This skeleton is used to explicitly restrict the search space of directed graphical structures to promising regions. Then, a search algorithm is used to traverse the restricted space to find a high-quality network. Our approach is general in the sense that it can be used with various different dependence tests, quality measures, and search algorithms. We have demonstrated the feasibility of our approach by means of experiments with two specific instances. These instances have shown good performance on datasets of various sizes generated from two real-world Bayesian networks.

The good performance of even a simple hillclimber within our approach suggests that the restriction of the search space of graphical structures by means of a skeleton is safe, in the sense that it is not likely to prune high-quality networks. To corroborate this observation, we have compared the computed skeletons against the true skeleton in varying situations. Using virtual datasets for the well-known Alarm network, skeletons without any missing edges and with up to twenty extra edges have been found, using very small thresholds for the dependence tests. Except for its three weakest dependences, also all edges from the true skeleton of the Oesoca network have been recovered with small thresholds; the numbers of extra edges for the skeletons computed from virtual datasets of various sizes range between 20 and 70. We conclude that the use of a skeleton provides for a careful balance of accuracy and a tractable search space.

To conclude, we would like to note that the Oesoca network includes a dependence that is weak in general yet becomes very strong for patients in whom a relatively rare condition is found. This dependence is important from the point of view of the application domain, although for the learning task it is indistinguishable from the numerous irrelevant dependences found in the data. Since there is always a tradeoff between considering weak dependences that may be important and the computational resources one is willing to spend, we feel that learning a Bayesian network from data should always be performed in close consultation with a domain expert.

References

W. L. Buntine. Classifiers: A theoretical and empirical study. In R. L. de Mantaras and D. Poole, editors, *Proceedings of the International Joint Conference on Artificial Intelligence*, pages 638–644. Morgan Kaufmann, 1991.

J. Cheng, R. Greiner, J. Kelly, D. A. Bell, and W. Liu. Learning Bayesian networks from data: An information-theory based approach. *Artificial Intelligence*, 137(1–2):43–90, 2002.

G. F. Cooper and E. Herskovits. A Bayesian method for the induction of probabilistic networks from data. *Machine Learning*, 9:309–347, 1992.

R. G. Cowell, A. P. Dawid, S. L. Lauritzen, and D. J. Spiegelhalter. *Probabilistic Networks and Expert Systems*. Statistics for Engineering and Information Science. Springer-Verlag, 1999.

L. M. de Campos and J. F. Huete. A new approach for learning belief networks using independence criteria. *International Journal of Approximate Reasoning*, 24(1):11–37, 2000.

D. Heckerman, D. Geiger, and D. M. Chickering. Learning Bayesian networks: The combination of knowledge and statistical data. *Machine Learning*, 20(3):197–243, 1995.

W. Lam and F. Bacchus. Using causal information and local measures to learn Bayesian networks. In D. Heckerman and A. Mamdani, editors, *Proceedings of the Ninth Conference on Uncertainty in Artificial Intelligence*, pages 243–250. Morgan-Kaufmann, 1993.

G. Rebane and J. Pearl. The recovery of causal poly-trees from statistical data. In L. N. Kanal et al., editors, *Proceedings of the Third Conference on Uncertainty in Artificial Intelligence*, pages 175–182. Elsevier, 1987.

P. Spirtes and C. Glymour. An algorithm for fast recovery of sparse causal graphs. *Social Science Computer Review*, 9(1):62–73, 1991.

H. Steck and V. Tresp. Bayesian belief networks for data mining. In *Proceedings of the Second Workshop on Design and Management of Data Warehouses*, pages 145–154, 1999.

S. van Dijk, D. Thierens, and L. C. van der Gaag. Building a GA from design principles for learning Bayesian networks. In E. Cantú-Paz et al., editors, *Lecture Notes in Computer Science, Volume 2723: Proceedings of the Genetic and Evolutionary Computation Conference*, pages 886–897. Springer-Verlag, 2003.

M. L. Wong, S. Y. Lee, and K. S. Leung. A hybrid data mining approach to discover Bayesian networks using evolutionary programming. In W. B. Langdon et al., editors, *Proceedings of the Genetic and Evolutionary Computation Conference*. Morgan-Kaufmann, 2002.

On Decision Boundaries of Naïve Bayes in Continuous Domains

Tapio Elomaa and Juho Rousu

Department of Computer Science, University of Helsinki, Finland
{elomaa,rousu}@cs.helsinki.fi

Abstract. Naïve Bayesian classifiers assume the conditional independence of attribute values given the class. Despite this in practice often
violated assumption, these simple classifiers have been found efficient,
effective, and robust to noise.
Discretization of continuous attributes in naïve Bayesian classifiers has
achieved a lot of attention recently. Continuous attributes need not necessarily be discretized, but it unifies their handling with nominal attributes
and can lead to improved classifier performance.
We show that optimal partitioning results from decision tree learning
carry over to Naïve Bayes as well. In particular, it sets decision boundaries on borders of segments with equal class frequency distribution. An
optimal univariate discretization with respect to the Naïve Bayes rule
can be found in linear time but, unfortunately, optimal multivariate optimization is intractable.

1 Introduction

The naïve Bayesian classifier, or Naïve Bayes, is surprisingly effective in classification tasks. Therefore, even if it does not belong to state-of-the-art methods, it
plays an important role — alongside decision tree learning — as standard baseline
methods of inductive algorithms. Naïve Bayesian classifiers have been studied
extensively over the years [18,19,7].

In Naïve Bayes numerical attributes can be handled without explicit discretization of the value range [7,15] unlike in, e.g., decision tree induction. An
often made assumption is that within each class the data is generated by a
single Gaussian distribution. To model actual distributions more faithfully one
can abandon the normality assumption and, rather, use nonparametric density
estimation [7,15].

Treating numerical attributes by density estimation, Gaussian or other, indicates that numerical and discrete attributes are handled differently. Furthermore,
discretization has been observed to increase the prediction accuracy and make
the method more efficient [2,6]. There are discretization methods that are specific
to Naïve Bayes [2,6,25,4,26] as well as general approaches that are often used
with naïve Bayesian classifiers [13]. A particularly interesting fact is that Naïve
Bayes permits overlapping discretization [16,27] unlike many other classification
learning algorithms.

N. Lavrač et al. (Eds.): PKDD 2003, LNAI 2838, pp. 144–155, 2003.
© Springer-Verlag Berlin Heidelberg 2003

In decision tree setting the line of research stemming from Fayyad and Irani's [12] seminal work on optimal discretizations for evaluation functions of ID3 has led to more efficient preprocessing approaches and a better understanding of the necessary and sufficient preprocessing needed to guarantee finding optimal partitions with respect to common evaluation functions [8,9,10,11]. In this paper we show that this type of analysis carries over to naïve Bayesian classifiers despite the difference of univariate inspection in decision trees and multivariate one in naïve Bayesian classifiers. The *decision boundaries* separating decision regions — class prediction changes — of naïve Bayesian classifiers fall exactly on the so-called segment borders. No other cut point candidates need to be considered in order to find the error-minimizing discretization.

We show that with respect to one numerical attribute, a partition that optimizes the naïve Bayes rule can be found in linear time using the same algorithm as in connection with decision trees. However, simultaneously satisfying the optimality with respect to more than one attribute, unfortunately, has recently turned out to be NP-complete [23]. This does not leave us with possibilities to solve the problem efficiently.

In Sect. 2 we first recapitulate the basics on naïve Bayesian classification. In Sect. 3 the optima-preserving preprocessing of numerical value ranges is reviewed. In Sect. 4 we prove that the same line of analysis applies to naïve Bayesian classifiers as well. We also briefly consider multivariate discretization in this section. Finally, Sect. 5 concludes this article by summarizing the work and discussing further research possibilities.

2   Naïve Bayes

Naïve Bayes gives an instance $x = \langle a_1, \ldots, a_n \rangle$ the label

$$\arg \max_{c \in C} \mathbf{Pr} \left(c \mid x \right), \tag{1}$$

where C is the set of classes. In other words, the classifier assigns for the given instance the class that is most probable. The computation of the conditional probability $\mathbf{Pr} \left(c \mid x \right)$ is based on the Bayes rule

$$\mathbf{Pr} \left(c \mid x \right) = \frac{\mathbf{Pr} \left(c \right) \mathbf{Pr} \left(x \mid c \right)}{\mathbf{Pr} \left(x \right)}$$

and the (naïve) assumption that the attributes $A_1, \ldots, A_n$ are independent of each other given the class, which indicates that

$$\mathbf{Pr} \left(x \mid c \right) = \prod_{i=1}^{n} \mathbf{Pr} \left(A_i = a_i \mid c \right).$$

The denominator $\mathbf{Pr} \left(x \right)$ of the Bayes rule is the same for all classes in C. Therefore, it is convenient to consider the quantity

$$\arg \max_{c \in C} \mathbf{Pr} \left(c \cap x \right) = \arg \max_{c \in C} \mathbf{Pr} \left(c \mid x \right) \mathbf{Pr} \left(x \right)$$

instead of (1). The two formulas, of course, always predict the same label.

Probability estimation is based on a training set of classified examples $E = \{\langle x_i, y_i \rangle\}_{i=1}^{m}$, where $y_i \in C$ for all i. Let m_c denote the number of instances from class c in E. Then, the data prior for class c is $\widehat{P}(c) = m_c/m$. Discrete attributes are easy to handle: we just estimate $\mathbf{Pr}(x \mid c)$ based on the training set by estimating the conditional marginals $\widehat{P}(A_i = a_i \mid c)$ by counting the fraction of occurrences of each value $A_i = a_i$ in m_c.

Unless discretized, continuous values are harder to take care of and require a different strategy. It is common to assume that within each class c the values of numeric attributes are normally distributed. Then by estimating from the training set the mean μ_c and standard deviation σ_c of the continuous attribute given c, one can compute the probability of the observed value. After obtaining μ_c and σ_c for an attribute A_i the estimation boils down to calculating the probability density function for a Gaussian distribution:

$$\mathbf{Pr}(A_i = a_i \mid c) = \frac{1}{\sqrt{2\pi}\sigma_c} \exp\left(-\frac{(a_i - \mu_c)^2}{2\sigma_c^2}\right).$$

Using Dirichlet prior, or more generally Bayesian estimation methods [17,4], and kernel density estimation [15] are some alternatives to the straightforward normality assumption for estimating a model for the distribution of the continuous attribute. In this paper, however, we are only concerned with probability estimates computed as data priors $\widehat{P}(\cdot)$.

Despite the unrealistic attribute independence assumption underlying Naïve Bayes it is a very successful classifier in practical situations. Some explanations have been offered by Domingos and Pazzani [5], who showed that Naïve Bayes may be globally optimal even though the attribute independence assumption is violated. It was shown that, under 0–1 loss, Naïve Bayes is globally optimal for the concept classes conjunctions and disjunctions of literals. Gama [14] discusses Naïve Bayes and quadratic loss.

It is well-known that the naïve Bayesian classifier is equivalent to a linear machine and, hence, for nominal attributes its decision boundary is a hyperplane [7,21,5]. Thus, Naïve Bayes can only be globally optimal for linearly separable concept classes. Ling and Zhang [20] consider the representational power of Naïve Bayes and more general Bayesian networks further. They characterize the representational power through the maximum XOR contained in a function.

3   Discretizing Continuous Attributes

The dominating discretization techniques for continuous attributes in Naïve Bayes are unsupervised equal-width binning [24] and the greedy top-down approach of Fayyad and Irani [13]. These straightforward heuristic approaches have also been offered some analytical backing [4]. However, in other classifier learners — decision trees in particular — analysis of discretization has been taken much further. In the following we recapitulate briefly the line of analysis

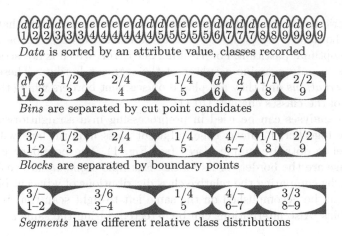

Data is sorted by an attribute value, classes recorded

Bins are separated by cut point candidates

Blocks are separated by boundary points

Segments have different relative class distributions

Fig. 1. The original set of 27 examples (top) can only be partitioned at bin borders (second from top) where the value of the attribute changes. Class uniform bins can be combined into blocks (second from bottom). Block borders are the boundary points of the numerical range. Furthermore, partitions can only happen between blocks with different relative class distribution. Thus, we may arrange the data into segments (below). Segment borders are a subset of boundary points

initiated by Fayyad and Irani [12]. The goal is to reduce the number of examined cut points without losing the possibility to recover optimal partitions.

In decision tree learning the processing of a numerical value range usually starts with sorting of the data points [1,22]. If one could make its own partition interval out of each data point in the sorted sequence, this discretization would have zero training error. However, only those points that differ in their value can be separated from each other. Therefore, we can preprocess the data into *bins*, one bin for each existing data point value. Within each bin we record the class distribution of the instances that belong to it (see Fig. 1). The class distribution information suffices to evaluate the goodness of the partition; the actual data set does not need to be maintained.

The sequence of bins attains the minimal misclassification rate. However, the same rate can usually be obtained with a smaller number of intervals. The analysis of the entropy function by Fayyad and Irani [12] has shown that cut points embedded into class-uniform intervals need not be taken into account, only the end points of such intervals — the *boundary points* — need to be considered to find the optimal discretization. Elomaa and Rousu [8] showed that the same is true for several commonly-used evaluation functions.

Subsequently, a more general property was also proved for some evaluation functions [9]: *segment borders* — points that lie in between two adjacent bins with different relative class distributions — are the only points that need to be taken into account. It is easy to see that segment borders are a subset of boundary points.

For strictly convex evaluation functions it was shown later that examining segment borders is necessary as well as sufficient in order to be able to discover the optimal partition. For Training Set Error, which is not strictly convex, it suffices to only examine a subset of the segment borders. These points are called *alternations* and they are placed on segment borders where the frequency ordering of the classes changes [10,11].

These analyses can be used in preprocessing in a straightforward manner: we merge together, in linear time, adjacent class uniform bins with the same class label to obtain example *blocks* (see Fig. 1). The boundary points of the value range are the borders of its blocks. Example *segments* are easily obtained from bins by comparing the relative class distributions of adjacent bins (see Fig. 1). This can be accomplished on the same left-to-right scan that is required to identify bins. Also alternations can be detected during the same scan.

4   Decision Boundaries of Naïve Bayes

We show now that segments in the domain of a continuous attribute are the locations where Naïve Bayes changes its class prediction, i.e., its decision boundaries. We start from undiscretized domains, go on to error-minimizing discretizations, and finally consider optimal partitions with respect to several attributes.

4.1   Decision Boundaries for Undiscretized Attributes

We start by examining the decision boundaries that Naïve Bayes sets when the continuous attribute is not discretized, but each numerical value is treated separately. When we consider the decision boundaries from the point of view of one attribute, we assume an arbitrary fixed value setting for the other attributes. In the following, notation $\widehat{P}(\cdot)$ is used to denote the probability estimates computed by Naïve Bayes, to distinguish them from true probabilities.

Theorem 1. *The decision boundaries of the naïve Bayesian classifier are situated on segment borders.*

Proof. Let A_i be a numerical attribute and let V' and V'' be two adjacent intervals in its range separated by cut point v_i. For any other attribute A_j, $i \neq j$, let V_j be an arbitrary subset of the values of A_j. Let us denote by

$$T = V_1 \times \cdots \times V_{i-1} \times V_{i+1} \times \cdots \times V_n$$

the Cartesian product of these subsets. We assume that the prediction of naïve Bayesian classifier within $V' \times T$ is $c' \in C$, and the prediction within $V'' \times T$ is $c'' \in C$. In other words, looking at the situation only from the point of view of A_i and taking all other attributes to have an arbitrary (but fixed) value combination, the decision boundary is set between intervals V' and V''. Then

$$\widehat{P}(c' \mid V' \times T) = \widehat{P}(c') \, \widehat{P}(V' \mid c') \prod_{j \neq i} \widehat{P}(V_j \mid c')$$
$$> \widehat{P}(c'') \, \widehat{P}(V' \mid c'') \prod_{j \neq i} \widehat{P}(V_j \mid c'') = \widehat{P}(c'' \mid V' \times T).$$

By reorganizing the middle inequality we get

$$\frac{\widehat{P}(V' \mid c')}{\widehat{P}(V' \mid c'')} > \frac{\widehat{P}(c'') \prod_{j \neq i} \widehat{P}(V_j \mid c'')}{\widehat{P}(c') \prod_{j \neq i} \widehat{P}(V_j \mid c')}. \tag{2}$$

On the other hand, within $V'' \times T$ we obtain, by similar manipulation,

$$\frac{\widehat{P}(V'' \mid c')}{\widehat{P}(V'' \mid c'')} < \frac{\widehat{P}(c'') \prod_{j \neq i} \widehat{P}(V_j \mid c'')}{\widehat{P}(c') \prod_{j \neq i} \widehat{P}(V_j \mid c')}. \tag{3}$$

Put together, (2) and (3) imply

$$\frac{\widehat{P}(V'' \mid c')}{\widehat{P}(V'' \mid c'')} < \frac{\widehat{P}(c'') \prod_{j \neq i} \widehat{P}(V_j \mid c'')}{\widehat{P}(c') \prod_{j \neq i} \widehat{P}(V_j \mid c')} < \frac{\widehat{P}(V' \mid c')}{\widehat{P}(V' \mid c'')}.$$

By using the Bayes rule to the conditional probabilities and canceling out equal factors we get

$$\frac{\widehat{P}(c' \mid V'')}{\widehat{P}(c'' \mid V'')} < \frac{\widehat{P}(c' \mid V')}{\widehat{P}(c'' \mid V')}.$$

Hence, the relative class distributions must be strictly different within the intervals V' and V'' making v_i thus a segment border.

The above result does not, of course, mean that all segment borders would be places for class prediction change. However, the class prediction changes of an undiscretized domain are confined to segment borders. Consequently, no loss is incurred in grouping the examples in segments of equal class distribution. On the contrary, we expect to benefit from the more accurate probability estimation.

4.2 Decision Boundaries in k-Interval Discretization

Let us now turn to the case where the continuous range has been discretized into k intervals. We will prove that in this case too segment borders are the only potential points for the decision boundaries.

The following proof has the same setting as the proofs in connection with decision trees [9]. The sample contains three subsets, P, Q, and R, with class frequency distributions

$$p = \sum_{j=1}^{m} p_j, \quad q = \sum_{j=1}^{m} q_j, \quad \text{and } r = \sum_{j=1}^{m} r_j,$$

where p is the number of examples in P and p_j is the number of instances of class j in P. Furthermore, m is the number of classes. The notation is similar also for Q and R.

We consider the k-ary partition $\{ S_1, \ldots, S_k \}$ of the sample, where subsets S_h and S_{h+1}, $1 \leq h \leq k - 1$, consist of the set $P \cup Q \cup R$, so that the split point

$S_1, \ldots, S_{h-1}$ $S_h(\ell)$ $S_{h+1}(\ell)$ $S_{h+2}, \ldots, S_k$

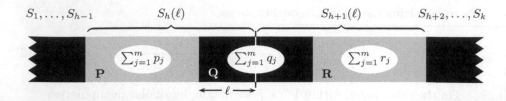

Fig. 2. The following proofs consider partitioning of the example set $P \cup Q \cup R$ into two subsets S_h and S_{h+1} within Q. No matter where, within Q, the cut point is placed, equal class distributions result

is inside Q, on the border of P and Q, or that of Q and R (see Fig. 2). Let ℓ be a real value, $0 \le \ell \le q$ [1]. Let $S_h(\ell)$ denote partition interval S_h that contains P and the ℓ first examples from Q. In the same situation $S_{h+1}(\ell)$ denotes the interval S_{h+1}. We assume that splitting the set Q so that ℓ examples belong to $S_h(\ell)$ and $q - \ell$ to $S_{h+1}(\ell)$ results in identical class frequency distributions for both subsets of Q regardless of the value of ℓ.

Let T again be the Cartesian product of the (arbitrary) subsets in dimensions other than the one under consideration. In this setting we can prove the following result which will be put to use later.

Lemma 1. *The sum* $\max_{c \in C} \widehat{P}(c \cap S_h(\ell) \times T) + \max_{c \in C} \widehat{P}(c \cap S_{h+1}(\ell) \times T)$ *is convex over* $\ell \in [0, q]$.

Proof. Let $\ell_1, \ldots, \ell_{r-1}$ be the class prediction change points within $[0, q]$. Without loss of generality, let us denote by c_i, $1 \le i \le r-1$, the class predicted within $S_h(\ell)$, $\ell \in]\ell_i, \ell_{i+1}]$. The probability of instances of class c within $S_h(\ell) \times T$ can be expressed as

$$\widehat{P}(c \cap S_h(\ell) \times T) = \frac{p_c \cdot \widehat{P}(T \mid c)}{n} + \ell \cdot \frac{q_c/q \cdot \widehat{P}(T \mid c)}{n},$$

which describes a line with offset $(p_c/n)\widehat{P}(T \mid c)$ and slope $((q_c/q)/n)\widehat{P}(T \mid c)$ (see Fig. 3). Now, it must be that the offsets satisfy

$$\frac{p_{c_1} \cdot \widehat{P}(T \mid c_1)}{n} \ge \cdots \ge \frac{p_{c_{r-1}} \cdot \widehat{P}(T \mid c_{r-1})}{n}$$

and the slopes of the lines satisfy

$$\frac{q_{c_1}/q \cdot \widehat{P}(T \mid c_1)}{n} \le \cdots \le \frac{q_{c_{r-1}}/q \cdot \widehat{P}(T \mid c_{r-1})}{n}.$$

Interpreting the situation geometrically, we see that $\max_c \widehat{P}(c \cap S_h(\ell) \times T)$ forms a convex curve (Fig. 3). By symmetry, $\max_c \widehat{P}(c \cap S_{h+1}(\ell) \times T)$ also is convex, and the claim follows by the convexity of the sum of convex functions.

---

[1] No harm is done considering splitting Q in other points than those corresponding to integral number of examples, since we are proving absence of local extrema.

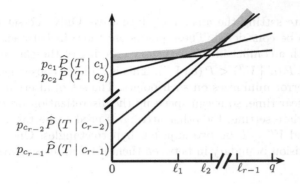

Fig. 3. The maxima of the sum of the most probable classes in $S_h(\ell)$ and $S_{h+1}(\ell)$ forms a convex curve over $[0, q]$

The following proof shows that a cut point in between two adjacent subsets S_h and S_{h+1} in one dimension is on a segment border, regardless of the context induced by the other attributes. Due to the additivity of the error, the result also holds in the multisplitting case, where a number of cut points are chosen in each dimension.

Theorem 2. *The error-minimizing cut points of Naïve Bayes are located on segment borders.*

Proof. Let $c_L(\ell) = \arg\max_{c \in C} \widehat{P}(c \cap S_h(\ell) \times T)$ be the most probable class for $S_h(\ell)$ according to Naïve Bayes criterion in this situation. In other words,

$$\widehat{P}(c_L(\ell) \cap S_h(\ell) \times T) = \max_{c \in C} \widehat{P}(c \cap S_h(\ell)) \, \widehat{P}(T \mid c).$$

Similarly, let $c_R(\ell)$ denote the most probable class in $S_{h+1}(\ell)$.

The minimum-error partition is the one that has the smallest combined error in the subsets $S_h(\ell)$ and $S_{h+1}(\ell)$. Thus, we want to optimize

$$\min_{\ell \in [0,q]} (\widehat{P}(S_h(\ell) \times T) - \widehat{P}(c_L(\ell) \cap S_h(\ell) \times T)$$

$$+ \widehat{P}(S_{h+1}(\ell) \times T) - \widehat{P}(c_R(\ell) \cap S_{h+1}(\ell) \times T)),$$

which is equal to

$$\min_{\ell \in [0,q]} (\widehat{P}(S \times T) - (\widehat{P}(c_L(\ell) \cap S_h(\ell) \times T) + \widehat{P}(c_R(\ell) \cap S_{h+1}(\ell) \times T))),$$

where $S = P \cup Q \cup R$. By Lemma 1 this is a concave function of $\ell \in [0, q]$. Hence, it minimizes at one of the extreme values of ℓ, which are the locations of the segment borders. Thus, we have proved the claim.

In principle it might be possible to reduce the number of examined points by leaving some segment borders without attention. Can we identify such a subset efficiently?

In univariate setting the answer is affirmative: Only the set of *alternation points* need to be considered. These points are those in between adjacent bins V' and V'' with a conflict in the frequency ordering of the classes: $\widehat{P}(c \mid V') < \widehat{P}(c' \mid V')$ and $\widehat{P}(c' \mid V'') < \widehat{P}(c \mid V'')$. This is a direct consequence of the fact that training error minimizes on such points. The set of alternation points can be found in linear time, so it can speed up the discretization process [10].

In multivariate setting, the other attributes need to be taken into account. Let $V' \times T$ and $V'' \times T$ be two adjacent hyperrectangles. One can show that there is no decision boundary in between them if for all class pairs c' and c'' we have either

1. $\widehat{P}(c' \mid V' \times T) \leq \widehat{P}(c \mid V' \times T)$ and $\widehat{P}(c' \mid V'' \times T) \leq \widehat{P}(c \mid V'' \times T)$, or
2. $\widehat{P}(c' \mid V' \times T) \geq \widehat{P}(c \mid V' \times T)$ and $\widehat{P}(c' \mid V'' \times T) \geq \widehat{P}(c \mid V'' \times T)$.

The problem with using this criterion to prune the set of candidate cut points is that the definition depends on the context T, and there is an exponential number of such contexts. So, even if all segment borders are not useful, deciding which of them can be discarded seems difficult.

Thus, in practice, finding a linear-time preprocessing scheme to reduce the set of potential cut points to a proper subset of segment borders is difficult.

4.3   Decision Boundaries of Naïve Bayes in Multiple Dimensions

It is well known that in the discrete (two-class) case the decision boundary is a (single) hyperplane in the input space [7,21]. In case of continuous attributes the situation is much more difficult: The decision regions and their boundaries may have arbitrary shape [7]. However, from preceding results we know that decision boundaries in reality can only occur at segment borders of each continuous attribute. Therefore, we actually can consider discretized ranges instead of truly continuous attributes.

In Fig. 4 the example set of Fig. 1 has been augmented with another (arbitrary) dimension. The segments of these two dimensions divide the input space (a plane) into a 6×5 grid, where each grid cell gets assigned a class label. Class uniform rows and columns get a uniform labeling but otherwise one cannot determine the labeling of grid cells based only on one dimension. Values of both attributes are needed to determine the class label. For example, when the attribute depicted on the y-axis of Fig. 4 has a value in its last segment, depending on the value of the attribute along the x-axis, there are two segments where the most probable prediction would be d and two segment where it would be e.

In general the discretized input space is divided into hyperrectangular cells, each assigned the class label according to the relevant segment statistics.

4.4   On Finding Optimal Discretizations for Naïve Bayes

Theorem 2 tells us that the decision boundaries of Naïve Bayes are always located on segment borders, which makes it possible to preprocess the data into

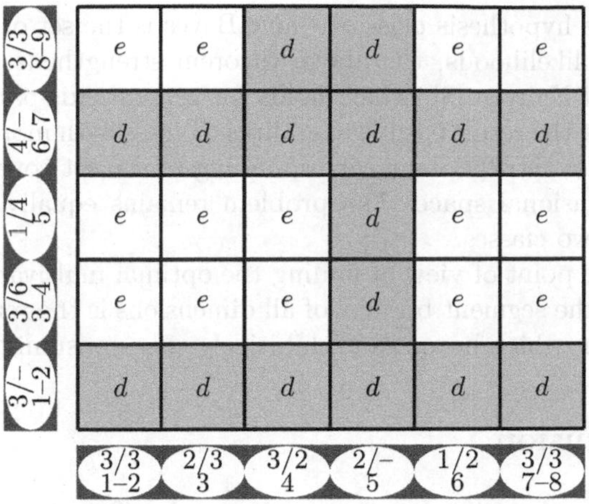

$\frac{3/3}{8-9}$	e	e	d	d	e	e
$\frac{4/-}{6-7}$	d	d	d	d	d	d
$\frac{1/4}{5}$	e	e	e	d	e	e
$\frac{3/6}{3-4}$	e	e	e	d	e	e
$\frac{3/-}{1-2}$	d	d	d	d	d	d

$\frac{3/3}{1-2}$ $\frac{2/3}{3}$ $\frac{3/2}{4}$ $\frac{2/-}{5}$ $\frac{1/2}{6}$ $\frac{3/3}{7-8}$

Fig. 4. The segments of two continuous attributes divide the input space into a rectangular grid. All grid cells are assigned the class label determined by the residual sums of the corresponding segments

segments prior to discretization. By this result, one can use the same linear-time optimization algorithm to find the univariate Naïve Bayes optimal multisplits as in the case of decision trees [9,11]. This so-called Auer-Birkendorf algorithm is based on dynamic programming. During a left-to-right scan over the segments, one can maintain the information required to decide into how many intervals should the data be split and where to locate the interval borders to obtain as good value as possible for the partition.

However, the situation in the multivariate setting is much more difficult. Even with the data preprocessed into segments we may still have a daunting amount of possible discretizations: $O(2^T)$ to be exact, where $T = \sum_{i=1}^{n} T_i$ and T_i is the number of cut points candidates along the i-th dimension. Could there, nevertheless, exist an efficient algorithm for optimal discretization? Unfortunately, we have to answer in the negative, as shown by the next theorem [23].

Theorem 3. *Finding the Naïve Bayes optimal discretization of the real plane* $\mathbb{R}^2$ *is NP-complete.*

This can be proved by a reduction from *Minimum Set Cover* using a similar construction as Chlebus and Nguyen [3] to show that already optimal consistent splitting of the real plane $\mathbb{R}^2$ is NP-complete. We construct a configuration of points in the 2D plane corresponding to the set covering instance and show two properties [23]:

1. The plane can be consistently discretized with k cut lines if and only if there is a set cover of size k for the given set cover instance.
2. The optimal Naïve Bayes discretization coincides with the consistent discretization.

Since the hypothesis class of Naïve Bayes is the set of product distributions of marginal likelihoods, the above theorem strengthens the negative result of Chlebus and Nguyen [3], which holds for general axis-parallel partitions of $\mathbb{R}^2$. Observe that the result easily generalizes to cases with more than two dimensions by embedding the 2D plane corresponding to the set covering instance into the higher dimensional space. The problem remains equally hard when there are more than two classes.

From the point of view of finding the optimal multivariate splits, exhaustive search over the segment borders of all dimensions is the remaining possibility for optimization, which becomes prohibitively time-consuming on larger datasets.

5   Conclusion

Examining segment borders is necessary and sufficient in searching for the optimal partition of a value range with respect to a strictly convex evaluation function [11]. The same set of cut point candidates is relevant for Naïve Bayes: Their decision boundaries (in disjoint partitioning) fall exactly on segment borders.

On the other hand, it seems that for an algorithm to rule out some segment borders from among the decision boundary candidates, it would have to examine too many contexts to be efficient. Therefore, preprocessing the value ranges of continuous attributes into segments appears necessary if one wants to detect all class prediction changes. Such preprocessing, naturally, is sufficient.

As future work we leave the empirical evaluation of the usefulness of segment borders and their accuracy in probability estimation as well as studying possibilities to approximate optimal multivariate discretization and the utility of segment borders therein.

References

1. Breiman, L., Friedman, J.H., Olshen, R.A., Stone, C.J.: Classification and Regression Trees. Wadsworth, Pacific Grove, CA (1984)
2. Catlett, J.: On changing continuous attributes into ordered discrete attributes. In: Proc. Fifth European Working Session on Learning. Lecture Notes in Computer Science, Vol. 482. Springer-Verlag, Berlin Heidelberg New York (1991) 164–178
3. Chlebus, B.S., Nguyen, S.H.: On finding optimal discretizations for two attributes. In: Polkowski. L., Skowron, A. (eds.): Rough Sets and Current Trends in Computing, Proc. First International Conference. Lecture Notes in Artificial Intelligence, Vol. 1424, Springer-Verlag, Berlin Heidelberg New York (1998) 537–544
4. Chu, C.-N., Huang, H.-J., Wong, T.-T.: Why discretization works for naïve Bayesian classifiers. In: Langley, P. (ed.): Proc. Seventeenth International Conference on Machine Learning. Morgan Kaufmann, San Francisco, CA (2000) 399–406
5. Domingos, P., Pazzani, M.: On the optimality of the simple Bayesian classifier under zero-one loss. Mach. Learn. **29** (1997) 103–130
6. Dougherty, J., Kohavi, R., Sahami, M.: Supervised and unsupervised discretization of continuous features. In: Proc. Twelfth International Conference on Machine Learning. Morgan Kaufmann, San Francisco, CA (1995) 194–202

7. Duda, R.O., Hart, P.E., Stork, D.G.: Pattern Classification, Second Edition. John Wiley & Sons, New York (2001)
8. Elomaa, T., Rousu, J.: General and efficient multisplitting of numerical attributes. Mach. Learn. **36** (1999) 201–244
9. Elomaa, T., Rousu, J.: Generalizing boundary points. In: Proc. Seventeenth National Conf. on Artificial Intelligence. MIT Press, Cambridge, MA (2000) 570–576
10. Elomaa, T., Rousu, J.: Fast minimum error discretization. In: Sammut, C., Hoffmann, A. (eds.): Proc. Nineteenth International Conference on Machine Learning. Morgan Kaufmann, San Francisco, CA (2002) 131–138
11. Elomaa, T., Rousu, J.: Necessary and sufficient pre-processing in numerical range discretization. Knowl. Information Systems **5** (2003) in press
12. Fayyad, U.M., Irani, K.B.: On the handling of continuous-valued attributes in decision tree generation. Mach. Learn. **8** (1992) 87–102
13. Fayyad, U.M., Irani, K.B.: Multi-interval discretization of continuous-valued attributes for classification learning. In: Proc. Thirteenth International Joint Conference on Artificial Intelligence. Morgan Kaufmann, San Francisco (1993) 1022–1027
14. Gama, J.: Iterative Bayes. Theor. Comput. Sci. **292** (2003) 417–430
15. John, G.H., Langley, P.: Estimating continuous distributions in Bayesian classifiers. In: Proc. Eleventh Annual Conference on Uncertainty in Artificial Intelligence, Morgan Kaufmann, San Francisco, CA (1995) 338–345
16. Kononenko, I.: Naive Bayesian classifier and continuous attributes. Informatica **16** (1992) 1–8
17. Kontkanen, P., Myllymöki, P., Silander, T., Tirri, H.: A Bayesian approach to discretization. In: European Symposium on Intelligent Techniques. ELITE Foundation, Aachen (1997) 265–268
18. Langley, P., Iba, W., Thompson, K.: An analysis of Bayesian classifiers. In: Proc. Tenth National Conference on Artificial Intelligence. MIT Press, Cambridge, MA (1992) 223–228
19. Langley, P., Sage, S.: Tractable average-case analysis of naive Bayesian classifiers. In: Bratko, I., Džeroski, S. (eds.): Proc. Sixteenth International Conference on Machine Learning. Morgan Kaufmann, San Francisco, CA (1999) 220–228
20. Ling, C.X., Zhang, H.: The representational power of discrete Bayesian networks. J. Mach. Learn. Res. **3** (2002) 709–721
21. Peot, M.A.: Geometric implications of the naive Bayes assumption. In: Horvitz, E., Jensen, F. (eds.): Proc. Twelfth Annual Conference on Uncertainty in Artificial Intelligence. Morgan Kaufmann, San Francisco, CA (1996) 414–419
22. Quinlan, J.R.: C4.5: Programs for Machine Learning, Morgan Kaufmann, San Mateo, CA (1993)
23. Rousu, J.: Optimal multivariate discretization for naive Bayesian classifiers is NP-hard. Tech. Rep. C-2003-8, Dept. of Computer Science, Univ. of Helsinki (2003)
24. Wong, A.K.C., Chiu, D.K.Y.: Synthesizing statistical knowledge from incomplete mixed-mode data. IEEE Trans. Pattern Anal. Mach. Intell. **9** (1987) 796–805
25. Wu, X.: A Bayesian discretizer for real-valued attributes. Computer J. **39** (1996) 688–691
26. Yang, Y., Webb, G.I.: Proportional k-interval discretization for naive-Bayes classifiers. In: De Raedt, L., Flach, P. (eds.): Proc. Twelfth European Conference on Machine Learning. Lecture Notes in Artificial Intelligence, Vol. 2167, Springer-Verlag, Berlin Heidelberg New York (2001) 564–575
27. Yang, Y., Webb, G.I.: Non-disjoint discretization for naive-Bayes classifiers. In: Sammut, C., Hoffmann, A. (eds.): Proc. Nineteenth International Conference on Machine Learning. Morgan Kaufmann, San Francisco, CA (2002) 666–673

Application of Inductive Logic Programming to Structure-Based Drug Design

David P. Enot and Ross D. King

Computational Biology Group, Department of Computer Science,
University of Wales Aberystwyth, SY23 3DB, UK
{dle,rdk}@aber.ac.uk

Abstract. Developments in physical and biological technology have resulted in a rapid rise in the amount of data available on the 3D structure of protein-ligand complexes. The extraction of knowledge from this data is central to the design of new drugs. We extended the application of Inductive Logic Programming (ILP) in drug design to deal with such structure-based drug design (SBDD) problems. We first expanded the ILP pharmacophore representation to deal with protein active sites. Applying a combination of the ILP algorithm Aleph, and linear regression, we then formed quantitative models that can be interpreted chemically. We applied this approach to two test cases: Glycogen Phosphorylase inhibitors, and HIV protease inhibitors. In both cases we observed a significant ($P < 0.05$) improvement over both standard approaches, and use of only the ligand. We demonstrate that the theories produced are consistent with the existing chemical literature.

1 Introduction

Most drugs are small molecules (ligands) that bind to proteins [19]. When knowledge of the 3D structure of the target protein is used in the drug design process, the term structure-based drug design (SBDD) is used. Knowledge of the co-crystallized protein-ligand complex structure is particularly important as it shows how a drug interacts with its target. The binding of the ligand to its target can be regarded as a key (ligand) fitting a lock (active site) (figure 1). To ensure this complementarity, a potential candidate must be the right size for the binding site, must have the correct binding groups to form a variety of weak interactions and must have these binding groups correctly positioned to maximize such interactions. These interactions are primarily hydrophobic and electrostatic (hydrogen bonds, interactions between groups of opposite charges). They are individually weak, but they lead if in sufficient number, to a strong overall interaction (*binding energy*) enabling the ligand to bind to the target site (also referred as activity). These general principles of drug interactions are now well understood, but specific relations between molecular structure and function are still too complex to be delineated from physico-chemical theory and semi-empirical approaches are necessary. From the computational side[14], SBDD involves two main sub-problems to design new active compounds: the prediction of

N. Lavrač et al. (Eds.): PKDD 2003, LNAI 2838, pp. 156–167, 2003.
© Springer-Verlag Berlin Heidelberg 2003

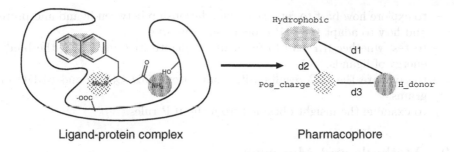

Fig. 1. Schematic representation of a ligand binding a protein illustrating the complementarity of shape and property (left). Example of a three elements pharmacophore (right) derived from the 3D structure of the ligand in the known ligand-protein complex (left).

the most likely ligand mode binding conformation (docking) and the estimation of the relative binding energy of a protein-ligand complex (scoring)[9].

The Protein Data Bank (PDB)[3] is the single worldwide repository for the processing and distribution of 3D biological macromolecular structure data and the number of co-crystallized protein-ligand complexes is rising exponentially over the years. The state-of-the-art in SBDD is to use general propositional regression functions that are designed to be applicable to any active sites (although parameterized using only a small subset of the PDB). Predictions are not generally tuned for specific active sites [20]. Here we describe an Inductive Logic Programming (ILP)/ Relational Data Mining (RDM) approach for SBDD based on generalizing over examples of ligands bound to a specific active site.

The structural nature of many chemical structure-function/property relationships has proven to be well suited to Inductive Logic Programming (ILP) [17]. We take the name of ILP to generalise all work in ILP and the related field of Relational Data Mining. In drug design, ILP has been successfully applied to model structure-activity relationships (SAR). Here the task was to obtain rules that could predict biological activity or toxicity of compounds from their chemical structure[12,16]. ILP is based on *logical relations* and differs from standard chemoinformatics approaches that use *attributes* (molecular descriptor, molecular field, etc) to encode the chemical information. For such problems, logic provides a unified way of representing the relations between objects (atoms and bonds). ILP systems have progressively been shown to be capable of handling 1[12], 2[16] and 3 dimensional[8,21] descriptions of the molecular structures, allowing the development of compact and comprehendable theories. Moreover, ILP has achieved the same predictive power or has significantly improved the traditional QSAR (Quantitative SAR) built using standard propositional learners and statistical methods[15,16].

We take the next natural step in developing the ILP approach to drug design by extending it to SBDD. The aim of this study is four-fold:

- to explore how best to represent the relationship between ligand and protein and how to adapt the ILP tools to suit our study.
- to test whether ILP can form accurate quantitative models of the binding energy of ligands.
- to compare the ILP results with conventional 3D QSAR and SBDD programs.
- to examine the insight obtained from the ILP rules.

2   Methods and Materials

This section describes the complete process we employed to address our problem. The methodology adopted for this study is organized as follows: 1) collect 3D structural data from the PDB and their corresponding biological activities in the literature; 2) transform the molecular structures into facts from a molecular modelling package and extract the features of interest to build the background knowledge; 3) form 3D structural features (pharmacophores) using ILP; 4) form regression models using the pharmacophores and assess their predictive power.

2.1   Datasets

A complete description of the protein-ligand series is reported in section 3. While the PDB gathers most of the structural data of biomolecular systems, there is no unified way to distribute biological activities and structures directly to analysis methods. A preprocessing step is necessary to clean the PDB files: isolation of the ligand, addition of missing atoms or residues, removal of useless information, etc. Despite the fact that the way ILP encodes chemical information is less sensitive to the initial preparation of the complexes than other SBDD methods (protonation state for example), extra care was required to form the proper assignment of the atom types before building the Datalog program.

2.2   Background Knowledge and Its Representation

ILP systems use background knowledge to further describe problems. The background knowledge comprises our statements about the most relevant features to explain the biological activity. This mainly involves using the most comprehensive and the most declarative representation to encode domain-dependant information. The content of the background knowledge used for this study is illustrated in figure 2.

In our representation, the three dimensional information is expressed in terms of distances between atoms or structural groups (*building blocks*) giving the final rule a *pharmacophore* like form. The concepts of (3D)pharmacophore and pharmacophore elements are very important in medicinal chemistry: a pharmacophore is an arrangement of atoms or groups of atoms which influence drastically the activity at a target receptor[19]. Pharmacophore representation expresses the potential activity in a language familiar to medicinal chemists and

Predicates related to the ligand (all arity 3):
hacc,hdon,alcohol,equiv_ether,six_ring,hetero_non_ar_6_ring,amide,
carbonyl,amine_Oh,methyl,lipo_seg,ar_6c_ring,halogen,five_ring.
    Predicates related to the protein (all arity 4):
prot_backc2,prot_cooh,prot_alcohol,prot_negcharge,prot_poscharge,
prot_amide,prot_guadinium,prot_lipo_seg.
    Hydrogen bonding predicate: hb/4.
    Water position predicate: water/3.

Fig. 2. General chemical knowledge defined in the background knowledge.

is easily convertible for searching compounds in chemical databases. A pharmacophore usually refers to the ligand only but, in the following, we apply this definition to the active site as well.

The Prolog implementation requires facts that store the location of particular groups and a predicate *dist/4* which states the Euclidean distance in 3D space between two groups. For example, the following conjunction,

hdonor(l10,1,A),methyl(l10,1,B),dist(A,B,6.3,1.0)

represents the fact that in the compound *l10* in its conformation labelled *1*, there are a methyl group *A* and hydrogen bond donor *B* separated by 6.3 ± 1.0 Angstroms.

Pharmacophore mapping with ILP avoids the need of traditional 3D QSAR and pharmacophore learning methods to prealign and superpose all the ligands to a common extrinsic coordinate system. The requirement is forced by the propositional nature of the traditional approaches[19]. ILP has the advantage that it can directly use the intrinsic coordinate system of each complex.

Some ligands may also have more than one conformation (3D structure). This is the problem which first highlighted the multiple instance problem, and most propositional machine learning algorithms require major changes to deal with it [13]. ILP has the advantage that it can naturally deal with multiple instance problems.

Only a brief summary of the predicates used for this study is presented here. A Prolog example of generating building blocks facts from molecular structure is illustrated in [21]. The pharmacophore elements, available for the present SAR analysis (figure 2) can be divided in the following two categories:

– Ligand related predicates state the position in 3D space of simple or complex chemical groups providing, for example, the definition of methyl group or aromatic rings. They can also encode some important physico-chemical properties of the atoms or the building blocks, such as their ability to form hydrogen bonds.
– The active site is described by integrating specific chemical knowledge related to a number of important amino acids and water molecules as well as representing hydrogen bonds(*hb/4*) explicitly.

2.3 Constructing Theories with Aleph

The learning algorithm used for this study is the ILP system Aleph[25]. This algorithm follows the classic ILP search engine framework[6]: given a background knowledge (i.e. relations describing the molecular structures), a set of examples (i.e. training data) and a language specification for hypotheses, an ILP system will attempt to find a set of rules that explain the examples using the background knowledge provided. We chose Aleph because it can be easily tuned to suit our learning system which proceeds by iterating through the following basic algorithm:

- The training data is formed by dichotomising the data into two sets (positive and negative) based on their biological activity. Because there is not a natural cut-off to the predictor, an example is chosen from the training data and the positive set comprises the molecules with the closest activity (1/3 of the training data are used in this study). The rest of the examples (2/3 of the molecules) are considered as negative examples.
- The most specific clause (*bottom clause*) that entails the above example is then constructed within the language restrictions provided[22]. This is known as the saturation step. The bottom clause prunes the search before it begins by identifying all the potential clauses explaining the activity of the selected molecule.
- The search is a refinement graph search: it proceeds along the space of clauses (partially ordered by Θ-subsumption) between the specific hypothesis (*bottom clause*) and the most general clause (*empty body*)[25]. We require a complete search in order to find all the possible pharmacophores consistent with the data.
- The new clause is added to the theory and the search is repeated until all the examples are saturated once. Pharmacophores are, thus, learnt for both highly and less active compounds. This contrasts with the usual ILP framework where all examples made redundant are removed (*cover removal step*[25]). Our aim is to use the rules as indicator variables to build quantitative models and the compactness will be assured by the model rather than by the ILP process.

2.4 Building QSAR Models

To combine the ILP pharmacophore into a regression model we used a vanilla in-house multiple linear regression program. The predictive power of the model was evaluated using leave-one-out cross-validation (involving the ILP and regression steps). The results are presented using the squared correlation coefficient (R_{cv}^2) between the actual and the predicted value of the activity. This is the standard measure in drug design. In the following, the activity is evaluated in logarithm units of the inhibition constant ($log(1/Ki)$).

We compare the results of the ILP models with the use of two conventional drug design approaches, CoMFA and a SBDD scoring function.

CoMFA (Comparative Molecular Field Analysis[5]) is the most commonly used 3D QSAR ligand-based approach[18]. The basic idea of CoMFA is to superimpose ligands onto a common 3D grid, and then sample their electronic structure at regular points (voxols). This has the benefit of transforming the data into a propositional form, but relies on the (often false) assumption that every molecule in the series interacts with the target molecule and in the same way (*common receptor assumption*)[18]. It can also be difficult to know how best to superimpose molecules that do share much common structure. CoMFA also has the drawback of producing thousands of correlated attributes which requires the powerful PLS regression approach to avoid overfitting. In CoMFA, neighbouring voxol attributes are generally highly correlated, yet this information is thrown away. PLS can be used to partially regenerate this correlated structure. In the following, we present the CoMFA analysis using the observed ligand conformation in the protein-ligand complex (common receptor assumption) within an optimized molecular field (superposition/translation).

As no general scoring function has been reported to date that is able to predict binding affinities with a high degree of accuracy[10], we present results with the most accurate approach, for each series under study, among five functions available in the CScore module of Sybyl[1] to compare models including information on the active site.

3   Results and Discussion

We report results obtained from our approach on two protein targets: the glycogen phosphorylase *b* (GP) and the human immunodeficiency virus protease (HIV-PR) enzymes. Chemical structures, inhibition data and predicted biological activities can be accessed from
http://www.aber.ac.uk/compsci/Research/bio/dss/.

We chose to study GP and HIV-PR because: a significant amount of 3D information is available on them in the PDB, allowing an accurate validation of the method; they have already been extensively studied, giving us the opportunity to verify the meaning of the rules found by Aleph, and comparable published models; the two datasets stand at two extreme points in SBDD problems. The GP dataset is an homogeneous series of 3D structures with only slight modifications of the structure of ligands. This contrasts with the HIV-PR dataset where the structures of the inhibitor, and to a lesser extent the protein sequence, exhibit dramatic changes from one complex to the next.

3.1   Glycogen Phosphorylase *b*

The set of 51 co-crystallized inhibitors of the glycogen phosphorylase *b* has been taken from the same SBDD project[23]. In this case, the chemical structure of the GP inhibitors is homogeneous; meeting then the usual requirements of traditional 2D/3D QSAR (common receptor assumption). However, the CoMFA[5] analysis on the 51 inhibitors leads to a poor predictive power (r_{cv}^2=0.46, table 1). One would have thought that we should have been able to derive more physical properties characterising ligand-receptor interaction but the best structure-based binding energy function accuracy is only r_{cv}^2=0.34 (FlexX[24], table 1).

Table 1. Models accuracies from the GP dataset.

Id.	Method	Accuracy (r_{cv}^2)
1	CoMFA	0.46
2	FlexX	0.36
3	ILP: Ligand only	0.66
4	ILP: Ligand + *water/3*	0.74
	+ H-bonds involving ligand and water	

In the case of GP, ligands bind at the catalytic site buried deeply from the surface of the enzyme and they stabilize an inactive form of the protein mainly through specific hydrophilic interactions with the protein and some water molecules. Water molecules are well known to play a significant role in stabilizing protein-ligand complexes but they remain a challenge for many QSAR analyses as their mobility violates the common receptor assumption. Table 1 also shows a comparison between results where the background knowledge contains facts only related to the ligand and where the background knowledge also contains facts related to the water molecule position and all the possible hydrogen bonds between the ligand, the active site and the molecules of water.

The results show that our ILP approach outperforms CoMFA and FlexX ($P < 0.005$ for both cases). Addition of more informative knowledge regarding the active site improves the predictive power of the model ($P < 0.025$). The results demonstrate the need to explicitly include hydrophilic interactions in forming a good predictor. The addition of the protein and water interaction also makes the interpretation of the model easier, as they highlight the most important features involved in the binding (see below). The resulting theory and QSAR model are reported in figure 3. The first three (pharmacophores) rules P1, P2 and P3 are overlaid with a highly active ligand to illustrate the main features found by the hypothesis on the same figure. Taking into account the relative homogeneity of the inhibitors, a close inspection of the rules found by Aleph in experiment 3 found that all the key chemical groups are involved in the final model. As shown in figure 3, ILP globally simplifies the interpretation. Insight into the binding mechanism is outlined in two points:

- The amide group in the region **2** is a constant in the three rules (*amide/3*), acting, though, as the basis for the construction of the three pharmacophores. This not surprising as this group is associated with the high activity of the series. Due to the high number of possible interactions in the region **1** and **3**, the theory involves OH groups (*alcohol/3*, rules P2 and P3) rather than explicit hydrogen bonds.
- The most surprising feature denoted by our method is related to the distal part (region **4**) of the active site. Most rules involve either the position of

```
P1 : active(A) :-
hb(A,B,C,D),carbonyl(A,B,E),amide(A,B,F),dist(A,F,E,1.35,1.0),
dist(A,C,E,9.47,1.0),dist(A,D,E,10.77,1.0),dist(A,C,F,10.74,1.0),
dist(A,D,F,11.95,1.0).
P2 : active(A) :-
water(A,B,C),alcohol(A,B,D),alcohol(A,B,E),amide(A,B,F),
dist(A,C,D,14.56,1.0),dist(A,C,E,13.12,1.0),dist(A,D,E,5.98,1.0),
dist(A,F,D,4.63,1.0),dist(A,F,E,3.00,1.0),dist(A,C,F,11.12,1.0).
P3 : active(A) :-
water(A,B,C),water(A,B,D),alcohol(A,B,E),amide(A,B,F),
dist(A,C,D,4.83,1.0),dist(A,C,E,13.69,1.0),dist(A,D,E,14.38,1.0),
dist(A,F,E,4.80,1.0),dist(A,C,F,9.29,1.0),dist(A,D,F,9.75,1.0).
P4 : active(A) :-
water(A,B,C),alcohol(A,B,D),methylen(A,B,E),equiv_ether(A,B,F),
dist(A,C,D,12.50,1.0),dist(A,E,D,4.42,1.0),dist(A,C,E,8.72,1.0),
dist(A,C,F,10.03,1.0),dist(A,D,F,3.01,1.0),dist(A,E,F,1.84,1.0).
```

QSAR model : log(1/Ki) = 2.43 + 0.76*P1 + 0.91*P2 + 0.35*P3 - 0.49*P4

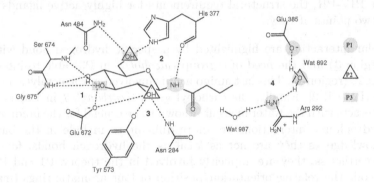

Fig. 3. Theory from experiment 4, table 1 (top). 2D representation of the interaction involved in the binding of the ligand (numbered **26** in [23]) found by our ILP approach (bottom). Shaded circles/rectangles and open triangle outline the pharmacophore elements involved in the theory. Intermolecular interactions between the inhibitor and the binding site are represented with dashed lines.

two water molecules or an explicit hydrogen bond interaction with Arg292 (*water/3* and *hb/4*). How could these interactions be involved in the binding process? We found that [4] suggested that the presence of water overlapping this region could explain a high inhibitory effect with a strong stabilization of the enzyme in the 280's loop.

3.2 Human Immunodeficiency Virus Protease

The second set concerns a series of inhibitors of the well studied human immunodeficiency virus protease. In this case, we are dealing with a series of diverse ligands, some inhibitors are present in two conformations and some residues in

Table 2. Models accuracies from the HIV-PR dataset.

Id.	Method	Accuracy (r_{cv}^2)
1	CoMFA	0.58
2	ChemScore	0.35
3	ILP: Ligand only	0.62
4	ILP: Ligand + Active site	0.75
	+ H-bonds involving the ligand + *water/3*	

the protein may be mutated (i.e. the sequence of amino-acids can differ from one structure to the next). The same process as for GP is reported in table 2.

In this case, the ILP structure based model (r_{cv}^2=0.75) improves on the CoMFA (r_{cv}^2=0.58, $P < 0.05$) and the scoring function ChemScore[7] (r_{cv}^2=0.35, $P < 0.001$) prediction of the binding energy. The theory from experiment 4 (table 2) is reported in figure 4. The first three rules P1, P2 and P3 are mapped onto the highest active inhibitor (PDB code: **1hvj**).

For HIV-PR, the structural requirements for highly active ligands can seen upon two points of view:

- Polar interaction are highlighted by a specific hydrogen bond with Asp29 (region **3**) and the need of a group (*alcohol/3* in P3) able to interact with Asp25 (region **1**). This last amino acid is involved in the catalytic mechanism of HIV-PR[2]. Finally, the carbonyl group (*carbonyl/3* in P3) in region **2** interacts with the water molecule known to be crucial for the binding process.
- Hydrophobic interactions are more difficult to include in the background knowledge as they are not as local as the hydrogen bonds, for example. Nevertheless, they are implicitly involved in the theory. P1 and P2 largely encode the relative orientation/position of four aromatic rings (mapped by *lipo_seg/3* and *six_ring/3*). The hydrophobic behaviour (*prot_lipo_seg/3*) of the residues 81 and 84 (regions **4** and **5**) are revealed to be important to ensure these non polar contacts.

4   Conclusions

We have presented a new procedure for the formulation of accurate and easily interpretable QSARs to predict binding energy within a series of protein-ligand complexes. This extends the application of ILP in drug design to problems where the structure of the binding protein is known. To form the models we used a relational description of the molecular structure to find rules in the form of pharmacophores, and linear regression to combine the pharmacophores into a predictive model. We consider that the ILP approach was effective for the following reasons:

- the logical formalism is an effective representation for the diverse types of knowledge required.

```
P1 : active(A) :-
hb(A,B,C,D),lipo_seg(A,B,E),six_ring(A,B,F),dist(A,C,E,5.49,1.0),
dist(A,C,F,5.31,1.0),dist(A,D,E,7.22,1.0),dist(A,D,F,7.77,1.0).
P2 : active(A) :-
lipo_seg(A,B,C),prot_lipo_seg(A,B,84,D),six_ring(A,B,E),
dist(A,C,D,5.93,1.0),dist(A,C,E,4.88,1.0),dist(A,D,E,9.19,1.0).
P3 : active(A) :-
alcohol(A,B,C),carbonyl(A,B,D),prot_lipo_seg(A,B,81,E),
dist(A,C,D,5.30,1.0),dist(A,C,E,11.54,1.0),dist(A,D,E,8.67,1.0).
P4 : active(A) :-
carbonyl(A,B,C),pos_charge(A,B,D),prot_negcharge(A,B,29,E),
dist(A,C,D,9.23,1.0),dist(A,C,E,6.09,1.0),dist(A,D,E,9.61,1.0).
```

QSAR model : log(1/Ki) = 8.00 + 0.81*P1 + 0.43*P2 + 0.58*P3 - 0.90*P4

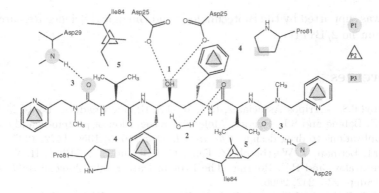

Fig. 4. Theory from experiment 4, table 2 (top). 2D representation of the interaction involved in the binding of **1hvj** found by our ILP approach. The same notation as in figure 3 is adopted.

- the coordinates of molecular structures can be used directly without the superposition or prealignment prior to some traditional approaches.
- ILP deals naturally with the multiple instances problem and can find all possible pharmacophore consistent with the background.
- the theories generated are compact and comprehensible in a language familiar to scientists.

We have tested this approach on two qualitatively different datasets. In both examples, the ILP models outperformed and yet were of equal complexity to the results of traditional SBDD approaches. The ILP models were directly interpretable by mapping the learned pharmacophore onto selected examples, and these interpretations were consistent with previous reported analysis. The derivation of so-called receptor-based pharmacophore does not only improve the predictive power of the models but allows the identification of key interaction *hotspots*. In the case of GP, ILP has brought an unexpected insight into the binding mechanism. Analysis of HIV-PR hypotheses shows that our approach could deal with heterogeneous series of protein-ligand. Here, we used direct information from the experimentally resolved structure of a similar protein-ligand complex to give the clues to whereabouts in the active site the ligand binds and in what conformation. Work is in progress to evaluate the applicability of our approach when such information is unavailable or insufficient. Flexible docking techniques can be used to explore the conformational space of the ligand within the active site leading to a highly diverse docking solution set: either our ILP models can be used to restrict the search space[11] or pharmacophores can be learnt from the docking set.

Acknowledgments

DPE was supported by the Biotechnology and Biological Science Research Council (grant no 2/B11471).

References

1. Sybyl 6.8. - Tripos Associates, Inc., 1699 S. Hanley Road, St. Louis, MO.
2. R.E. Babine and S.L. Bender. Molecular recognition of protein-ligand complexes: Applications to drug design. *Chemical Reviews*, 97(5):1359–1472, 1997.
3. H.M. Berman, J. Westbrook, Z. Feng, G. Gilliland, T.N. Bhat, H. Weissig, I.N. Shindyalov, and P.E. Bourne. The Protein Data Bank. *Nucleic Acids Research*, 28(Supp):235–242, 2000.
4. C.J.F. Bichard, E.P. Mitchell, M.R. Wormald, K.A. Watson, L.N. Johnson, S.E. Zographos, D.D. Koutra, N.G. Oikonomakos, and Fleet G.W.J. Potent inhibition of glycogen phosphorylase by a spirohydantoin of glucopyranose: first pyranose analogues of hydantocidin. *Tetrahedron Letters*, 36:2145–2148, 1995.
5. R.D. Cramer, D.E. Patterson, and J.D. Bunce. Comparative molecular field analysis (CoMFA). 1. Effect of shape on binding of steroids to carrier proteins. *Journal of the American Chemical Society*, 110(18):5959–5967, 1988.
6. S. Dzeroski and N. Lavrac. An introduction to inductive logic programming. In Dzeroski S. and Lavrac N., editors, *Relational Data Mining*, pages 28–73. Springer-Verlag, 2001.
7. M.D. Eldridge, C.W. Murray, T.R. Auton, G.V. Paolini, and R.P. Mee. Empirical scoring functions: I. The development of a fast empirical scoring function to estimate the binding affinity of ligands in receptor complexes. *Journal of Computer-Aided Molecular Design*, 11(5):425–445, 1997.
8. P.W. Finn, S. Muggleton, D. Page, and A. Srinivasan. Pharmacophore discovery using the inductive logic programming system PROGOL. *Machine Learning*, 30(2-3):241–270, 1998.
9. I. Halperin, B. Ma, H. Wolfson, and R. Nussinov. Principles of docking: An overview of search algorithms and a guide to scoring functions. *Proteins*, 47(4):409–443, 2002.
10. S. Ha, R. Andreani, A. Robbins, and I. Muegge. Evaluation of docking/scoring approaches: a comparative study based on MMP3 inhibitors. *Journal of Computer-Aided Molecular Design*, 14(5):435–448, 2000.

11. S.A. Hindle, M. Rarey, C. Buning, and T. Lengauer. Flexible docking under pharmacophore type constraints. *Journal of Computer-Aided Molecular Design*, 16(2):129–149, 2002.
12. J.D. Hirst, R.D. King, and M.J.E. Sternberg. Quantitative structure-activity relationships by neural networks and inductive logic programming. I. The inhibition of dihydrofolate reductase by pyriminides. *Journal of Computer-Aided Molecular Design*, 8(4):405–420, 1994.
13. A.N. Jain, K. Koile, and D. Chapman. Compass: predicting biological activities from molecular surface properties. performance comparisons on a steroid benchmark. *Journal of Medicinal Chemistry*, 37(15):2315–2327, 1994.
14. D. Joseph-McCarthy. Computational approaches to structure-based ligand design. *Pharmacology and Therapeuthics*, 84(2):179–191, 1999.
15. R.D. King, S. Muggleton, R. Lewis, and M.J.E Sternberg. Drug design by machine learning: The use of inductive logic programming to model the structure-activity relationships of trimethoprim analogues binding to dihydrofolate reductase. *Proceedings of the National Academy of Sciences of the USA*, 89(23):11322–11326, 1992.
16. R.D. King, S. Muggleton, A. Srinivasan, and M.J.E. Sternberg. Structure-activity relationships derived by machine learning: The use of atoms and their bond connectivities to predict mutagenicity by inductive logic programming. *Proceedings of the National Academy of Sciences*, 93:438–442, 1996.
17. R.D. King and A. Srinivasan. Relating chemical activity to structure: An examination of ILP successes. *New Generation Computing Special issue on Inductive Logic Programming*, 13(3-4):411–434, 1995.
18. H. Kubinyi. *3D QSAR in drug design. Theory methods and application.* Kluwer, Dordrecht, 1997.
19. T. Liljefors and I. Pettersson. Computer-aided development and use of three dimensional pharmacophore. In P. Krogsgaard-Larsen, U. Madsen, and T. Liljefors, editors, *A Textbook of Drug Design and Development*, pages 86–116. Taylor and Francis, London, 2002.
20. A. Logean, A. Sette, and D. Rognan. Customized versus universal scoring functions: application to class I MHC-peptide binding free energy predictions. *Bioorganic Medicinal Chemistry Letters*, 11(5):675–679, 2001.
21. N. Marchand-Geneste, K.A. Watson, B.K. Alsberg, and R.D. King. New approach to pharmacophore mapping and QSAR analysis using inductive logic programming. Application to thermolysin inhibitors and glycogen phosphorylase *b* inhibitors. *Journal of Medicinal Chemistry*, 44(18):2861–2864, 2001.
22. S. Muggleton. Inverse entailment and Progol. *New Generation Computing, Special issue on Inductive Logic Programming*, 13(3-4):245–286, 1995.
23. M. Pastor, G. Cruciani, and K.A. Watson. A strategy for the incorporation of water molecules present in a ligand-binding site into a 3D-QSAR analysis. *Journal of Medicinal Chemistry*, 40(25):4089–4102, 1997.
24. M. Rarey, B. Kramer, T. Lengauer, and G. Klebe. A fast flexible docking method using an incremental construction algorithm. *Journal of Molecular Biology*, 261(3):470–489, 1996.
25. A. Srinivasan. Aleph: A Learning Engine for Proposing Hypotheses http://web.comlab.ox.ac.uk/oucl/research/areas/machlearn/Aleph/aleph.pl.

Visualizing Class Probability Estimators

Eibe Frank and Mark Hall

Department of Computer Science
University of Waikato
Hamilton, New Zealand
{eibe,mhall}@cs.waikato.ac.nz

Abstract. Inducing classifiers that make accurate predictions on future data is a driving force for research in inductive learning. However, also of importance to the users is how to gain information from the models produced. Unfortunately, some of the most powerful inductive learning algorithms generate "black boxes"—that is, the representation of the model makes it virtually impossible to gain any insight into what has been learned. This paper presents a technique that can help the user understand why a classifier makes the predictions that it does by providing a two-dimensional visualization of its class probability estimates. It requires the classifier to generate class probabilities but most practical algorithms are able to do so (or can be modified to this end).

1 Introduction

Visualization techniques are frequently used to analyze the input to a machine learning algorithm. This paper presents a generic method for visualizing the output of classification models that produce class probability estimates. The method has previously been investigated in conjunction with Bayesian network classifiers [5]. Here we provide details on how it can be applied to other types of classification models.

There are two potential applications for this technique. First, it can help the user understand what kind of information an algorithm extracts from the input data. Methods that learn decision trees and sets of rules are popular because they represent the extracted information in intelligible form. This is not the case for many of the more powerful classification algorithms. Second, it can help machine learning researchers understand the behavior of an algorithm by analyzing the output that it generates. Standard methods of assessing model quality—for example, receiver operating characteristic (ROC) curves [4]—provide information about a model's predictive performance, but fail to provide any insight into why a classifier makes a particular prediction.

Most existing methods for visualizing classification models are restricted to particular concept classes. Decision trees can be easily visualized, as can decision tables and naive Bayes classifiers [8]. In this paper we discuss a general visualization technique that can be applied in conjunction with any learning algorithm for

N. Lavrač et al. (Eds.): PKDD 2003, LNAI 2838, pp. 168–179, 2003.

classification models as long as these models estimate class probabilities. Most learning algorithms fall into this category[1].

The underlying idea is very simple. Ideally we would like to know the class probability estimate for each class for every point in instance space. This would give us a complete picture of the information contained in a classification model. When there are two attributes it is possible to plot this information with arbitrary precision for each class in a two-dimensional plot, where the color of the point encodes the class probability (e.g. black corresponds to probability zero and white corresponds to probability one). This can easily be extended to three dimensions. However, it is not possible to use this simple visualization technique if there are more than three attributes.

This paper presents a data-driven approach for visualizing a classifier regardless of the dimension of the instance space. This is accomplished by projecting its class probability estimates into two dimensions. It is inevitable that some information will be lost in this transformation but we believe that the resulting plotting technique is a useful tool for data analysts as well as machine learning researchers. The method is soundly based in probability theory and aside from the classifier itself only requires an estimate of the attributes' joint density (e.g. provided by a kernel density estimator).

The structure of the paper is as follows. Section 2 describes the visualization technique in more detail. Section 3 contains some experimental results. Section 4 discusses related work, and Section 5 summarizes the paper.

2   Visualizing Expected Class Probabilities

The basic idea is to visualize the information contained in a classification model by plotting its class probability estimates as a function of two of the attributes in the data. The two attributes are user specified and make up the x and y axes of the visualization. In this paper we only consider domains where all the attributes are numeric. We discretize the two attributes so that the instance space is split into disjoint rectangular regions and each region corresponds to one pixel on the screen. The resulting rectangles are open-sided along all other attributes. Then we estimate the expected class probabilities in each region by sampling points from the region, obtaining class probability estimates for each point from the classification model, and averaging the results. The details of this method are explained below.

The probability estimates for each region are color coded. We first assign a color to each class. Each of these colors corresponds to a particular combination of RGB values. Let (r_k, g_k, b_k) be the RGB values for class k, i.e. if class k gets probability one in a given region, this region is colored using those RGB values. If no class receives probability one, the resulting color is computed as a linear combination of all the classes' RGB values. Let $\hat{e}_k$ be the estimated expected

---

[1] Note that the technique can also be used in conjunction with clustering algorithms that produce cluster membership probabilities.

probability for class k in a particular region. Then the resulting RGB values are computed as follows:

$$r = \sum_k \hat{e}_k \times r_k \qquad g = \sum_k \hat{e}_k \times g_k \qquad b = \sum_k \hat{e}_k \times b_k \qquad (1)$$

This method smoothly interpolates between pure regions—regions where one class obtains all the probability mass.

The class colors can be chosen by the user based on a standard color chooser dialog. For example, when there are only two classes, the user might choose black for one class, and white for the other, resulting in a grey-scale image. Note that the colors will not necessarily uniquely identify a probability vector. In a four-class problem setting the corresponding colors to $(1,0,0), (0,1,0), (0,0,1)$, and $(0,0,0)$ will result in a one-to-one mapping. However, with other color settings and/or more classes there may be clashes. To alleviate this problem our implementation shows the user the probability vector corresponding to a certain pixel on mouse over. It also allows the user to change the colors at any time, so that the situation in ambiguous regions can be clarified.

We now discuss how we estimate the expected class probabilities for each pixel (i.e. each rectangular region in instance space). If the region is small enough we can assume that the density is approximately uniform within the region. In this case we can simply sample points uniformly from the region, obtain class probability estimates for each point from the model, and average the results. However, if the uniformity assumption does not hold we need an estimate $\hat{f}$ of the density function—for example, provided by a kernel density estimator [2]—and sample or weight points according to this estimate. Using the density is crucial when the method is applied to instance spaces with more than two dimensions (i.e. two predictor attributes) because then the uniformity assumption is usually severely violated.

Given a kernel density estimator $\hat{f}$ we can estimate the expected class probabilities by sampling instances from a region using a uniform distribution and weighting their predicted class probabilities $\hat{p}_k$ according to $\hat{f}$. Let $S = (\mathbf{x}_1, ..., \mathbf{x}_l)$ be our set of l uniformly distributed samples from a region. Then we can estimate the expected class probability $\hat{e}_k$ of class k for that region as follows:

$$\hat{e}_k = \frac{\sum_{\bullet \in S} \hat{f}(\mathbf{x}) \hat{p}_k(\mathbf{x})}{\sum_{\bullet \in S} \hat{f}(\mathbf{x})}. \qquad (2)$$

If there are only two dimensions this method is quite efficient. The number of samples required for an accurate estimate could be determined automatically by computing a confidence interval for it, but the particular choice of l is not critical if the screen resolution is high enough (considering the limited resolution of the color space and the sensitivity of the human eye to local changes in color).

Unfortunately this estimation procedure becomes very inefficient in higher-dimensional instance spaces because most of the instances in S will receive a very low value from the density function: most of the density will be concentrated in specific regions of the space. Obtaining an accurate estimate would require a very large number of samples. However, it turns out that there is a more efficient sampling strategy for estimating $\hat{e}_k$. This strategy is based on the kernel density estimator that we use to represent $\hat{f}$.

A kernel density estimator combines local estimates of the density based on each instance in a dataset. Assuming that there are n instances $\mathbf{x}_i$ it consists of n kernel functions:

$$\hat{f}(\mathbf{x}) = \frac{1}{n} \sum_{i=1}^{n} k_i(\mathbf{x}) \tag{3}$$

where k_i is the kernel function based on instance $\mathbf{x}_i$:

$$k_i(\mathbf{x}) = \prod_{j=1}^{m} k_{ij}(x_j). \tag{4}$$

This is a product of m component functions, one for each dimension. We use a Gaussian kernel, for which the component functions are defined as:

$$k_{ij}(x_j) = \frac{1}{\sqrt{2\pi}\sigma_{ij}} \exp\left(-\frac{(x_j - x_{ij})^2}{2\sigma_{ij}^2}\right). \tag{5}$$

Each k_{ij} is the density of a normal distribution centered on attribute value j of instance $\mathbf{x}_i$. The parameter σ_{ij} determines the width of the kernel along dimension j. In our implementation we use $\sigma_{ij} = (max_j - min_j) \times d_i$, where max_j and min_j are the maximum and minimum value of attribute j, and d_i is the Euclidean distance to the k-th neighbor of $\mathbf{x}_i$ after all the attributes' values have been normalized to lie between zero and one. The value of the parameter k is user specified. Alternatively it could be determined by maximizing the cross-validated likelihood of the data [7].

Based on the kernel density estimator we can devise a sampling strategy that produces a set of instances Q by sampling a fixed number of instances from each kernel function. This can be done by sampling from the kernel's normal distributions to obtain the attribute values for each instance. The result is that the instances in Q are likely to be in the populated areas of the instance space. Given Q we can estimate the expected class probability for a region R as follows:

$$\hat{e}_k = \frac{1}{|\mathbf{x} \in R \wedge \mathbf{x} \in Q|} \sum_{\bullet \in R \wedge \bullet \in Q} \hat{p}_k(\mathbf{x}). \tag{6}$$

Unfortunately this is not the ideal solution: for our visualization we want accurate estimates for every pixel, not only the ones corresponding to populated parts of the instance space. Most regions R will not receive any samples. The solution is to split the set of attributes into two subsets: the first set containing the two attributes our visualization is based on, and the second set containing the remainder. Then we can fix the values of the attributes in the first set so that we are guaranteed to get an instance in the area that we are interested in (corresponding to the current pixel), sample values for the other attributes from a kernel, and use the fixed attributes to weight the resulting instance according to the density function.

Let us make this more precise by assuming that our visualization is based on the first two attributes in the dataset. For these two attributes we fix values x_1 and x_2 in the region corresponding to the pixel that we want to plot. Then we obtain an instance $\mathbf{x}_i$ from kernel k_i by sampling from the kernel's normal distributions to obtain attribute values $x_{i3}, ..., x_{im}$ and setting $x_{i1} = x_1$ and $x_{i2} = x_2$. We can then estimate the class probability $p_k(x_1, x_2)$ for location x_1, x_2 as follows:

$$\hat{p}_k(x_1, x_2) = \frac{\sum_{i=1}^n \hat{p}_k(\mathbf{x}_i) k_{i1}(x_1) k_{i2}(x_2)}{\sum_{i=1}^n k_{i1}(x_1) k_{i2}(x_2)}. \tag{7}$$

This is essentially the likelihood weighting method used to perform probabilistic inference in Bayesian networks [6]. The $p_k(\mathbf{x}_i)$ are weighted by the $k_{i1}(x_1) k_{i2}(x_2)$ to take the effect of the kernel on the two fixed dimensions into account. The result of this process is that we have marginalized out all dimensions apart from the two that we are interested in.

One sample per kernel is usually not sufficient to obtain an accurate representation of the density and thus an accurate estimate of $p_k(x_1, x_2)$, especially in higher-dimensional spaces. In our implementation we repeat the sampling process r^{m-2} times, where r is a user-specified parameter, evaluate Equation 7 for each resulting set of instances, and take the overall average as an estimate of $p_k(x_1, x_2)$. A more sophisticated approach would be to compute a confidence interval for the estimated probability and to stop the sampling process when a certain precision has been attained.

Note that the running time can be decreased by first sorting the $\mathbf{x}_i$ according to their weights $k_{i1}(x_{i1}) k_{i2}(x_{i2})$ and then sampling from the corresponding kernels in decreasing order until the cumulative weight exceeds a certain percentage of the total weight (e.g. 99%). Usually only a small fraction of the kernels need to be sampled from as a result of this filtering process.

To obtain the expected class probability $\hat{e}_k$ for a region corresponding to a particular pixel we need to repeat this estimation process for different locations x_{l1}, x_{l2} within the pixel and compute a weighted average of the resulting probability estimates based on the density function:

$$\hat{e}_k = \frac{\sum_l \hat{f}(x_{l1}, x_{l2}) \hat{p}_k(x_{l1}, x_{l2})}{\sum_l \hat{f}(x_{l1}, x_{l2})}, \tag{8}$$

where

$$\hat{f}(x_{l1}, x_{l2}) = \frac{1}{n} \sum_{i=1}^n k_{i1}(x_{l1}) k_{i2}(x_{l2}). \tag{9}$$

This weighted average is then plugged into Equation 1 to compute the RGB values for the pixel.

3   Some Example Visualizations

In the following we visualize class probability estimators on three example domains. We restrict ourselves to two-class domains so that all probability vectors
can be represented by shades of grey. Some color visualizations are available
online at http://www.cs.waikato.ac.nz/ml/weka/bvis. Note that our implementation is included in the Java package weka.gui.boundaryvisualizer as
part of the Weka machine learning workbench [9] Version 3.3.6 and later[2].

For each result we used two locations per pixel to compute the expected
probability, set the parameter r to two (i.e. generating four samples per kernel
for a problem with four attributes), and used the third neighbor ($k = 3$) for
computing the kernel width.

The first domain is an artificial domain with four numeric attributes. Two
of the attributes are relevant (x_1 and x_2) and the remaining ones (x_3 and x_4)
are irrelevant. The attribute values were generated by sampling from normal
distributions with unit variance. For the irrelevant attributes the distributions
were centered at zero for both classes. For the relevant ones they were centered
at -1 for class one and $+1$ for class two. We generated a dataset containing 100
instances from each class.

We first built a logistic regression model from this data. The resulting model
is shown in Figure 1. Note that the two irrelevant attributes have fairly small coefficients, as expected. Figure 2 shows the results of the visualization procedure
for three different pairs of attributes based on this model. The points super-
imposed on the plot correspond to the actual attribute values of the training
instances in the two dimensions visualized. The color (black or white) of each
point indicates the class value of the corresponding instance.

Figure 2a is based on the two relevant attributes (x_1 on the x axis and
x_2 on the y axis). The linear class boundary is clearly defined because the two
visualization attributes are the only relevant attributes in the dataset. The lower
triangle represents class one and the upper triangle class two. Figure 2b shows
the result for x_1 on the x axis, and x_3 on the y axis. It demonstrates visually
that x_1 is relevant while x_3 is not. Figure 2c displays a visualization based on
the two irrelevant attributes. It shows no apparent structure—as expected for
two completely irrelevant attributes.

Figure 4 shows visualizations based on the same pairs of attributes for the
decision tree from Figure 3. The tree is based exclusively on the two relevant
attributes, and this fact is reflected in Figure 4a: the area is divided into rectangular regions that are uniformly colored (because the probability vectors are
constant within each region). Note that the black region corresponds to three
separate leafs and that one of them is not pure. The difference in "blackness" is
not discernible.

Figure 4b shows the situation for attributes x_1 (relevant) and x_3 (irrelevant).
Attribute x_1 is used twice in the tree, resulting in three distinct bands. Note that

---

[2] Available from http://www.cs.waikato.ac.nz/ml/weka.

$$p(class = one|\mathbf{x}) = \frac{1}{1+e^{4.77x_1+4.21x_2-0.15x_3-0.14x_4-1.05}}$$

Fig. 1. The logistic regression model for the artificial dataset.

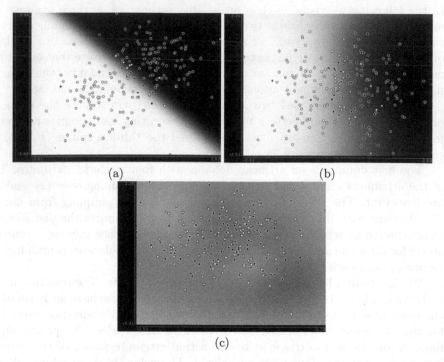

(a)   (b)

(c)

Fig. 2. Visualizing logistic regression for the artificial data using (a) the two relevant attributes, (b) one relevant and one irrelevant attribute, and (c) the two irrelevant attributes.

the three bands are (almost) uniformly colored, indicating that the attribute on the y axis (x_3) is irrelevant.

Figure 4c is based on the two irrelevant attributes. The visualization shows no structure and is nearly identical to the corresponding result for the logistic regression model shown in Figure 2c. Minor differences in shading compared to Figure 2c are due to differences in the class probability estimates that are caused by the two relevant attributes (i.e a result of the differences between Figures 4a and 2a).

For illustrative purposes Figure 6 shows a visualization for a two-class version of the iris data (using the 100 instances pertaining to classes iris-virginica and iris-versicolor) based on the decision tree in Figure 5. The iris data can be obtained from the UCI repository [1]. In Figure 6a the petallength attribute is shown on the x axis and the petalwidth attribute on the y axis. There are four uniformly colored regions corresponding to the four leaves of the tree. In Figure 6b petallength is on the x axis and sepallength on the y axis. The influence of sepallength is clearly visible in the white area despite

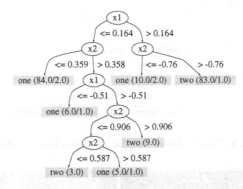

Fig. 3. The decision tree for the artificial dataset.

Fig. 4. Visualizing the decision tree for the artificial data using (a) the two relevant attributes, (b) one relevant and one irrelevant attribute, and (c) the two irrelevant attributes.

this attribute not being used in the tree. Figure 6c is based on `sepallength` (x) and `sepalwidth` (y). Although these attributes are not used in the tree the visualization shows a clear trend going from the lower left to the upper right, and a good correlation of the probability estimates with the actual class values of the training instances. This is a result of correlations that exist between `sepallength` and `sepalwidth` and the two attributes used in the tree.

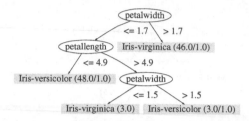

Fig. 5. The decision tree for the two-class iris dataset.

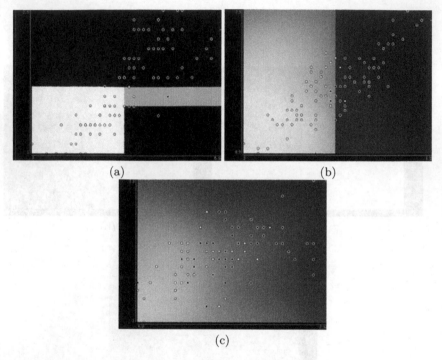

Fig. 6. Visualizing the decision tree for the two-class iris data using (a) `petallength` and `petalwidth`, (b) `petallength` and `sepallength`, and (c) `sepallength` and `sepalwidth` (with the first attribute on the x axis and the second one on the y axis).

This particular example shows that the pixel-based visualization technique can provide additional information about the structure in the data even when used in conjunction with an interpretable model like a decision tree. In this case it shows that the decision tree implicitly contains much of the information provided by the `sepallength` and `sepalwidth` attributes (although they are not explicitly represented in the classifier).

To provide a more realistic example Figure 8 shows four visualizations for pairs of attributes from the `pima-indians diabetes` dataset [1]. This dataset has eight attributes and 768 instances (500 belonging to class `tested_negative`

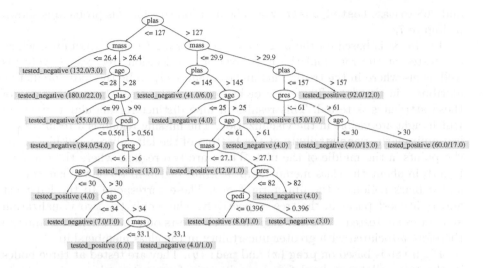

Fig. 7. The decision tree for the diabetes dataset.

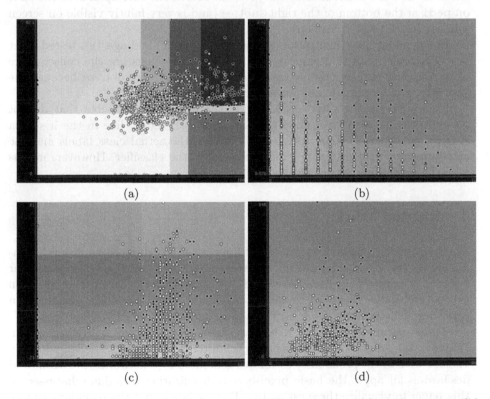

Fig. 8. Visualizing the decision tree for the diabetes data using (a) `plas` and `mass`, (b) `preg` and `pedi`, (c) `pres` and `age`, and (d) `skin` and `insu` (with the first attribute on the x axis and the second one on the y axis).

and 268 to class `tested_positive`). The decision tree for this problem is shown in Figure 7.

Figure 8a is based on the `plas` (x axis) and `mass` (y axis) attributes, which are tested at the root and the first level of the decision tree respectively (as well as elsewhere in the tree). This makes them likely to be the most predictive attributes in the data. There are eight nodes in the tree where either one of these attributes is tested. This results in nine distinct rectangular regions, of which eight are visible in the visualization. The missing region is a result of the last split on the `mass` attribute at the bottom of the left subtree and hidden by the points in the middle of the plot. There are two regions where the classifier is certain about the class membership: a white region in the lower left corner, and a black region in the upper right one. These correspond to the left-most and right-most paths in the tree respectively, where only the two visualization attributes are tested. All other regions involve tests on other attributes and are therefore associated with greater uncertainty in the class membership.

Figure 8b is based on `preg` (x) and `pedi` (y). They are tested at three nodes in the tree—all below level four—resulting in four rectangular regions. Only three of these are discernible in the plot; the fourth one corresponds to the split on `pedi` at the bottom of the right subtree (and is very faintly visible on screen but not in the printed version).

In Figure 8c the visualization is based on `pres` (x) and `age` (y), tested eight times in total. Note that some of the rectangular regions are the consequence of overlapping regions originating from splits in different subtrees because the subtrees arise by partitioning on non-visualized attributes.

Figure 8d visualizes the model using the only two attributes that do not occur in the decision tree: `skin` (x) and `insu` (y). Again, like in the iris data (Figure 6b), there is some correlation between the actual class labels and the attributes not explicitly taken into account by the classifier. However, in this case the correlation is very weak.

4   Related Work

There appears to have been relatively little attention devoted to general techniques for the visualization of machine learning models. Methods for particular types of classifiers have been developed, for example, for decision trees, decision tables, and naive Bayesian classifiers [8], but in these cases the visualization procedure follows naturally from the model structure.

The structure of Bayesian networks can be visualized as a directed graph. However, the graph-based visualization is limited because it does not provide any visualization of the probability estimates generated by a network. Rheingans and desJardins [5] apply the basic pixel-based visualization technique discussed in this paper to visualize these estimates. They indicate that the technique can be used in conjunction with other types of class probability estimators but do not provide details on how this can be done. Inference methods for Bayesian networks directly provide estimates of conditional probabilities based on evidence variables (in this case the two attributes used in the visualization), and this means a separate density estimator and sampling strategy is not required. Rheingans and desJardins also investigate a visualization technique that maps the instance space into two dimensions using a self-organizing map (SOM) [3]. However, this makes it difficult to relate a pixel in the visualization to a point in the original instance space.

5 Conclusions

This paper has presented a generic visualization technique for class probability estimators. The basic method is not new and has been investigated in the context of Bayesian network classifiers before. Our contribution is that we have provided details on how to generalize it to arbitrary classification models that produce class probability estimates. We have provided some example visualizations based on logistic regression and decision trees that demonstrate the usefulness of this method as a general tool for analyzing the output of learning algorithms. Potential applications are two fold: practitioners can use this tool to gain insight into the data even if a learning scheme does not provide an interpretable model, and machine learning researchers can use it to explore the behavior of a particular learning technique.

Acknowledgments

Many thanks to Len Trigg for pointing out that the method is applicable to probabilistic clustering algorithms, and to Geoff Holmes and Bernhard Pfahringer for their comments. This research was supported by Marsden Grant 01-UOW-019.

References

1. C.L. Blake and C.J. Merz. UCI repository of machine learning databases, 1998. [www.ics.uci.edu/~mlearn/MLRepository.html].
2. Trevor Hastie, Robert Tibshirani, and Jerome Friedman. *The Elements of Statistical Learning: Data Mining, Inference, and Prediction*. Springer-Verlag, 2001.
3. T. Kohonen. *Self-Organizing Maps*. Springer-Verlag, 1997.
4. Foster J. Provost and Tom Fawcett. Analysis and visualization of classifier performance: Comparison under imprecise class and cost distributions. *Knowledge Discovery and Data Mining*, pages 43–48, 1997.
5. Penny Rheingans and Marie desJardins. Visualizing high-dimensional predictive model quality. In *Proceedings of IEEE Visualization 2000*, pages 493–496, 2000.
6. Stuart Russell and Peter Norvig. *Artificial Intelligence*. Prentice-Hall, 1995.
7. P. Smyth. Model selection for probabilistic clustering using cross-validated likelihood. *Statistics and Computing*, pages 63–72, 2000.
8. Kurt Thearling, Barry Becker, Dennis DeCoste, Bill Mawby, Michel Pilote, and Dan Sommerfield. *Information Visualization in Data Mining and Knowledge Discovery*, chapter Visualizing Data Mining Models. Morgan Kaufmann, 2001.
9. Ian H. Witten and Eibe Frank. *Data Mining: Practical Machine Learning Tools and Techniques with Java Implementations*. Morgan Kaufmann, 2000.

Automated Detection of Epidemics from the Usage Logs of a Physicians' Reference Database

Jaana Heino[1] and Hannu Toivonen[2]

[1] National Public Health Institute, Helsinki, Finland
Department of Computer Science, University of Helsinki
jaana.heino@cs.helsinki.fi
[2] Department of Computer Science, University of Helsinki
hannu.toivonen@cs.helsinki.fi

Abstract. Epidemics of infectious diseases are usually recognized by an observation of an abnormal cluster of cases. Usually, the recognition is not automated, and relies on the alertness of human health care workers. This can lead to significant delays in detection. Since real-time data from the physicians' offices is not available. However, in Finland a Web-based collection of guidelines for primary care exists, and increases in queries concerning certain disease have been shown to correlate to epidemics. We introduce a simple method for automated online mining of probable epidemics from the log of this database. The method is based on deriving a smoothed time series from the data, on using a flexible selection of data for comparison, and on applying randomization statistics to estimate the significance of findings. Experimental results on simulated and real data show that the method can provide accurate and early detection of epidemics.

1 Introduction

The usual way of recognizing an infectious disease epidemic is through an observation of clustering (temporal, geographical, or both) of new cases by someone involved either with the diagnosis of patients or a disease registry. When the epidemic is widely spread – either spatially, temporally, or both – it might be very difficult or outright impossible for a single individual on the field to notice the change, and registries only receive notification after the patient has met a physician, laboratory samples have been analyzed, and reports filed, which causes delay.

In certain diseases, however, early detection would be very desirable. This is the case in, for instance, rare diseases whose etiology is not clear, in order to begin epidemiological studies as early as possible. Another example of the benefits of early detection comes from the detection of food-borne epidemics, where control measures are often much easier to conduct in the early phases of the epidemic.

If the registries could utilize diagnostic hypotheses made by physicians as soon as the patient has met the doctor, the delay would be eliminated. Unfortunately, such data is not available. However, in Finland a database exists that

N. Lavrač et al. (Eds.): PKDD 2003, LNAI 2838, pp. 180–191, 2003.
© Springer-Verlag Berlin Heidelberg 2003

might be a way around this problem. The Physician's Reference Database [1] is a collection of medical guidelines for the primary care, from which physicians often seek information about infectious diseases [2]. According to a preliminary study, an increase in the rate of database searches about a certain disease sometimes correlates to the onset of an epidemic [3].

In this work we develop a simple method for automatic detection and evaluation of such increases.

What Is an Epidemic? Incidence, in epidemiology, is defined as the number of new cases per unit of time. *Prevalence* refers to the number of the people with the certain characteristic (e.g. a certain disease) in the population at any given moment; disease prevalence is thus dependent on the incidence of the disease and the duration of the disease.

Incidence and prevalence are well-defined concepts. An *epidemic*, on the other hand, is more difficult to define precisely. In mathematical modeling of infectious diseases (e.g. [4]) an epidemic is said to occur if the introduction of an infectious agent to a population causes people (other than the first introducer) to get sick through people transmitting the disease to each other.

This definition is, however, useful only in theoretical modeling, or when the actual incidence in the absence of an epidemic is zero. Many diseases have a certain *baseline incidence*: when a steady incidence and prevalence are considered the normal situation, even if the transfer happens person-to-person. Often, we also call a cluster of cases an "epidemic" even though no actual person-to-person transmission happens. What "an epidemic" is thus depends on the observer's subjective goals and estimates of local conditions.

Many infectious diseases have a seasonal cycle cycle: the incidence of the disease increases and decreases with a certain, steady interval. Some diseases, like influenza, have yearly peaking epidemics. Some have simply a slightly higher incidence during a certain time of year: for instance, food- and water-borne diseases are more common in warm weather. Some diseases have a longer cycle: for example, a major Pogosta disease epidemic occurs about every seven years in Finland.

A generic goal of automated detection of epidemics is to discover the beginning of any epidemic, whether cyclic or not. For cyclic diseases, an alternative goal is to consider the cyclic variation normal and to detect incidences that are exceptionally high given the normal cycle.

Requirements for Online Surveillance of Epidemics. It is of course important that a detection system achieves high sensitivity (proportion of epidemics detected), so that epidemics are not overlooked. It is also important that epidemics are detected as soon as possible after their onset. However, false alarms severly undermine the credibility of the warning system, which might result in the users no longer taking the warnings seriously. Thus, high specificity (proportion of non-epidemic periods classified as non-epidemic), leading to a higher positive prediction value (probability of epidemic given that the system outputs an alarm), should be a priority even at the cost of some reduction in the detection speed.

With a good system, the user can specify the period to which the present moment is compared: for instance, it should be possible to ask "is this week different from the previous n months", "is this month different from the same month during previous years", and several other questions like that, with the same method.

The method we introduce allows such flexibility and different treatments of cyclic diseases. We will analyze the sensitivity, specificity and positive prediction value of the method on both synthetic and real data.

2   Physician's Reference Database and Data Preprocessing

The Physician's Reference Database is a web-based collection of medical guidelines used by physicians. The database consists of thousands of articles, each described by a number of keywords (typically names of diseases, symptoms and findings). Reading events are recorded in a usage log, allowing one to mine for physicians' active interests in different diseases.

Let A be the set of all articles in the database, and $i = 1, 2, ..., n$ the sequence of days under surveillance. The raw data consists of the number of reading events in a day, $D(a)_i$, for each article $a \in A$. Each of the articles has associated keywords. Let $A(k)$ be the set of articles that contain disease k among its keywords. For each day i and each keyword k, the *daily total count of events of all relevant articles* is

$$D(k)_i = \sum_{a \in A(k)} D(a)_i.$$

Usage data from the reference database is available from October 1st 2000 onward; in the analysis in this work we use data until September 30th 2002. On average, there were 1465 reading events (a user viewing an article) per day, by all users total. The event counts show a notable upward trend: during the first 100 days the average was 633.9 events and during the 100 last days 2696.1.

As trends related to the changing usage of the database are not relevant, it is necessary is to "normalize" daily counts in relation to the overall database usage. We divide the daily event count per keyword by the total event count, giving the basic unit of our data, the *proportional daily event count*:

$$d(k)_i = \frac{D(k)_i}{\sum_{a \in A} D(a)_i}.$$

There are some potential problems with this normalization approach: something that is likely to affect the keyword-specific event counts is also likely to affect the total count. This might cause artefacts, ie. trends or peaks that are not present in the original data, or "dilution" of a smaller epidemic by a bigger one. At the moment we do not try to counteract such possible side effects.

3   The Method

The goal is to detect possible epidemics, observable as exceptionally high proportional daily event counts, and to output an alarm as soon as possible after the beginning of an epidemic. We develop a simple randomization-based framework to recognize significant increases in event counts.

The series is first *smoothed* to remove some of the daily variations while retaining most of the trends. Two smoothing methods, namely sliding average and sliding linear regression, are used (see below).

A *null hypothesis period*, a sequence of days from the past to which the present moment is compared, is chosen. The way the null hypothesis period is designated determines the exact question we are trying to answer. When the present moment is compared to all past non-epidemic times, the question is "is there an epidemic now". Other examples include the last n days ("has the situation changed for the worse recently"), the same months during previous years ("is this June different from what is typical for previous Junes"), and all previous epidemic times ("is there an epidemic now that is worse than typical epidemics of this disease").

We assume that in the absence of an epidemic the proportional daily event counts for a given disease are independent and identically distributed, but note that the independence assumption is most likely not completely true. A person that has read a lot about some disease recently is less likely to review that information than he would be if he knew nothing about the subject. Modeling such dependencies would be tedious, though, and we hope that this dependency averages out among all users.

As we do not know the true distribution behind the data, we cannot obtain a p-value or other comparison based on that. Instead, we use *randomization statistics*. The null hypothesis period is sampled with replacement for samples the size of the smoothing function's window[3], and the smoothed value is calculated for each of these samples. The resulting empirical distribution is used as the distribution of the smoothed values under the null hypothesis of no epidemic, and the p-value of the day in question is taken to be the proportion of the sampled sequences having the same or a higher smoothed value. (For more information on randomization statistics, see for instance [6].)

Testing every day like this could cause a bad positive prediction value for two reasons. First, if the proportion of negatives to positives in the set of objects tested rises, even a good specificity causes bad prediction values eventually. Second, the detection method itself includes randomization, and thus will eventually err if run repeatedly.

To avoid this problem, we require that the smoothed value of a day is both high (when compared with other observed values) and statistically significant (tested with randomization). The first requirement is fulfilled by checking if the smoothed value of that day is higher than a certain percentile of smoothed

---

[3] Note that here we sample new sequences, that is, individual points until we have w points, not windows of size w from the original series.

values of the original series in the null hypothesis period. Thus, the tested set of days is limited to a subset of all days having a high value of the statistic under surveillance. The use of a cut-off value can also be seen as a crude and quick estimate of our statistical test.

To put this together, given a time series, a window length $w > 1$, a cutoff value $c \in [50, 100[$, a smoothing function f from a window in the series (a run of consecutive points) to a real number, a null hypothesis period, and a p-value, the method works as follows:

> (1) for the null hypothesis period, calculate the w-day smoothed values according to f, store these in S_1
> (2) check if today's smoothed value exceeds the c'th percentile of all smoothed values in S_1, and if it does:
> > (3) resample samples of size w from the null hypothesis period of the original non-smoothed series, and calculate the smoothed value for each of these windows, store these in S_2
> > (4) determine the proportion of values in S_2 that are the same or higher than the value for today, and if that proportion is lower than p:
> > > (5) output an alarm, together with the proportion.

The (non-weighted) w-day *sliding average* is calculated simply by replacing each data point with an average of w days. This can be done either using time points on both sides of the day, or using the previous $w - 1$ days. As in this problem future data is not available, we use the latter method:

$$SA_i = \sum_{j=1}^{w} d_{i-j+1}$$

Sliding linear regression, on the other hand, works by fitting *a line* with the least squares over the w days $[i - w + 1, i]$, and then taking the smoothed value at i to be the value of this linear function at that point. The benefit of this smoothing method is that it reacts faster to abrupt changes in the series; in a way the sliding linear function exaggerates the linear tendencies in the series, while the sliding average tries to smooth them out.

Naturally, the longer the window, the less short-term changes affect the smoothed series. However, as the window stretches only backward in time, as opposed to backward and forward, this causes a lag in the smoothed curves' reaction to changes. See Figure 1 for an example series and smoothings on it.

4   Test Results

4.1   Results on Artificial Test Data

Forty artificial test series were constructed to test the performance of the algorithm with different parameters. Each series is 700 time points long. 20 are constructed from an exponential distribution, another 20 from a normal distribution with values below zero replaced by zero. 10 of each type of series had no epidemics; in the other 20 datasets timepoints [351, 450] were replaced by samples drawn from a similar distribution as the main series, but with a higher mean. Parameters for the distributions were chosen based on the means and variations in the real life test data. See Figure 1 for an example series.

In all the tests on artificial data, the null hypothesis period for timepoint i is $[1, i - w]$ for the non-epidemic series and $[1, i - w] \setminus [351, 450]$ for the epidemic ones. As we know for certain which days are epidemic and which are not, we can calculate sensitivity, specificity and delays exactly. No epidemic was completely missed, giving an *epidemics-wise* sensitivity of exactly 1 for all settings. Below, we have explored the sensitivity and specificity *day-wise*, that is, the algorithm is expected to mark each day either belonging to an epidemic or not. We also examine the detection delay, defined as the number of false negatives from the first day of an epidemic (time point 351) until the first true positive during the epidemic.

In practice, 20,000 samples were enough to produce steady results on the randomization tests; 80,000 were used for the sliding average and 20,000 for the sliding linear, due to the first one's Matlab implementation being so much faster that the extra certainty was worthwhile. Unless mentioned otherwise, the p-value is 0.01. When calculating the performance statistics, the first one hundred days are ignored.

Figure 2.a shows the day-wise specificity of the algorithm for all the test series with different parameter values. Note how specificity drops in the sliding average tests. This is mostly due to the fact that the longer the window, the longer it takes after the epidemic period before the smoothed values return to the baseline. As a faster-reacting function, sliding linear regression does not suffer from similar problems, but the distribution of the specificities widens when the window grows: more series are detected with 100 % specificity, but some epidemic periods are also less well detected than they would have a shorter window.

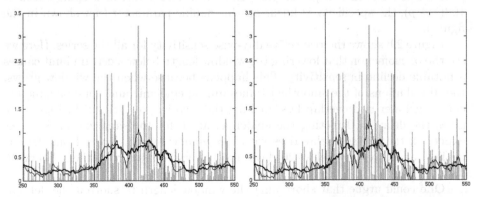

Fig. 1. A portion of an artificial time series (gray bars) with an "epidemic" in the middle of it (bordered by the vertical black lines). On the left; a 14-day sliding average (thin line) and a 30-day sliding average (thick line). On the right, a 30-day sliding average (again the thick line) and a 30-day sliding linear regression (thin line).

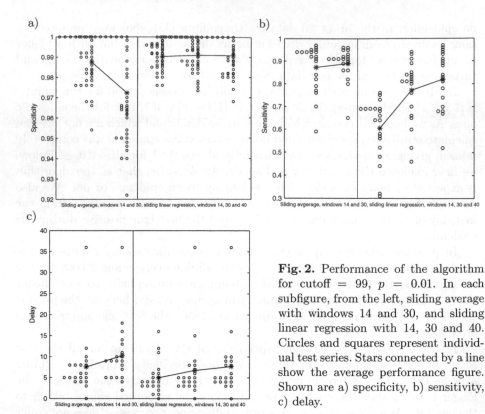

Fig. 2. Performance of the algorithm for cutoff = 99, $p = 0.01$. In each subfigure, from the left, sliding average with windows 14 and 30, and sliding linear regression with 14, 30 and 40. Circles and squares represent individual test series. Stars connected by a line show the average performance figure. Shown are a) specificity, b) sensitivity, c) delay.

As one would expect, specificity is dependent on the parameters cutoff and p. If window length is kept constant and cutoff and p varied, keeping cutoff $= 100(1 - p)$, the specificity is linear on this double parameter (not shown in the figure).

Figure 2.b shows the respective day-wise sensitivity for all the series. Here we see the phenomenon that lowering the window length below a certain limit causes a notable decline in sensitivity. This happens because when the window grows, the distributions of the smoothed values during epidemic and non-epidemic become narrower and overlap less, making them easier to distinguish. Figure 2.c gives the delay in detecting the epidemic, again for the same parameters. As expected, the average delay rises with the window length. Apart from the one outlier, all delays are below 20 days and the vast majority of them below 10 days.

One could argue that shortening the window length to shorten the delay is worthwhile, since as the epidemic still is detected the lower sensitivity does not matter. To some extent this is true. However, one important way to tell a real epidemic from a false positive is whether the situation *persists*. False positives appear singly or in short clusters, real epidemics last for several days or weeks.

While we can afford to lose some sensitivity in order to gain a shorter delay, we cannot let it go altogether.

Experimenting on the effect of window length, keeping other parameters constant, revealed that when the window length shortens, sensitivity stays reasonably good (that is, around 80-90 %) up to a point, and then drops steeply. This drop happens around window length 15 for the sliding average and around window length 40 for the sliding linear regression (at least on the data used in this work). Comparing the specificity and sensitivity of the smoothing methods on these lengths, it can be seen see that while there are some differences on certain series between the methods, their overall performance is close to equal, and that there are no systematic differences depending on the type of the series. (Test results not shown.)

4.2 Results on Real Life Data

Ten diseases were selected as test targets, namely hepatitis A, influenza, diphtheria, legionellosis, Pogosta disease, polio, parotitis, tularemia, varicella and measles. The keyword for each is the (Finnish) name of the disease, and each test series is 729 days long. Parameters used are cutoff $= 99, p = 0.01$, and window lengths 14 (sliding average) and 40 (sliding linear regression).

Unlike with artificial data, we do not now have conclusive knowledge of which days are epidemic and which are not; the only definite example of an epidemic interesting from the public health point of view is the Pogosta epidemic of 2002, which began in August (Figure 3.a). So we count negatives after August 1st 2002 (day 668) in that series as false negatives, and exclude timepoints from day 680 onward from the null hypothesis period. In other series the null hypothesis period is $[1, i - w]$.

Delay of Detection. The Pogosta epidemic was detected on day 665 by the sliding average, and on day 666 by the sliding linear regression method (see Figure 3.a). Compared to the shape of the actual epidemic (the thick line at the bottom), we can see that this is remarkably early. The epidemic curve is drawn based on the day the diagnostic blood sample was taken, which is likely to be the same day as the day after the patient's physician first suspected the disease. Even if we had data straight from the physician's office, we could not have seen the epidemic much earlier. (Data for the epidemic curve is based on cases later notified to the Infectious Diseases Registry of the National Public Health Institute.)

On the sliding average there was a short alarm peak also at time-points 636-639. It is unclear whether this is a true first alarm of the epidemic, or a sequence of false positives; in the following it is counted as the latter. No days were falsely classified as negative during the epidemic.

Sensitivity, Specificity and Positive Prediction Value. Over all the series, in the sliding average, if we consider only the Pogosta epidemic of 2002 as true positives, specificity was 99.02 %, and the positive prediction value 51.2 %. The sliding linear regression had specificity 99.00 % and positive prediction value of 50.5 %.

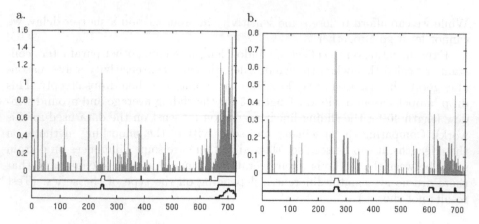

Fig. 3. The output of the algorithm in some situations. a. The Pogosta series with the output of both smoothing methods. Thin line, sliding average; thick line in the middle, sliding linear regression. The null hypothesis period is the whole preceding period minus timepoints 680 and onward. The bottom-most thick line shows the shape of the actual epidemic: weekly incidence according to the day the first diagnostic blood sample was taken, which probably is the same day or close to the day that the patient's physician first suspected the disease. b. The tularemia series with the output of the sliding average smoothing method. Thin line, the null hypothesis period is the whole preceding series; thick line, it is the preceding 180 days.

With the two methods, continuous runs of "false positives" happened in four series (legionella, mumps series, Pogosta, and tularemia) in about the same places. Looking at the series, and bearing in mind that the detection is based only on time previous to those time-points, these four periods of alarms seem reasonable and even desirable. See Figure 3 for two of the cases.

If we count the positives during these periods as true positives, we get a positive predictive value of 85.3 % for the sliding average and 90.1 % for the sliding linear regression. (Stating specificity would require arbitrarily determining which days, if any, around these positives are also positives.) In the real situation, where the series under surveillance and the null hypothesis periods are chosen by an epidemiologist, the positive prediction value (defined through the usefulness of the alarm) will probably be somewhere between these estimates of 50 and 90 %.

4.3   Changing the Null Hypothesis Period

Figure 3.b demonstrates the effect of the null hypothesis period. The thin line shows the output when the null hypothesis period is the whole preceding series; the thick line when it is last 180 days. The yearly tularemia epidemic during the second year does not cause an alarm when the epidemic during the previous year is included in the null hypothesis, and does cause an alarm when it is not.

Another interesting feature is also visible in the lower output. Looking at the second predicted epidemic (from timepoint 600 onward), we can see that there first is an alarm period, but when the alarm has been on for some time, it ceases. When the epidemic rises again, new alarms are put out. This happens because the epidemic time is now not excluded from the null hypothesis. Thus "epidemics inside epidemics" can be detected.

5   Technical Comments

The time requirement of the method for the check on one day is linear to the number of iterations i performed in the randomization and the window length w, as taking a random sample and calculating a mean or fitting a line are all linear. Thus, running the check for m diseases (which is what would normally be done daily) with i iterations is $O(wmi)$, and calculating the results for n days is $O(nwmi)$. Since typically $w << i$, these are close to $O(mi)$ and $O(nmi)$.

Currently, data arrives from the system administrator of the actual database daily, as a text file. The data is read into an (Oracle RDB) database, and the reading events per keyword are calculated and stored. The raw article counts are also stored, to make it simpler to add a new disease keyword to the base (without the need to reread the data files). When the amount of data grows too large, article counts might be preserved for perhaps two years, and only keyword-specific values stored for a longer period, enabling the beginning of a new surveillance with some data, but not requiring too much space.

The test versions of the algorithms as explained in this work were implemented on Matlab. Conversion to a Java program for end users is planned.

6   Related Work

Basically, most methods of online detection of changes in a time series fall into two categories. In the first, we compare each value, at the time of its arrival, to some baseline value calculated from previous data, and decide if the new value is "different enough" to be considered abnormal [7,8]. In the second, we fit some model, often a curve – constant [9], linear [10,11], or more complex [12] – piecewise to the data, searching for the change-points.

Other than these two approaches, data mining of time series data has been studied mainly from the point of view of mining inter-series relations in either patterns (slopes, peaks) or in concurrently occurring different events (see, for example, [13,14]), which approaches are not directly related to the problem of this study.

Piatetsky-Shapiro and Matheus were among the first data miners to investigate deviations in time series data [7]. Their basic concept is a deviation, defined as a difference between an observed value and a normative value. In addition, they categorize the deviations based on their interestingness, defined as the utility of the finding to a user.

Stern and Lightfoot describe a system for detecting clusters of human infection with enteric pathogens [8]. In it, a smoothing technique is used to determine normal baseline incidences for each pathogen, area and time-of-year, based on data from several previous years. Then weekly counts of cases are compared to a threshold calculated from the base incidence. The system achieves great sensitivity, over 90 %, but the positive predictive value remains at about 60 % or lower - meaning that almost every other alarm is false.

The problem with these approaches is that calculating the normative values requires data from several years, and that they give accurate results only if the difference between normal and abnormal values is rather sharp. Another problem is that the calculation of normative values requires some knowledge or assumption about the shape of the distribution of the values; in our case such knowledge is not readily available.

Ogden and Sugiura [10] describe test statistics for determining whether a time series has undergone a change. The change is defined as a linear change in the parameters of the underlying distribution: the parameter vector is θ from the beginning of the series to some timepoint t_i, then changes in a linear way until it reaches $\theta + \delta$ at $t_j, j > i$. The null hypothesis tested is $\delta = 0$. However, the tests cannot be applied online to decide if there has been a change recently, and require information on the distribution of the data.

Keogh et al. [11] explore segmenting of a time series into piecewise linear representation, relative to an error criterion stating whether a line fit is "good enough" (for example, the total error must not exceed a certain value). They describe three basic greedy approaches, only one of them online, and a combination online algorithm that performed well. Guralnik and Srivastava [12] suggest not restricting the function to be fitted to the segments to lines (for instance, one could allow the algorithm to choose the best fit of 0-3 degree polynomials, instead). In his thesis [9] Marko Salmenkivi introduced methods for intensity modeling; that is, assuming a sequence of independent events in time, finding a piece-wise constant function describing the intensity of the Poisson-process producing that sequence.

Most of the above methods is directly suitable for online detection of change points. We experimented with the online algorithm of Keogh et al., trying to use them to detect epidemics. The idea was to segment the series, and then look at the slope of the last segments. Unfortunately, we were unable to calibrate an error criterion that would be both specific enough and produce a sort enough delay, and unable to adapt the method to answer several surveillance questions (for instance, comparing this month's situation to the same months two previous years proved impractical). Similar problems apply to the other change-point detection/segmentation approaches.

7   Conclusions

A method was developed to automatically detect epidemics from an online time-series. Despite its simplicity, the method works reliably. In all the test data, all epidemics were correctly detected. Even when calculated day-wise instead of epidemics-wise, we achieved specificity over 99 % and sensitivity over 80 %. Also the results on real-life data were very encouraging.

A nice feature of the method is the adaptability that is achieved by changing the null hypothesis period. The same method can readily answer several kinds of questions of interest such as detecting acute short-term changes and comparing epidemics.

However, caution must be used before widely applying this – or any – method of online surveillance. It must be kept in mind that all surveillance requires the capacity to deal with both true and false alarms; surveillance is useless unless personnel exist to work on the alarms. A separate prospective study will be necessary to establish the actual benefits of surveillance.

References

1. Physician's Desk Reference and Database (in Finnish). Kustannus Oy Duodecim. Yearly revised edition.
   URL:http://www.duodecim.fi/kustannus/cdrom/index.html (14.3.2002).
2. Jousimaa. J., Kunnamo. I., Physicians' patterns of using a computerized collection of guidelines of primary health care. *Int J Technol Assess Health Care*. 14:484-93. 1998.
3. Jormanainen. V., Jousimaa. J., Kunnamo. I., Ruutu. P., Physicians' Database Searches as a Tool for Early Detection of Epidemics. *Emerging Infectious Diseases*. 7(3). 2001.
4. Diekman, O., Heesterbek, J. A. P, *Mathematical Epidemiology of Infectious Diseases*. Wiley Series in Mathematical and Computational Biology, John Wiley & Sons Ltd, 2000.
5. Farmer, R. D. T., Miller D. L., Lawrenson R., *Lecture Notes on Epidemiology and Public Health Medicine*. Blackwell Scientific, Oxford, 1996.
6. Manly, B. F. J. *Randomization, Bootstrap and Monte Carlo Methods in Biology*, Second Edition. Volume 41 in the series "Texts in Statistical Science Series". CRC Press, 1997.
7. Piatetsky-Shapiro G., Matheus, C. J., The Interestingness of Deviations. *AAAI-94 Knowledge Discovery in Databases Workshop*, 1994.
8. Stern, L., Lightfoot, D., Automated Outbreak Detection: a quantitative retrospective analysis. *Epidemiol. Infect.*, 122, 103-110, 1999.
9. Salmenkivi. S., Computational Methods for Intensity Models. Department of Computer Science. University of Helsinki. Series of Publications A-2001-2. 2001.
10. Ogden. S., Sugiura. N., Testing Change-points with Linear Trend. *Communications in Statistics B: Simulation and Computation*. 23:287-322. 1994.
11. Keogh, E., Chu, S., Hart, D., Pazzani. M., An Online Algorithm for Segmenting Time Series. *IEEE International Conference on Data Mining. 2001*.
12. Guralnik. V., Srivastava. J., Event Detection from Time Series Data. *Proc. Fifth ACM SIGKDD*, 1999.
13. Last, M., Klein, Y., Kandel, A., Knowledge discovery in Time Series Databases. Correspondence in *IEEE Transactions on Systems, Man and Cybernetics - Part B: Cybernetics*, 2001.
14. Das. G., Lin. K., Mannila. H., Renganathan. G., Smyth. P., Rule discovery from time series. *Proc. KDD'98*, 16-22, 1998.

An Indiscernibility-Based Clustering Method with Iterative Refinement of Equivalence Relations

Shoji Hirano and Shusaku Tsumoto

Department of Medical Informatics, Shimane Medical University, School of Medicine
89-1 Enya-cho, Izumo, Shimane 693-8501, Japan
shirano@ieee.org, tsumoto@computer.org

Abstract. In this paper, we present an indiscernibility-based clustering method that can handle relative proximity. The main benefit of this method is that it can be applied to proximity measures that do not satisfy the triangular inequality. Additionally, it may be used with a proximity matrix – thus it does not require direct access to the original data values. In the experiments we demonstrate, with the use of partially mutated proximity matrices, that this method produces good clusters even when the employed proximity does not satisfy the triangular inequality.

1 Introduction

Clustering is a powerful tool for revealing underlying structure of the data. A number of methods, for example, hierarchical, partial, and model-based methods, have been proposed and have produced good results on both artificial and real-life data [1].

In order to assess the quality of clusters being produced, most of the conventional clustering methods employ quality measures that are associated with centroids of clusters. For example, the internal homogeneity of a cluster can be measured as the sum of differences from objects in the cluster to their centroid, and it can be further used as a component of the total quality measure for assessing a clustering result. Such centroid-based methods work well on datasets in which the proximity of objects satisfies the natures of distance that are, positivity ($d(x, y) \geq 0$), identity ($d(x, y) = 0$ iff $x = y$), symmetry ($d(x, y) = d(y, x)$), and triangular inequality ($d(x, z) \leq d(x, y) + d(y, z)$), for any objects x, y and z. However, they have a potential weakness in handling relative proximity. Relative proximity is a class of proximity measures that is suitable for representing subjective similarity or dissimilarity such as the degree of likeness between people. It may not satisfy the triangular inequality because the proximity $d(x, z)$ of x and z is allowed to be independent of y. Usually, the centroid c of objects x, y and z is expected to be in their convex hull. However, if we use relative proximity, the centroid can be out of x, y, and z's convex hull because proximity between c and other objects can be far greater (if we use dissimilarity as proximity) or smaller (if we use similarity) than $d(x, y)$, $d(y, z)$ and $d(x, z)$. Namely, a centroid does

N. Lavrač et al. (Eds.): PKDD 2003, LNAI 2838, pp. 192–203, 2003.

not hold its geometric properties under these conditions. Thus another criterion should be used for evaluating the quality of the clusters.

In this paper, we present a new clustering method based on the indiscernibility degree of objects. The main benefit of this method is that it can be applied to proximity measures that do not satisfy the triangular inequality. Additionally, it may be used with a proximity matrix – thus it does not require direct access to the original data values.

2   The Method

2.1   Overview

Our method is based on iterative refinement of N binary classifications, where N denotes the number of objects. First, an equivalence relation, that classifies all the other objects into two classes, is assigned to each of N objects by referring to the relative proximity. Next, for each pair of objects, the number of binary classifications in which the pair is included in the same class is counted. This number is termed the indiscernibility degree. If the indiscernibility degree of a pair is larger than a user-defined threshold value, the equivalence relations may be modified so that all of the equivalence relations commonly classify the pair into the same class. This process is repeated until class assignment becomes stable. Consequently, we may obtain the clustering result that follows a given level of granularity, without using geometric measures.

2.2   Assignment of Initial Equivalence Relations

When dissimilarity is defined relatively, the only information available for object x_i is the dissimilarity of x_i to other objects, for example to x_j, $d(x_i, x_j)$. This is because the dissimilarities for other pairs of objects, namely $d(x_j, x_k)$, $x_j, x_k \neq x_i$, are determined independently of x_i. Therefore, we independently assign an initial equivalence relation to each object and evaluate the relative dissimilarity observed from the corresponding object.

Let $U = \{x_1, x_2, ..., x_N\}$ be the set of objects we are interested in. An equivalence relation R_i for object x_i is defined by

$$U/R_i = \{P_i,\ U - P_i\},\tag{1}$$

where

$$P_i = \{x_j|\ d(x_i, x_j) \leq Th_{di}\},\ \ \forall x_j \in U.\tag{2}$$

$d(x_i, x_j)$ denotes dissimilarity between objects x_i and x_j, and Th_{di} denotes an upper threshold value of dissimilarity for object x_i. The equivalence relation, R_i classifies U into two categories: P_i, which contains objects similar to x_i and $U - P_i$, which contains objects dissimilar to x_i. When $d(x_i, x_j)$ is smaller than Th_{di}, object x_j is considered to be indiscernible to x_i. U/R_i can be alternatively written as $U/R_i = \{\{[x_i]_{R_i}\}, \{\overline{[x_i]_{R_i}}\}\}$, where $[x_i]_{R_i} \cap \overline{[x_i]_{R_i}} = \phi$ and $[x_i]_{R_i} \cup \overline{[x_i]_{R_i}} = U$ hold.

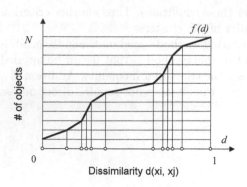

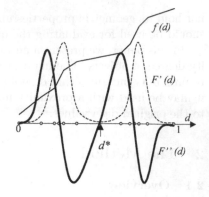

Fig. 1. An example of function $f(d)$ generated by $d(x_i, x_s)$.

Fig. 2. Relations between $f(d)$ and its smoothed first- and second-order derivatives $F'(d)$ and $F''(d)$.

Definition of the dissimilarity measure $d(x_i, x_j)$ is arbitrary. If all the attribute values are numerical, ordered, and independent of each other, conventional Minkowski distance

$$d(x_i, x_j) = \left(\sum_{a=1}^{N_a} |x_{ia} - x_{ja}|^p \right)^{\frac{1}{p}}, \tag{3}$$

where N_a denotes the number of attributes, x_{ia} denotes the a-th attribute of object x_i, and p denotes a positive integer, is a reasonable choice since it has been successfully applied to many areas and its mathematical properties have been well investigated. More generally, any type of dissimilarity measure can be used regardless of whether or not the triangular inequality is satisfied among objects.

Threshold of dissimilarity Th_{di} for object x_i is automatically determined based on the spatial density of objects. The procedure is summarized as follows.

1. Sort $d(x_i, x_j)$ in ascending order. For simplicity, we denote the sorted dissimilarity using the same representation $d(x_i, x_s)$, $1 \le s \le N$.
2. Generate a function $f(d)$ that represents the cumulative distribution of d. For a given dissimilarity d, function f returns the number of objects whose dissimilarity to x_i is smaller than d. Figure 1 shows an example. Function $f(d)$ can be generated by linearly interpolating $f(d(x_i, x_s)) = n$, where n corresponds to the index of x_s in the sorted dissimilarity list.
3. Obtain the smoothed second-order derivative of $f(d)$ as a convolution of $f(d)$ and the second-order derivative of Gaussian function as follows.

$$F''(d) = \int_{-\infty}^{\infty} f(u) \frac{-(d-u)}{\sigma^3 \sqrt{2\pi}} e^{-(d-u)^2/2\sigma^2} du, \tag{4}$$

where $f(d) = 1$ and $f(d) = N$ are used for $d < 0$ and $d > 1$ respectively. The smoothed first-order derivative $F'(d)$ of $f(d)$ represents spatial density of objects because it represents increase or decrease velocity of the objects induced by the change of dissimilarity. Therefore, by calculating its further derivative as $F''(d)$, we find a sparse region between two dense regions. Figure 2 illustrates relationship between $f(d)$ and its smoothed derivatives. The most sparse point d^* should take a local minimum of the density where the following conditions are satisfied.

$$F''(d^* - \Delta d) < 0 \text{ and } F''(d^* + \Delta d) > 0. \tag{5}$$

Usually, there are some d^*s in $f(d)$ because $f(d)$ has multiple local minima. The value of σ in the above Gaussian function can be adjusted to eliminate meaningless small minima.

4. Choose the smallest d^* and object x_{j*} whose dissimilarity is the closest to but not larger than d^*. Finally, the dissimilarity threshold Th_{di} is obtained as $Th_{di} = d(x_i, x_{j*})$.

2.3   Refinement of Initial Equivalence Relations

Suppose we are interested in two objects, x_i and x_j. In indiscernibility-based classification, they are classified into different categories regardless of other relations, if there is at least one equivalence relation that has an ability to discern them. In other words, the two objects are classified into the same category only when all of the equivalence relations commonly regard them as indiscernible objects. This strict property is not acceptable in clustering because it will generate many meaningless small categories, especially when global associations between the equivalence relations are not taken into account. We consider that objects should be classified into the same category when most of, but not necessarily all of, the equivalence relations commonly regard the objects as indiscernible. In the second stage, we perform global optimization of initial equivalence relations so that they produce adequately coarse classification to the objects. The global similarity of objects is represented by a newly introduced measure, the *indiscernibility degree*. Our method takes a threshold value of the indiscernibility degree as an input and associates it with the user-defined granularity of the categories. Given the threshold value, we iteratively refine the initial equivalence relations in order to produce categories that meet the given level of granularity.

Now let us assume $U = \{x_1, x_2, x_3, x_4, x_5\}$ and classifications of U by $\mathbf{R} = \{R_1, R_2, R_3, R_4, R_5\}$ is given as follows.

$$\begin{aligned}
U/R_1 &= \{\{x_1, x_2, x_3\}, \{x_4, x_5\}\}, \\
U/R_2 &= \{\{x_1, x_2, x_3\}, \{x_4, x_5\}\}, \\
U/R_3 &= \{\{x_2, x_3, x_4\}, \{x_1, x_5\}\}, \\
U/R_4 &= \{\{x_1, x_2, x_3, x_4\}, \{x_5\}\}, \\
U/R_5 &= \{\{x_4, x_5\}, \{x_1, x_2, x_3\}\}.
\end{aligned} \tag{6}$$

This example contains three types of equivalence relations: $R_1 (= R_2 = R_5)$, R_3 and R_4. Since each of them classifies U slightly differently, classification of U

by the family of equivalence relations $\mathbf{R}$, $U/\mathbf{R}$, contains four very small, almost independent categories.

$$U/\mathbf{R} = \{\{x_1\}, \{x_2, x_3\}, \{x_4\}, \{x_5\}\}. \tag{7}$$

In the following we present a method to reduce the variety of equivalence relations and to obtain coarser categories.

First, we define an *indiscernibility degree*, $\gamma(x_i, x_j)$, for two objects x_i and x_j as follows.

$$\gamma(x_i, x_j) = \frac{\sum_{k=1}^{|U|} \delta_k^{indis}(x_i, x_j)}{\sum_{k=1}^{|U|} \delta_k^{indis}(x_i, x_j) + \sum_{k=1}^{|U|} \delta_k^{dis}(x_i, x_j)}, \tag{8}$$

where

$$\delta_k^{indis}(x_i, x_j) = \begin{cases} 1, \text{ if } (x_i \in [x_k]_{R_k} \wedge x_j \in [x_k]_{R_k}) \\ 0, \text{ otherwise.} \end{cases} \tag{9}$$

and

$$\delta_k^{dis}(x_i, x_j) = \begin{cases} 1, \text{ if } (x_i \in [x_k]_{R_k} \wedge x_j \notin [x_k]_{R_k}) \text{ or} \\ \quad \text{ if } (x_i \notin [x_k]_{R_k} \wedge x_j \in [x_k]_{R_k}) \\ 0, \text{ otherwise.} \end{cases} \tag{10}$$

Equation (9) shows that $\delta_k^{indis}(x_i, x_j)$ takes 1 only when the equivalence relation R_k regards both x_i and x_j as indiscernible objects, under the condition that both of them are in the same equivalence class as x_k. Equation (10) shows that $\delta_k^{dis}(x_i, x_j)$ takes 1 only when R_k regards x_i and x_j as discernible objects, under the condition that either of them is in the same class as x_k. By summing $\delta_k^{indis}(x_i, x_j)$ and $\delta_k^{dis}(x_i, x_j)$ for all $k(1 \leq k \leq |U|)$ as in Equation (8), we obtain the percentage of equivalence relations that regard x_i and x_j as indiscernible objects. Note that in Equation (9), we excluded the case when x_i and x_j are indiscernible but not in the same class as x_k. This is to exclude the case where R_k does not significantly put weight on discerning x_i and x_j. As mentioned in Section 2.2, P_k for R_k is determined by focusing on similar objects rather than dissimilar objects. This means that when both of x_i and x_j are highly dissimilar to x_k, their dissimilarity is not significant for x_k, when determining the dissimilarity threshold Th_{dk}. Thus we only count the number of equivalence relations that certainly evaluate the dissimilarity of x_i and x_j.

For example, the indiscernibility degree $\gamma(x_1, x_2)$ of objects x_1 and x_2 in the above case is calculated as follows.

$$\begin{aligned} \gamma(x_1, x_2) &= \frac{\sum_{k=1}^{5} \delta_k^{indis}(x_1, x_2)}{\sum_{k=1}^{5} \delta_k^{indis}(x_1, x_2) + \sum_{k=1}^{5} \delta_k^{dis}(x_1, x_2)} \\ &= \frac{1 + 1 + 0 + 1 + 0}{(1 + 1 + 0 + 1 + 0) + (0 + 0 + 1 + 0 + 0)} = \frac{3}{4}. \end{aligned} \tag{11}$$

Let us explain this example with the calculation of the numerator $(1+1+0+1+0)$. The first value 1 is for $\delta_1^{indis}(x_1, x_2)asshown$. Since x_1 and x_2 are in the same class of R_1 and obviously, they are in the same class to x_1, $\delta_1^{indis}(x_1, x_2) = 1$

Table 1. Degree γ for objects in Eq. (6).

	$x.$	$x.$	$x.$	$x.$	$x.$
$x.$	3/3	3/4	3/4	1/5	0/4
$x.$		4/4	4/4	2/5	0/5
$x.$			4/4	2/5	0/5
$x.$				3/3	1/3
$x.$					1/1

Table 2. Degree γ after the first refinement.

	$x.$	$x.$	$x.$	$x.$	$x.$
$x.$	3/3	3/4	3/4	2/4	1/5
$x.$		4/4	4/4	3/4	0/5
$x.$			4/4	3/4	0/5
$x.$				3/3	1/5
$x.$					1/1

Table 3. Degree γ after the second refinement.

	$x.$	$x.$	$x.$	$x.$	$x.$
$x.$	4/4	4/4	4/4	4/4	0/5
$x.$		4/4	4/4	4/4	0/5
$x.$			4/4	4/4	0/5
$x.$				4/4	0/5
$x.$					1/1

holds. The second value is for $\delta_2^{indis}(x_1, x_2)$, and analogously, it becomes 1. The third value is for $\delta_3^{indis}(x_1, x_2)$. Since x_1 and x_2 are in the different classes of R_3, it becomes 0. The fourth value is for $\delta_4^{indis}(x_1, x_2)$ and it obviously, becomes 1. The last value is for $\delta_5^{indis}(x_1, x_2)$. Although x_1 and x_2 are in the same class of R_5, their class is different to that of x_5. Thus $\delta_5^{indis}(x_1, x_2)$ returns 0.

Indiscernibility degrees for all of the other pairs in U are tabulated in Table 1. Note that the indiscernibility degree of object x_i to itself, $\gamma(x_i, x_i)$, will always be 1.

From its definition, a larger $\gamma(x_i, x_j)$ represents that x_i and x_j are commonly regarded as indiscernible objects by the large number of the equivalence relations. Therefore, if an equivalence relation R_l discerns the objects that have high γ value, we consider that it represents excessively fine classification knowledge and refine it according to the following procedure (note that R_l is rewritten as R_i below for the purpose of generalization).

Let $R_i \in \mathbf{R}$ be an initial equivalence relation on U. A refined equivalence relation $R_i' \in \mathbf{R}'$ of R_i is defined as

$$U/R_i' = \{P_i', \ U - P_i'\}, \tag{12}$$

where P_i' denotes a set of objects represented by

$$P_i' = \{x_j | \gamma(x_i, x_j) \geq T_h\}, \quad \forall x_j \in U. \tag{13}$$

and T_h denotes the lower threshold value of the indiscernibility degree above, in which x_i and x_j are regarded as indiscernible objects. It represents that when $\gamma(x_i, x_j)$ is larger than T_h, R_i is modified to include x_j into the class of x_i.

Suppose we are given $Th = 3/5$ for the case in Equation (6). For R_1 we obtain the refined relation R_1' as

$$U/R_1' = \{\{x_1, x_2, x_3\}, \{x_4, x_5\}\}, \tag{14}$$

because, according to Table 1, $\gamma(x_1, x_1) = 1 \geq T_h = 3/5$, $\gamma(x_1, x_2) = 3/4 \geq 3/5$, $\gamma(x_1, x_3) = 3/4 \geq 3/5$, $\gamma(x_1, x_4) = 1/5 \leq 3/5$, $\gamma(x_1, x_5) = 0/5 \leq 3/5$ hold. In the same way, the rest of the refined equivalence relations are obtained as follows.

$$U/R_2' = \{\{x_1, x_2, x_3\}, \{x_4, x_5\}\},$$

$$U/R_3' = \{\{x_1, x_2, x_3\}, \{x_4, x_5\}\},$$
$$U/R_4' = \{\{x_4\}, \{x_1, x_2, x_3, x_5\}\},$$
$$U/R_5' = \{\{x_5\}, \{x_1, x_2, x_3, x_4\}\}. \tag{15}$$

Then we obtain classification of U by the refined family of equivalence relations $\mathbf{R}'$ as follows.

$$U/\mathbf{R}' = \{\{x_1, x_2, x_3\}, \{x_4\}, \{x_5\}\}. \tag{16}$$

In the above example, R_3, R_4 and R_5 are modified so that they include similar objects into the equivalence class of x_3, x_4 and x_5, respectively. Three types of the equivalence relations remain, however, the categories become coarser than those in Equation (7) by the refinement.

2.4   Iterative Refinement of Equivalence Relations

It should be noted that the state of the indiscernibility degrees could also be changed after refinement of the equivalence relations, since the degrees are re-calculated using the refined family of equivalence relations $\mathbf{R}'$.

Suppose we are given another threshold value $T_h = 2/5$ for the case in Equation (6). According to Table 1, we obtain $\mathbf{R}'$ after the first refinement, as follows.

$$U/R_1' = \{\{x_1, x_2, x_3\}, \{x_4, x_5\}\},$$
$$U/R_2' = \{\{x_1, x_2, x_3, x_4\}, \{x_5\}\},$$
$$U/R_3' = \{\{x_1, x_2, x_3, x_4\}, \{x_5\}\},$$
$$U/R_4' = \{\{x_2, x_3, x_4\}, \{x_1, x_5\}\},$$
$$U/R_5' = \{\{x_5\}, \{x_1, x_2, x_3, x_4\}\}. \tag{17}$$

Hence

$$U/\mathbf{R}' = \{\{x_1\}, \{x_2, x_3\}, \{x_4\}, \{x_5\}\}. \tag{18}$$

The categories in $U/\mathbf{R}'$ are exactly the same as those in Equation (7). However, the state of the indiscernibility degrees are not the same because the equivalence relations in $\mathbf{R}'$ are different from those in $\mathbf{R}$. Table 2 summarizes the indiscernibility degrees, recalculated using $\mathbf{R}'$. In Table 2, it can be observed that the indiscernibility degrees of some pairs of objects, for example $\gamma(x_1, x_4)$, increased after the refinement, and now they exceed the threshold $th = 2/5$. Thus we perform refinement of equivalence relations again using the same T_h and the recalculated γ. Then we obtain

$$U/R_1' = \{\{x_1, x_2, x_3, x_4\}, \{x_5\}\},$$
$$U/R_2' = \{\{x_1, x_2, x_3, x_4\}, \{x_5\}\},$$
$$U/R_3' = \{\{x_1, x_2, x_3, x_4\}, \{x_5\}\},$$
$$U/R_4' = \{\{x_1, x_2, x_3, x_4\}, \{x_5\}\},$$
$$U/R_5' = \{\{x_5\}, \{x_1, x_2, x_3, x_4\}\}. \tag{19}$$

Hence

$$U/\mathbf{R}' = \{\{x_1, x_2, x_3, x_4\}, \{x_5\}\}. \tag{20}$$

After the second refinement, the number of the equivalence relations in $\mathbf{R}'$ are reduced from 3 to 2, and the number of categories are also reduced from 4 to 2. We further update the state of the indiscernibility degrees according to the equivalence relations after the second refinement. The results are shown in Table 3. Since no new pairs, whose indiscernibility degree exceeds the given threshold appear, refinement process may be halted and the stable categories may be obtained, as in Equation (20).

As shown in this example, refinement of the equivalence relations may change the indiscernibility degree of objects. Thus we iterate the refinement process using the same T_h until the categories become stable. Note that each refinement process is performed using the previously 'refined' set of equivalence relations.

3   Experimental Results

We applied the proposed method to some artificial numerical datasets and evaluated its clustering ability. Note that we used numerical data, but clustered them without using any type of geometric measures.

3.1   Effects of Iterative Refinement

We first examined the effects of refinement of the initial equivalence relations. A two-dimensional numerical dataset was artificially created using Neyman-Scott method [2]. The number of seed points was set to 5. Each of the five clusters contained approximately 100 objects, and a total of 491 objects were included in the data. We evaluated validity of the clustering result based on the following measure:

$$\text{Validity } v. \ (C) = \min \left(\frac{|X. \cap C|}{|X. |}, \frac{|X. \cap C|}{|C|} \right),$$

where $X.$ and C denote the clusters obtained by the proposed method and the expected clusters, respectively. The threshold value for refinement T_h was set to 0.2, meaning that if two objects were commonly regarded as indiscernible by 20% of objects in the data, all the equivalence relations were modified to regard them as indiscernible objects.

Without refinement, the method produced 461 small clusters. Validity of the result was 0.011, which was the smallest value assigned to this dataset. This was because the small size of the clusters produced very low coverage, namely, amount of overlap between the generated clusters and their corresponding expected clusters was very small compared with the size of the expected clusters.

By performing refinement one time, the number of clusters was reduced to 429, improving validity to 0.013. As the refinement proceeds, the small clusters merged as shown in Figures 3 and 4. Validity of the results continued to increase. Finally, clusters became stable at the 6th refinement, where 10 clusters were

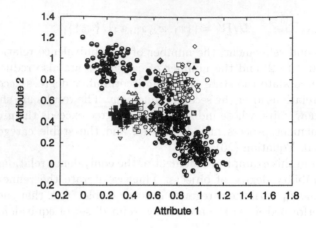

Fig. 3. Clusters after 4th refinement.

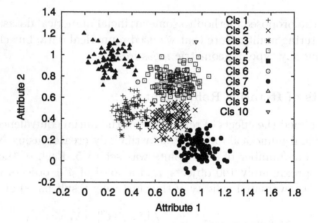

Fig. 4. Clusters after 6th refinement.

formed as shown in Figure 4. Validity of the clusters was 0.927. One can observe that a few small clusters, for example, clusters 5 and 6, were formed between the large clusters. These objects were classified into independent clusters because of the competition of the large clusters containing almost the same populations. Aside from this, the results revealed that the proposed method automatically produced good clusters that have high correspondence to the original ones.

3.2   Capability of Handling Relative Proximity

In order to validate the method's capability of handling relative proximity, we performed clustering experiments with another dataset. The data was originally generated on the two-dimensional Euclidean space likewise the previous dataset; however, in this case we randomly modified distances between data points in

Table 4. Comparison of the clustering results

Mutation Ratio[%]	0	10	20	30	40	50
AL-AHC	0.990	0.688 ± 0.011	0.670 ± 0.011	0.660 ± 0.011	0.633 ± 0.013	0.633 ± 0.018
CL-AHC	0.990	0.874 ± 0.076	0.792 ± 0.093	0.760 ± 0.095	0.707 ± 0.098	0.729 ± 0.082
Our method	0.981	0.980 ± 0.002	0.979 ± 0.003	0.980 ± 0.003	0.977 ± 0.003	0.966 ± 0.040

order to make the induced proximity matrix not fully satisfy the triangular inequality.

The dataset was prepared as follows. First, we created a two-dimensional data set by using the Neyman-Scott method [2]. The number of seed points was set to three, and a total of 310 points were included in the dataset. Next, we calculated the Euclidean distances between the data points and constructed a 310 × 310 proximity matrix. Then we randomly selected some elements of the proximity matrix and mutated them to zero. The ratio of elements to be mutated was set to 10%, 20%, 30%, 40%, and 50%. For each of these mutation ratio, we created 10 proximity matrices in order to include enough randomness. Consequently, we obtained a total of 50 proximity matrices.

We took each of the proximity matrices as an input and performed clustering of the dataset. Parameters used in the proposed method were manually determined to $\sigma = 15.0$ and $T_h = 0.3$. Additionally, we employed average-linkage and complete-linkage agglomerative hierarchical clustering methods (for short, AL-AHC and CL-AHC respectively) [3] for the purpose of comparison. Note that we partly disregarded the original data values and took the mutated proximity matrix as input of the clustering methods. Therefore, we did not employ clustering methods that require direct access to the data value.

We evaluated validity of the clustering results using the same measures as in the previous case. Table 4 shows the comparison results. The first row of the table represents the ratio of mutation. For example, 30 represents 30% of the elements in the proximity matrix were mutated to zero. The next three rows contain the validity obtained by AL-AHC, CL-AHC and the proposed method, respectively. Except for the cases in zero mutation ratio, validity is represented in the form of 'mean ± standard deviation', summarized from the 10 randomly mutated proximity matrices.

Without any mutation, the proximity matrix exactly corresponded to the one obtained by using the Euclidean distance. Therefore, both of AL-AHC and CL-AHC could produce high validity over 0.99. The proposed method also produced the high validity over 0.98. However, when mutation had occurred, the validity of clusters obtained by AL-AHC and CL-AHC largely reduced to 0.688 and 0.874, respectively. They kept decreasing moderately following the increase of mutation. The primary reason for inducing decrease of the validity was considered as follows. When the distance between two objects was forced to be mutated into zero, it brought a kind of local warp to the proximity of the objects. Thus the two objects could become candidates of the first linkage. If the two objects were originally belonged to the different clusters, these clusters were merged at

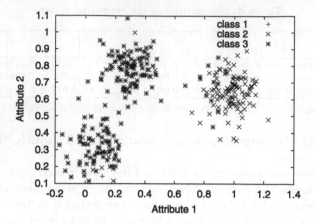

Fig. 5. Clustering results by AL-AHC. Ratio of mutation was 40%. Linkage was terminated when three clusters were formed.

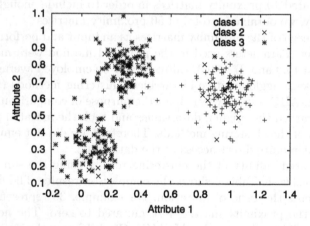

Fig. 6. Clustering results by CL-AHC. Ratio of mutation was 40%. Linkage was terminated when three clusters were formed.

an early stage of the merging process. Since both of AL-AHC and CL-AHC do not allow inverse of the cluster hierarchy, these clusters would never be separated. Consequently, inappropriately bridged clusters were obtained as shown in Figures 5 and 6.

On the contrary, the proposed method produced high validity even when the mutation ratio approached to 50%. In this method, effects of a mutation was very limited. The two concerning objects would consider themselves as indiscernible objects, however, the majority of other objects never change their classification. Although the categories obtained by the initial equivalence relations could be distorted, they could be globally adjusted through iterative refinement of the equivalence relations. Consequently, good clusters were obtained as shown in

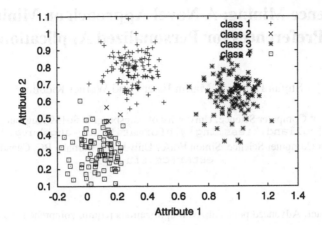

Fig. 7. Clustering results by the proposed method. Ratio of mutation was 40%. Iteration terminated at the fourth cycle.

Figure 7. This demonstrates the capability of the method for handling locally distorted proximity matrix that do not satisfy the triangular inequality.

4 Conclusions

In this paper, we have presented an indiscernibility-based clustering method, which clusters objects according to their relative proximity. Experimental results from the artificially created numerical datasets demonstrated that this method could produce good clusters even when the proximity of the objects did satisfy the triangular inequality. Future work include reduction of the computational complexity of the method and empirical evaluation of its clustering ability on large and complex real-life databases.

Acknowledgment

This work was supported in part by the Grant-in-Aid for Scientific Research on Priority Area (B)(No.759) by the Ministry of Education, Culture, Science and Technology of Japan.

References

1. P. Berkhin (2002): Survey of Clustering Data Mining Techniques. Accrue Software Research Paper. URL: http://www.accrue.com/products/researchpapers.html.
2. J. Neyman and E. L. Scott (1958): "Statistical Approach to Problems of Cosmology," *Journal of the Royal Statistical Society*, Series B20: 1–43.
3. B. S. Everitt, S. Landau, and M. Leese (2001): Cluster Analysis Fourth Edition. Arnold Publishers.

Preference Mining: A Novel Approach on Mining User Preferences for Personalized Applications

Stefan Holland[1], Martin Ester[2], and Werner Kießling[1]

[1] Institute of Computer Science, University of Augsburg, D-86159 Augsburg, Germany
{holland,kiessling}@informatik.uni-augsburg.de
[2] School of Computer Science, Simon Fraser University, Burnaby BC, Canada V5A 1S6
ester@cs.sfu.ca

Abstract. Advanced personalized e-applications require comprehensive knowledge about their user's likes and dislikes in order to provide individual product recommendations, personal customer advice and custom-tailored product offers. In our approach we model such preferences as strict partial orders with "A is better than B" semantics, which has been proven to be very suitable in various e-applications. In this paper we present novel Preference Mining techniques for detecting strict partial order preferences in user log data. The main advantage of our approach is the semantic expressiveness of the Preference Mining results. Experimental evaluations prove the effectiveness and efficiency of our algorithms. Since the Preference Mining implementation uses sophisticated SQL statements to execute all data-intensive operations on database layer, our algorithms scale well even for large log data sets. With our approach personalized e-applications can gain valuable knowledge about their customers' preferences, which is essential for a qualified customer service.

1 Introduction

The enormous growth of web content and web-based applications leads to an unsatisfactory behavior for users: search engines retrieve a huge number of results and they are left on their own to find interesting web sites or preferred products. Such a behavior leads not only to frustrated users but also to a reduction of turnover in commercial businesses because customers who are willing to buy cannot do it since they do not find the right product even if it is available. In recent years, several techniques have been developed to build user adaptive web sites and personalized web applications. For instance, e-commerce applications use link personalization to recommend items based on the customer's buying history or some categorization of customers based on ratings and opinions [11]. Another technique is content personalization: web pages present different information to different users based on their individual needs. Thereby, the user can indicate his preferences explicitly using the predefined tools of the underlying portal or the preferences may be inferred automatically from his profile.

State-of-the-art personalization techniques suffer from some drawbacks. Manually customizing web sites is not very feasible to the customer since it is a very time-

N. Lavrač et al. (Eds.): PKDD 2003, LNAI 2838, pp. 204–216, 2003.

consuming task to select relevant content from the huge repertoire provided by the web portal. Personalizing products or web content automatically is a more promising approach. However, the current approaches of automatic personalization lack of preference models with limited expressiveness. State-of-the-art techniques either use scores to describe preferences [7] or just distinguish between liked and disliked values [9]. Thus, complex *"I like A more than B"*-relationships as well as *preferences for numeric attributes* cannot be expressed in a natural way. Furthermore, these approaches are not able to handle *dependencies among preferences*. For example, two preferences can be of equally importance to a customer or one preference can be preferred to another one.

A very expressive and mathematically well-founded framework for preferences has recently been introduced [5]. Customer preferences are modeled as strict partial orders with "A is better than B"-semantics, where negative, numeric and complex preferences form special cases. This approach has been proven to be very suitable for modeling user preferences of almost any complexity. Standard query languages like SQL and XPATH were extended by such preferences [6] in order to deal carefully with user wishes. In this paper, we present algorithms for automatically mining such strict partial order preferences from user log data. Basic categorical and numerical preferences are discovered based on the frequencies of the different attribute values in the user log. These base preferences are then combined to detect complex preferences.

The rest of the paper is organized as follows: After a survey of related work in section 2 we describe the underlying preference model and Preference Mining requirements in section 3. In section 4, we present algorithms for mining categorical, numerical and complex preferences. Section 5 summarizes the results of an extensive experimental evaluation of the accuracy and efficiency of the proposed algorithms. We conclude our paper with a summary and outlook in section 6. All proves are omitted her, but can be found in the extended version of this paper [3].

2   Related Work

Several research groups have studied the usage of log data analysis for personalized applications. In particular, web log mining is a commonly used approach of analyzing web log data with data mining techniques for the discovery of sequential patterns, association rules or user clusters. Such mining techniques have been applied to provide personalized link recommendations to web users [10]. Thereby the user profile of the current user is matched against one or more previously discovered usage profiles.

Lin et al. applied association rule mining algorithms for collaborative recommendations [9]. They mapped user ratings (liked/disliked/not rated) of articles into transactions and used their algorithms in order to detect association rules within these transactions. Such gained rules can be used for the recommendation of articles to the users.

Beeferman and Berger analyzed query log data of search engines [1]. They developed clustering algorithms in order to find groups of URLs that match various keywords given by the user. This approach is not only helpful for delivering better search results but also for the construction of web categories and the generation of ontologies. In [4], Joachims analyzed clickthrough data to improve the results of search engines. He uses the search results that are chosen by the user as additional information. He argues that selected items are better in the opinion of the user and applies this knowledge to find better rankings for future search results.

Our Preference Mining techniques can work either on web logs or query logs. The main advantage of our approach is the semantic expressiveness of the Preference Mining results. Our algorithms compute no scores to distinguish between liked and disliked values but detect intuitive preferences like positive or negative preferences, numerical preferences or complex preferences. Personalized web applications [11] can gain significant improvements by using detailed knowledge about user preferences.

3   User Preferences in Log Data

In this section we revisit those aspects of the preference model of [5] that are relevant for the scope of this paper. We also define requirements on the user log data for mining such preferences.

3.1   Preferences as Strict Partial Orders

A preference P is defined as a strict partial order $P = (A, <_p)$, where $A = \{A_1, ..., A_k\}$ denotes a set of attributes with corresponding domains $dom(A_i)$. The domain of A is defined as Cartesian product of the $dom(A_i)$, $<_p \subseteq dom(A) \times dom(A)$ and $x <_p y$ is interpreted as "y is better than x". A set of intuitive preference constructors for base and complex preferences is defined.

The constructors for base preferences on categorical domains are POS(A, POS-set), NEG(A, NEG-set), POS/NEG(A, POS-set; NEG-set), POS/POS(A, POS1-set; POS2-set) and EXP(A, E-graph). The POS-set $\subseteq dom(A)$ of a POS preference defines a set of values that are better than all other values of dom(A). Analogously, the NEG-set of a NEG preference describes disliked values. The POS/NEG preference is a combination of the previous preferences and in a POS/POS preference optimal values (POS1-set) and alternative values (POS2-set) can be specified. In E-graph of an EXPLICIT preference a user can specify any better-than relationships.

The preference constructors for numerical domains include AROUND(A, z), BETWEEN(A, [low, up]), LOWEST(A) and HIGHEST(A). In an AROUND preference the desired value is z, but if this it not available values with nearest distance apart from z are best alternatives. For a BETWEEN preference the values within [low, up] are optimal. For LOWEST (HIGHEST) preferences lower (higher) values are better.

Preferences can inductively be combined with complex preference constructors. A Pareto preference $P = P_1 \otimes P_2$ treats the underlying preferences as equally important and a Prioritized preference $P = P_1 \& P_2$ treats P_1 as more important than P_2. For instance, $P = POS(author, \{Douglas\ Adams,\ Edgar\ Wallace\})\ \&\ NEG(binder, \{paperback\})$ denotes a POS preference for the authors Douglas Adams and Edgar Wallace, and a NEG preference for paperbacks, whereby the latter preference is less important.

This definition of preference constructors has been proven to be appropriate to describe complex user wishes. Preference engineering examples are shown in [5]. Our Preference Mining developments should be consistent to this preference model. Therefore, not only all base and complex preferences should be detectable by the Preference Miner but also preference properties like preference hierarchies or preference algebra laws (see [5] for details) should be valid for the detected preferences.

3.2 Requirements on User Log Data in Web Applications

Data mining benefits from the availability of a huge amount of data since having many records ensures the statistical significance of patterns [8]. Log data of user transactions can have several sources like web server log-files or transaction logging on an application server.

Web server logs are generated by the web server when a user is visiting a web site. Such files can comply with standardized formats like the Common Logfile Format[1]. The log data includes the IP of the client host, the current timestamp and the URL (uniform resource locator) he is visiting. Valuable information about a user's wishes is stored in the URL, since it contains not only the address but also requested keywords or preferred product properties the user inserted into a web form. For example, if the user requests the book "The Raven" in the e-shop Barnes & Noble[2] the logged URL is *http://search.barnesandnoble.com/booksearch/results.asp?WRD=The+Raven*. But web server logs also have some disadvantages, especially for e-commerce applications [8]. Events like "add to cart" or "change item" are not available in web logs. Furthermore, a user can deactivate cookies in his browser, so no session information or user identification is available. Preference Mining on web server logs requires some data preprocessing. User input like "The Raven" in the above example has to be extracted from the logged URL and has to be stored in a relational database since our Preference Mining algorithms work on database relations. Furthermore, user identification is required to detect preferences for each customer separately.

Application server logs can handle user transactions much better [8]. User and session identification can be accomplished with a login and logout mechanism. Another advantage is the capability to detect business events like "add to cart" or "buy items". For example, an e-commerce application server can record queries, search results, selected items and bought products for each customer separately. Furthermore, application server log data can be stored in databases and therefore huge amount of log

---

[1] http://www.w3.org/Daemon/User/Config/Logging.html
[2] http://www.barnesandnoble.com

data can be managed by using database technology. The Preference Mining algorithms can work directly on these log relations without any data preprocessing. For instance, analyzing the properties of bought products can lead to preferences about liked and disliked features, price preferences and dependencies between such preferences.

While browsing or shopping in an online environment, a customer has typically several different types of input fields for interacting with the underlying system. Text fields allow the input of keywords and choices allow the selection of static or dynamic predefined values of an attribute. To describe these different situations we define the closed world assumption and the difference between static and dynamic domains.

Definition 1 (Closed world assumption (CWA))
The assumption that a customer knows all possible values of an attribute is called Closed World Assumption or CWA. If this assumption doesn't hold we abbreviate it with ¬CWA.

Definition 2 (Static and dynamic domains)
If a domain of attribute values is constant over time, we call it a static domain otherwise we call it a dynamic domain.

The CWA is required for the detection of negative preferences since only if the user knows all possible values we can assume dislike for values he never selected. Otherwise (¬CWA), we can't decide whether he doesn't know or doesn't like such values. For instance, in a book shop the customer knows all possible values for binder (paperback or hardcover) but doesn't know all available authors. After submitting a search query, a customer gets a set of results and chooses one or more of them as his preferred products. Such search results define dynamic domains and can lead to valuable clickthrough data, which can be used to get information about explicit user preferences since the clicked items of the query result are preferred by the user [4].

4   Preference Mining Algorithms

In this section we present algorithms for mining the strict partial order preferences introduced in section 3.1. Our methods work on log relations as described in section 3.2 and use appropriate data mining and statistical methodologies in order to detect the right preference and correct additional information like POS-sets. To detect base preferences we use the frequencies of the different values in the log relation.

Definition 3 (Frequency of a value)
Let A be an attribute of a log relation R and $x \in$ dom(A). The number of entries of x in R(A) is called frequency of x or $freq_A(x)$. If dom(A) is numerical, $freq_A([x_1, x_2])$ denotes the number of entries of all values between x_1 and x_2 ($x_1 \leq x_2$).

We have introduced the concept of user-defined preferences $P = (A, <_p)$. The actual user preferences shall be predicted from the implicit preferences hidden in the user log data. To that purpose, we define data-driven preferences denoted by $P_D = (A, <_{PD})$.

Definition 4 (Data-driven preference)

- For categorical domains dom(A) a data-driven preference $P_D = (A, <_{PD})$ is defined as: $x <_{PD} y$ iff $freq_A(x) < freq_A(y)$.
- For numerical domains dom(A) a data-driven preference $P_D = (A, <_{PD})$ is defined as: $x <_{PD} y$ iff $\exists \varepsilon > 0$: $freq_A([x-\varepsilon, x+\varepsilon]) < freq_A([y-\varepsilon, y+\varepsilon])$.

As it can easily be shown data-driven preferences define strict partial orders. Depending on the design of the log data, values can be products (e.g. search results) or just product properties like color or price. If the frequency of a value x is zero, a customer has never selected the according value. If CWA holds, $freq_A(x) = 0$ means that a customer doesn't like the property x because he never selected it although he knows it. Otherwise, if CWA doesn't hold, the customer may either not like the property x or may never have heard of it. The relation $freq_A(x) < freq_A(y)$ shows that the corresponding customer has selected y more often than x. In this sense the relation $x <_{PD} y$ denotes a preference.

Numeric domains need a slightly different approach to data-driven preferences. For instance, an attribute A may have the real numbers as domain (dom(A) = $\mathbb{R}$) and we want to test, if a user has a data-driven LOWEST(A) preference, i.e. lower values are better and should occur with higher frequencies. Since $\mathbb{R}$ consists of an infinity number of different values, the log relation only contains some of them and typically each value occurs only a few times in the log relation. Therefore, we use frequencies of intervals. E.g. for a data-driven LOWEST preference the relation $freq_A([x-\varepsilon, x+\varepsilon])$ < $freq_A([y-\varepsilon, y+\varepsilon])$ for $y < x$ must hold for some ε.

4.1 Mining Categorical Preferences

Based on $P_D = (A, <_{PD})$ we can define data-driven preferences for categorical domains.

Definition 5 (Data-driven preferences for categorical data)
Let A be a categorical attribute of a log relation R and POS-set, NEG-set, POS1-set, POS2-set, $E \subseteq$ dom(A).

- There is a data-driven POS preference, iff $\forall x \in$ POS-set, $\forall y \notin$ POS-set: $y <_{PD} x$.
- There is a data-driven NEG preference, iff $\forall x \in$ NEG-set, $\forall y \notin$ NEG-set: $x <_{PD} y$.
- There is a data-driven POS/POS preference, iff $\forall x \in$ POS1-set, $\forall y \in$ POS2-set, $\forall z \notin$ (POS1-set $\cup$ POS2-set): $y <_{PD} x$ and $z <_{PD} y$.
- There is a data-driven POS/NEG preference, iff $\forall x \in$ POS-set, $\forall y \in$ NEG-set, $\forall z \notin$ (POS-set $\cup$ NEG-set): $z <_{PD} x$ and $y <_{PD} z$.
- Let $<_E$ be a strict partial order on E. A data-driven EXPLICIT preference holds, iff (1) $\forall x, y \in$ E with $x <_E y$: $x <_{PD} y$, and (2) $\forall u \in$ E, $\forall v \notin$ E: $v <_{PD} u$.

For a data-driven POS preference the values in the POS-set must occur more often than the other values and in a data-driven NEG preference the other values must occur more often than the values in the NEG-set. POS/POS and POS/NEG run analogously. A data-driven EXPLICIT preference with underlying E-graph exists, if a value y occurs more often than any successor x in E-graph. Values outside the E-graph occur with lowest frequencies.

The main task for an algorithm for mining categorical preferences is the detection of proper POS-sets, NEG-sets, etc. Consider the following example of frequencies for an attribute author (CWA doesn't hold, the domain is static):

Table 1. Example of frequencies for an attribute "author"

Douglas Adams	Edgar Wallace	Natalie Angier	Agatha Christie	John Grisham
50	49	2	3	2

The set {Douglas Adams} is a correct POS-set for a data-driven POS preference. But intuitively, the set {Douglas Adams, Edgar Wallace} denotes are more reasonable POS-set since these two values occurred much more frequently than Natalie Angier, Agatha Christie and John Grisham. The following algorithm for mining categorical preferences uses cluster techniques in order to detect such proper sets.

Algorithm 1: Miner for categorical preferences in static domains
INPUT: log relation R, attribute A, dom(A)
(1) Compute for each value x_i the frequency in the log relation $freq_A(x_i)$.
(2) Compute a clustering of the x_i with $freq_A(x_i) \geq 1$ by using a clustering technique.
(3) Depending on the clustering results we have the following possibilities:
   (a) There is only one cluster C_1 and CWA holds. Here we have a NEG(A, $\{x \in dom(A) |\ freq_A(x) = 0\}$) preference.
   (b) There are two clusters C_1 and C_2, where $\forall c_1 \in C_1, \forall c_2 \in C_2$: $freq_A(c_2) <$ $freq_A(c_1)$.
   (b1) If ¬CWA, we have a POS(A, C_1) preference.
   (b2) If CWA, there is a POS/NEG(A, C_1; $\{x \in dom(A)|\ freq_A(x) = 0\}$) preference.
   (c) There are three clusters C_1, C_2 and C_3, where $\forall c_1 \in C_1,\ \forall c_2 \in C_2, \forall c_3 \in C_3$: $freq_A(c_3) < freq_A(c_2) < freq_A(c_1)$. Here we have a POS/POS(A, C_1; C_2) preference.
   (d) There are more than three clusters C_1, ..., C_n, where $\forall c_1 \in C_1, \forall c_2 \in C_2$, ..., $\forall c_n \in C_n$: $freq_A(c_n) < ... < freq_A(c_2) < freq_A(c_1)$. Here we have an EXPLICIT preference EXP(A, $<_E$) with $c_n <_E ... <_E c_2 <_E c_1$, $\forall c_1 \in C_1, \forall c_2 \in C_2, ..., \forall c_n \in C_n$.
   (e) In all other situations there is no data-driven preference.
OUTPUT: the detected preference or that no preference was found

Complexity: If n denotes the number of tuples in the log relation and k is the number of different values, the k-means clustering needs $O(k^2)$ [2] leading to the overall complexity $O(n + k^2)$. Typically, we have k << n and with it the complexity O(n).

By using a state-of-the-art clustering technique like k-means – and silhouettes for getting the right number of clusters, see [12] – this algorithm detects two clusters $C_1 = $ {Douglas Adams, Edgar Wallace} and $C_2 = $ {Natalie Angier, Agatha Christie, John Grisham} leading to a POS(author, {Douglas Adams, Edgar Wallace}) preference in the above example. The data-driven preferences constructed in algorithm 1 are correct since the frequencies match to our requirements stated in definition 5. Data-driven NEG preferences can only be detected, if the user knows all possible values (CWA).

In dynamic domains the CWA holds, because the user must know the varying values for his decisions. By selecting or clicking on one or more of the available values, the user provides preference knowledge since he prefers the selected items to the other available values. The following algorithm for mining such EXPLICIT preferences requires an advanced structure of the log relation. We assume we have the information (query_id, value, selected) within the log relation, whereby "value" contains a value available for the user, "selected" ($\in \{0,1\}$) denotes whether the according value was selected or not and "query_id" specifies which values belong to one search query. The ability of a low-cost construction of such log data has been shown in [4].

Algorithm 2 (Miner for EXPLICIT preferences in dynamic domains)
INPUT: log data in the format (query_id, value, selected)
(1) Compute the k occurring values $(x_1, ..., x_k)$ in the log relation. Initialize the better-than graph with E-graph = $\emptyset$.
(2) FOR(i = 1, ..., k) and FOR(j = i + 1, ..., k) DO:
    (a) Consider the query ids, whose according values contain x_i and x_j.
    (b) Compute the number s of query ids, where x_i was selected and x_j wasn't.
    (c) Compute the number t of query ids, where x_j was selected and x_i wasn't.
    (d1) If s > t and there is no path from x_j to x_i in E-graph, set E-graph = E-graph $\cup$ (x_j, x_i). Otherwise, if a path from x_j to x_i exists, remove it.
    (d2) If s < t and there is no path from x_i to x_j in E-graph, set E-graph = E-graph $\cup$ (x_i, x_j). Otherwise, if a path from x_i to x_j exists, remove it.
    (d3) If s = t remove within E-graph all direct and transitive connections from x_i to x_j and vice versa.
OUTPUT: the detected EXPLICIT preference based on E-graph as better-than graph.

Complexity: In the first step the n entries in the log relation have to be considered (O(n)). The two nested FOR-loops have the complexity $O(k^2)$. Within the loops all the tuples in the log relation need to be analyzed O(n) and a path between two given vertices ($O(k^2)$ by using Dijkstra's algorithm) has to be computed. Thus we have $O(n + k^2(n + k^2))$ as complexity. A main effort lies in the detection of inconsistencies in the user's shopping and browsing behavior. If we assume a consistent user behavior, we can avoid the path detection. The resulting complexity would be $O(n + k^2n)$.

For two values x_i and x_j the algorithm computes the query ids that have both values in the result set. Now x_i is better than x_j, if the user selected it more often. In step (2d) cycles are removed. Therefore we check if there is a path from x_j to x_i in E-graph before inserting (x_j, x_i) and vice versa. Cycles can occur, if the browsing or shopping behavior of the user has inconsistencies like blue $<_p$ red $<_p$ green $<_p$ blue. In such situations the preferences of the customer are not clear and therefore we leave out such relations. If s = t, the user is indifferent between x_i and x_j and therefore existing preference relations between x_i and x_j have to be removed.

4.2   Mining Numerical Preferences

The distribution of numerical log data defines a statistical density function $\varphi(x)$. Properties of this density function provide information about data-driven preferences. For instance, if $\varphi(x)$ has a unique maximum at z and the gradient is positive for $x < z$ and negative for $x > z$, there is an AROUND preference with around-value z. This approach is consistent to the definition of numeric data-driven preferences (def. 4) because an increasing density guarantees $freq_A([x-\varepsilon, x+\varepsilon]) < freq_A([y-\varepsilon, y+\varepsilon])$ for $x <$ y and a decreasing density implicates $freq_A([x-\varepsilon, x+\varepsilon]) > freq_A([y-\varepsilon, y+\varepsilon])$. Thus, in the above example values around z are requested most frequently and the frequency decreases with increasing distance to z. Since the density is usually unknown, it has to be estimated using the underlying numerical log data. In our implementation we use histograms as an easy to use and efficient density estimation technique [3].

4.3   Mining Complex Preferences

Definition 6 (Data-driven Prioritized preference)
Let $P_D = (A, <_{PD})$ and $Q_D = (B, <_{QD})$ be two data-driven preferences and $x = (x_1, x_2)$, y $= (y_1, y_2) \in dom(A) \times dom(B)$. A data-driven Prioritized preference $P_D \& Q_D = (\{A, B\}, <_{PQ-D})$ is defined as: $x <_{PQ-D} y$ iff $x_1 <_{PD} y_1 \vee (x_1 = y_1 \wedge x_2 <_{QD} y_2)$.

Data-driven Pareto preferences can be handled analogously [3]. In order to detect such complex data-driven preferences we need the definition of associate values.

Definition 7 (Associate Values)
Consider a log relation R(A, B, ...). For $a \in \pi_A(R)$ the associate values in B are defined as $asv_{A, B}(a) = \pi^*_B(\sigma_{A=a}(R))$.

Thereby π^* denotes the relational projection without removing duplicates.

Algorithm 3 (Miner for Prioritized preferences)
INPUT: log relation R(A, B, ...) and a data-driven preference P_D on A
(1) Compute the set M of maximal values of P_D and for all $a_i \in M$ the set of associate values $asv_{A, B}(a_i)$.
(2) If there is the same preference Q_D in all sets $asv_{A, B}(a_i)$ and P_D does not occur in the associate values of the maxima of Q_D, there is a Prioritized preference $P = P_D \& Q_D$.
(3) Otherwise there is no Prioritized preference.
OUTPUT: the detected Prioritized preference or that no preference was found

Complexity: If n denotes the number of tuples, k_1 and k_2 the effort for mining P_D and Q_D, respectively, above algorithm has the complexity $O(n^2 + nk_1 + nk_2)$, since in maximal n values in A, the associate values in B are computed leading to $O(n^2)$. Furthermore, in maximal n sets the existence of the preference Q_D has to be tested ($O(nk_2)$) and vice versa ($O(nk_1)$).

A data-driven Prioritized preference $P = P_D \& Q_D$ exists, if, firstly, there is a data-driven preference P_D, and, secondly, in those tuples, which have equal values in A, there is a data-driven preference Q_D in B. Thereby, we consider only the maximal values of P since users often don't care about a second-level preference, if the prioritized preference isn't fulfilled optimal. If P_D also occurs in the maximal values of Q_D, a Pareto preference has been found. Therefore, we have to eliminate this situation here.

In our previous example the preference P_D = POS(author, {Douglas Adams, Edgar Wallace}) was detected. If above algorithm detects Q_D = NEG(binder, {paperback}) (dom(binder) = {hardcover, paperback}) in the associate values of Douglas Adams and Edgar Wallace and, furthermore, P_D is not detected within the hardcover books, a Prioritized preference P = P_D & Q_D is found.

5   Experimental Evaluation

In this section we present test results and performance measurements of an efficient database-driven implementation of a Preference Miner prototype.

5.1   Preference Mining Test Results

We performed an analysis of the Preference Mining algorithms on the log data of the COSIMA application [3]. Over five hundred users queried the COSIMA comparison shop almost four thousand times. COSIMA offers shopping in the three categories books, cds and computer products. The application server records for each query the timestamp, the shop category, the preferred price interval and – depending on the category – title and author in the book shop, title and performer in the cd shop and product name and product group in the computer hardware category. We applied the Preference Mining algorithms on the COSIMA log data, whereby we analyzed the log-data for each user separately. The Preference Miner detected lots of POS preferences and also one POS/POS preference for the shop category. Quite a few LOWEST preferences for price were detected by analyzing the lower price limit. NEG preferences and even complex preferences were also detected with the Preference Miner. Mining preferences works very fast: on a PC with 1,3 ghz and 1,5 gb main memory the Preference Miner needed less than one second to detect the preferences of a user.

Though these test results on real data show the practical usability of our techniques we cannot prove the correctness with it since the customer preferences are a priori unknown. Therefore we created synthetic log data using simulated users with predefined preference profiles. We defined 35 profiles, where each profile contains between two and six preferences, e.g. {P1 = POS(color, {blue}, P2 = LOWEST(price), P3 = P1 & P2}. In our simulation each user queries the product database between 25 to 50 times. In each query a preference of the considered user is chosen and a product database is requested with it using Preference SQL [6]. The results are stored in a log database. Afterwards we use the Preference Mining algorithms to detect preferences within the log data. A comparison of the detected preference profiles with the predefined user preferences will show the effectiveness of the Preference Mining algorithms. To assess the quality of our results we define preference precision and preference recall.

Definition 8 (Precision and recall for preferences)

Preference precision and preference recall are defined as

$$precision = \frac{\text{number of correctly detected preferences of user i}}{\text{number of all detected preferences of user i}}$$

$$recall = \frac{\text{number of correctly detected preferences of user i}}{\text{number of all preferences of user i}}$$

The algorithms for mining base preferences lead to a 60 % precision and a 39 % recall averaged over all users. This means that 60 % of the detected preferences occur exactly in the predefined preference profiles and 39 % of the predefined preferences are detected with our algorithms. Since a preference is regarded as correct only if all underlying information (POS-set, NEG-set, around-value, etc.) is detected correctly, a 60 % precision and a 39 % recall are very promising. Mining complex preferences yield to 55 % precision and 15 % recall. The poor recall here is caused by dependencies between preferences: if a base preference of a complex preference is not detected, it is not possible to detect the complex preference itself. Note, that we filled the log relation with the search results. In real-life applications even better Preference Mining results can be achieved, if the selected results or query information is used.

5.2   Performance Measurements

In this section we analyze the efficiency of the Preference Miner prototype for large data sets. The underlying database system is an Oracle 8i database server on an AMD CPU with 1,3 ghz and 1,5 gigabyte main memory. For our tests, we created relations with 10,000, 20,000, 30,000, 40,000 and 50,000 tuples of synthetic data. Categorical attributes contain 20 different categories. Numerical attributes have a data range of 200 (maximal minus minimal value). For mining complex preferences we assume one categorical and one numerical attribute. Fig. 1 reports the average runtimes for detecting a single preference for the different preference types w.r.t. the number of tuples.

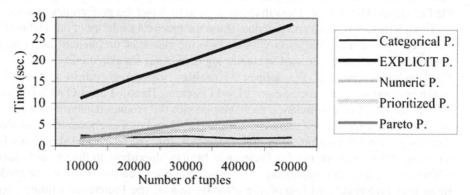

Fig. 1. Runtimes for detecting a single preference for the different preference types

Mining numerical preferences is the fastest task, since histograms can be computed very efficiently in the database layer. The miner for categorical data needs more effort since clustering is a more expensive iterative process. Mining Prioritized and Pareto preferences needs about 5 seconds in the average. The most expensive algorithm is the miner for EXPLICIT preferences (algorithm 2). The cost-intensive part is the cycle test and leads to a performance which depends linearly to the number of tuples. The efficiency of our Preference Mining algorithms allows their usage for *online Preference Mining*: while interacting with a customer an e-application can check online his preferences and react flexible to his wishes during the sales process.

6   Summary and Outlook

In this paper we have presented a novel approach for mining preferences from user log data based on the concept of strict partial order preferences. We presented several algorithms for the detection of categorical, numerical and complex preferences. Our prototype implementation executes all data-intensive operations on the database server and exhibits excellent efficiency. Our experimental results also demonstrate promising precision and recall of the detected user preferences.

Our next steps include the integration of user situations into preferences. Situations can be described with a set of parameters like current time and location, the user's role or physical and psychological condition of the user. Some preferences may only be relevant under specific situations; for example, in a bookshop a user may have different preferred categories whether he is at work or at home. A major task is the adaptation of our Preference Mining algorithms in order to detect *situated preferences*.

Another research task is the design of an appropriate storage structure for preferences. Such a *Preference Repository* should not only be able to record preferences detected with the Preference Miner but also preferences defined with Preference SQL or Preference XPATH. The integration of situations should also be possible as well as user identifiers to assign users and user groups. Finally, the Preference Repository shall also include a set of appropriate access operations for inserting, deleting and updating preferences. It can also be used to find users with similar preferences and with it product recommendations based on preferences can be offered. Therefore the Preference Repository is also a major step towards advanced personalized applications.

Acknowledgements

This work is partially supported by the German Research Foundation DFG within the research group "Efficient Electronic Coordination in the Service Sector".

References

1. D. Beeferman and A. Berger: *Agglomerative Clustering of a Search Engine Query Log.* Proc. ACM SIGKDD 2000, p. 407-416, Boston, Massachusetts, USA, 2000.
2. V. Estivill-Castro and M. E. Houle: *Robust Distance-Based Clustering with Applications to Spatial Data Mining.* In 3rd Pacific-Asia Conference on Knowledge Discovery and Data Mining, p. 327–337, Beijing, China, 1999.
3. S. Holland, M. Ester and W. Kießling: *Preference Mining: A Novel Approach on Mining User Preferences for Personalized Applications.* Technical Report 2003-5, Institute of Computer Science, University of Augsburg, May 2003. http://www.informatik.uni-augsburg.de/nav/forschung.
4. T. Joachims: *Optimizing Search Engines using Clickthrough Data.* Proc. ACM SIGKDD 2002, Edmonton, Alberta, Canada, 2002.
5. W. Kießling: *Foundations of Preferences in Database Systems.* Proc. VLDB 2002, p. 311-322, Hong Kong, China, 2002.
6. W. Kießling and G. Köstler: *Preference SQL - Design, Implementation, Experiences.* Proc. VLDB 2002, p. 990-1001, Hong Kong, China, 2002.
7. S.-J. Ko, J.-H. Lee: *User Preference Mining through Collaborative Filtering and Content Based Filtering in Recommender System.* Proc. of the 3rd Intern. Conf. on E-Commerce and Web Technologies (EC-Web 2002), p. 244-253, Aix-en-Provence, France, 2002.
8. R. Kohavi: *Mining E-Commerce Data: The Good, the Bad, and the Ugly.* Proc. ACM SIGKDD 2001, p. 8-13, San Francisco, California, USA, 2001.
9. W. Lin, S. A. Alvarez, and C. Ruiz: *Efficient Adaptive-Support Association Rule Mining for Recommender Systems.* In DMKD Journal, vol. 6 (1), p. 83-105, 2002.
10. B. Mobasher, R. Cooley and J. Srivastava: *Automatic Personalization Based on Web Usage Mining.* In Communications of the ACM, vol. 43 (8), p. 142-151, August, 2000.
11. G. Rossi, D. Schwabe and R. Guimaraes: *Designing Personalized Web Applications.* Proc. 10th World Wide Web Conference (WWW 2001), p. 275-284, Hong Kong, China, 2001.
12. P. J. Rousseeuw: *Silhouettes: A Graphical Aid to the Interpretations and Validation of Cluster Analysis.* Journal of Computational and Applied Mathematics, 20, 53–65, 1987.

Explaining Text Clustering Results
Using Semantic Structures

Andreas Hotho, Steffen Staab, and Gerd Stumme

Institute of Applied Informatics and Formal Description Methods AIFB
University of Karlsruhe, D–76128 Karlsruhe, Germany
{hotho,staab,stumme}@aifb.uni-karlsruhe.de
http://www.aifb.uni-karlsruhe.de/WBS

Abstract. Common text clustering techniques offer rather poor capabilities for explaining to their users why a particular result has been achieved. They have the disadvantage that they do not relate semantically nearby terms and that they cannot explain how resulting clusters are related to each other. In this paper, we discuss a way of integrating a large thesaurus and the computation of lattices of resulting clusters into common text clustering in order to overcome these two problems. As its major result, our approach achieves an explanation using an appropriate level of granularity at the concept level as well as an appropriate size and complexity of the explaining lattice of resulting clusters.

1   Introduction

Clustering is an important task that is performed as part of many text mining and information retrieval systems. It can be used for efficiently finding the nearest neighbors of a document [1], for improving the precision or recall in information retrieval systems [15,11], for aid in browsing a collection of documents [3,8], as well as for the organization [19] and personalization of search engine results [13].

Most current document clustering approaches are based on the vector-space model (also called bag of words model or word space), the dimensions of the vector space are constituted by the important words of the document collection. The respective term or word frequencies (TF) in a given document constitute the vector describing this document. In order to discount frequent words with little discriminating power, each word can additionally be weighted based on its Inverse Document Frequency (IDF) in the document collection. Once the documents are mapped into the vector space, they can be clustered according to the distances between the vectors. However, what is neglected in these approaches are the explanations of why particular clusters have been formed and how the different clusters are related to each other.

To elaborate on this, we build on a finding by Karypis and Han. They have shown [10] that words occurring with high weights in the centroid of a cluster can be used to summarize the content of the cluster. They observed that "prevalent terms of the various centroids often contain terms that act as synonyms within the context of the topic they describe." Common text clustering algorithms lack the capabilities

N. Lavrač et al. (Eds.): PKDD 2003, LNAI 2838, pp. 217–228, 2003.

(A) to recognize such synonyms in order to improve text clustering quality;

(B1) to use such synonymity in order to improve the quality of the explanation of why a cluster has been formed (e.g., just state 'this cluster is about Volkswagen' instead of stating 'this cluster is about Volkswagen and about VW');

(B2) to exploit semantic hierarchies of words in order to abstract an explanation (e.g., instead of 'this cluster is about pork and beef and veal' state 'this cluster is about meat');

(C) to give an account of how resulting clusters are related (e.g. 'cluster1 is about the same topics as cluster2, but additionally about meat').

With regard to *(A)*, we have investigated how a large thesaurus like WordNet [5] may help to improve text clustering results exploiting synonymity and other semantic relationships[1].

With regard to *(B1)* and *(B2)*, we use the synonymity of words and the hierarchy of their corresponding concepts as defined in a thesaurus[2] to come up with more concise and abstracting explanations. We extract explanations from the centroid representation of resulting clusters, but we also use thesaurus concepts instead of words only.

Finally, with regard to *(C)*, we have explored the use of lattice theory. Formal concept analysis (FCA) [6] computes the place of an object representation (e.g. the representation of a text or the representation of a text cluster) in a lattice according to its vector representation. Unfortunately, formal concept analysis (and similar means of analysis) are not suited to directly relate vector representations of large collections of texts. Tests with text samples have revealed that even homogeneous text sets have relatively few joint word occurrences leading to large and complex lattices with unsatisfying explanatory power[3].

In this paper, we address the challenges listed above. Here, we will specifically discuss the challenges *(B)* and *(C)*. Our approach proceeds along the following lines:

1. It represents text documents by a vector model that exploits the hierarchy of the concepts in the WordNet thesaurus (cf. Section 2);
2. it uses a common text clustering algorithm, BiSec–k–Means, to aggregate texts without supervision into a pre-defined number of clusters. Moreover it extracts a representation of each resulting cluster (cf. Section 3);
3. it computes a lattice from the resulting cluster representations to relate, *(i)* words from the thesaurus hierarchy with the different clusters and, *(ii)* to compare the different cluster representations (cf. Section 4);
4. eventually, it visualizes (parts of) the resulting lattice structure(s) allowing the user to explore explanations of how and why clustering results have been produced (cf. Section 5).

---

[1] A comprehensive empirical investigation on how a thesaurus may improve text clustering is reported in [9].

[2] What is typically called a 'concept' in ontologies or thesauri is a very close match to what is called a 'synset' in WordNet.

[3] Result size complexity of FCA may become a problem if the lattice is computed on several thousands of texts, as the number of nodes in the lattice may grow exponentially with the number of objects. In practice runs, however, we have observed that formal concept analysis scaled quite well with regard to the number of texts — much better than worst case.

We discuss our approach along the Reuters–21578 text collection. Section 6 provides an overview of related work. In particular, we emphasize that while the individual algorithms of steps 1 through 4 are well-known and to some extent interchangable with likewise approaches, the combination we describe here is unique and original, while it serves frequently arising objectives in text clustering.

2   Representing Texts

This section describes the representation of texts in an exemplary manner drawing from the Reuters-21578 dataset on which we performed many of our experiments. The basic idea of this section is the extension of the common text representation as a vector in a word space towards a representation in a word/concept space.

The Reuters-21578 Dataset We selected the Reuters-21578 [4] text collection for our experiments. The corpus consists of 21578 documents. This corpus is especially interesting for evaluation, as part of it comes along with a (hand-crafted) classification. It contains 135 so-called topics. To be more general, we will refer to them as 'classes' in the sequel. For allowing evaluation, we restrict ourselves to the 12344 documents which have been classified manually by Reuters. Some of them could not be assigned by the experts to one of the predefined classes; we collect them in an additional class 'defnoclass'. Reuters assigns some of its documents to multiple classes, but we consider only the first assignment. After these steps, we obtain our final corpus $\mathcal{D}$ for evaluation. It consists of the 12344 documents, distributed over 82 Reuters topics.

Preprocessing the Document Set For the preprocessing of the documents, we used the text mining system developed at AIFB within the KAON[5] framework. We performed the following steps on the selected corpus: First we lowered the letters of all words and removed stopwords. We used a stopword list with 571 entries which removed 416 stopwords from the documents. We also dropped all words with less than 30 occurrences over the whole corpus. 17917 words were removed in total. After these steps, 2657 different words remained in our list, with a total occurrence of 784434.

WordNet as Background Knowledge Instead of using a bag-of-word model directly, we additionally enriched the text representation with background knowledge. The basic idea is to replace the words by concepts and their broader concepts as defined in a given thesaurus, in order to capture similarities at various levels of generalization. For this purpose we needed a resource suitable for the Reuters corpus. We choose WordNet[6] as our background knowledge. WordNet consists of so-called synsets, together with a hypernym/hyponym hierarchy[7].

To modify the existing word vector representations of text, we have first replaced all nouns that appeared in the documents and that were known by WordNet by the corresponding concept ('synset') identifiers from WordNet. At this point, we had several choices of, *(i)*, how to deal with terms not known by WordNet (delete or keep), *(ii)*, how to deal with ambiguity (one word in the document, like 'bank', may correspond

---

[4] http://www.daviddlewis.com/resources/testcollections/ reuters21578/
[5] http://kaon.semanticweb.org
[6] http://www.cogsci.princeton.edu/~wn/
[7] See http://www.cogsci.princeton.edu/~wn/man1.7.1/wngloss.7WN.html for a glossary.

to several concepts in WordNet), and, *(iii)*, how many generalizations of a concept to consider to use for the text representation. We have elaborated on these choices in [9] and present here only a simple, but quite effective combination.

In the simplest case, *(i)*, we have ignored all words that either were not nouns or that were not known by WordNet. *(ii)*, we have used a disambiguation method provided by WordNet. WordNet has a ranking of what is the 'most common' meaning for a word in English. We here use only this static ranking to map a word onto corresponding concepts. *(iii)*, we have mapped a word occurrence in a document to its most highly ranked concept in WordNet as well as to the four most specific generalizations of this concept. For instance, the occurrence of 'bank' in a document would increase the vector entry corresponding to 'banking company' (the concept) as well as the vector entries corresponding to 'financial institution', 'institution', 'organization' and 'social group' as these are the four most specific generalizations of 'banking company'. The concepts that were assigned to at least one document formed then the new set $\mathcal{T}$ of terms used for describing the documents, i. e., they constitute the dimensions of the vector space for the new text representation.

Enriching the term vectors with concepts from WordNet has two benefits. First it resolves synonyms; and second it introduces more general concepts which help identifying related topics. For instance, a document about 'bank' may not be related to a document about 'insurer' by the cluster algorithm if there are only 'bank' and 'insurer' in the term vector. But if the more general concept 'financial institution' is added to both documents, their semantical relationship is revealed.

In the remainder of this paper, we will use the expression 'term' both for words and for concepts (synsets) of the thesaurus for sake of simplicity. If we talk about one of them specifically, we will mention it explicitly.

Building the Term Vectors Based on the work done so far, we built a term vector for each document $d \in \mathcal{D}$. For each document, the terms $t \in \mathcal{T}$ are weighted by *tfidf* (term frequency × inverse document frequency) [16], which is defined as follows: $tfidf(d, t) = tf(d, t) \times \log \left(\frac{|\mathcal{D}|}{|\mathcal{D}_t|} \right)$, where $tf(d, t)$ is the frequency of term t in document d, and $\mathcal{D}_t \subseteq \mathcal{D}$ is the set of all documents containing term t. The term vector for document d is then the tuple $\mathbf{w}_d := (tfidf(d, t))_{t \in \mathcal{T}}$.

Tfidf weighs the frequency of a term in a document with a factor that discounts its importance when it appears in almost all documents. Therefore terms that appear too rarely or too frequently are ranked lower than terms that hold the balance and, hence, are expected to be better able to contribute to clustering results.

After this first step, we have thus obtained a description of all documents, which is enriched by background knowledge, and which will also allow to relate semantically close (but syntactically different) documents.

3   Text Clustering and Feature Extraction

In this section, we show how to cluster our, so far uncommon, text representations with state-of-the-art methods (cf. [17]). The major output of this section is an explanation of the text clusters achieved by known means, like [10], which is then used as an input for analysis and explanations in subsequent sections.

The reader may note that while we present a specific, effective and efficient, approach for text clustering and extraction of cluster representations here, this approach might be replaced by other methods that digest similar input and produce similar output without changing our principal approach.

3.1   Text Clustering with BiSec–k–Means

On the preprocessed data we applied BiSec–k–Means [17], a 'bisecting' variant of k–Means, using the cosine similarity: the similarity between two documents $d_1, d_2 \in \mathcal{D}$ is calculated as the cosine of the angle between their term vectors $\boldsymbol{w}_{d_1}$ and $\boldsymbol{w}_{d_2}$.

For this clustering step we need a fast algorithm (such as k–Means), which is able to deal with large datasets, which should also provide a reasonable accuracy. Instead of a slow agglomerative clustering technique with a good accuracy we choose BiSec–k–Means which tends to give better results as k–Means and is sometimes also better as agglomerative clustering, while it is as fast as k–Means (cf. the seminal paper [17]).

BiSec–k–Means is based on the k–Means algorithm. It repeatedly splits the largest cluster (using k–Means) until the desired number of clusters is obtained. As input, it takes the list $(\boldsymbol{w}_d)_{d \in \mathcal{D}}$ of document descriptions and the number k of desired clusters. As output, it provides a partitioning $\mathcal{C}$ of of the set $\mathcal{D}$ of documents (i. e., a set $\mathcal{C}$ of k disjoint subsets of $\mathcal{D}$ with $\bigcup_{C \in \mathcal{C}} C = \mathcal{D}$). Each cluster $C \in \mathcal{C}$ is represented by its centroid $\boldsymbol{w}_C$.

3.2   Extracting Cluster Descriptions

For a good explanation of results, it is necessary to detect important terms that are concise about the explanation created. The basic idea of mechanisms like latent semantic indexing [4] or concept indexing [10] is that the 'importance' of a component can be derived from the weight it receives by an analysis (be it singular value decomposition or k-means clustering, respectively). Correspondingly, we here rank the importance of terms for explaining clustering results based on the weights they have in the cluster centroids.

In order to be able to control how many terms remain to describe the clusters and, hence, be concise, we discretize the term ranks into three descriptions 'very important', 'important', or 'uninteresting' by two thresholds θ_1, θ_2. In our running example, we set θ_1 to 7 % and θ_2 to 20 % of the maximal value. We can then explain the clustering results by considering the terms that are at least 'important' for a resulting cluster.

3.3   Examining BiSec–k–Means Explanations on the Reuters-21578 Dataset

Table 1 shows the highest ranked terms from the centroids of ten (out of 100) resulting clusters on the Reuters-21578 dataset together with their value in the respective centroid. All listed values are above the lower threshold $\theta_1 = 7\%$ (i.e., they are at least 'important'). In general the set of terms that exceed the threshold is much larger (e.g., up to 50 terms of a cluster centroid exceed θ_1) than the set that can be listed here.

A general overview of these results reveals that it is hard to understand the results. While some part of the difficulty stems from the simple, tabular way in which it is presented to the user, quite a substantial part of the difficulty comes from the sheer fact that there are only few meaningful structures that can be represented to the user at all. To

Table 1. The highest ranked terms in the first ten out of 100 clusters resulting from a BiSec–k–Means run on the Reuters-21578 dataset (ordered by their values in the respective centroids).

Cluster 0		Cluster 1		Cluster 2		Cluster 3		Cluster 4	
amount	0,12	depository financial instit	0,09	loss	0,34	Irani, Iranian, Persian'	0,14	indebtedness, liability, fin	0,12
billion, one million million	0,11	financial institution, finan	0,09	failure	0,33	Iran, Islamic Republic of	0,13	obligation	0,12
large integer'	0,11	rate, charge per unit'	0,09	nonaccomplishment, non	0,32	gulf	0,13	debt	0,12
integer, whole number'	0,11	charge	0,09	Connecticut, Nutmeg Sta	0,28	vessel, watercraft'	0,12	written agreement'	0,1
insufficiency, inadequacy	0,1	institution, establishment	0,09	ten, 10, X, tenner, decad	0,24	ship	0,12	agreement, understandin	0,08
deficit, shortage, shortfall	0,1	loss	0,08	American state'	0,23	craft	0,12	creditor	0,08
number	0,09	monetary unit'	0,07	state, province'	0,22	Asian, Asiatic'	0,10	lender, loaner'	0,08
excess, surplus, surplus	0,09	central, telephone exchai	0,07	system, unit'	0,19	person of color, person o	0,10	statement	0,07
overabundance, overmuc	0,09	financial loss'	0,06	network, net, mesh, mesi	0,19	Asian country, Asian nati	0,10	billion, one million million	0,06
abundance, copiousness	0,09	outgo, expenditure, outla	0,06	September, Sep, Sept'	0,18	oil tanker, oiler, tanker, ta	0,10	large integer'	0,05

Cluster 5		Cluster 6		Cluster 7		Cluster 8		Cluster 9	
text, textual matter'	0,15	loss	0,34	gross sales, gross reven	0,11	tender, legal tender'	0,15	metric weight unit, weight	0,15
matter	0,15	failure	0,33	sum, sum of money, amd	0,09	offer, offering'	0,14	metric ton, MT, tonne, t'	0,15
letter, missive'	0,15	nonaccomplishment, non	0,32	income	0,09	medium of exchange, mo	0,11	mass unit'	0,14
sign, mark'	0,13	common fraction, simple	0,22	financial gain'	0,09	speech act'	0,1	palm, thenar'	0,14
clue, clew, cue'	0,13	fraction	0,22	gain	0,09	indicator	0,1	area, region'	0,12
purpose, intent, intention	0,11	rational number'	0,22	enterprise	0,05	standard, criterion, meas	0,1	unit of measurement, unit	0,10
evidence	0,11	real number, real'	0,22	business, concem, busin	0,05	reference point, point of r	0,09	organic compound'	0,10
indication, indicant'	0,11	complex number, comple	0,22	assets	0,05	signal, signaling, sign'	0,08	oil	0,10
goal, end'	0,1	one-half, half'	0,22	division	0,05	acquisition	0,06	lipid, lipide, lipoid'	0,10
writing, written material,	0,07	revolutions per minute, r	0,22	army unit'	0,05	giant	0,06	compound, chemical com	0,08

substantiate this claim, let us investigate in detail what kind of structures are available for subsequent explanation in a visualization tool and which are not.

E.g., one may recognize from Table 1 that clusters 2 and 6 are *similar* because they both are about 'loss', 'failure' and 'non-accomplishment'. Also, the human observer may *interpret* a cluster description like the one of cluster 1, in order to guess that the list 'depository financial institution', 'financial institution', 'rate', 'charge', 'institution', 'loss', 'monetary unit', 'financial loss', 'expenditure' probably means that this cluster is about financial transaction with loss (which is also correct when investigating the corresponding Reuters news documents). While these structures are not that easy to find in the tables, it is not hard to imagine a user interface to facilitate their discovery.

However, there are meaningful structures that are more difficult to find. For instance, the occurrence of 'oil' relates Cluster 3 (ranked further down in the list) with Cluster 9 and several other clusters (from the set of clusters 10 to 99). Along similar lines, it would be nice to see how switching from a general concept like 'chemical compound' to a more specific one like 'oil' switches the set of associated clusters. Eventually, one would like to find that a particular type of oil or a term like 'palm' is a unique property of Cluster 9 as compared against all other clusters. Such structural dependencies require a further analysis as we propose in the following sections.

Eventually, we want to summarize the problems encountered by extracting explaining terms from cluster centroids: This model, if used on its own, assumes that the ranking of terms adequately reflects the importance of terms, which is often not the case (e.g., for Cluster 6 it remains unclear what type of 'loss' is encountered). In fact, importance frequently depends on what terms help to explain *commonalities* or *differences* between clusters — an analysis provided by the next step.

4   Computing the Lattice of Cluster Representations

The clusters obtained by the previous step have the advantage that they cluster similar documents. However, the clustering does not give a description of how the clusters are related to each other, i.e. an explicit account of what their commonalities and differences are. Formal Concept Analysis derives a lattice that incorporates this account.

4.1   Formal Concept Analysis

Formal Concept Analysis (FCA) was introduced for modeling the concept 'concept'[8] in terms of lattice theory. We recall the basics of FCA as far as needed for this paper. An extensive overview is given in [6]. To allow a mathematical description of concepts as being composed of extensions and intensions, FCA starts with a *formal context*:

Definition: A *formal context* is a triple $\mathbb{K} := (G, M, I)$, where G is a set of *objects*, M is a set of *attributes*, and I is a binary relation between G and M (i.e. $I \subseteq G \times M$). $(g, m) \in I$ is read *"object g has attribute m"*.

A straightforward way of modeling our problem in FCA would be to let the set of objects consist of all clusters determined in the previous step, i. e., $G := \mathcal{C}$ and let the set of attributes consists of all terms which remain from the step described in Section 3.3, i. e., $M := \mathcal{T}_c$. In order to obtain a more fine-grained view, we additionally apply *conceptual scaling*. We use the two thresholds of our example and impose an ordinal scale on the object set with two thresholds θ_1 and θ_2. The formal context (G, M, I) is then composed as follows: $G := \mathcal{C} \times \{\theta_1, \theta_2\}$, $M := \mathcal{T}_c$, and $((C, \theta_i), t) \in I :\iff (t_C)_t \geq \theta_i$. The relation I, applied to a pair (C, θ_i), returns thus the set $\{(C, \theta_i)\}'$ of all attributes which are more or less (i. e., with threshold θ_i) relevant for cluster C. From a formal context, a concept hierarchy, called *concept lattice*, can then be derived:

Definition: For $A \subseteq G$, we define $A' := \{m \in M \mid \forall g \in A : (g, m) \in I\}$ and, for $B \subseteq M$, we define $B' := \{g \in G \mid \forall m \in B : (g, m) \in I\}$.

A *formal concept* of a formal context (G, M, I) is defined as a pair (A, B) with $A \subseteq G$, $B \subseteq M$, $A' = B$ and $B' = A$. The sets A and B are called the *extent* and the *intent* of the formal concept (A, B). The *subconcept–superconcept relation* is formalized by

$$(A_1, B_1) \leq (A_2, B_2) :\iff A_1 \subseteq A_2 \quad (\iff B_1 \supseteq B_2) .$$

The set of all formal concepts of a context $\mathbb{K}$ together with the partial order $\leq$ is always a complete lattice[9], called the *concept lattice* of $\mathbb{K}$ and denoted by $\mathfrak{B}(\mathbb{K})$.

The resulting concept lattice can also be interpreted as a concept hierarchy directly on the documents, as it is isomorphic to the concept lattice of the context $\mathbb{K}' := (G', M', I')$ with $G' := \mathcal{D}$, $M' := \mathcal{T}_c$, and $(d, t) \in I'$ iff $d \in C$ and $(w_C)_t \geq \theta$ for some cluster $C \in \mathcal{C}$. This context is an approximation of the descriptions of the documents by term vectors, with the property that all documents in one cluster obtain exactly the same description.

Clustering the objects before applying FCA is an abstraction that might be considered a loss of information. However, it is predominantly beneficial for the following reasons. Firstly, it reduces the number of objects such that FCA becomes more efficient. Secondly,

---

[8] 'concept' in FCA is a different notion than 'concept' in a thesaurus or ontology.
[9] I. e., for each set of formal concepts, there exists always a unique greatest common subconcept and a unique least common superconcept.

the technique is robust with regard to upcoming documents: A new document is first assigned to the cluster with the closest centroid, and then finds its place within the concept lattice. If on the contrary the document would be considered directly for computing the concept lattice, it could not be guaranteed that the structure of the lattice would not change. Finally and most importantly, formal concept analysis applied directly on all documents suffers from the low co-occurrence of terms. The application of FCA on the Reuters-21578 dataset has shown that hardly any two texts are placed into a common node of the lattice. Thus, the lattice became large, unwieldy and hard to understand for the human user. Therefore, in our approach we first cluster a large number of texts (e.g. 10^5) into a more manageable number of clusters (e.g. 10^2). Only then we compute the lattice, allowing for abstraction from some randomness in the joint occurrences of terms.

4.2   The Lattice of the Reuters Clusters

In the Reuters setting, we obtain from the representation computed in the previous section a list of over hundred formal concepts. Each of them groups together clusters of the previous steps. This grouping indicates the conceptual similarity of the clusters. E.g., we obtain a formal concept, which we here refer to by (*), that has {CL 3 (m), CL 9 (m), CL 23 (m), CL 79 (m), CL 85 (m), CL 95 (m)} as extent, and {organic compund, oil, 'lipid, lipide, lipoid', 'compound, chemical compound'} as intent[10]. This formal concept indicates the commonalities ('conceptual similarity' when calling it by FCA terminology) of these clusters: the majority of documents within these clusters are about oil.

The formal concept (*) has three direct subconcepts: the first has {CL 3 (m)} as extent, and the attributes from above plus some attributes like 'oil tanker' and 'Iranian' as intent. The second has {CL 9 (m)} as extent, and the attributes from above plus some attributes like 'area', 'palm', and 'metric ton' as intent. The third subconcept has {CL 23 (m), CL 79 (m), CL 85 (m), CL 95 (m)} as extent, and the attributes from above plus 'substance, matter' as intent. These three subconcepts of (*) show what distinguishes the clusters grouped together in the formal concept (*). The majority of documents in Cluster 3 are about transport of oil (from Iran), those in Cluster 9 about (packaging of) palm oil, and those in the remaining clusters about crude oil.

This example shows that the lattice computed on the resulting clusters can in fact provide meaningful explanations about the commonalities and differences of the set of clusters — beyond what could be provided in Section 3.3. Since it remains inconvenient to figure out these structures just from the list of formal concepts, we furthermore exploit techniques for visualizing the computed lattice in the next step.

5   Visualizing the Concept Lattice

We make use of *Hasse diagrams* for visualizing the concept lattice. They follow the conventions for the visualization of hierarchical concept systems as established in the international standard ISO 704. Figure 1 highlights a part of the concept lattice of our

---

[10] (m) stands here for the important and (h) for the very important terms with the higher threshold.

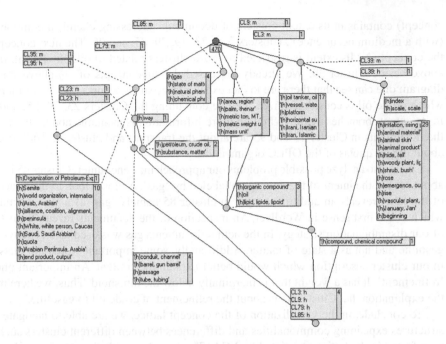

Fig. 1. The resulting conceptual clustering of the text clusters (visualized for the clusters related to chemical compounds.

context by a Hasse diagram. It will be explained in detail below. The lattice was computed and visualized using the Cernato software of NaviCon Gmbh[11]. It shows all clusters where the value of the synset 'compound, chemical compound' in the centroid is above the threshold $\theta_1 = 7\%$.

In a Hasse diagram, each node represents a formal concept. Due to technical reasons, we reverse the usual reading order: A concept $c_1 \in \mathfrak{B}(\mathbb{K})$ is a subconcept of a concept $c_2 \in \mathfrak{B}(\mathbb{K})$ if and only if there is a path of descending(!) edges from the node representing c_1 to the node representing c_2.

The name of an object g is always attached to the node representing the most specific concept (i. e., the smallest concept with respect to $\leq$) with g in its extent (i. e., in our figure, the highest such node); dually, the name of an attribute m is always attached to the node representing the most general concept with m in its intent (i. e., the lowest such node in the diagram). We can always read the context relation from the diagram, since an object g has an attribute m if and only if the concept labeled by g is a subconcept of the one labeled by m. The extent of a concept consists of all objects whose labels are attached to subconcepts, and, dually, the intent consists of all attributes attached to superconcepts.

For example, the concept in the lower middle of the diagram labeled by 'oil' is the concept (*) that we encountered above. In the diagram, we can see that it is part of a chain of concepts with increasing specificity. The most general of them (beside the top

---

[11] http://www.navicon.de

concept) contains in its extent clusters of documents addressing chemical compounds (with a medium occurrence) : Clusters 3, 9, 23, 39, 79, 85, and 95. The next concept is the concept (*). Its extent is restricted to those clusters related to oil: all clusters from above beside Cluster 39. We already discussed the subconcepts of (*) above. In the diagram one can see that there are in fact exactly three subconcepts. The one of them in which crude oil is considered, i. e., the one containing the Clusters 23, 79, 85, and 95 in its extent, branches again: While no more information is available about cluster 79, the documents in Cluster 23 and 95 are about the transport and cluster 95 additionally about the oil quotas of the OPEC organization.

Let us also analyze possible problems our approach may encounter: Cluster 85 is also about oil but the intent of the concept is labeled by 'gas'. A closer look to the concepts of the label reveals an additional topic in cluster 85 namely 'gas' as a state of matter which is the first sense in WordNet. An inspection of the documents reveals a mistake of our disambiguation strategy. In the actual documents gas was used as a synonym of gasoline and not as a state of matter. Additionally, some important words are missing in our cluster description which would better explain the content. An important one is 'refinement'. It has a weight that is marginally below the threshold. Thus, we here miss the explanation that Cluster 85 is about the refinement of crude oil to gasoline.

To conclude, in the visualization of the concept lattice, we are able to navigate the structures explaining commonalities and differences between different clusters such as manifested in the lattice computed by FCA. The lattice extends the set of meaningful, explanatory structures by means that relate clusters to each other and that exploit the hierarchy defined in the thesaurus for this purpose as a side effect. On the other hand, without BiSec–k–Means as a preceeding step, the FCA step would not have produced a lattice of reasonable and understandable size, because individual texts are too volatile what concerns the joint occurrence of relevant terms. As shown, our approach thus combines the thorough analysis of FCA with the reduction of term and document space to a concise, but relevant basis.

6   Related Work

As just summarized, the orginality of our approach is not so much based on the individual algorithms used, as vector representations, BiSec–k–Means, Formal Concept Analysis and Hasse Diagramms are all well known, but on their original integration. This integration serves the purpose to achieve a careful balance with regard to the *granularity* of information used for explanation in three dimensions. First, it automatically finds the adequate level of generalization of concepts in the thesaurus (e.g., 'financial institution' instead of 'bank', whereby only the latter actually appears in the texts). Second, it restricts the term space to a subspace. Thereby, the major components in the cluster centroids are terms that are particularly able to group and discriminate larger text subsets. Third, our approach restricts the document space to a subspace. The subspace abstracts from outlying non-occurrences of individual terms (e.g., one document being about 'financial losses in business acquisitions', but only exhibiting the 'loss' information and not 'company B burned money').

Our experiments have shown, that in order to come up with a concise, but elaborate description, one must carefully balance these dimensions before computing a lattice-based, hence expressive, explanation. Against this background we may compare paradigms related to our approach.

Hierarchical text clustering, by agglomerative or by partitional algorithms, may be used to derive a tree of cluster representations. One document is in general not only found in one cluster, but it is assigned to a hierarchy of clusters. The hierarchy with its clusters at different levels of representations may be used to describe how clusters are similar or different. Unlike FCA, however, the tree-like hierarchy does not allow multiple assignment of categories, as is common for text documents, e.g. Reuters news.

We have explored explanations produced by applying common rule/decision tree learners like Ripper [2] or C4.5 [14] on resulting clusters. While the result that these algorithms produce may be very good to classify into the categories they learn, they tend to produce a larger number of rules to explain a single resulting cluster. The explanation for 100 clusters appears to be rather unmanageable for a human user.

We have mentioned that conceptual clustering techniques might be applied directly on the text representations instead of on the text cluster representations. We have also mentioned that there arise problems because of the large term and document space. There are means to reduce the term and document space based on counting support for formal concepts (e.g., Titanic [18]). However, this type of algorithm is rather new and we don't know about work that would have applied it to texts in any way.

Latent Semantic Indexing (LSI; cf. [4,7]) constitutes a paradigm that groups words into 'concepts' based on their cooccurrences in a given dataset. LSI then allows for text clustering or classification taking into account these 'concepts'. Compared against our approach, its biggest advantage is that a thesaurus is not needed, but the largest disadvantage of LSI is that the notion of 'concept' that LSI introduces cannot easily be explained to a common user. Also, an explanation by a more general concept (the 'first dimension' sketched above) is not possible.

Finally, Karypis and Han [10] have built on the principal idea of reducing term space from LSI, but introduce 'concepts' that are based on the automatic clustering of words. They achieve performance quality comparable to LSI, but their method is more accessible to a human user and might be integrated in the future with a manually defined core thesaurus. We see here a promising line of further research combining their idea of concept construction by clustering (also found in other areas like ontology learning [12]) with its immediate use in text classification and clustering as well as in explaining clustering results as we have proposed in this paper.

7   Conclusion

In this paper, we presented a novel combination of known techniques for text clustering. First, we extended the typical vector space representation of text by synsets of WordNet, in order to exploit its semantics. Then we clustered the documents with the BiSec–k–Means algorithm, using the cosine for measuring the similarity of documents. For each cluster, we extracted a conceptual description, which we used for arranging the clusters in a lattice using Formal Concept Analysis. This blend of known techniques has been shown to combine the benefits of each of the techniques involved: WordNet provides means to identify apparently different terms on a higher level of abstraction;

BiSec–k–Means structures the domain and reduces it to a manageable size; the extracted cluster descriptions help identifying the content of individual clusters; and FCA and its visualization means show up the relation between those clusters.

References

1. C. Buckley and A. Lewit. Optimizations of inverted vector searches. In *SIGIR-1985*, pages 97–110, 1985.
2. William W. Cohen. Fast effective rule induction. In *Proc. of ICML-95*, pages 115–123. Morgan Kaufmann, July 9–12, 1995.
3. D.R. Cutting, D. R. Karger, J. O. Pedersen, and J. W. Tukey. Scatter/gather: A cluster-based approach to browsing large document collections. In *SIGIR-1992*, pages 318–329, 1992.
4. S. C. Deerwester, S. T. Dumais, T. K. Landauer, G. W. Furnas, and R. A. Harshman. Indexing by latent semantic analysis. *Journal of the American Society of Information Science*, 41(6):391–407, 1990.
5. C. Fellbaum, editor. *WordNet: An Electronic Lexical Database*. The MIT Press, 1998.
6. B. Ganter and R. Wille. *Formal Concept Analysis: Mathematical Foundations*. Springer, Berlin–Heidelberg, 1999.
7. T. Hofmann. Probabilistic latent semantic indexing. In *Research and Development in Information Retrieval*, pages 50–57, 1999.
8. A. Hotho, A. Maedche, S. Staab, and R. Studer. SEAL-II — the soft spot between richly structured and unstructured knowledge. *Journal of Universal Computer Science (J.UCS)*, 7(7):566–590, 2001.
9. A. Hotho, S. Staab, and G. Stumme. Wordnet improves text document clustering. In *Proc. of the SIGIR 2003 Semantic Web Workshop*, 2003.
10. George Karypis and Eui-Hong Han. Fast supervised dimensionality reduction algorithm with applications to document categorization and retrieval. In *Proceedings of CIKM-00*, pages 12–19. ACM Press, New York, US, 2000.
11. G. Kowalski. *Information Retrieval systems-theory and implementations*. Kluwer Academic Publishers, 1997.
12. A. Maedche and S. Staab. Ontology learning for the semantic web. *IEEE Intelligent Systems*, 16(2):72 –79, 2001.
13. D. Mladenic. Text learning and related intelligent agents. *IEEE Expert*, July/August 1999.
14. J. R. Quinlan. *C4.5: Programs for Machine Learning*. Morgan Kaufmann, San Mateo, California, 1993.
15. C. Van Rijsbergen. *Information Retrieval*. Buttersworth, London, 1989.
16. G. Salton. *Automatic Text Processing: The Transformation, Analysis and Retrieval of Information by Computer*. Addison-Wesley, 1989.
17. M. Steinbach, G. Karypis, and V. Kumar. A comparison of document clustering techniques. In *KDD Workshop on Text Mining*, 2000.
18. G. Stumme, R. Taouil, Y. Bastide, N. Pasqier, and L. Lakhal. Computing iceberg concept lattices with Titanic. *J. on Knowledge and Data Engineering*, 42:189–222, 2002.
19. O. Zamir, O. Etzioni, O. Madani, and R.M. Karp. Fast and intuitive clustering of web documents. In *KDD-1997*, pages 287– 290, 1997.

Analyzing Attribute Dependencies

Aleks Jakulin and Ivan Bratko

Faculty of Computer and Information Science, University of Ljubljana
Tržaška 25, Ljubljana, Slovenia

Abstract. Many effective and efficient learning algorithms assume independence of attributes. They often perform well even in domains where this assumption is not really true. However, they may fail badly when the degree of attribute dependencies becomes critical. In this paper, we examine methods for detecting deviations from independence. These dependencies give rise to "interactions" between attributes which affect the performance of learning algorithms. We first formally define the degree of interaction between attributes through the deviation of the best possible "voting" classifier from the true relation between the class and the attributes in a domain. Then we propose a practical heuristic for detecting attribute interactions, called *interaction gain*. We experimentally investigate the suitability of interaction gain for handling attribute interactions in machine learning. We also propose visualization methods for graphical exploration of interactions in a domain.

1 Introduction

Many learning algorithms assume independence of attributes, such as the naïve Bayesian classifier (NBC), logistic regression, and several others. The independence assumption licenses the classifier to collect the evidence for a class from individual attributes separately. An attribute's contribution to class evidence is thus determined independently of other attributes. The independence assumption does not merely simplify the learning algorithm; it also results in robust performance and in simplicity of the learned models.

Estimating evidence from given training data with the independence assumption is more robust than when attribute dependencies are taken into account. The evidence from individual attributes can be estimated from larger data samples, whereas the handling of attribute dependencies leads to fragmentation of available data and consequently to unreliable estimates of evidence. This increase in robustness is particularly important when data is scarce, a common problem in many applications. In practice these unreliable estimates often cause inferior performance of more sophisticated methods.

Methods like NBC that consider one attribute at a time are called "myopic." Such methods compute evidence about the class separately for each attribute (independently from other attributes), and then simply "sum up" all these pieces of evidence. This "voting" does not have to be an actual arithmetic sum (for example, it can be the product, that is the sum of logarithms, as in NBC). The

N. Lavrač et al. (Eds.): PKDD 2003, LNAI 2838, pp. 229–240, 2003.

aggregation of pieces of evidence coming from individual attributes does not depend on the relations among the attributes. We will refer to such methods as "voting methods;" they employ "voting classifiers."

A well-known example where the myopia of voting methods results in complete failure, is the concept of exclusive OR: $C = XOR(X, Y)$, where C is a Boolean class, and X and Y are Boolean attributes. Myopically looking at attribute X alone provides no evidence about the value of C. The reason is that the relation between X and C critically depends on Y. For $Y = 0, C = X$; for $Y = 1, C \neq X$. Similarly, Y alone fails. However, X *and* Y *together* perfectly determine C. We say that there is a *positive interaction* between X and Y with respect to C. In the case of a positive interaction the evidence from jointly X and Y about C is greater than the sum of the evidence from X alone and evidence from Y alone.

The opposite may also happen, namely that the evidence from X and Y jointly is worth less than the sum of the individual pieces of evidence. In such cases we say that there is a *negative interaction* between X and Y w.r.t. C. A simple example is when attribute Y is (essentially) a duplicate of X. For example, the length of the diagonal of a square duplicates the side of the square. Voting classifiers are confused by negative interactions as well by positive ones.

2   Attribute Interactions

Let us first define the concept of interaction among attributes formally. Let there be a supervised learning problem with class C and attributes $X_1, X_2, \ldots$. Under conditions of noise or incomplete information, the attributes need not determine the class values perfectly. Instead, they provide some "degree of evidence" for or against particular class values. For example, given an attribute-value vector, the degrees of evidence for all possible class values may be a probability distribution over the class values given the attribute values.

Let the *evidence function* $f(C, X_1, X_2, \ldots, X_k)$ define some chosen "true" degree of evidence for class C in the domain. The task of machine learning is to induce an approximation to function f from training data. In this sense, f is the target concept for learning. In classification, f (or its approximation) would be used as follows: if for given attribute values $x_1, x_2, \ldots, x_k : f(c_1, x_1, x_2, \ldots, x_k) > f(c_2, x_1, \ldots, x_k)$, then the class c_1 is more likely than c_2.

We define the presence, or absence, of interactions among the attributes as follows. If the evidence function can be written as a ("voting") sum:

$$f(C, X_1, X_2, \ldots, X_k) = v \left(\sum_{i=1, \ldots, k} e_i(C, X_i) \right) \qquad (1)$$

for some voting function v, and myopic predictor functions $e_1, e_2, \ldots, e_k$, then there is no interaction between the attributes. Equation (1) requires that the joint evidence of all the attributes can essentially be reduced to the sum of the pieces of evidence $e_i(C, X_i)$ from individual attributes. The function e_i is a predictor that investigates the relationship between an attribute X_i and the class C.

If, on the other hand, no such functions $v, e_1, e_2, \ldots, e_k$ exist for which (1) holds, then there *are* interactions among the attributes. The strength of interactions IS can be defined as

$$IS := f(C, X_1, X_2, \ldots, X_k) - v \left(\sum_i e_i(C, X_i) \right). \qquad (2)$$

IS greater than some positive threshold would indicate a positive interaction, and IS less than some negative threshold would indicate a negative interaction. Positive interactions indicate that a holistic view of the attributes unveils new evidence. Negative interactions are caused by multiple attributes providing the same evidence, while the evidence should count only once.

Many classifiers are based on the linear form of (1): naïve Bayesian classifier, logistic and linear regression, linear discriminants, support vector machines with linear kernels, and others. Hence, interaction analysis is relevant for all these methods. All we have written about relationships between attributes also carries over to relationships between predictors in an ensemble.

3 Interaction Gain: A Heuristic for Detecting Interactions

The above definition of an interaction provides a "golden standard" for deciding, in principle, whether there is interaction between two attributes. The definition is, however, hard to use as a procedure for detecting interactions in practice. Its implementation would require combinatorial optimization.

We will not refine the above definition of interactions to make it applicable in a practical learning setting. Instead, we propose a heuristic test, called *interaction gain*, for detecting positive and negative interactions in the data, in the spirit of the above definition. Our heuristic will be based on information-theoretic notion of entropy as the measure of classifier performance, joint probability distribution as the predictor, and the chain rule as the voting function. Entropy has many useful properties, such as linear additivity of entropy with independent sources. We will consider discriminative learning, where our task is to study the class probability distribution. That is why we will always investigate relationships between an attribute and the class, or between attributes with respect to the class.

Interaction gain is based on the well-known idea of information gain. *Information gain* of a single attribute X with respect to class C, also known as *mutual information* between X and C, measured in bits:

$$\text{Gain}_C(X) = I(X; C) = \sum_x \sum_c P(x, c) \log \frac{P(x, c)}{P(x)P(c)}. \qquad (3)$$

Information gain can be regarded as a measure of the strength of a 2-way interaction between an attribute X and the class C. In this spirit, we can generalize it to 3-way interactions by introducing the *interaction gain* [1]:

$$I(X;Y;C) := I(X,Y;C) - I(X;C) - I(Y;C). \tag{4}$$

Interaction gain is also measured in bits, and can be understood as the difference between the actual decrease in entropy achieved by the joint attribute XY and the expected decrease in entropy with the assumption of independence between attributes X and Y. The higher the interaction gain, the more information was gained by joining the attributes in the Cartesian product, in comparison with the information gained from single attributes. When the interaction gain is negative, both X and Y carry the same evidence, which was consequently subtracted twice.

To simplify our understanding, we can use the entropy $H(X)$ to measure the uncertainty of an information source X through the identity $I(X;Y) = H(X) + H(Y) - H(X,Y)$. It it then not difficult to show that $I(X;Y;C) = I(X;Y|C) - I(X;Y)$. Here, $I(X;Y|C) = H(X|C) + H(Y|C) - H(X,Y|C)$ is conditional mutual information, a measure of dependence of two attributes given the *context* of C. $I(X;Y)$ is an information-theoretic measure of dependence or "correlation" between the attributes X and Y regardless of the context.

Interaction gain (4) describes the change in a dependence of a pair of attributes X, Y by introducing context C. It is quite easy to see that when interaction gain is negative, context decreased the amount of dependence. When the interaction gain is positive, context increased the amount of dependence. When the interaction gain is zero, context did not affect the dependence between the two attributes. Interaction gain is identical to the notion of interaction information [2] and mutual information among three random variables [3,4].

4   Detecting and Resolving Interactions

A number of methods have been proposed to account for dependencies in machine learning, in particular with respect to the naïve Bayesian classification model [5,6,7], showing improvement in comparison with the basic model. The first two of these methods, in a sense, perform feature construction; new features are constructed from interacting attributes, by relying on detection of interactions. On the other hand, tree augmentation [7], merely makes the dependence explicit, but this is more a syntactic distinction. In this section we experimentally investigate the relevance of interaction gain as a heuristic for guiding feature construction.

The main questions addressed in this section are: Is interaction gain a good heuristic for detecting interactions? Does it correspond well to the principled definition of interactions in Section 2?

The experimental scenario is as follows:

1. We formulate an operational approximation to our definition of interaction. This is a reasonable and easy to implement special case of formula (1) as follows: the degree of evidence is a probability, and the formula (1) is instantiated to the naïve Bayesian formula. It provides an efficient test for interactions. We refer to this test as BS.

2. In each experimental data set, we select the most interacting pair of attributes according to (a) BS, (b) positive interaction gain (PIG), and (c) negative interaction gain (NIG).

3. We build naïve Bayesian classifiers (NBC) in which the selected interactions are "resolved." That is, the selected pair of most interacting attributes is replaced in NBC by its Cartesian product. This interaction resolution is done for the result of each of the three interaction detection heuristics (BS, PIG and NIG), and the performance of the three resulting classifiers is compared.

We chose to measure the performance of a classifier with Brier score (described below). We avoided classification accuracy as a performance measure for the following reasons. Classification accuracy is not very sensitive in the context of probabilistic classification: it usually does not matter for classification accuracy whether a classifier predicted the true class with the probability of 1 or with the probability of, e.g., 0.51. To account for the precision of probabilistic predictions, we employed Brier score. Given two probability distributions, the predicted class probability distribution $\hat{p}$, and the actual class probability distribution p, where the class can take N values, the Brier score [8] of the prediction is:

$$b(\hat{p}, p) := \frac{1}{N} \sum_{i=1}^{N} (\hat{p}_i - p_i)^2 \tag{5}$$

The larger the Brier score, the worse a prediction. Error rate is a special case of Brier score for deterministic classifiers, while Brier score could additionally reward a probabilistic classifier for better estimating the probability. In a practical evaluation of a classifier given a particular testing instance, we approximate the actual class distribution by assigning a probability of 1 to the true class of the testing instance. For multiple testing instances, we compute the average Brier score.

We used two information-theoretic heuristics, based on interaction gain: the interaction with the maximal positive magnitude (PIG), and the interaction with the minimal negative magnitude (NIG). We also used a wrapper-like heuristic: the interaction with the maximum improvement in the naïve Bayesian classifier performance after merging the attribute pair, as measured with the Brier score or classification accuracy on the training set, $b(NBC(C|X)(C|Y)) - b(NBC(C|X,Y))$, the first term corresponding to the independence-assuming naïve Bayesian classifier and the second to the Bayesian classifier assuming dependence. This heuristic (BS) is closely related to the notion of mutual conditional information, which can be understood as the Kullback-Leibler divergence between the two possible models.

As the basic learning algorithm, we have used the naïve Bayesian classifier. After the most important interaction was determined outside the context of other attributes, we modified the NBC model created with all the domain's attributes by taking the single most interacting pair of attributes and replacing them with their Cartesian product, thus eliminating that particular dependence. All the numerical attributes in the domains were discretized beforehand, and missing values represented as special values. Evaluations of the default NBC model and of its modifications with different guiding heuristics were performed with 10-fold cross-validation. For each fold, we computed the average score. For each domain, we computed the score mean and the standard error over the 10 fold experiments. We performed all our experiments with the Orange toolkit [9].

Table 1. The table lists Brier scores obtained with 10-fold cross validation after resolving the most important interaction, as assessed with different methods. A result is set in bold face if it is the best for the domain, and checked if it is within the standard error of the best result for the domain. We marked the artificial[†] domains.

domain	NB	PIG	NIG	BS	domain	NB	PIG	NIG	BS
lung	0.230√	**0.208**	0.247	0.243	soy-large	0.008√	**0.007**	0.008√	0.008√
soy-small	**0.016**	**0.016**	**0.016**	**0.016**	wisc-canc	0.024√	**0.023**	0.024√	0.026√
zoo	**0.018**	0.019√	**0.018**	**0.018**	austral	0.120√	0.127	**0.114**	0.116√
lymph	0.079√	0.094	0.077√	**0.075**	credit	0.116√	0.122	**0.111**	0.115√
wine	**0.010**	0.010√	0.015	0.014	pima	0.159√	0.159√	**0.158**	0.159√
glass	**0.070**	0.071√	0.071√	0.073√	vehicle	0.142	0.136	0.138	**0.127**
breast	**0.212**	0.242	0.212√	0.221√	heart	**0.095**	0.098	0.095√	0.095√
ecoli	**0.032**	0.033√	0.039	0.046	german	**0.173**	0.175√	0.174√	0.175√
horse-col	0.108√	0.127	0.106√	**0.104**	cmc	0.199√	**0.194**	0.195√	0.198√
voting	0.089	0.098	0.089	**0.063**	segment	0.016	0.017	0.017	**0.015**
monk3[†]	0.042	**0.027**	0.042	**0.027**	krkp[†]	0.092	0.077√	0.088	**0.076**
monk1[†]	0.175	**0.012**	0.176	**0.012**	mushroom	0.002	0.006	**0.002**	**0.002**
monk2[†]	0.226√	**0.223**	0.224√	0.226√	adult	0.119	0.120	**0.115**	0.119

In Table 1 we sorted 26 of the UCI KDD archive [10] domains according to the number of instances in the domain, from the smallest on the top left to the largest on the bottom right, along with the results obtained in the above manner. We can observe that in two domains resolution methods matched the original result. In 6 domains resolution methods worsened the results, in 10 domains the original performance was within a standard error of the best, and in 8 domains, the improvement was significant beyond a standard error. We can thus confirm that accounting for interactions in this primitive way did help in ~70% of the domains.

If comparing different resolution algorithms, the Brier-score driven interaction detection was superior to either of the information-based heuristics PIG or NIG, achieving the best result in 11 domains. However, in only two domains, 'voting' and 'segment,' neither of PIG and NIG was able to improve the result while BS did. Thus, information-theoretic heuristics are a reasonable and effective choice for interaction detection, providing competitive results even if the BS heuristic had the advantage of using the same evaluation function as the final classifier evaluation. PIG improved the results in 5 natural domains, and in 4 artificial domains. This confirms earlier intuitions that the XOR-type phenomena occur more often in synthetic domains. NIG provided an improvement in 7 domains, all of them natural. Negative interactions are generally very frequent, but probabilistic overfitting cannot always be resolved to a satisfactory extent by merely resolving the strongest interaction because the result is dependent on the balance between multiple negatively interacting attributes.

Table 2. A summary of results shows that BS-driven interaction resolution provides the most robust approach, while AIG follows closely behind.

times	NB	PIG	NIG	BS	AIG
best	8	8	7	11	10
good √	10	7	10	11	7
bad	8	11	9	4	9

In Table 2, we summarize the performance of different methods, including AIG as a simple approach deciding between the application of NIG and PIG in a given domain. AIG suggests resolving the interaction with the largest absolute interaction gain. We can also observe that success is more likely when the domain contains a large number of instances. It is a known result from statistical literature that a lot of evidence is needed to show the significance of higher-order interactions [11].

5   Visualization of Interactions

The analysis of attribute relationships can be facilitated by methods of information visualization. We propose two methods for visualization of attribute interactions. Interaction dendrogram illustrates groups of mutually interacting attributes. Interaction graph provides detailed insight into the nature of attribute relationships in a given domain.

5.1   Interaction Dendrograms

Interaction dendrogram illustrates the change in dependence between pairs of attributes after introducing the context. The direction of change is not important: we will distinguish this later. If we bind proximity in our presentation to the change in level of dependence, either positive or negative, we can define the distance d_m between two attributes X, Y as:

$$d_m(X, Y) := \begin{cases} |I(X;Y;C)|^{-1} & \text{if } |I(X;Y;C)|^{-1} < 1000, \\ 1000 & \text{otherwise.} \end{cases} \tag{6}$$

Here, 1000 is a chosen upper bound as to prevent attribute independence from disproportionately affecting the graphical representation. To present the function d_m to a human analyst, we tabulate it in a dissimilarity matrix and apply

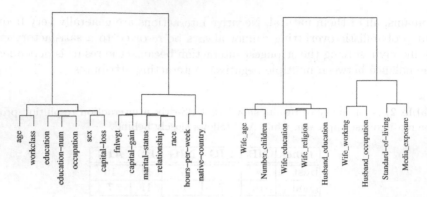

Fig. 1. An interaction dendrogram illustrates which attributes interact, positively or negatively, in the 'census/adult' (left) and 'cmc' (right) data sets. We used the Ward's method for agglomerative hierarchical clustering [12].

the techniques of hierarchical clustering or multi-dimensional scaling. Dependent attributes will hence appear close to one another; independent attributes will appear far from one another. This visualization is an approach to variable clustering, which is normally applied to numerical variables outside the context of supervised learning. Diagrams, such as those in Fig. 1, may be directly useful for feature selection: the search for the best model starts by only picking the individually best attribute from each cluster. We must note, however, that an attribute's membership in a cluster merely indicates its average relationship with other cluster members.

5.2   Interaction Graphs

The analysis described in the previous section was limited to rendering the magnitude of interaction gains between attributes. We cannot use the dendrogram to identify whether an interaction is positive or negative, nor can we see the importance of each attribute. An interaction graph presents the proximity matrix better. To reduce clutter, only the strongest N interactions are shown, usually $5 \leq N \leq 20$. With an interactive method for graph exploration, this trick would not be necessary. We also noticed that the distribution of interaction gains usually follows a Gaussian-like distribution, with only a few interactions standing out from the crowd, either on the positive or on the negative side.

Each node in the interaction graph corresponds to an attribute. The information gain of each attribute is expressed as a percentage of the class entropy (although some other uncertainty measure, such as the error rate or Brier score, could be used in the place of class entropy), and written below the attribute name. There are two kinds of edges, bidirectional arrows and undirected dashed arcs. Arcs indicate negative interactions, implying that the two attributes provide partly the same information. The amount of shared information, as a percentage of the class entropy, labels the arc. Analogously, the amount of novel

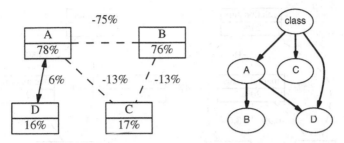

Fig. 2. The four most informative attributes were selected from a real medical domain. In the interaction graph (left), the most important attribute A alone eliminates 78% of class entropy. The second most important attribute B alone eliminates 76% of class entropy, but A and B interact negatively (dashed arc), and share 75% of class entropy. So B reduces class entropy by only 76-75=1% of its truly own once we have accounted for A: but if we leave B out in feature subset selection, we are giving this information up. Similarly, C provides 4% of its own information, while the remaining 13% is contained in both, A and B. Attribute D provides 'only' 16% of information, but if we account for the positive interaction between A and D (solid bidirectional arrow), we provide for 78+16+6=100% of class entropy. Consequently, only attributes A and D are needed, and they should be treated as dependent. A Bayesian network [14] learned from the domain data (right) is arguably less informative.

information labels the arrow, indicating a positive interaction between a pair of attributes. Figure 2 explains the interpretation of the interaction graph, while Figs. 3 and 4 illustrate two domains. We used the 'dot' utility [13] for generating the graph.

6   Implications for Classification

In discussing implications of interaction analysis for classification, there are two relevant questions. The first is the question of significance: when is a particular interaction worth considering. The second is the question of how to treat negative and positive interactions between attributes in the data.

In theory, we should react whenever the conditional mutual information $I(X; Y|C)$ deviates sufficiently from zero: it is a test of conditional dependence. In practice, using a joint probability distribution for XY would increase the complexity of the classifier, and this is often not justified when the training data is scarce. Namely, introducing the joint conditional probability distribution $P(X, Y|C)$ in place of two marginal probability distributions $P(X|C)P(Y|C)$ increases the degrees of freedom of the model, thus increasing the likelihood that the fit was accidental. In the spirit of Occam's razor, we should increase the complexity of a classifier only to obtain a significant improvement in classification performance. Hence, the true test in practical applications is improvement in generalization performance, measured with devices such as the training/testing set separation and cross-validation. When the improvement after accounting for an interaction is significant, the interaction itself is significant.

238   Aleks Jakulin and Ivan Bratko

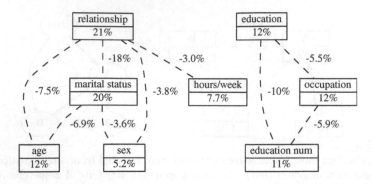

Fig. 3. An interaction graph for the 'census/adult' domain confirms our intuitions about natural relationships between the attributes. All interactions in this graph are negative, but there are two clusters of them.

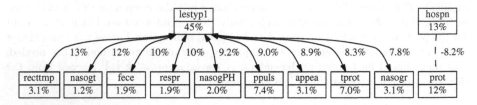

Fig. 4. In this illustration of the 'horse colic' domain, one attribute appears to moderate a number of other attributes' relationships with the class. There is a separate and independent negative interaction on the right.

6.1 Negative Interactions

If X and Y are interacting negatively, they both provide the same information. If we disregard a negative interaction, we are modifying the class probability distribution with the same information twice. If the duplicated evidence is biased towards one of the classes, this may shift the prediction. The estimated class probabilities may become excessively confident for one of the classes, another case of *overfitting*. Unbalanced class probability estimates by themselves do not necessarily bother several classification performance measures, such as the classification accuracy and the ROC, because they do not always change the classifications.

Even if the naïve Bayesian classifier sometimes works optimally in spite of negative interactions, it often useful to resolve them. The most frequent method is feature selection. However, we observe that two noisy measurements of the same quantity are better than a single measurement, so other approaches may be preferable. One approach is assigning weights to attributes, such as feature weighting or least-squares regression. Alternatively, a latent attribute L can be inferred, to provide evidence for all three attributes: X, Y and C. The trivial approach to latent attribute inference is the introduction of the Cartesian product between attributes, the technique we applied in our experiments, but methods like factor analysis, independent component analysis are also applicable here. This kind of attribute dependence is a simple and obvious explanation for conditional dependence. Negative interactions allow us to simplify the model in the sense of reducing the quantity of evidence being dealt with.

6.2   Positive Interactions

The second cause of independence assumption violation is when two attributes together explain more than what we estimated from each attribute individually. There could be some unexplained moderating effect of the first attribute onto the second attribute's evidence for C. There could be a functional dependence involving X, Y and C, possibly resolved by feature construction. Such positive interactions can be inferred from a positive value of interaction gain. Positive interactions are interesting subjects for additional study of the domain, indicating complex regularities. They too can be handled with latent attribute inference. Positive interactions indicate a possible benefit of complicating the model.

If we disregard a positive interaction by assuming attribute independence, we are not taking advantage of all the information available: we are *underfitting*. Of course, one should note that the probabilities, on the basis of which entropy is calculated, might not be realistic. Since the probabilities of a joint probability distribution of two values of two attributes and the class are computed with fewer supporting examples than those computed with only one attribute value and the class, the 3-way interaction gains are less trustworthy than 2-way interaction gains. Consequently, positive interaction gains may in small domains indicate only a coincidental regularity. Taking accidental dependencies into consideration is a well-known cause of *overfitting*, but there are several ways of remedying this probability estimation problem, e.g. [15].

7   Conclusion

In this paper we studied the detection and resolution of dependencies between attributes in machine learning. First we formally defined the degree of interaction between attributes through the deviation of the best possible "voting" classifier from the true relation between the class and the attributes in a domain. Then we proposed the *interaction gain* as a practical heuristic for detecting attribute interactions. We experimentally investigated the suitability of interaction gain (IG) for handling attribute dependencies in machine learning. Experimental results can be summarized as follows:

- IG as a heuristic for detecting interactions performs similarly as the BS criterion (a heuristic that was directly derived from the principled formal definition of attribute interaction), and enables the resolution of interactions in classification learning with similar performance as BS.
- IG enables the distinction between *positive* and *negative* interactions while BS does not distinguish between these two types of interactions. Here, IG can explain the reason for a certain recommendation of BS.

- According to empirical results in real-world domains, strong positive interactions are rare, but negative interactions are ubiquitous.
- In typical artificial domains, strong interactions are more frequent, particularly positive interactions. The IG heuristic reliably detects them.

We also presented visualization methods for graphical exploration of interactions in a domain. These are useful tools that should help expert's understanding of the domain under study, and could possibly be used in constructing a predictive model. Problems for future work include: handling n-way interactions where $n > 3$; building learning algorithms that will incorporate interaction detection facilities, and provide superior means of resolving these interactions when building classifiers.

References

1. Jakulin, A.: Attribute interactions in machine learning. Master's thesis, University of Ljubljana, Faculty of Computer and Information Science (2003)
2. McGill, W.J.: Multivariate information transmission. Psychometrika **19** (1954) 97–116
3. Han, T.S.: Multiple mutual informations and multiple interactions in frequency data. Information and Control **46** (1980) 26–45
4. Yeung, R.W.: A new outlook on Shannon's information measures. IEEE Transactions on Information Theory **37** (1991) 466–474
5. Kononenko, I.: Semi-naive Bayesian classifier. In Kodratoff, Y., ed.: European Working Session on Learning - EWSL91. Volume 482 of LNAI., Springer Verlag (1991)
6. Pazzani, M.J.: Searching for dependencies in Bayesian classifiers. In: Learning from Data: AI and Statistics V. Springer-Verlag (1996)
7. Friedman, N., Goldszmidt, M.: Building classifiers using Bayesian networks. In: Proc. National Conference on Artificial Intelligence, Menlo Park, CA, AAAI Press (1996) 1277–1284
8. Brier, G.W.: Verification of forecasts expressed in terms of probability. Weather Rev **78** (1950) 1–3
9. Demšar, J., Zupan, B.: Orange: a data mining framework. http://magix.fri.uni-lj.si/orange (2002)
10. Hettich, S., Bay, S.D.: The UCI KDD archive http://kdd.ics.uci.edu. Irvine, CA: University of California, Department of Information and Computer Science (1999)
11. McClelland, G.H., Judd, C.M.: Statistical difficulties of detecting interactions and moderator effects. Psychological Bulletin **114** (1993) 376–390
12. Struyf, A., Hubert, M., Rousseeuw, P.J.: Integrating robust clustering techniques in S-PLUS. Computational Statistics and Data Analysis **26** (1997) 17–37
13. Koutsofios, E., North, S.C.: Drawing Graphs with dot. (1996) Available on research.att.com in dist/drawdag/dotguide.ps.Z.
14. Myllymaki, P., Silander, T., Tirri, H., Uronen, P.: B-Course: A web-based tool for Bayesian and causal data analysis. International Journal on Artificial Intelligence Tools **11** (2002) 369–387
15. Cestnik, B.: Estimating probabilities: A crucial task in machine learning. In: Proc. 9th European Conference on Artificial Intelligence. (1990) 147–149

Ranking Interesting Subspaces
for Clustering High Dimensional Data*

Karin Kailing, Hans-Peter Kriegel, Peer Kröger, and Stefanie Wanka

Institute for Computer Science
University of Munich
Oettingenstr. 67, 80538 Munich, Germany
{kailing,kriegel,kroegerp,wanka}@dbs.informatik.uni-muenchen.de

Abstract. Application domains such as life sciences, e.g. molecular biology produce a tremendous amount of data which can no longer be managed without the help of efficient and effective data mining methods. One of the primary data mining tasks is clustering. However, traditional clustering algorithms often fail to detect meaningful clusters because of the high dimensional, inherently sparse feature space of most real-world data sets. Nevertheless, the data sets often contain clusters hidden in various subspaces of the original feature space. We present a pre-processing step for traditional clustering algorithms, which detects all interesting subspaces of high-dimensional data containing clusters. For this purpose, we define a quality criterion for the interestingness of a subspace and propose an efficient algorithm called RIS (*R*anking *I*nteresting *S*ubspaces) to examine all such subspaces. A broad evaluation based on synthetic and real-world data sets empirically shows that RIS is suitable to find all relevant subspaces in large, high dimensional, sparse data and to rank them accordingly.

1 Introduction

The tremendous amount of data produced nowadays in various application domains such as molecular biology can only be fully exploited by efficient and effective data mining tools. One of the primary data mining tasks is clustering which is the task of partitioning objects of a data set into distinct groups (clusters) such that two objects from one cluster are similar to each other, whereas two objects from distinct clusters are not.

Considerable work has been done in the area of clustering. Nevertheless, clustering real-world data sets often raises problems, since the data space is usually a high dimensional feature space. A prominent example is the application of cluster analysis to gene expression data. Depending on the goal of the application, the dimensionality of the feature space can be up to 10^2 when clustering

---

* The work is supported in part by the German Ministery for Education, Science, Research and Technology (BMBF) under grant no. 031U112F within the BFAM (Bioinformatics for the Functional Analysis of Mammalian Genomes) project which is part of the German Genome Analysis Network (NGFN).

N. Lavrač et al. (Eds.): PKDD 2003, LNAI 2838, pp. 241–252, 2003.

the genes and can be in the range of 10^3 to more than 10^4 when clustering the samples. In general, most of the common clustering algorithms fail to generate meaningful results because of the inherent sparsity of the data space. In such high dimensional feature spaces data does not cluster anymore. But usually, there are clusters in lower dimensional subspaces. In addition, objects can often be clustered differently in varying subspaces, i.e. objects may be grouped with different objects when subspaces vary. Again, gene expression data is a prominent example. When clustering the genes to detect co-regulated genes, one has to cope with the problem, that usually the co-regulation of the genes can only be detected in subsets of the samples (attributes). In other words, different subsets of the samples are responsible for different co-regulations of the genes. When clustering the samples this situation is even worse. As different phenotypes are hidden in varying subsets of the genes, the samples could usually be clustered differently according to various phenotypes, i.e. in varying subspaces.

1.1   Related Work

A common approach to cope with the curse of dimensionality for data mining tasks are dimensionality reduction or methods. In general, these methods map the whole feature space onto a lower-dimensional subspace of relevant attributes, using e.g. principal component analysis (PCA) and singular value decomposition (SVD). However, the transformed attributes often have no intuitive meaning any more and thus the resulting clusters are hard to interpret. In some cases, dimensionality reduction even does not yield the desired results (e.g. [1] presents an example where PCA does not reduce the dimensionality). In addition, using dimensionality reduction techniques, the data is clustered only in a particular subspace. The information of objects clustered differently in varying subspaces is lost. This is also the case for most common feature selection methods.

A second approach for coping with clustering high-dimensional data is projected clustering, which aims at computing k pairs $(C_i, S_i)_{(0 \leq i \leq k)}$ where C_i is a set of objects representing the i-th cluster, S_i is a set of attributes spanning the subspace in which C_i exists (i.e. optimizes a given clustering criterion), and k is a user defined integer. Representative algorithms include the k-means related PROCLUS [2], ORCLUS [3] and the density-based approach OPTIGRID [4]. While the projected clustering approach is more flexible than dimensionality reduction, it also suffers from the fact that the information of objects which are clustered differently in varying subspaces is lost. Figure 1(a) illustrates this problem using a feature space of four attributes A,B,C, and D. In the subspace AB the objects 1 and 2 cluster together with objects 3 and 4, whereas in the subspace CD they cluster with objects 5 and 6. Either the information of the cluster in subspace AB or in subspace CD will be lost.

The most informative approach for clustering high-dimensional data is subspace clustering which is the task of automatically identifying (in general several) subspaces of a high dimensional data space that allow better clustering of the data objects than the original space [1]. One of the first approaches to subspace clustering is CLIQUE [1], a grid-based algorithm using an *Apriori*-like method

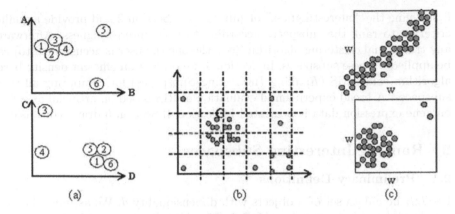

Fig. 1. Drawbacks of existing approaches (see text for explanation).

to recursively navigate through the set of possible subspaces in a bottom-up way. The dataspace is first partitioned by an axis-parallel grid into equi-sized blocks of width ξ called *units*. Only units whose densities exceed a threshold τ are retained. Both ξ and τ are the input parameters of CLIQUE. A cluster is defined as a maximal set of connected dense units. Successive modifications of CLIQUE include ENCLUS [5] and MAFIA [6]. But the information gain of these approaches is also sub-optimal. As they only provide clusters and not complete partitionings of some subspaces, we do not get the information in which subspaces the whole dataset clusters best. Another drawback of these methods is caused by the use of grids. In general, grid-based approaches heavily depend on the positioning of the grids. Clusters may be missed if they are inadequately oriented or shaped. Figure 1(b) illustrates this problem for CLIQUE: Each grid by itself is not dense, if $\tau > 4$, and thus, the cluster C is not found. On the other hand if $\tau = 4$, the cell with four objects in the lower right corner just above the x-axis is reported as a cluster.

Another recent approach called DOC [7] proposes a mathematical formulation for the notion of an optimal projected cluster, regarding the density of points in subspaces. DOC is not grid-based but as the density of subspaces is measured using hypercubes of fixed width w, it has similar problems drafted in Figure 1(c). If a cluster is bigger than the hypercube, some objects may be missed. Furthermore, the distribution inside the hypercube is not considered, and thus it need not necessarily contain only objects of one cluster.

1.2   Contributions

In this paper, we propose a new approach which eliminates the problems mentioned above and enables the user to gain all the clustering information contained in high-dimensional data. We present a preprocessing step, which selects all interesting subspaces using a density-connected clustering notion. Thus we are able to detect all subspaces containing clusters of arbitrary size and shape. We first define the "interestingness" of subspaces in Section 2 and provide a quality criterion to rank the subspaces according to their interestingness. Afterwards any traditional clustering algorithm (e.g. the one the user is accustomed to) can be applied to these subspaces. In Section 3, we present an efficient density-based algorithm called RIS (*R*anking *I*nteresting *S*ubspaces) for computing all those subspaces. A broad experimental evaluation of RIS based on artificial as well as on gene expression data is presented in Section 4. Section 5 draws conclusions.

2   Ranking Interesting Subspaces

2.1   Preliminary Definitions

Let DB be a data set of n objects with dimensionality d. We assume, that DB is a database of feature vectors ($DB \subseteq I\!R^d$). All feature vectors have normalized values, i.e. all values fall into $[0, attrRange]$ for a fixed $attrRange \in I\!R^+$. Let $\mathcal{A} = \{a_1, \ldots, a_d\}$ be the set of all attributes a_i of DB. Any subset $S \subseteq \mathcal{A}$, is called a subspace. The projection of an object o into a subspace $S \subseteq \mathcal{A}$ is denoted by $\pi_S(o)$. The distance function is denoted by $dist$. We assume that $dist$ is one of the L_p-norms. The ε-neighborhood of an object o is defined by $\mathcal{N}_\varepsilon(o) = \{x \in DB \mid dist(o, x) \leq \varepsilon\}$. The ε-neighborhood of an object in a subspace $S \subseteq \mathcal{A}$ is denoted by $\mathcal{N}_\varepsilon^S(o) := \{x \in DB \mid dist(\pi_S(o), \pi_S(x)) \leq \varepsilon\}$.

2.2   Interestingness of a Subspace

Our approach to rate the interestingness of subspaces is based on a density-based notion of clusters. This notion is a common approach for clustering used by various clustering algorithms such as DBSCAN [8], DENCLUE [9], and OPTICS [10]. All these methods search for regions of high density in a feature space that are separated by regions of lower density. We adopt the notion of [8] to define "dense regions" by means of core-objects:

Definition 1. *(Core-Object)*
Let $\varepsilon \in I\!R$ *and* $MinPts \in I\!N$. *An object* o *is called* core object *if* $|\mathcal{N}_\varepsilon(o)| \geq MinPts$.

The core-object property is the key concept of the formal density-connected clustering notion in [8]. This property can also be used for deciding about the interestingness of a subspace. Obviously, if a subspace contains no core-object, it contains no dense region (cluster) and therefore contains no relevant information.

Observation 1. *The number of core-objects of a dataset* DB *(wrt.* ε *and* $MinPts$*) is proportional to the number of different clusters in* DB *and/or the size of the clusters in* DB *and/or the density of clusters in* DB.

This observation can be used to rate the interestingness of subspaces. However, simply counting all the core objects for each subspace delivers not enough information. Even if two subspaces contain the same number of core-objects the quality may differ a lot. Dense regions contain objects which are no core-objects but lie within the ε-neighborhood of a core-object and are thus a vital part of the dense region. Therefore, it is not only interesting how many core-objects a subspace contains but also how many objects lie within the ε-neighborhood of these core-objects. In the following the variable $count[S]$ denotes the sum of all points lying in the ε-neighborhood of all core-objects in the subspace S. The number of core-objects of S is denoted by $core[S]$. If we measure the interestingness of a subspace according to its $count[S]$ value and rank all subspaces according to this quality value, two problems are not adressed. Since naturally with each dimension the number of expected objects in the ε-neighborhood of an object decreases, this naive quality value favors lower dimensional subspaces over higher dimensional ones. To overcome this problem we introduce a scaling coefficient that takes the dimensionality of the subspace into account. We take the ratio between the $count[S]$ value and the $count[S]$ value we would get if all data objects were uniformly distributed in S. For that purpose, we compute the volume of a d-dimensional ε-neighborhood denoted by Vol_ε^d and the number of objects lying in Vol_ε^d assuming uniform distribution.

Definition 2. *The quality of a subspace S, measuring the interestingness of S is defined by:*

$$\text{QUALITY}(S) = \frac{count[S]}{n \cdot \frac{Vol_\varepsilon^{dim[S]} \cdot n}{attrRange^{dim[S]}}}$$

If $dist$ is the L_∞-norm, Vol_ε^d is a hypercube and can be computed by $Vol_\varepsilon^d = (2\varepsilon)^d$, or if $dist$ is the Euclidian distance (L_2-norm) Vol_ε^d is a hypersphere and can be computed as given below:

$$Vol_\varepsilon^d = \frac{\sqrt{\pi^d}}{\Gamma(d/2 + 1)} \cdot \varepsilon^d$$

where $\Gamma(x + 1) = x \cdot \Gamma(x)$, $\Gamma(1) = 1$ and $\Gamma(\frac{1}{2}) = \sqrt{\pi}$.

The second problem is the phenomenon that in high-dimensional spaces more and more points are located on the boundary of the data space. The ε-neighborhoods of these objects are smaller because they exceed the borders of the data space. In [11] the authors show that the average volume of the intersection of the data space and a hypersphere with radius ε can be expressed as the integral of a piecewise defined function integrated over all possible positions of the ε-neighborhood, i.e the core-objects. For our implementation we choose a less complex heuristics to eliminate this effect based on periodical extensions of the data space (cf. Section 3.2 for details).

For two arbitrary subspaces $U, V \in I\!\!R^d$ this quality criterion has two complementary effects which are summerized in the following observation:

Observation 2. *Let $U \supset V$. Then the following inequalities hold:*

1. *$core[U] \le core[V]$ and $count[U] \le count[V]$.*
2. *If $core[U] = core[V]$ and $count[U] = count[V]$ then $\text{QUALITY}(U) > \text{QUALITY}(V)$.*

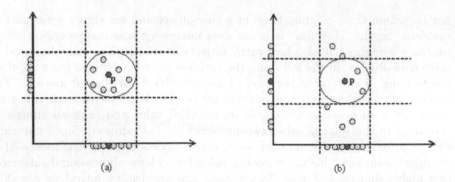

Fig. 2. Visualisation of Lemma 1 for *MinPts* = 5 (2D feature space).

The first observation states that, while navigating through the subspaces bottom-up, at a certain point the core-objects loose their core-object property due to the addition of irrelevant features and thus the quality decreases. On the other hand, as long as this is not the case, the features are relevant for the cluster and the quality increases.

2.3   General Idea of Finding Interesting Subspaces

A straightforward approach would be to examine all possible subspaces (e.g. bottom-up). The problem is, that the number of subspaces is 2^d. Basically all subspaces that do not contain any core-object can be dropped since they cannot contain any clusters. Furthermore, the core-object condition is decreasing strictly monotonic:

Lemma 1. *(Monotonicity of Core-Object Condition)*
Let $o \in DB$ and $S \subseteq A$ be an attribute subset. If o is a core-object in S, then it is also a core-object in any subspace $T \subseteq S$ wrt. ε and MinPts, formally:

$$\forall T \subseteq S : |\mathcal{N}_\varepsilon^S(o)| \geq MinPts \Rightarrow |\mathcal{N}_\varepsilon^T(o)| \geq MinPts.$$

Proof. $\forall x \in \mathcal{N}_\varepsilon^S(o)$ the following holds:

$$dist(\pi_S(o), \pi_S(x)) \leq \varepsilon \Rightarrow \sqrt[p]{\sum_{a_i \in S} (\pi_{a_i}(o) - \pi_{a_i}(x))^p} \leq \varepsilon \overset{T \subseteq S}{\Rightarrow}$$

$$\sqrt[p]{\sum_{a_i \in T} (\pi_{a_i}(o) - \pi_{a_i}(x))^p} \leq \varepsilon \Rightarrow dist(\pi_T(o), \pi_T(x)) \leq \varepsilon \Rightarrow x \in \mathcal{N}_\varepsilon^T(o)$$

It follows that $|\mathcal{N}_\varepsilon^T(o)| \geq |\mathcal{N}_\varepsilon^S(o)| \geq MinPts$ □

The Lemma is visualized in Figure 2(a). The reverse conclusion of Lemma 1 is illustrated in Figure 2(b) and states: If an object o is not a core-object in T, then o is also not a core-object in any super-space $S \supset T$.

The next sections will present in detail, how this property helps to eliminate a lot of subspaces in the process of generating all relevant subspaces in a bottom-up process.

```
RIS(SetOfObjects, Eps, MinPts)
    Subspaces := emptySet;
    FOR i FROM 1 TO SetOfObjects.size() DO
        Object := SampleObjects.get(i);
        RelevantSubspaces := GenerateSubspaces(Object,SetOfObjects);
        Subspaces.add(RelevantSubspaces);
    END FOR
    Subspaces.prune();
    Subspaces.sort();
END //RIS
```

Fig. 3. The RIS algorithm.

3 Implementation of RIS

3.1 Algorithm

Given a set of objects DB and density parameters ε and $MinPts$, RIS finds all interesting subspaces and presents them to the user ordered by relevance. For each object, RIS computes a set of relevant subspaces. All these sets are then merged. A pruning and sorting procedure is applied to the resulting set of subspaces. The pseudocode of the algorithm RIS is given in Figure 3. For each object $o \in DB$, all subspaces in which the core-object condition holds for o, are computed. This step will be described in detail in Section 3.2. Let us note that the algorithm can also be applied to a sample of DB, e.g. for performance reasons (cf. Section 4.3). For each detected subspace, statistical data is accumulated. The detected subspaces are pruned according to certain criteria. In Section 3.3, these criteria will be discussed. Finally, the subspaces are sorted for a more comprehensible user presentation. The clustering in these subspaces can then be done by any clustering algorithm.

3.2 Efficient Generation of Subspaces

For a given object $o \in DB$, the method `GenerateSubspaces` finds all subspaces S in which the core-object condition holds wrt. ε and $MinPts$. Formally, it computes the following set: $K_o := \{T \subseteq \mathcal{A} \mid |\mathcal{N}_\varepsilon^T(o)| \geq MinPts\}$.

The problem of finding the set K_o is equivalent to the problem of determining all frequent itemsets in the context of mining association rules [12] when using the L_∞-norm as distance function and thus can be computed rather efficiently[1]:
For each $x \in DB$ a transaction $T_x \subseteq \mathcal{A}$ is defined, such that,

$$a_i \in T_x \Leftrightarrow |\pi_{a_i}(x) - \pi_{a_i}(o)| \leq \varepsilon \quad \text{for all } i \in \{1, \ldots, d\}.$$

---

[1] Let us note that the use of L_∞-norm is no serious constraint. The only difference is that by using the L_∞ norm we may find additional core-objects and thus additional subspaces. However, these additional subspaces get low quality values anyway.

Lemma 2.

$$K_o = \{T \subseteq \mathcal{A} \mid \mathrm{Supp}_{DB}(T) \geq \frac{MinPts}{|DB|}\}$$

$$where\ \mathrm{Supp}_{DB}(T) = \frac{|\{x \in DB \mid T \subseteq T_x\}|}{|DB|}$$

Proof. $T \subseteq \mathcal{A} \wedge |N_\varepsilon^T(o)| \geq MinPts$

$\Leftrightarrow T \subseteq \mathcal{A} \wedge |\{x \in DB \mid dist_{L_\infty}(\pi_T(o), \pi_T(x)) \leq \varepsilon\}| \geq MinPts$

$\Leftrightarrow T \subseteq \mathcal{A} \wedge$

$\quad |\{x \in DB \mid \forall i \in \{1,\ldots,d\} : a_i \in T \Rightarrow |\pi_{a_i}(o) - \pi_{a_i}(x)| \leq \varepsilon\}| \geq MinPts$

$\Leftrightarrow T \subseteq \mathcal{A} \wedge |\{x \in DB \mid T \subseteq T_x\}| \geq MinPts \Leftrightarrow T \subseteq \mathcal{A} \wedge \mathrm{Supp}_{DB}(T) \geq \frac{MinPts}{|DB|}$

$\hfill\square$

The method `GenerateSubspaces` extends the familar *Apriori* [12] algorithm in accumulating the statistical information for measuring the subspace quality using the monotonicity of the core-object condition (cf. Lemma 1). As mentioned before, we are extending the data space periodically to ensure that all ε-neighborhoods have the same size. This can be done very easily by changing the way the transactions are defined. Instead of only checking if $|\pi_{a_i}(x) - \pi_{a_i}(o)| \leq \varepsilon$ we have to check if $|\pi_{a_i}(x) - \pi_{a_i}(o)| \leq \varepsilon$ or $|\pi_{a_i}(x) - \pi_{a_i}(o)| \geq attrRange - \varepsilon$.

3.3   Pruning of Subspaces

As we are only interested in the subspaces which provide the most information, we can perform the following downward pruning step to eliminate redundant subspaces: If there exists a $(k + 1)$-dimensional subspace S, with higher quality than the k-dimensional subspace T ($S \supset T$), we delete T.

For the second pruning, we assume, that for a given data set the k-dimensional subspace S reflects the clustering in that special data set in a best possible way. Thus, its quality value and the quality values of all its $(k - 1)$-dimensional subspaces $T_1, \ldots, T_m$ is high. On the other hand, if we combine one of these $(k - 1)$-dimensional subspaces $T_1, \ldots, T_m$ with another 1-dimensional subspace with lower quality, the quality of the resulting k-dimensional subspace can still be good. But as we know that it does not reflect the clustering in a best possible way, we are not interested in this k-dimensional subspace. The following heuristic upward pruning eliminates such subspaces. Let S be a k-dimensional attribute space and $S_{k-1} := \{T \mid T \subset S \wedge \dim[T] = k - 1\}$ be the set of all $(k - 1)$-dimensional subspaces of S. Let $\overline{count}$ be the mean count value of all $T \in S_{k-1}$ and $\overline{s}$ be the standard deviation. Let $maxdiff := \max_{T \in S_{k-1}} (|count[T] - \overline{count}|)$ be the maximum deviation of the count-values of all $T \in S_{k-1}$ from the mean count-value. Then, the so-called *bias*-value can be computed as follows: $bias = \frac{\overline{s}}{maxdiff}$. If this bias-value falls below a certain threshhold, we prune the k-dimensional subspace S. Experimental evaluations indicate that 0.56 is a good value for this bias-criterion.

3.4   Determination of Density Parameters

A heuristic method, which is experimentally shown to be sufficient, suggests $MinPts \approx \ln(n)$ where n is the size of the database. Then, ε must be picked depending on the value of $MinPts$. In [8] a simple heuristics is presented to determine the ε of the "thinnest" cluster in the database (for a given $MinPts$). But as we do not know beforehand in which subspaces clusters will be found, we cannot determine ε to find a single subspace with one particular clustering. Quite the contrary, we want to choose the parameters such that RIS detects subspaces which might have clusters of different density and different dimensionality.

However, we can determine an upper bound for ε for a given value of $MinPts$. If we take uniform distribution as worst case, the ε-neighborhood of an object should not contain more than $MinPts - 1$ objects in the full-dimensional space. Otherwise all objects are core-objects. In case of the L_∞-norm an upper bound for ε can be computed as follows:

$$n \cdot \frac{Vol_\varepsilon^d}{attrRange^{dim}} < MinPts \quad \overset{L_\infty}{\Longrightarrow} \quad \varepsilon < \frac{attrRange}{2} \cdot \sqrt[dim]{\frac{MinPts}{n}}$$

where $dim = d$. If we have any knowledge about the dimensionality of the subspaces we want to find, we can further decrease the upper bound by setting dim to the highest dimension of such a subspace.

This upper bound is very rough. Nevertheless, it provides a good indication for the choice of ε. Indeed, it empirically turned out, that $upperbound/4$ is a reasonable choice for ε. Experiments on synthetic data sets show, that our suggested criteria for the choice of the density parameters are sufficient to detect the relevant subspaces containing clusters.

4   Performance Evaluation

We tested RIS using several synthetic as well as a real-world data set. The experiments were run on a workstation with a 1.7 GHz CPU and 2 GB RAM.

The synthetic data sets were generated by a self-implemented data generator. It permits to control the size and structure of the generated data sets through parameters such as number and dimensionality of subspace clusters, dimensionality of the feature space and density parameters for the whole data set as well as for each cluster. In a subspace that contains a cluster the average density of data points in that cluster is much larger than the density of points not belonging to the cluster in this subspace. In addition, it is ensured, that none of the synthetically generated data sets can be clustered in full dimensional space.

The real world data set is the well-studied gene expression data set of Spellman et al. [13] analyzing the yeast mitotic cell cycle. We only chose the data of the cdc15 mutant and eliminated all genes having missing attribute values. The resulting test data set consists of approximately 4400 genes expressed at 24 different time spots.

A subsequent clustering of the data sets in the detected subspaces was performed for each experiment using the above mentioned algorithm OPTICS to validate the interestingness of the subspaces computed by RIS.

4.1 Effectiveness Evaluation

Synthetic Data Sets. We evaluated the effectiveness of RIS using several synthetic data sets of varying dimensionality. The data sets contained between two and five overlapping clusters in varying subspaces. In all experiments, RIS detected the correct subspaces in which clusters exist and assigned the highest quality values to them. All higher dimensional subspaces which were generated, were removed by the upward pruning procedure.

Gene Expression Data. We also applied RIS to the above described gene expression data set. A clustering using OPTICS in the two top-ranked subspaces provided several clusters. The first subspace spanned by the time spots 90, 110, 130, and 190 contains three biologically relevant clusters with several genes playing a central role during mitosis[2]. For example, cluster 1 consists of the genes CDC25 (starting point for mitosis), MYO3 and NUD1 (known for an active role during mitosis) and various other transcription factors (e.g. CHA4, ELP3) necessary during the cell cycle. Cluster 2 contains the gene STE12, identified by [13] as an important transcription factor for the regulation of the cell cycle. In addition, the genes CDC27 and EMP47 which have possible STE12-sites and are most likely co-regulated with STE12 are in that cluster. The cluster is completed by several transcription factors (e.g. XBP1, SSL1). Cluster 3 also consists of several genes which are known to play a role during the cell cycle such as DOM34, CKA1, CPA1, and MIP6. The second subspace is spanned by the time spots 190, 270 and 290 and consists of three clusters that have similar characteristics to those of the first subspace. In addition, a fourth cluster contains several mitochondrion related genes which have similar functions and are therefore most likely co-regulated, indeed. For example, the genes MRPL17, MRPL31, MRPL32, and MRPL33 are four mitochondrial large ribosomal subunits, the genes UBC1 and UBC4 are subunits of a certain protease, the genes SNF7 and VPS4 are direct interaction partners, and several other genes that code for mitochondrial proteins (e.g. MEF1, PHB1, CYC1, MGE1, ATP12). This indicates a higher mitochindrial activity at these time spots, which could be explained by a higher demand of biological energy during the cell cycle (the energy metabolism is located in mitochondrions). In summary, RIS detects two subspaces containing several biologically relevant co-regulations.

4.2 Efficiency Evaluation

The results of the efficiency evaluation are depicted in Figure 4. This evaluation is based on several synthetic data sets. The experiments were run with $MinPts = \ln(n)$ and ε choosen as suggested in Section 3.4. All run times are in seconds.

---

[*] The analysis of the clusters is partly based on the Saccharomyces Genome Database (SGD), available at: http://genome-www.stanford.edu/Saccharomyces/

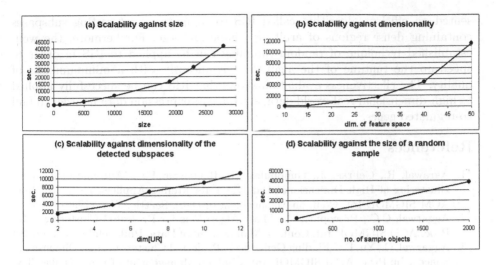

Fig. 4. Efficiency evaluation.

RIS scales well to the dimensionality of the relevant subspaces. With increasing dimensionality of the relevant subspaces, the runtime of RIS grows with a linear factor. On the other hand, the scalability of RIS to the size n and the dimensionality d of the input data set is not linear. With increasing n and d, the runtime of RIS grows with an at least quadratic factor for rather large n and d, respectively. The reason for this scalability vs. the size n is that RIS performs multiple range-queries without any index support, due to the fact that the ε-neighborhoods of all points in arbitrary subspaces have to be computed. However, there is no index structure to efficiently support range queries in arbitrary subspaces. The observed scalability with respect to d can be explained by the *Apriori*-like navigation through the search space of all subspaces.

4.3   Speed-up for Large Data Sets

Since the runtime of RIS is rather high especially for large data sets, we applied random sampling to accelerate our algorithm. Figure 4 shows that for a large data set of $n = 750,000$ data objects, sampling yields a rather good speed-up. The data set contained two overlapping four-dimensional subspace clusters, containing approximately 400,000 and 350,000 points. Even using only 100 sample points, RIS had no problem to detect the subspaces of these two clusters. For all sample sizes, these subspaces had by far the highest quality values. Further experiments empirically show, that random sampling can be successfully applied to RIS in order to speed-up the runtime of this algorithm paying a minimum loss of quality.

5   Conclusions

In this paper, we introduced a preprocessing step for clustering high-dimensional data. Based on a quality criterion for the interestingness of a subspace, we presented an efficient algorithm called RIS to compute all interesting subspaces containing dense regions of arbitrary shape and size. Furthermore, the well-established technique of random sampling can be applied to RIS in order to speed-up the runtime of the algorithm significantly with a minimum loss of quality. The effectiveness evaluation shows that RIS can be succesfully applied to high-dimensional real-world data, e.g. on gene expression data in order to find co-regulated genes.

References

1. Agrawal, R., Gehrke, J., Gunopulos, D., Raghavan, P.: "Automatic Subspace Clustering of High Dimensional Data for Data Mining Applications". In: Proc. ACM SIGMOD Int. Conf. on Management of Data, Seattle, WA. (1998)
2. Aggarwal, C.C., Procopiuc, C.: "Fast Algorithms for Projected Clustering". In: Proc. ACM SIGMOD Int. Conf. on Management of Data, Philadelphia, PA. (1999)
3. Aggarwal, C., Yu, P.: "Finding Generalized Projected Clusters in High Dimensional Space". In: Proc. ACM SIGMOD Int. Conf. on Management of Data, Dallas, TX. (2000)
4. Hinneburg, A., Keim, D.: "Optimal Grid-Clustering: Towards Breaking the Curse of Dimensionality in High-Dimensional Clustering". In: Proc. 25th Int. Conf. on Very Large Databases, Edinburgh, Scotland. (1999)
5. Cheng, C.H., Fu, A.C., Zhang, Y.: "Entropy-Based Subspace Clustering for Mining Numerical Data". In: Proc. ACM SIGKDD Int. Conf. on Knowledge Discovery in Databases, San Diego, FL. (1999)
6. Goil, S., Nagesh, H., Choudhary, A.: "MAFIA: Efficiant and Scalable Subspace Clustering for Very Large Data Sets". Tech. Report No. CPDC-TR-9906-010, Center for Parallel and Distributed Computing, Dept. of Electrical and Computer Engineering, Northwestern University (1999)
7. Procopiuc, C.M., Jones, M., Agarwal, P.K., Murali, T.M.: "A Monte Carlo Algorithm for Fast Projective Clustering". In: Proc. ACM SIGMOD Int. Conf. on Management of Data, Madison, WI. (2002) 418–427
8. Ester, M., Kriegel, H.P., Sander, J., Xu, X.: "A Density-Based Algorithm for Discovering Clusters in Large Spatial Databases with Noise". In: Proc. 2nd Int. Conf. on Knowledge Discovery and Data Mining, Portland, OR. (1996) 291–316
9. Hinneburg, A., Keim, D.A.: "An Efficient Approach to Clustering in Large Multimedia Databases with Noise". In: Proc. 4th Int. Conf. on Knowledge Discovery and Data Mining, New York City, NY. (1998) 224–228
10. Ankerst, M., Breunig, M.M., Kriegel, H.P., Sander, J.: "OPTICS: Ordering Points to Identify the Clustering Structure". In: Proc. ACM SIGMOD Int. Conf. on Management of Data, Philadelphia, PA. (1999) 49–60
11. Berchtold, S., Böhm, C., Keim, D.A., Kriegel, H.P.: "A Cost Model For Nearest Neighbor Search in High-Dimensional Data Space". In: Proc. ACM PODS Symp. on Principles of Database Systems, Tucson, AZ. (1997) 78–86
12. Agrawal, R., Srikant, R.: "Fast Algorithms for Mining Association Rules". In: Proc. ACM SIGMOD Int. Conf. on Management of Data, Minneapolis, MN. (1994) 94–105
13. Spellman, P., Sherlock, G., Zhang, M., Iyer, V., Anders, K., Eisen, M., Brown, P., Botstein, D., Futcher, B.: "Comprehensive Identification of Cell Cycle-Regulated Genes of the Yeast Saccharomyces Cerevisiae by Microarray Hybridization.". Molecular Biolology of the Cell 9 (1998) 3273–3297

Efficiently Finding Arbitrarily Scaled Patterns in Massive Time Series Databases

Eamonn Keogh

University of California - Riverside
Computer Science & Engineering Department
Riverside, CA 92521,USA
eamonn@cs.ucr.edu
www.cs.ucr.edu/~eamonn/

Abstract. The problem of efficiently finding patterns in massive time series databases has attracted great interest, and, at least for the Euclidean distance measure, may now be regarded as a solved problem. However in recent years there has been an increasing awareness that Euclidean distance is inappropriate for many real world applications. The limitations of Euclidean distance stems from the fact that it is very sensitive to distortions in the time axis. A partial solution to this problem, Dynamic Time Warping (DTW), aligns the time axis before calculating the Euclidean distance. However, DTW can only address the problem of *local* scaling. As we demonstrate in this work, *uniform* scaling may be just as important in many domains, including applications as diverse as bioinformatics, space telemetry monitoring and motion editing for computer animation. In this work, we demonstrate a novel technique to speed up similarity search under uniform scaling. As we will demonstrate, our technique is simple and intuitive, and can achieve a speedup of 2 to 3 orders of magnitude under realistic settings.

1 Introduction

The problem of efficiently finding patterns in massive time series databases has attracted great interest in the database and data mining communities, and, at least for the Euclidean distance measure, may now be regarded as a solved problem [2, 5, 11, 12]. However in recent years there has been an increasing awareness that Euclidean distance is inappropriate for many real world applications [1, 6]. The limitations of Euclidean distance stems from the fact that it is very sensitive to distortions in the time axis. A partial solution to this problem, Dynamic Time Warping (DTW), essentially aligns the time axis before calculating the Euclidean distance. Because of its well-documented lethargy, DTW was deemed impractical for large databases until a recent breakthrough demonstrated that DTW can be indexed [10]. DTW can only address the problem of *local* scaling, however uniform scaling may be just as important in many domains, including applications as diverse as bioinformatics, space telemetry monitoring and motion editing for computer animation.

N. Lavrač et al. (Eds.): PKDD 2003, LNAI 2838, pp. 253–265, 2003.

There exists a handful of techniques that can support similarity search under uniform scaling if the scaling factor is known in advance [3, 9]; however, in most domains it is unlikely that we know the scaling factor. In such instances we must resort to multiple queries, one for each possible scaling factor. Clearly, this is untenable for even moderately large databases. What we really need is a technique that can perform a single efficient query to retrieve all qualifying time series with *any* scaling. This is exactly the contribution of this paper.

The rest of this paper is organized as follows. Section 2 carefully motivates the need for similarity search under uniform scaling, and reviews related work. In Section 3 we introduce our approach to the problem. Section 4 contains an extensive empirical evaluation on 5 real world datasets. Finally, Section 5 contains conclusions and directions for future work.

2   Motivating the Need for Uniform Scaling

In addition to the classic Euclidean and Dynamic Time Warping distance measures, the last decade has seen the introduction of dozens of new similarity measures for time series. Recent empirical studies, however, suggest that the majority of these measures are of dubious utility for real world problems [13]. We will therefore take the time to motivate the absolute need for uniform scaling in several real world applications.

2.1   Space Shuttle Telemetry Monitoring

The Space Shuttle transmits thousands of sensor readings to Earth at 1mhz or greater during flight. With over 100 missions, averaging 8.6 days in orbit, this massive repository of data constitutes a potential goldmine for engineers wishing understand and predict in-flight anomalies [4]. Consider an engineer wishing to discover all occurrences of a "dipping" event. This event consists of a sudden positive change in yaw, followed by an auto correction by the Shuttle's onboard flight guidance system. Such events can easily be visually located in a small time series, as they form a 'V' pattern. However, in a massive dataset we must resort to a computerized similarity search.

If we create a 'V' shaped query that is 4 minutes long, and search using the Euclidean distance, we correctly find one true event as shown in Fig. 1 A. However, the second and third best matches fail to find the other two "dips". In contrast, if we issue a query for all 'V' shaped patterns in the range of 4 minutes to 6 minutes, we can correctly discover all three such events as shown in Fig. 1 B.

2.2   Gene Expression Data

Recent advances in bioinformatics technology have resulted in an explosion of gene expression data to be analyzed [1]. Several of the most important tasks, such as clustering, classification and missing value reconstruction, require similarity matching as a first step. Both Euclidean distance and DTW are used; however, we argue that uniform scaling may be more useful for some tasks and datasets. Consider the two sequences shown in Fig. 2.

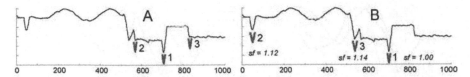

Fig. 1. Eight hours of STS-57 Space Shuttle Inertial Sensor Data: A) A 'V' shaped query correctly matches one steep valley in the data, but the second and third best matches fail to find the two other valleys because they happen more slowly. B) A 'V' shaped query that is allowed to rescale itself by up to 50% correctly finds the three valleys. The second and third best matches have a scaling factor (*sf*) of 1.12 and 1.14 respectively

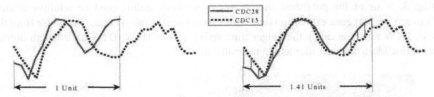

Fig. 2. Two yeast cell-cycle gene expression time series, from genes known to be functionally related. (*Left*) Using the original scale, the genes appear to be a poor match. (*Right*) If the shorter time series is rescaled by a scaling factor of 1.41, it becomes a high quality match to the "prefix" of the longer time series

Although the two genes are known to be functionally related [1], the raw time series subjectively appear to be a poor match. Simply rescaling the shorter time series by a factor of 1.41 allows the underlying similarity to be more readily discovered.

We considered other approaches for this problem. Euclidean distance is a very commonly used technique, but it is only defined for time series of the same length. One solution is to normalize the lengths with interpolation; another is to truncate the longer time series. Although DTW is defined for time series of different lengths, interpolation and truncation can also be useful here. In Fig. 3. we show all combinations of possibilities, none of them succeeds in capturing the underlying similarity of the data.

2.3   Motion Capture Editing

Motion capture data is increasingly used in video games, movie special effects and gait analysis [6]. The following is a classic problem in this domain. Given two examples of a human performing a task, once slowly, and once quickly, interpolate the motion at any desired speed [20]. Figure 4 shows an example. The problem is non-trivial because of non-linear effects in human dynamics. Nevertheless correctly aligning the two time series from each instance is a critical first step in solving the problem. This can be achieved manually for a simple movie special effect, but for real time video games, or complex effect shots (i.e, the battle scenes in The Lord of the Rings), automation is required.

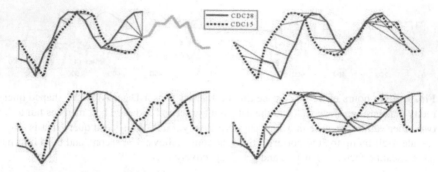

Fig. 3. None of the published alternatives to uniform scaling produce intuitive alignments between the two gene expression time series introduced in this section. Clockwise from the top left, DTW after truncating the longer time series, classic DTW, DTW after length normalization, Euclidean distance after length normalization

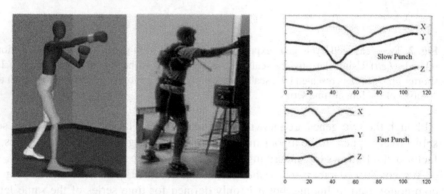

Fig. 4. (*left*) A computer animation of a boxer, driven by a motion capture system (*center*). Given that we have captured an example of a fast moment and a slow movement (*right*), an important problem in motion capture editing is to interpolate the movement at any desired speed. Aligning the signals with uniform scaling is a important first step in this process

Having motivated the need for uniform scaling in several domains, we will next consider related work.

2.4   Related Work

The past decade has seen literally hundreds of papers on similarity search using the Euclidean distance [2, 5, 11, 12]; useful surveys can be found in [8] and [17]. However recent years have seen an increasing awareness that the Euclidean distance may be unsuitable for many applications [1, 10, 18, 19].

Many non Euclidean distance measures for time series have been introduced, however, a recent empirical study suggests that most of them are of questionable utility [10]. The only non-Euclidean distance measure that has been forcefully shown to be superior to Euclidean distance is DTW, it's utility has been demonstrated in domains as diverse as bioinformatics [1], chemical engineering, gait analysis [6], speech recognition, meteorology, and robotics. However DTW only considers local stretching and shrinking of the time axis. As we demonstrated in the previous section, uniform scaling may be equally important in many domains.

The utility of uniform scaling has been noted before [9, 14, 15]. However, all previous work has focused on speeding up similarity search, when the scaling factor is *known*. For example, there are systems that can index data of length 200, and support queries of any length from 150 to 200. However the user must specify what length query they wish to run, perhaps a query of length 175. If the user wishes to find the best matching time series, at any length from 150 to 200, they would have to run every possible query, of length 150, 151 ,…, 200 to find the answer. This is clearly untenable. As all these systems claim about one order of magnitude speed up, placing them in a loop and running them 50 times is clearly going to be self defeating. The feature that differentiates our work from all the rest is that we allow a user to issue a single query, and find the best match at *any* scaling. Our proposed technique is unique in this aspect.

3   Uniform Scaling

We begin by formally defining the uniform scaling problem.

Suppose we have two time series, a query Q and a candidate match C, of length n and m respectively, where:

$$Q = q_1, q_2, \ldots, q_i, \ldots, q_n \tag{1}$$
$$C = c_1, c_2, \ldots, c_j, \ldots, c_m \tag{2}$$

For clarity of presentation we will assume that $n \le m$, that is to say, C is always longer than or equal to Q, and thus we are only interested in stretching the query to match some prefix of C. This assumption is only to simplify notion and does not preclude matching a time series by shrinking, since we can always reverse the roles of the sequences.

If we wish to compare the two time series, and it happens that $n = m$, we can use the ubiquitous Euclidean distance:

$$D(Q, C) \equiv \sqrt{\sum_{i=1}^{n} (q_i - c_i)^2} \tag{3}$$

Since the square root function is monotonic and concave, we can remove the square root step and get identical rankings, clustering and classifications. This measure is called the squared Euclidean distance:

$$D(Q,C) \equiv \sum_{i=1}^{n} (q_i - c_i)^2 \qquad (4)$$

In addition to the utility of slightly speeding up the calculations, working with this distance measure makes other optimizations possible [13].

If n is smaller than m, then the distance measures introduced above are not defined. To compare the two time series in this case, we have several choices; we can truncate C, and compare Q to $[c_1, c_2, \dots, c_n]$, or we can somehow *stretch* Q to be of length m, or more generally we can *stretch* Q to be of length p, $(n \le p \le m)$, truncate off the last $m-p$ values of Q, then use squared Euclidean distance. The informal idea behind *stretching* can be captured in the more formal definition of scaling. To scale time series Q to produce a new time series QP of length p, the formula is:

$$QP_j = Q_{\lceil j * n/p \rceil}, 1 \le j \le p \qquad (5)$$

Note that we can quickly obtain any scaling in $O(p)$ time. We call the ratio p/n the *scaling factor* or *sf*. Slightly different definitions of scaling do exist, but they do not affect the results that follow. Fig. 5. visually summarizes the above definitions.

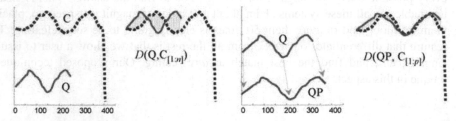

Fig. 5. A visual summary of the notation introduced in this section. From (*left*) to (*right*) A candidate time series C, and a shorter query Q. The squared Euclidean distance between Q and the first n datapoints in C can be visualized as the sum of the squared lengths of the gray hatch lines. The query Q can be stretched to length p, producing a new time series QP. In this case, QP is a good match to the first p datapoints in C

3.1   Brute Force Search under Uniform Scaling

If we wish to find the best scaled match between Q and C, we can simply test all possible scalings, as illustrated in Table 1.

Table 1. An algorithm to find the best scaled match between two time series

```
Algorithm: Test_All_Scalings(Q,C)
   best_match_val      = inf;
   best_scaling_factor = null;
   for p = n to m
       QP = rescale(Q,p);
       distance = squared_Euclidean_distance(QP, C[1..p]);
       if distance <  best_match_val
           best_match_val = distance;
           best_scaling_factor = p/n;
       end;
   end;
return(best_match_val, best_scaling_factor)
```

The algorithm takes only $O(p*(m-n))$ time and seems unworthy of any optimization effort. However, when mining real world datasets, rather than having a single candidate time series C, we are typically confronted with massive collection of possible candidate time series, which will denote as **C**. As a motivating example, the MACHO dataset, a collection of star light curve microlensing events, has over 40 million time series [7]. To find the best scaled match to a query Q, in data collection **C**, we can use a brute force algorithm as shown in Table 2.

Note that the time complexity for this algorithm is $O(|C| * (m-n))$, this is simply untenable for large datasets.

Table 2. An algorithm to find the best scaled match to query from a set of possible matches

```
Algorithm: Search_Database_for_Scaled_Match(Q,C)
    overall_best_time_series = null;
    overall_best_match_val   = inf;
    overall_best_scaling     = null;
    for i = 1 to number_of_time_series_in_(C)
        [dist, scale] = Test_All_Scalings(Q,Ci)
        if dist <  overall_best_match_val
                overall_best_time_series = i;
                overall_best_match_val   = dist;
                overall_best_scaling     = scale;
        end;
    end;
    return(overall_best_time_series, overall_best_match_val, overall_best_scaling)
```

3.2   Speeding up Search with Lower Bounding

To speed up matching under uniform scaling we will rely on the classic idea of lower bounding. The intuition is this: given some technique for quickly calculating the minimum possible distance between the query and a candidate sequence at any possible scaling, we can prune off many calculations. In more detail, we maintain a variable that contains the distance of the best-scaled match encountered thus far. Before calling the subroutine `Test_All_Scalings` on the next candidate time series, we first perform the quick lower bounding test. If the lower bound distance between the candidate and the query is greater than the distance of the best-scaled match already seen, we can simply discarded the candidate from consideration. For clarity, the idea is formalized in Table 3, although the algorithm differs from the algorithm in Table 2 only in the addition of the lower bounding test as a precondition to the subroutine `Test_All_Scalings`.

There are only two important properties of a lower bounding measure:

- It must be fast to compute. A measure that takes as long to compute as `Test_All_Scalings` is of little use. We would like the time complexity to be at most linear in the length of the time series.
- It must be a relatively tight lower bound. A function can achieve a trivial lower bound by always returning zero as the lower bound estimate. However, in order for the algorithm in Table 3 to be effective, we require a method that tightly bounds the value of the best match.

Table 3. A modified algorithm for searching for the best match under uniform scaling

```
Algorithm: Faster_Search_Database_for_Scaled_Match(Q,C)
    overall_best_time_series = null;
    overall_best_match_val   = inf;
    overall_best_scaling     = null;
    for i = 1 to number_of_time_series_in_(C)
        if lower_bound_distance(Q,Ci) < overall_best_match_val
                [dist, scale] = Test_All_Scalings(Q,Ci)
                    if dist <  overall_best_match_val
                        overall_best_time_series = i;
                        overall_best_match_val   = dist;
                        overall_best_scaling     = scale;
                end;
        end;
    end;
    return(overall_best_time_series, overall_best_match_val, overall_best_scaling)
```

The idea of speeding up search using lower bounding is not new; in fact, it is the cornerstone of virtually every time series similarity search algorithm. However, while dozens of lower bounding measures are known for Euclidean distance [2, 5, 9, 11, 12], and 3 lower bounding measures known for DTW [10], there are no lower bounding measures in the literature for uniform scaling. In the next section we introduce the first such measure.

3.3   Lower Bounding Uniform Scalings

To create a lower bounding distance measure for uniform scaling we will generate a bounding envelope. Bounding envelopes were introduced in [10] to lower bound DTW, and since then they have sparked a flurry of research activity [16, 18, 19]. While the principle is the same here, the definitions of the envelope are very different. In particular, we create two sequences U and L, such that:

$$U_i = \max(\, c_{\lfloor (i-1)*m/n \rfloor +1}, \ldots, c_{\lfloor i*m/n \rfloor}\,) \tag{6}$$

$$L_i = \min(\, c_{\lfloor (i-1)*m/n \rfloor +1}, \ldots, c_{\lfloor i*m/n \rfloor}\,) \tag{7}$$

These sequences can be visualized as bounding the first n points of the time series C. Fig. 6. shows some examples.

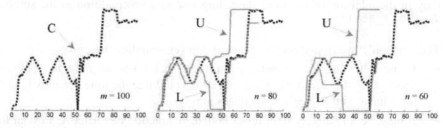

Fig. 6. (*Left*) A time series C of length 100. (*Center*) The time series shrouded by upper and lower envelopes U and L with lengths 80. (*Right*) The same time series shrouded by upper and lower envelopes U and L with lengths 60

Having defined the U and L, we can now introduce the lower bounding function, it was originally introduced in [10] for the problem of DTW.

$$LB\_Keogh(Q,C) = \sum_{i=1}^{n} \begin{cases} (q_i - U_i)^2 & \text{if } q_i > U_i \\ (q_i - L_i)^2 & \text{if } q_i < L_i \\ 0 & otherwise \end{cases} \tag{8}$$

This function can be visualized as the squared Euclidean distance between any part of the query time series not falling within the envelope and the nearest (orthogonal) corresponding section of the envelope. Fig. 7. illustrates the idea.

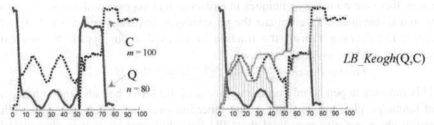

Fig. 7. (*Left*) A time series C and a shorter query Q. (*Right*) A visualization of the lower-bounding function *LB_Keogh*(Q,C). Note that any part of query time series Q that falls inside the bounding envelope is ignored. Otherwise the distance corresponds to the sum of the squared straight line distances from the query to the nearest point in the envelope (the gray hatch lines)

We have claimed that *LB_Keogh*(Q,C) lower bounds the squared Euclidean distance between any scaling of Q, and the appropriate prefix of C. The proof is straightforward, we omit it brevity.

3.4   Further Optimizations

While *LB_Keogh*(Q,C) is the optimal lower bound for uniform scaling, given only U and L, several further optimizations are possible in the context of similarity search. We will give one such example here, using concrete numbers for clarity. Suppose we are using the algorithm in Table 3 for similarity search, with $n = 100$, and $m = 200$. Further suppose that the best matching time series encountered thus far is at a distance of 10. If we test the lower bound of the next candidate time series and we find it to be 11, we can prune it from the search space. However, if the lower bound is 9 we must call the `Test_All_Scalings` subroutine.

We can observe, however, that although the lower bounding test did fail for the fairly drastic scaling factor of 2 (i.e. 200/100), it would be less likely to do so for smaller scaling factor, say 3/2. We could rescale the query to length 150, rebuild U and L and apply the lower bounding test again. If it happens that the lower bound is now 10 or greater, we could prune all possible scalings from length 150 to 200 from consideration, and only examine the scalings from 100 to 149. Of course, we could apply the above logic recursively to the scalings from 100 to 149, and more generally this suggests doing a binary search over all the scalings. We call this algorithm Bi-

`nary_Test_All_Scalings`, but omit a detailed description since it is rather obvious. Note that we cannot use binary search to speed up the brute force algorithm, since the squared Euclidean distance does not vary monotonically with the scaling factor (in general). We use this optimization in all our experiments below.

4   Experimental Results

In this section we test our proposed approach with a comprehensive set of experiments. We compare only to the brute force search algorithm defined in Table 2, because there are no other techniques in existence that support uniform scaling queries, with a single query. To eliminate the possibility of implementation bias [13], we will report the *Pruning Power*, the fraction of times that our approach must call the squared Euclidean distance function.

$$Pruning\ Power = \frac{Number\ of\ calls\ to\ distance\ function\ by\ proposed\ approach}{Number\ of\ calls\ to\ distance\ function\ by\ brute\ force\ search}\qquad(13)$$

This measure depends only on the tightness of the lower bounds, and is independent of language, platform, caching or any other implementation details. As an additional sanity check we also measured the CPU time, however since it is almost perfectly correlated with the *Pruning Power*, we will omit it for brevity.

It has been forcefully demonstrated that the quality of lower bounding measures, and therefore the speed of search, can vary greatly depending on the data [13]. We therefore tested our approach on a variety of datasets. Fig 8. shows a sample of each.

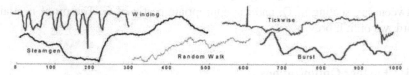

Fig. 8. Randomly extracted samples of the time series datasets

Since the speed-up obtained for our approach clearly depends on range of scaling factors and the length of the time series, we will test our approach for the cross product of scaling factors = {1.05, 1.10, 1.15, 1.20, 1.25} and time series candidate lengths of {16, 32, 64, 128, 256}.

We conducted our experiments as follows. We randomly removed a subsequence of the appropriate length from the data to use as a query, then we randomly choose 5,000 other subsequences to act as the database. We then searched for the best scaled match, noting the pruning power. We repeated this 100 times for every combination of scaling factors and candidate lengths. Fig 9. shows the results.

The results are quite impressive, the worst case is a single order of magnitude speed-up, more generally two to three orders of magnitude speedup are observed. Note that, the pruning power seems independent of the candidate time series lengths, but does get worse as the scaling factor increases. This is to be expected, since for large scaling factors the *LB_Keogh* function has relatively little information with which to calculate the lower bound.

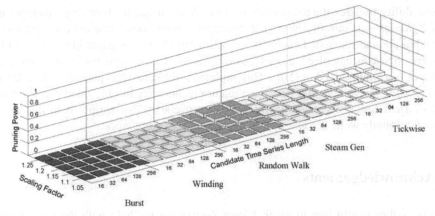

Fig. 9. The pruning power of *LB_Keogh* of 5 different datasets, over a range of scaling factors and candidate lengths

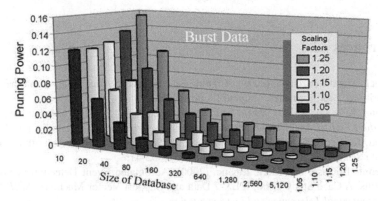

Fig. 10. The pruning power of *LB_Keogh* on the burst dataset, over a range of scaling factors and database sizes. Note the scale of the Z-axis is different from that of Fig. 9

As with many indexing techniques, the pruning power of our approach improves with the size of the dataset. The intuition behind this effect is that the larger the dataset, the more likely we are to find a very close match early on in the search, and thus derive the maximum benefit from the lower bound pruning test (the outermost `if` statement in Table 3). To demonstrate this, we repeated the previous experiment for different size datasets. The results for just the burst dataset are shown in Fig. 10.

The results clearly show that as the database size increases, the pruning power improves. This is a very desirable property when mining larger datasets.

5   Discussion and Conclusions

We have shown how to dramatically speed up similarity search under uniform warping, however, we have not considered *indexing* under uniform warping. Fortunately the ability to index the data comes for free! A technique for indexing envelopes under *LB_Keogh* was introduced in [10]. Since then, many other researchers have used this technique and suggested extensions [16, 18, 19] (Note that paper [19] claims to *introduce* the "concept of envelopes", *introduce* must be a typo for *review*, since envelopes were introduced in [10]). This explosion of interest has ensured that indexing of time series envelopes has become a mature technology in only one year. We omitted empirical testing of indexing for brevity and clarity; we simply note that it works exceptionally well. We leave a full discussion for future work.

Acknowledgements

The author would like to thank Victor Zordan for his help with the motion capture example, Dennis DeCoste at JPL for contributing the Space Shuttle data, and Jessica Lin and Michalis Vlachos for their suggestions.

References

1. Aach, J. and Church, G. (2001). Aligning gene expression time series with time warping algorithms. Bioinformatics. Volume 17, pp 495-508
2. Chan, K. & Fu, A. W. (1999). Efficient time series matching by wavelets. In *proceedings of the 15th IEEE Int'l Conference on Data Engineering*. Sydney, Australia. pp 126-133
3. Chu, K., Lam., S. & Wong, M. (1998) An Efficient Hash-Based Algorithm for Sequence Data Searching. The Computer Journal 41 (6): 402-415
4. Dennis DeCoste and Marie Levine. (2000). Automated Event Detection in Space Instruments: A Case Study Using IPEX-2 Data and Support Vector Machines. *SPIE Conference Astronomical Telescopes and Instrumentation.*
5. Faloutsos, C., Ranganathan, M., & Manolopoulos, Y. (1994). Fast subsequence matching in time-series databases. *In Proc. ACM SIGMOD Conf.*, Minneapolis. pp. 419-429
6. Gavrila, D. M. & Davis,L. S.(1995). Towards 3-d model-based tracking and recognition of human movement: a multi-view approach. In *International Workshop on Automatic Face-and Gesture-Recognition*
7. Hegland, M., Clarke, W. & Kahn, M. (2002). Mining the MACHO dataset, Computer Physics Communications, Vol 142(1-3), December 15. pp. 22-28
8. Hetland, M. (2003). A Survey of Recent Methods for Efficient Retrieval of Similar Time Sequences. To appear in an Edited Volume, *Data Mining in Time Series Databases*. Published by the World Scientific Publishing Company
9. Kahveci, T. & Singh, A. (2001). Variable length queries for time series data. In *proceedings of the 17th Int'l Conference on Data Engineering*. Heidelberg, Germany, pp 273-282
10. Keogh, E. (2002). Exact indexing of dynamic time warping. In 28th *International Conference on Very Large Data Bases*. Hong Kong. pp 406-417
11. Keogh, E,. Chakrabarti, K,. Pazzani, M. & Mehrotra (2000). Dimensionality reduction for fast similarity search in large time series databases. Journal of Knowledge and Information Systems. pp 263-286

12. Keogh, E,. Chakrabarti, K,. Pazzani, M. & Mehrotra (2001) Locally adaptive dimensionality reduction for indexing large time series databases. In *Proc of ACM SIGMOD Conference on Management of Data*. pp 151-162
13. Keogh, E. and Kasetty, S. (2002). On the Need for Time Series Data Mining Benchmarks: A Survey and Empirical Demonstration. In *the 8ᵗʰ ACM SIGKDD International Conference on Knowledge Discovery and Data Mining*. Edmonton, Canada. pp 102-111.
14. Park, S., Chu, W. W., Yoon, J. & Hsu, C. (2000). Efficient searches for similar subsequences of different lengths in sequence databases. In *proceedings of the 16ᵗʰ Int'l Conference on Data Engineering*. San Diego, CA, pp 23-32
15. Perng, C., Wang, H., Zhang, S., & Parker, S. (2000). Landmarks: a new model for similarity-based pattern querying in time series databases. In *proceedings of 16ᵗʰ International Conference on Data Engineering*. pp 33-42
16. Rath, T. & Manmatha, R. (2002): Lower-Bounding of Dynamic Time Warping Distances for Multivariate Time Series. Tech Report MM-40, University of Massachusetts Amherst.
17. Roddick, J. F. and Spiliopoulou, M. (2001). A Survey of Temporal Knowledge Discovery Paradigms and Methods. *IEEE Tran's on Knowledge and Data Engineering*. pp. 750-767
18. Vlachos, M., Kollios, G., & Gunopulos, G. (2002). Discovering similar multidimensional trajectories. In *Proc 18ᵗʰ International Conference on Data Engineering*
19. Zhu, Y. & Shasha, D. (2003). Query by Humming: a Time Series Database Approach. To appear in SIGMOD 2003.
20. Zordan, V. B., Hodgins, J. K., (2002). Motion capture-driven simulations that hit and react, ACM SIGGRAPH Symposium on Computer Animation.

Using Transduction and Multi-view Learning to Answer Emails

Michael Kockelkorn[1], Andreas Lüneburg[1], and Tobias Scheffer[2]

. Tonxx, Berlin, Germany
{mk,alueneburg}@tonxx.com
. Humboldt Univerersity, School of Computer Science
Unter den Linden 6, 10099 Berlin, Germany
scheffer@iws.cs.uni-magdeburg.de

Abstract. Many organizations and companies have to answer large amounts of emails. Often, most of these emails contain variations of relatively few frequently asked questions. We address the problem of predicting which of several frequently used answers a user will choose to respond to an email. Our approach effectively utilizes the data that is typically available in this setting: inbound and outbound emails stored on a server. We take into account that there are no explicit links between inbound and corresponding outbound mails on the server. We map the problem to a semi-supervised classification problem that can be addressed by algorithms such as the transductive support vector machine and multi-view learning. We evaluate our approach using emails sent to a corporate customer service department.

1 Introduction

Companies allocate considerable economic resources to communication with their customers. A continuously increasing share of this communication takes place via email; marketing, sales and customer service departments as well as dedicated call centers have to process high volumes of emails, many of them containing repetitive routine questions. It appears overly ambitious to completely automate this process; however, any software support that leads to a significant productivity increase is already greatly beneficial. Our approach to support this process is to predict which answer a user will most likely send in reply to an incoming email, and to propose this answer to the user. The user, however, is free to modify – or to dismiss – the proposed answer.

Our approach is to learn a predictor that decides which of a small set of standard answers a user is most likely to choose in reply to a given inbound message. We learn such a predictor from the available data: inbound and outbound emails stored on an email server. We transform the email answering problem into a set of semi-supervised text classification problems. Contrasting studies that investigate identification of general subject areas of emails (*e.g.*, [8]), we explore whether text classification algorithms can identify instances of a specific frequently asked question.

N. Lavrač et al. (Eds.): PKDD 2003, LNAI 2838, pp. 266–277, 2003.
© Springer-Verlag Berlin Heidelberg 2003

Many approaches are known that learn text classifiers from data. The support vector machine (SVM) (*e.g.*, [10]) is generally considered to be one of the most accurate algorithms; this is supported, for instance, by the TREC filtering challenge [13]. The naive Bayes algorithm is also widely used for text classification.

Among the known algorithms that utilize unlabeled data, the transductive SVM and the multi-view framework apply for support vector learning. The transductive SVM [9] maximizes the distance between hyperplane and both, labeled and unlabeled data. In multi-view learning [3], two classifiers which use different attribute sets provide each other with labels for the unlabeled data.

The contribution of this paper is threefold. Firstly, we analyze the problem of answering emails, taking all practical aspects into account. Secondly, we present a case study on a practically relevant problem showing how well the naive Bayes algorithm, the support vector machine, the transductive support vector machine, and the co-training multi-view algorithm can identify instances of particular questions in emails. Thirdly, we describe how we integrated machine learning algorithms into a practical answering assistance system that is easy to use and provides immediate user benefit.

The rest of this paper is organized as follows. In Section 2, we analyze the problem setting. We discuss our general approach and our mapping of the email answering problem to a set of semi-supervised text classification problems in Section 3. In Section 4, we briefly describe the transductive SVM and the multi-view algorithm that we used for the case study that is presented in Section 5. In Section 6 we describe how we have integrated our learning approach into the Responsio email management system. Section 7 discusses related approaches.

2 Problem Setting

We consider the problem of predicting which of n (manually identified) standard answers $A_1, \ldots, A_n$ a user will reply to an email. In order to learn a predictor, we are given a repository $\{x_1, \ldots, x_m\}$ of inbound, and $\{y_1, \ldots, y_{m'}\}$ of outbound emails. Typically, these repositories contain at least hundreds, but often (at least) thousands of emails stored on a corporate email server.

Although both inbound and outbound emails are stored, it is not trivial to identify which outbound email has been sent in reply to a particular inbound email; neither the emails nor the internal data structures of the Outlook email client contain explicit links. When an outbound email does not *exactly* match one of the standard answers, this does *not* necessarily mean that none of the standard answers is the correct prediction. The user could have written an answer that is equivalent to one of the answers A_i but uses a few different words.

A characteristic property of the email answering domain is a non-stationarity of the distribution of inbound emails. While the likelihood $P(x|A_i)$ is quite stable over time, the prior probability $P(A_i)$ is not. Consider, for example, a server breakdown which will lead to a sharp increase in the probability of an answer like "we apologize for experiencing technical problems...."; or consider an advertising campaign for a new product which will lead to a high volume of requests for information on that product.

What is the appropriate utility criterion for this problem? Out goal is to assist the user by proposing answers to emails. Whenever we propose the answer that the user accepts, he or she benefits; whereas, when we propose a different answer, the user has to manually select or write an answer. Hence, the optimal predictor proposes the answer A_i which is most likely given x (i.e., maximizes $P(A_i|x)$), and thereby minimizes the probability of the need for the user to write an answer manually. Keeping these characteristics in mind, we can pose the problem which we want to solve as follows.

Problem 1. Given is a repository X of inbound emails and a repository Y of outbound emails in which instances of standard answers $A_1, \ldots, A_n$ occur. There is no explicit mapping between inbound and outbound mails and the prior probabilities $P(A_i)$ are non-stationary. The task is to generate a predictor for the most likely answer A_i to a new inbound email x.

3   Underlying Learning Problem

In this Section, we discuss our general approach that reduces the email answering problem to a semi-supervised text classification problem.

Firstly, we have to deal with the non-stationarity of the prior $P(A_i)$. In order to predict the answer that is most likely given x, we have to choose $\mathrm{argmax}_i P(A_i|x) = \mathrm{argmax}_i P(x|A_i)P(A_i)$ where $P(x|A_i)$ is the likelihood of question x given that it will be answered with A_i and $P(A_i)$ is the prior probability of answer A_i. Assuming that the answer will be exactly one of $A_1, \ldots, A_n$ we have $\sum_i P(A_i) = 1$; when the answer can be any subset of $\{A_1, \ldots, A_n\}$, then $P(A_i) + P(\bar{A}_i) = 1$ for each answer A_i.

We know that the likelihood $P(x|A_i)$ is stationary; only a small number of probabilities $P(A_i)$ has to be estimated dynamically. Equation 1 averages the time dependent priors (estimated by counting occurrences of the A_i in the outbound emails within time interval t) discounted over time.

$$\hat{P}(A_i) = \frac{\sum_{t=0}^{T} e^{-\lambda t} \hat{P}(A_i|t)}{\sum_{t=0}^{T} e^{-\lambda t}} \tag{1}$$

We can now focus on estimating the (stationary) likelihood $P(x|A_i)$ from the data. In order to map the email answering problem to a classification problem, we have to identify positive and negative examples for each answer A_i.

We use the following heuristic to identify cases where an outbound email is a response to a particular inbound mail. The recipient has to match the sender of the inbound mail, and the subject lines have to match up to a prefix ("Re:" for English or "AW:" for German email clients). Furthermore, either the inbound mail has to be quoted in the outbound mail, or the outbound mail has to be sent while the inbound mail was visible in one of the active windows. (We are able to check the latter condition because our email assistance system is integrated into the Outlook email client and monitors user activity.) Using this rule, we are able to identify *some* inbound emails as positive examples for $A_1, \ldots, A_n$.

We also need to identify negative examples. We can safely assume that no two different standard answers A_i and A_j are semantically equivalent. Hence, when an email has been answered by A_i, we can conclude that it is a negative example for all A_j, $j \neq i$. When the answer to an inbound email is different from all standard answers A_i, we cannot conclude that the inbound mail is a negative example for all standard answers because the response might have been semantically equivalent, or very similar, to one of the standard answers. Such emails are unlabeled examples in the resulting text classification problem.

For the same reason, we cannot obtain examples of inbound emails for which no standard answer is appropriate; hence, we cannot estimate $P(\text{no standard answer})$ or $P(\bar{A}_i)$ for any A_i. Thus, we have a small set of positive and negative examples for each A_i. Additionally, we have a large quantity of emails for which we cannot determine the appropriate answer.

Text classifiers typically return an uncalibrated decision function f_i for each binary classification problem; our decision on the answer to x has to be based on the $f_i(x)$ (Equation 2). We discriminate each class A_i against all other classes; that is, we have to assume that A_i is independent of all $f_j(x)$ for $i \neq j$. Since we have dynamic estimates of the non-stationary $P(A_i)$, Bayes' equation (Equation 3) provides us with a mechanism that combines n binary decision functions and the prior estimates optimally.

$$\mathrm{argmax}_i P(A_i|x) = \mathrm{argmax}_i P(A_i|f_1(x), \ldots, f_n(x)) \qquad (2)$$
$$\approx \mathrm{argmax}_i P(A_i|f_i(x)) = \mathrm{argmax}_i P(f_i(x)|A_i)P(A_i) \quad (3)$$

Equation 3 is only applicable for discrete $f_i(x)$, while the decision function values are really continuous. In order to estimate $P(f_i(x)|A_i)$ we have to fit a parametric model to the data. Following [2], we assume Gaussian likelihoods $P(f_i(x)|A_i)$ and estimate the μ_i, $\mu_{\bar{i}}$ and σ_i in a cross validation loop as follows. In each cross validation fold, we record the $f_i(x)$ for all held-out positive and negative instances. After that, we estimate μ_i, $\mu_{\bar{i}}$ and σ_i from the recorded decision function values of all examples. It is well known that Bayes' rule applied to a Gaussian likelihood yields a sigmoidal posterior; Equation 4 corresponds to Equation 3 for continuous $f_i(x)$ and Gaussian $P(f_i(x)|A_i)$.

$$P(A_i|f_i(x)) = \left(1 + e^{\frac{\mu_{\bar{i}} - \mu_i}{\sigma^2} f_i(x) + \frac{\mu_{\bar{i}}^2 - \mu_i^2}{2\sigma^2} + log\frac{1 - P(A_i)}{P(A_i)}}\right)^{-1} \qquad (4)$$

We have now reduced the email answering problem to a semi-supervised text classification problem. We have n binary classification problems for which few labeled positive and negative and many unlabeled examples are available. We need a text classifier that returns a (possibly uncalibrated) decision function $f_i : X \rightarrow real$ for each of the answers A_i.

We considered a Naive Bayes classifier and the support vector machine SVM^{light} [9]. Both classifiers use the bag-of-words representation which con-

siders only the words occurring in a document, but not the word order. As pre-
processing operation, we tokenize the documents but do not apply a stemmer.
For SVM^{light}, we calculate tf.idf vectors.

4   Using Unlabeled Data

We briefly sketch two approaches that allow to utilize unlabeled data for support
vector learning: the transductive SVM, and the co-training algorithm.

4.1   Transduction

In order to calculate the decision function for an instance x, the support vector
machine calculates a linear function $f(x) = wx + b$. Model parameters w and b
are learned from data $((x_1, y_1), \ldots, (x_m, y_m))$. Note that $\frac{w}{|w|}x_i + b$ is the distance
between plain (w, b) and instance x_i; this margin is positive for positive examples
($y_i = +1$) and negative for negative examples ($y_i = -1$). Equivalently, $y_i(\frac{w}{|w|}x_i +
b)$ is the positive margin for both positive and negative examples.

The optimization problem which the SVM learning procedure solves is to
find w and b such that $y_i(wx_i + b)$ is positive for all examples (all instances lie
on the "correct" side of the plain) and the smallest margin (over all examples)
is maximized. Equivalently to maximizing $y_i(\frac{w}{|w|}x_i + b)$, it is usually demanded
that $y_i(wx_i + b) \geq 1$ for all (x_i, y_i) and $|w|$ be minimized.

Optimization Problem 1 *Given data* $((x_1, y_1), \ldots, (x_m, y_m))$; *over all* w, b,
minimize $|w|^2$, *subject to the constraint* $\forall_{i=1}^m y_i(wx_i + b) \geq 1$.

The SVM^{light} software package [9] implements an efficient optimization algorithm which solves optimization problem 1. The transductive support vector
machine (TSVM) [10] furthermore considers unlabeled data. This unlabeled data
can (but need not) be new instances which the SVM is to classify. In transductive support vector learning, the optimization problem is reformulated such that
the margin between all (labeled and unlabeled) examples and hyperplain is maximized. However, only for the labeled examples we know on which side of the
hyperplain the instances have to lie.

Optimization Problem 2 *Given labeled data* $((x_1, y_1), \ldots, (x_m, y_m))$ *and un-
labeled data* $(x_1^*, \ldots, x_k^*)$; *over all* w, b, $(y_1^*, \ldots, y_k^*)$, *minimize* $|w|^2$, *subject to
the constraints* $\forall_{i=1}^m y_i(wx_i + b) \geq 1$ *and* $\forall_{i=1}^m y_i^*(wx_i^* + b) \geq 1$.

The TSVM algorithm which solves optimization problem 2 is related to the
EM algorithm. TSVM starts by learning parameters from the labeled data and
labels the unlabeled data using these parameters. It iterates a training step
(corresponding to the "M" step of EM) and switches the labels of the unlabeled
data such that optimization criterion 2 is maximized (resembling the "E" step).
The TSVM algorithm is described in [9].

Table 1. Co-training algorithm.

Given positive examples $(x_\cdot, x_\cdot, +)$, negative examples $(x_\cdot, x_\cdot, -)$ and unlabeled examples in two different views $V_\cdot$ and $V_\cdot$; number of iterations k.

1. Loop for k iterations
   (a) Train $f_\cdot$ and $f_\cdot$ using the labeled positive and negative examples.
   (b) Let $f_\cdot$ and $f_\cdot$ select the positive and negative example for which they make the most confident prediction. Remove the examples from the unlabeled data and add them to the labeled data.
2. Return the combined classifier $f(x) = f_\cdot(x_\cdot) + f_\cdot(x_\cdot)$.

4.2 Multi-view Learning

Blum and Mitchell [3] have proposed the multi-view approach to utilizing unlabeled data. In multi-view learning, the available attributes V are split into two subsets V_1 and V_2 such that $V_1 \cup V_2 = V$ and $V_1 \cap V_2 = \emptyset$. A labeled example (x, a) is then viewed as (x_1, x_2, a) where x_1 contains the values of the attributes in V_1 and x_2 the values of attributes in V_2.

The co-training algorithm is the most prominent multi-view algorithm. The idea of co-training is to learn two classifiers $f_1(x_1)$ and $f_2(x_2)$ which bootstrap each other by providing each other with labels for the unlabeled data. Co-training is applicable when either attribute set suffices to learn the target f – *i.e.*, there are classifiers f_1 and f_2 such that for all x: $f_1(x_1) = f_2(x_2) = f(x)$ (the *compatibility* assumption). When the views are furthermore *independent* given the class labels – $P(x_1|f(x), x_2) = P(x_1|f(x))$ – then co-training converts unlabeled examples into randomly drawn labeled examples [3].

As V_1, we use randomly drawn 50% of the words occurring in the training corpus; V_2 contains the remaining words. $f_1(x_1)$ and $f_2(x_2)$ are trained from the labeled examples. Now f_1 selects two examples from the unlabeled data that it most confidently rates positive and negative, respectively, and adds them to the labeled examples. If the representations in the two views are truly independent, then the new examples are randomly drawn positive and negative examples for f_2. Now f_2 selects two unlabeled examples, the two hypotheses are retrained, and the process recurs. The algorithm is presented in Table 1.

The compatibility and independence assumptions are usually violated in practice. However, empirical studies [14,11] show that co-training can nevertheless improve performance. In particular, text classification problems seem to be particularly suited for co-training [15]. In our experiments, we use co-training in association with SVM^{light}.

5 Case Study

The data used in this study was provided by the TELES European Internet Academy, an education provider that offers classes held via the internet. In order to evaluate the predictors, we manually labeled all inbound emails within a certain period with the matching answer. Table 2 provides an overview of the data statistics. Roughly 72% of all emails received can be answered by one of nine standard answers. The most frequent question "product inquiries" (requests for the information brochure) already covers 42% of all inbound emails.

Table 2. Statistics of the TEIA email data set.

Frequently answered question	emails	percentage
Product inquiries	224	42%
Server down	56	10%
Send access data	22	4%
Degrees offered	21	4%
Free trial period	15	3%
Government stipends	13	2%
Homework late	13	2%
TELES product inquiries	7	1%
Scholarships	7	1%
Individual questions	150	28%
Total	528	100%

We briefly summarize the basic principles of ROC analysis which we used to assess the decision functions [5,17]. The *receiver operating characteristic* (ROC) curve of a decision function plots the number of true positives against the number of false positives. By comparing the decision function against a decreasingly large threshold value we observe a trajectory of classifiers described by the ROC curve.

The area under the ROC curve is equal to the probability that, when we draw one positive and one negative example at random, the decision function assigns a higher value to the positive example than to the negative. Hence, the area under the ROC curve (the *AUC performance*) is a very natural measure of the ability of a decision function to separate positive from negative examples.

In order to estimate the AUC performance and its standard deviation for a decision function, we performed between 7 and 20-fold stratified cross validation and averaged the AUC values measured on the held out data. In order to plot the actual ROC curves, we also performed 10-fold cross validation. In each fold, we filed the decision function values of the held out examples into one global histogram for positives and one histogram for negatives. After 10 folds, we calculated the ROC curves from the resulting two histograms.

First, we studied the performance of a decision function provided by the Naive Bayes algorithm (which is used, for instance, in the commercial Autonomy Answer system) as well as the support vector machine SVM^{light} [9]. We use the default parameter settings for SVM^{light}. Figure 1 shows that the SVM impressively outperforms Naive Bayes in all cases except for one (TELES product inquiries). Remarkably, the SVM is able to identify even very specialized questions with as little as seven positive examples with between 80 and 95%

AUC performance. It has earlier been observed that the probability estimates of Naive Bayes approach zero and one, respectively, as the length of analyzed document increases [2]. This implies that Naive Bayes performs poorly when not all documents are equally long, as is the case here.

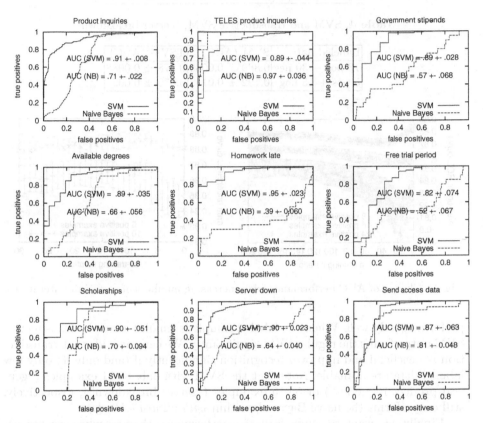

Fig. 1. ROC curves for nine most frequently asked questions of naive Bayes and the support vector machine.

In the next set of experiments, we observed how the transductive support vector machine improves performance by utilizing the available unlabeled data. We successively reduce the amount of labeled data and use the remaining data (with stripped class labels) as unlabeled and hold-out data (we use the same setting for the co-training experiments described in the following). We average five re-sampled iterations with distinct labeled training sets. We compare SVM performance (only the labeled data is used by the SVM) to the performance of the transductive SVM (using both labeled and unlabeled data). Table 3 shows the results for category "general product inqueries"; Table 4 for "server breakdown".

When the labeled sample is of size at least 24 + 33 for "general product inqueries", or 10 + 30 for "server breakdown", then SVM and transductive SVM

Table 3. SVM and transductive SVM, "general product inqueries".

Labeled data	SVM (AUC)	TSVM (AUC)
24 pos + 33 neg	0.87 ± 0.0072	0.876 ± 0.007
16 pos + 22 neg	0.855 ± 0.007	0.879 ± 0.007
8 pos + 11 neg	0.795 ± 0.0087	0.876 ± 0.0068

Table 4. SVM and transductive SVM, "server breakdown".

Labeled data	SVM (AUC)	TSVM (AUC)
10 pos + 30 neg	0.889 ± 0.0088	0.878 ± 0.0088
5 pos + 15 neg	0.792 ± 0.01	0.859 ± 0.009

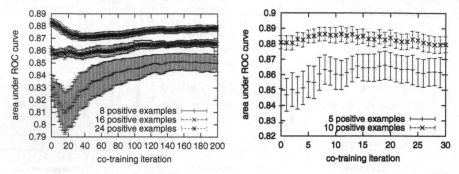

Fig. 2. Change of AUC performance with increasing numbers of co-training iterations.

perform equally well. When the labeled sample is smaller, then the transductive SVM outperforms the regular SVM significantly. We can conclude that transduction is beneficial and improves recognition significantly if (and only if) only few labeled data are available. Note that the SVM with only 8+11 examples ("general product inqueries") or 5+15 examples ("server breakdown"), respectively, still outperforms the naive Bayes algorithm *with all available data*.

Finally, we want to study how the performance changes when we use co-training in association with the Support Vector Machine. Figure 2 shows the AUC performance against the number of co-training iteration. The results resemble those obtained with the TSVM: Co-training improves performance only when at most 16 positive examples are available for product inqueries and when at most 5 positive examples are available for server breakdown. The benefit of both, co-training and transduction is greatest, when only few labeled data are available. Transduction outperforms co-training for product inqueries; transduction and co-training perform similar for server breakdown.

6   The Responsio Email Management System

We integrated the learning algorithms into an email assistance system. The key design principle is that, once the standard answers are entered, it does not require

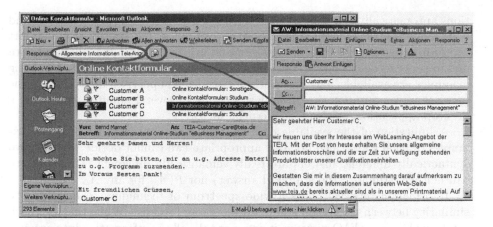

Fig. 3. When an email is read, the most likely answer is displayed in a special field in the Outlook window. On clicking the "Auto-Answer" button, a reply window with the proposed answer text is created.

any extra effort from the user. The system observes incoming emails and replies sent, but does not require explicit feedback. Responsio is an add-on to Microsoft Outlook. The control elements (Figure 3) are loaded as a COM object.

When an email is selected, the COM add-in sends the email body to a second process which identifies the language of the email, executes the language specific classifiers and determines the posterior probabilities of the configured answers. The classifier process notifies the COM add-in of the most likely answer which is displayed in the field marked. When the user clicks the "auto answer" button (circled in Figure 3), Responsio extracts first and last name of the sender, identifies the gender by comparing the first names against a list, and formulates a salutation line followed by the proposed standard answer. The system opens a reply window with the proposed answer filled in.

Whenever an email is sent, Responsio identifies whether the outbound mail is a reply to an inbound mail by matching recipient and subject line to sender and subject line of all emails that are visible in one of the Outlook windows. When the sent email includes one of the standard answers, the inbound mail is filed into the list of example mails for that answer. These examples can be viewed in the Responsio manager window. It is also possible to manually drag and drop emails into the example folders. Whenever an example list changes, the training unit starts a process with the learning algorithm.

7 Discussion and Related Results

We have discussed the problem of identifying instances of frequently asked questions in emails, using only stored inbound and outbound emails as training data. Our empirical data shows that identifying a relatively small set of standard questions automatically is feasible; we obtained AUC performance of between 80 and 95% using as little as seven labeled positive examples. The transductive support vector machine and the co-training algorithm utilize the available unlabeled data and improve recognition rate considerably if and only if only few labeled training examples are available. The drawback of both semi-supervised algorithms is the increase in computation time from few seconds to several minutes. For use in a desktop application, efficiency is a crucial factor.

A limitation of the available data sources is that we cannot determine examples of emails for which no A_i is appropriate (we cannot decide whether two syntactically different answers are really semantically different). Therefore, we can neither estimate $P(\text{no standard answer})$ nor $P(\bar{A}_i)$ for any A_i.

Information retrieval offers a wide spectrum of techniques to measure the similarity between a question and questions in an FAQ list. While this approach is followed in many FAQ systems, it does not take all the information into account that is available in the particular domain of email answering: emails received in the past. An FAQ list contains only one single instance of each question whereas we typically have many instances of each questions available that we can utilize to recognize further instances of these questions more accurately.

The domain of question answering [20] is rather loosely related to our email answering problem. In our application domain, a large fraction of incoming questions can be answered by very few answers. These answers can be pre-configured; the difficulty lies in recognizing instances of these frequently asked questions robustly, even in very ungrammatical emails. Question answering systems solve a problem that is in a way more difficult: selecting an answer sentence from a large corpus (such as an encyclopedia) for arbitrary questions.

Several email assistance systems have been presented. [8,4,19,7] use text classifiers in order to predict the correct folder for an email. In contrast to these studies, we study the feasibility of identifying instances of particular questions rather than general subject categories.

Related is the problem of filtering spam email. Keyword based approaches, Naive Bayes [1,16,18,12] and rule-based approaches [6] have been compared. Generating positive and negative examples for spam requires additional user interaction: the user might delete interesting emails just like spam after reading it. By contrast, our approach generates examples for the email answering task without imposing additional effort on the user.

Acknowledgment

We have received support from the German Science Foundation (DFG) under grant SCHE540/10-1, and from Hewlett Packard Consulting, Germany.

References

1. I. Androutsopoulos, J. Koutsias, K. Chandrinos, and C. Spyropoulos. An experimental comparison of naive bayesian and keaword based anti-spam filtering with personal email messsages. In *Proceedings of the International ACM SIGIR Conference*, 2000.

2. P. Bennett. Assessing the calibration of naive bayes' posterior estimates. Technical report, CMU, 2000.
3. A. Blum and T. Mitchell. Combining labeled and unlabeled data with co-training. In *Proceedings of the Workshop on Computational Learning Theory*, 1998.
4. T. Boone. Concept features in Re:Agent, an intelligent email agent. *Autonomous Agents*, 1998.
5. A. Bradley. The use of the area under the ROC curve in the evaluation of machine learning algorithms. *Pattern Recognition*, 30(7):1145–1159, 1997.
6. W. Cohen. Learnig rules that classify email. In *Proceedings of the IEEE Spring Symposium on Machine learning for Information Access*, 1996.
7. E. Crawford, J. Kay, and E. McCreath. IEMS - the intelligent email sorter. In *Proceedings of the International Conference on Machine Learning*, 2002.
8. C. Green and P. Edwards. Using machine learning to enhance software tools for internet information management. In *Proceedings of the AAAI Workshop on Internet Information Management*, 1996.
9. T. Joachims. Making large-scale svm learning practical. In B. Schölkopf, C. Burges, and A. Smola, editors, *Advances in Kernel Methods - Support Vector Learning*, 1999.
10. T. Joachims. Transductive inference for text classification using support vector machines. In *Proceedings of the International Conference on Machine Learning*, 1999.
11. S. Kiritchenko and S. Matwin. Email classification with co-training. Technical report, University of Ottawa, 2002.
12. A. Kolcz and J. Alspector. Svm-based filtering of e-mail spam with content-specific misclassification costs. In *Proceedings of the ICDM Workshop on Text Mining*, 2001.
13. D. Lewis. The trec-5 filtering track. In *Proceedings of the Fifth Text Retrieval Conference*, 1997.
14. I. Muslea, C. Kloblock, and S. Minton. Active + semi-supervised learning = robust multi-view learning. In *Proceedings of the International Conference on Machine Learning*, 2002.
15. K. Nigam and R. Ghani. Analyzing the effectiveness and applicability of co-training. In *Proceedings of Information and Knowledge Management*, 2000.
16. P. Pantel and D. Lin. Spamcop: a spam classification and organization program. In *Proceedings of the AAAI Workshop on Learning for Text Categorization*, 1998.
17. F. Provost, T. Fawcett, and R. Kohavi. The case against accuracy estimation in comparing classifiers. In *Proceedings of the International Conference on Machine Learning*, 1998.
18. M. Sahami, S. Dumais, D. Heckerman, and E. Horvitz. A bayesian approach to filtering junk email. In *Proceedings of the AAAI Workshop on Learning for Text Categorization*, 1998.
19. R. Segal and J. Kephart. Mailcat: An intelligent assistant for organizing mail. In *Autonomous Agents*, 1999.
20. E. Vorhees. The trec-8 question answering track report. In *Proceedings of TREC-8*, 1999.

Exploring Fringe Settings of SVMs for Classification

Adam Kowalczyk and Bhavani Raskutti

Telstra, 770 Blackburn Road, Clayton, Victoria 3168, Australia
{Adam.Kowalczyk,Bhavani.Raskutti}@team.telstra.com

Abstract. There are many practical applications where learning from single class examples is either, the only possible solution, or has a distinct performance advantage. The first case occurs when obtaining examples of a second class is difficult, e.g., classifying sites of "interest" based on web accesses. The second situation is exemplified by the one-class support vector machine which was the winning submission of the second task of the KDD Cup 2002.

This paper explores the limits of supervised learning using both positive and negative examples. To this end, we analyse the KDD Cup dataset using four classifiers (support vector machines and ridge regression) and several feature selection methods. Our analysis shows that there is a consistent pattern of performance differences between one and two-class learning for all algorithms investigated, and these patterns persist even with aggressive dimensionality reduction through automated feature selection. Using insight gained from the above analysis, we generate synthetic data showing similar pattern of performance.

1 Introduction

A standard approach for two class discrimination is to use examples from both classes to generate a model for discriminating them. This approach is so entrenched in machine learning that practitioners often will not consider data unless it contains examples of both classes. Moreover, many machine learning algorithms, such as decision trees, naive Bayes or multilayer perceptron, do not function unless the training data includes examples from two classes. However, there are many applications where obtaining examples of a second class is difficult, e.g., classifying sites of "interest" to a web surfer where the sole information that is available are the positive examples or sites that are of interest to the user. In such a case, learning from examples of one class is the only possible solution.

In addition, there are situations when the data has heavily unbalanced representatives of the two classes of interest, e.g., fraud detection and information filtering. A supervised algorithm applied to such a problem has to implement some form of balancing. In some situations, it may be beneficial to design rebalancing even more radically than warranted by unequal proportions, and ignore the large pool of negative examples and learn from positive examples only. A real life learning problem that has benefited from such an approach is the second task of the KDD Cup 2002 [4], where the winning submission learnt using just the positive examples which consisted of $< 3\%$ of the training data [8].

N. Lavrač et al. (Eds.): PKDD 2003, LNAI 2838, pp. 278–290, 2003.

This paper explores the limits of two-class learning and analyses situations when this discrimination learning may break down. This exploration begins with an analysis of the KDD Cup dataset using four different classifiers: support vector machines and ridge regression, in several different settings. We then study the performance of these classifiers in the one-class and two-class mode when the input feature space is significantly reduced using automatic feature selection methods. The consistently better performance of the one-class models in the above analysis, leads us to a systematic study of conditions when one-class learning is advantageous. This study using synthetic data and the four classifiers used in the earlier experiments shows that data with a certain combination of properties, e.g., the presence of label noise, sparsity of features and low proportion of minority class, lends itself to better performance with one-class learners.

The paper is organised as follows. Section 2 places our research in context of existing research. Section 3 introduces the basic support vector machines in the particular form used for this research, and describes the performance measure suitable for our task. We then present results of our experiments with the KDD 2002 Cup data in Section 4.1 and that for synthetic data in Section 4.2. Finally, we discuss the implications of our results in Section 5.

2   Related Research

The problem of discrimination of unbalanced classes is encountered in a large number of real life situations, e.g., detection of oil spills in satellite radar images [9], information retrieval and filtering [10] and biological domains [4,8]. Many solutions have been proposed to address the imbalance problem including sampling and weighting examples (cf. [7] for a thorough survey). However, they typically focus on cases when the imbalance ratio of minority to majority class is around 10:90. In this paper, we focus on extreme imbalance, where the minority class consists of around 1-3% of the data, and extend the sampling to situations when one of the classes is ignored completely and learning is accomplished using examples from a single class.

A possibility of single class learning with support vector machines (SVM) has been noticed previously. In particular, Schölkopf et al. [14] have suggested a method of adapting the SVM methodology to *one-class* learning by treating the origin as the only member of the second class. This methodology has been used for image retrieval [3] and for document classification [11]. In both cases, modelling is performed using examples from the positive class only, and the one-class models perform reasonably, although much worse than the *two-class* models learnt using examples from both classes. In contrast, in this paper, we show that for certain problems one-class models can perform better.

3   Classifiers and Performance Metrics

In this section we recall basic concepts of *kernel machines* in a form suitable for this paper. Given a training sequence (x_i, y_i) of binary n-vectors $x_i \in \{0,1\}^n \subset \mathbb{R}^n$ and bipolar labels $y_i \in \{\pm1\}$ for $i = 1, ..., m$. The case of prime interest here is when the target class, labelled $+1$, is much smaller than the background class (labelled -1), consisting of a minute fraction, $\approx 1 - 3\%$, of the data. Our aim is to find a "good" discriminating function $f : \{0,1\}^n \to \mathbb{R}$ that scores the target class instances higher than the background class instances. The solution will be given in a form of a kernel machine

$$f(x) = f^H(x) + b := \sum_{i=1}^{m} \beta_i k(x, x_i) + b \qquad (1)$$

where $k : \mathbb{R}^n \times \mathbb{R}^n \to \mathbb{R}$ is a kernel function of one of the forms specified below and $\beta_i, b \in \mathbb{R}$ are parameters to be defined for the given training set as the minimiser of the regularised risk of the form as follows.

$$(\beta_i, b) \mapsto \|f^H, b\|^2 + \sum_{i=1}^{m} C_{y_i} \phi\big(1 - y_i(f^H(x_i) + b)\big), \qquad (2)$$

where $C_{+1}, C_{-1} \geq 0$ are class dependent regularisation constants, $\phi : \mathbb{R} \to \mathbb{R}_+$ is a convex loss function penalising deviations of scores from allocated labels and $\|\cdot\|$ is a norm as specified below. Now we specify variations of the regularised risk (2) leading to four different cases of kernel machines used in this paper.

1. SVM^1: For the popular *support vector machine with linear penalty* we use the norm $\|f^H, b\|^2 := \|f^H\|_k^2$, where

$$\|f^H\|_k^2 := \sum_{i,j=1}^{m} \beta_i \beta_j k(x_i, x_j),$$

and the "hinge loss" $\phi(\theta) := \max(0, \theta)$, $\theta \in \mathbb{R}$ [5,15,16];

2. $hSVM^1$: Replacing the norm in the above definition by

$$\|f^H, b\|^2 := \|f^H\|_k^2 + b^2 \qquad (3)$$

we obtain *the homogeneous support vector machine with linear penalty*;

3. $hSVM^2$: For *the (homogeneous) support vector machine with quadratic penalty* [5] we use norm (3) and the squared hinge loss $\phi(\theta) := (\max(0, \theta))^2$ for $\theta \in \mathbb{R}$;

4. hRN^2: For *the regularisation network* [6,17] or ridge regression (c.f. [5,6,17] we use norm (3) and the ordinary square loss $\phi(\theta) := (\theta)^2$ for $\theta \in \mathbb{R}$.

If the kernel k satisfies the Mercer theorem assumptions [5,15,16] then for the minimiser of (2) we have $\beta_i = y_i \alpha_i$, where $\alpha_i \geq 0$ for $i = 1, ..., m$.

In our investigations we shall be using the popular polynomial kernel

$$k(x, x') = (x \cdot x')^d = (\sum_{i=1}^{n} \xi_i \xi_i')^d$$

for $x = (\xi_i)$ and $x' = (\xi_i')$ from $\{0,1\}^n$ and degree $d = 1, 2, 3$ and 4.

Note that $hSVM^1$, $hSVM^2$ and hRN^2 implement classifiers that correspond to separation of the data $(z_i, y_i) := (\Phi(x_i), 1, y_i) \in \mathbb{R}^N \times \mathbb{R} \times \{\pm 1\}$ by a hyperplane in the extended feature space passing through the $(0,0) \in \mathbb{R}^N \times \mathbb{R}$, $\Phi : \mathbb{R}^n \to \mathbb{R}^N$ is a *feature mapping* of *the observation space* $\mathbb{R}^n$ into an appropriate Euclidean space $\mathbb{R}^N$ (*the features* space). In particular such a solution is provided also if all data points belong to a single class, i.e. if $y_i = const$.

The geometrical meaning of the solution (2) can be most clearly illustrated in the limiting case of "hard margin", i.e. $C \to \infty$. In such a case, the optimal solution of (2) corresponds to the direction of the shortest vector to the convex shell spanned by all vectors $y_i z_i \in \mathbb{R}^N \times \mathbb{R}$, $i = 1, ..., m$.

3.1   Re-balancing of the Data

Two way of compensation for the imbalance in the training data will be investigated in this paper.

Hard balancing. We use all m_+ positive instances but randomly choose only m_- instances with the negative labels (the majority class) varying the *"mixture" ratio* $B_{-/+} = m_-/m_+$. We set the regularisation constants to $C_{-1} = C_{+1} = C/m_+$, where $C > 0$ is a chosen constant. In this form of balancing, $B_{-/+} = 0$ is the case of *positive 1-class learning*, and $B_{-/+} = 1$ represents the case of *balanced 2-class learning* when the same number of examples from both classes are used.

Soft balancing. We use all available training data but with different class regularisation constants: $C_{-1} = (1 - B)C/2m_-$ and $C_{-1} = (1 + B)C/2m_+$, where $C > 0$ and $-1 \leq B \leq +1$ is a balance parameter. Here $B = +1$ and $B = -1$ correspond to 1-class learning, and $B = 0$ is 2-class learning with both classes "balanced" according to their prior proportions.

The advantage of the hard balancing over the soft balancing is the speed of generation of a solution, as it typically uses a smaller training set. For this reason the hard balancing was used in the most of our experiments.

3.2   Centroids

Now we introduce the fifth and the simplest of the five algorithms considered here in terms of generation of the solution. In contrast to SVMs, it is inherently non-sparse and so can be complex to implement in the case of non-linear kernels.

Algorithm 5, $Cntr_B$: For the *centroid* classifier we set

$$f(x) = f_{Cntr}^H(x) := \frac{(1 + B) \sum_{i, y_i = +1} k(x_i, x)}{2 \max(1, m_+)} - \frac{(1 - B) \sum_{i, y_i = -1} k(x_i, x)}{2 \max(1, m_-)},$$

where $x \in \mathbb{R}^n$ and $-1 \leq B \leq +1$ is the balance factor.

In terms of the feature space, the centroid classifier implements the projection on the direction of a weighted differences between centroids of data from both class labels. For $B = -1$ it is the direction of the majority class, for $B = +1$ that of the minority class and for $B = 0$ that of the difference between centroids of both classes.

Now we formally link centroids and SVMs. Later we shall connect this result to some of our empirical findings. The formal proof, omitted here, can be easily derived from Karush-Khun-Tucker conditions for SVM solution.

Theorem 1 *If functions* $k(x_i, .)$, $i = 1, ..., m$ *are linearly independent, then*

$$\lim_{C \to 0^+} \frac{f_{hSVM^p}^H}{C} = \lim_{C \to 0^+} \frac{f_{hRN^2}^H}{C} = f_{Cntr}^H, \qquad (4)$$

for $p = 1, 2,$ *where* $f.^H$ *denotes the homogeneous part of the* soft balanced *SVM solution (1) for the appropriate machine. Moreover, if both classes are represented in training, i.e. for the soft balance factor* $-1 < B < +1,$ *then:*

$$\lim_{C \to 0^+} \frac{f_{SVM^1}^H}{C} = f_{Cntr}^H. \qquad (5)$$

3.3   Performance Measures

We have used *AROC*, the Area under the Receiver Operating Characteristic (ROC) curve as our main performance measure. In that we follow the steps of KDD 2002 Cup, but also, we see it as the natural metric of general goodness of classifier (as corroborated below) capable of meaningful results even if the target class is a tiny fraction of the data.

We recall that the ROC curve is a plot of the *true positive rate* or precision, $P(f(x_i) > \theta | y_i = 1)$, against the *false positive rate*, $P(f(x_i) > \theta | y_i = -1)$, as a decision threshold θ is varied. The concept of ROC curve originates in the military signal detection but these days it is widely used in many other areas, including data mining, psychophysics and medical diagnosis (cf. review [2]). In the latter case, *AROC* is viewed as a measure of general "goodness" of a test, formalised as a predictive model f in our context, with a clear statistical meaning as follows. According to Bamber's interpretation [1], $AROC(f)$ is equal to the probability of correctly answering the two-alternative-forced-choice problem: given two cases, one x_i from the negative and the other x_j from the positive class, allocate scores in the right order, i.e. $f(x_i) < f(x_j)$. Additional attraction of *AROC* as a figure of merit is its direct link to the well researched area of order statistics via U-statistics and Wilcoxon-Whitney-Mann test [1].

There are some ambiguities in the case of *AROC* estimated from a discrete set in the case of ties, i.e. when multiple instances from different classes receive the same score. Following [1] we implement in this paper the definition

$$AROC(f) = P(f(x_i) < f(x_j)| - y_i = y_j = 1)$$
$$+0.5P(f(x_i) = f(x_j)| - y_i = y_j = 1)$$

expressing *AROC* in terms of conditional probabilities, which can be re-formulated in terms the rank-ordered test sequence (where the rank is imposed by the scores allocated by f).

Note that the trivial uniform random predictor has *AROC* of 0.5.

4   Experiments

In order to understand the boundaries when the performance of two-class classifiers deteriorate, we have explored the following datasets: (1) Real life data in the form of the Aryl Hydrocarbon Receptor signalling pathway data provided for the second task of the 2002 KDD cup (henceforth referred to as the *AHR data*) (Section 4.1), and (2) Synthetic data created with some specific properties such as presence of noise in labels (Section 4.2).

4.1   Analysis of KDD Cup 2002 Data

In our main experiments we have used AHR-data set which is the combined training and test data sets used for task 2 of KDD Cup 2002. The data set is based on experiments by Guang Yao and Chris Bradfield of McArdle Laboratory for Cancer Research, University of Wisconsin. These experiments aimed at identification of yeast genes that, when knocked out, cause a significant change in the level of activity of the Aryl Hydrocarbon Receptor signalling pathway (cf. [4] for more details). In this paper we follow the setting of the "broad task" of the KDD Cup: the discrimination between 127 'positive' genes from the combined class encompassing the labels "change" and "control" and the remaining 4380 genes forming the 'negative' class. We note that the results for the first subtask, namely, learning "change" class are similar [8]. In our experiments this set has been repeatedly split into 70% for training and 30% for testing. All averages and standard deviations reported are for independent tests on 20 such random splits.

Each of the 4507 instances in the data set is described by a variety of information that characterise the gene associated with the instance, e.g., associated abstracts from scientific articles,genes whose encoded proteins physically interact with one another, information about the subcellular localisation and functional classes of the proteins encoded by various genes. For the experiments described in this paper, we convert all of the information from the different files to a sparse matrix containing 18330 binary features as described in [8].

Impact of Regularisation Constant. Figure 1 shows mean AROC as a function of C for four different linear kernel machines ($d = 1$) with the hard balancing (Figures A-D) and the soft balancing (Figures E-H). We use four different modes as follows: (*i*) positive 1-class ($B_{-/+} = 0$ and $B = +1$, solid line); (*ii*) negative 1-class ($B = -1$, dotted line); (*iii*) balanced 2-class ($B_{-/+} = 1$ and $B = 0$, dashed line); (*iv*) un-balanced 2-class ($B_{-/+} = 35 \approx 4380/127$) when all examples from both classes are used, the dash-dot line). The standard deviations are shown as vertical bars.

An inspection of plots brings a number of interesting observations:

1. The un-balanced 2-class machines (dash-dot lines) and negative 1-class machines (dot lines) have inferior performance relative to either positive 1-class machines or the balanced 2-class SVMs for most values of C (excepting very low C). Thus only the last two modes will be used in further research in this paper.

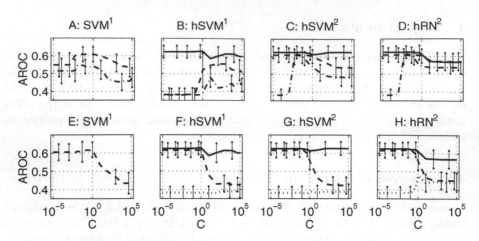

Fig. 1. Mean AROC for AHR-data as a function of the regularisation constant C for 1-class and 2-class SVMs with the hard balancing (figures A-D) and the soft balancing (figures E-H) for four linear ($d = 1$) SVMs and four different modes: (i) positive 1-class ($B_{-/.} = 0$ and $B = +1$, solid line); (ii) negative 1-class ($B = -1$, dotted line); (iii) balanced 2-class ($B_{-/.} = 1$ and $B = 0$, dashed line); (iv) un-balanced 2-class ($B_{-/.} = 35$, the dash-dot line). The standard deviations are shown as vertical bars.

2. All positive 1-class and balanced 2-class machines show a very good and roughly equal performance for very low values of C. Additionally, these values of mean AROC are equal to that for the positive 1-class centroid ($B = +1$, AROC $= 62.4 \pm 3.4$) and the balanced 2-class centroid $Cntr_0$ ($B = 0$, AROC $= 61.5 \pm 3.9$) trained on the whole data. In the case of soft balancing this can be inferred from Theorem 1 since it implies that the orders imposed on test data by the scores of the respective centroid and SVM classifiers coincide hence yield the same AROC (uniquely determined by such an order).

3. There are noticeable differences between the performance of different SVMs. For instance, note the differences between unbalanced 2-class $hSVM^1$ and SVM^1 (dash-dot lines in Figures 1A and 1B, respectively).

4. Positive 1-class $hSVM^2$ is very robust across the whole range of C values (cf. the solid line in Figure 1C). In particular, for high values of C, i.e, virtually the hard margin case [8], it performs better than any other SVM tested. This setting was used for the winning submission to KDD Cup 2002.

In summary, the top performance by SVMs is achieved at extremely high values of C, i.e. hard margin case, or at the limit of very low C. For low C the best SVMs are equivalent to respective centroid machines, for positive one-class ($B = 1$) or the balanced 2 class ($B = 0$). For the high Cs, the positive one-class consistently outperforms other settings. This motivates our restrictions on experimental settings for the rest of the paper as follows. We shall concentrate exclusively on hard balanced SVMs trained with high Cs and centroid classifiers with B set to 0 ($Cntr_0$) for a range of hard balance mixture ratios $B_{-/+}$.

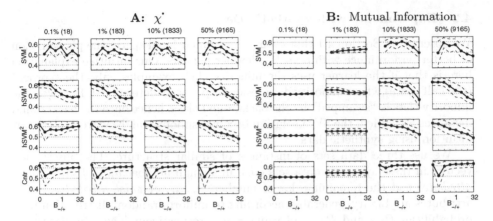

Fig. 2. Mean AROC with ± standard deviation envelopes for AHR-data as a function of the hard balanced data with mixture ratio $B_{-/.} \in \{0, 0.01, 0.1, 0.5, 1, 5, 10, 35\}$ for four different fractions of the original feature set (0.1%, 1%, 10% and 50% out of 18,330 features). The three linear ($d = 1$) kernel machines were trained with $C = 1000$, while the centroid classifiers were developed with $B = 0$.

Impact of Feature Selection. We have used several different feature selection strategies including the document frequency thresholding [13], χ^2 [18], mutual information [18], information gain [12], inverse document frequency – term frequency [13] and average discrimination scoring [13].

The results obtained have very similar trends, and due to space limitations we present in Figure 2 plots for only two selection methods: χ^2 and mutual information. The main thing to observe is the trend of the dropping performance by SVMs as the negative class examples are added. For $hSVM^1$ this is visible even with 18 features selected by χ^2 method. (This is also the case for all the other methods we have tested except mutual information.) The poor performance of mutual information at low fractions of features is exceptional among the techniques we have tested. It can be explained by a strong influence of this criterion by the marginal probability of features which tends to favour "rare" features rather than common ones.

The performance of SVM^1 is generally poorer than that of $hSVM^1$ or $hSVM^2$ at the same settings. Further, there are peaks and valleys at different mixture ratios that warrant further investigation into the behaviour of SVM^1.

The performance of the centroid classifiers $Cntr_0$ is quite different. There is a pronounced dip with high variance when very few negative examples are used ($B_{-/+} \leq 0.05$, which represents 1-6 negative examples only). However, as more negative examples are included, the performance improves and catches up with that of the positive 1-class classifier. Further, this pattern is consistent even at extremely low number of features.

4.2   Experiments with Synthetic Data

We observed in Section 4.1 that even in low dimensional space, the phenomenon of better performance with one-class learner persists. Our intuitive explanation here is that if the learner uses the minority class examples only, the "corner" (the half space) where minority data resides is properly determined. However, the minority class is "swamped" by the background class, hence once the background instances are added, the SVM solution becomes suboptimal. Now we explore this intuition using synthetic data.

We use three data sets of instances of similar structure. The observation vectors in these synthetic data sets contain a small number n_{inf} of *informative attributes* and the remaining, larger number, n_{noise}, of *noise attributes*. These attributes are binary, generated according to uniform random distribution with probabilities P_{inf} and P_{noise} of value $= +1$, respectively. The informative attributes determine the labels modulo the additional *label noise* which is the random reversal of certain proportions of labels, namely the proportions LN_+ of the positive and LN_- of the negative labels. In all sets, we generate $m = 9000$ instances of which $p_{y=+1} = 3\%$ have labels $y = +1$.

- S_1: For this data set we use $n = n_{inf} + n_{noise} = 1 + 999$ dimensions and $P_{noise} = 2\%$. The labels are generated as a random bipolar label vector $y \in \{\pm 1\}^{9000}$ with the proportion $p_{y=+1} = 3\%$ of positive examples. For the informative dimension we set $x_{inf} = (y + 1)/2 \in \{0, 1\}$ and then change randomly the proportion $LN_- = 20\%$ of 0s to 1s.
- S_2: In this case $n_{inf} = 10$, $n_{noise} = 990$, $P_{inf} = 5\%$, $P_{noise} = 2\%$. Having defined informative attributes $x_{inf,i} \in \mathbb{R}^{10}$ for $i = 1, ..., 9000$, we have randomly generated a vector $v \in \mathbb{R}^{10}$, then chosen a bias $b \in \mathbb{R}$ such that for 2004 ($\approx 22\%$) instances i we got the scores $x_{inf,i} \cdot v > b$. Of these 2004 instances, we randomly select 270 instances ($= 3\%$ of 9000) and label them $+1$ and the remaining 8730 instances we labelled -1.
- S_3: This set was designed to test the impact of non-linear kernels. It is generated as S_1 with the difference that only $n = n_{inf} + n_{noise} = 1 + 19 = 20$ dimensions are used and the random proportions LN_+ and LN_- of the both $+1$ and of 0 entries, respectively, are reversed in the second phase of the generation of the informative attribute x_{inf}.

In experiments, each set of 9000 instances generated as described above, was split randomly into 3000 training and 6000 test instances, with proportional sampling (without replacement) from both classes. All results reported are averages of 20 such random splits.

Figure 3 presents the results of experiments evaluating AROC as a function of the mixture ratio $B_{-/+}$, for the four kernel machines. For all three data sets, we show the results for the linear kernel (Figures 3A-3C), and for S_3 we show the impact of higher degree polynomial kernels (Figures 3D-3F).

The results, especially for $hSVM^1$, strikingly resemble those obtained for the AHR data (c.f Figure 2), with the consistent pattern of decreasing performance with increasing proportion of negative class instances. Note the 'collapse' of

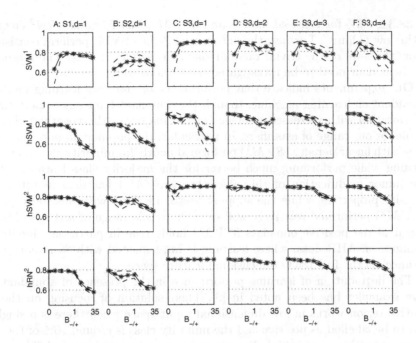

Fig. 3. Mean AROC with ± standard deviation envelopes as a function of the mixture ratio $B_{-/.}$ for four machines with $C = 1000$: SVM^*, $hSVM^*$, $hSVM^*$ and hRN^*. Plots A, B, C show results for linear kernels with $S.$, $S.$ and $S.$, respectively. Results with higher degree polynomial kernels ($d = 2, 3, 4$) are shown for $S.$ in plots D, E and F, respectively. $B_{-/.} = [0, 0.01, 0.1, 0.5, 1, 5, 10, 35]$.

SVM^1 algorithm when negative class proportion is very low and significantly better performance of $hSVM^1$ than SVM^1 for data dominated by the positive class ($B_{-/+} \approx 0$) and reverse of it for data dominated by the negative class ($B_{-/+} \approx 1$).

As kernel degree increases we observe the familiar pattern of decreasing performance with increasing dominance of negative class instances (Figures 3C-3F). Thus, the relatively low dimensional S_3 data set when used with higher degree polynomial kernels behaves in a way similar to that of the high dimensional datasets S_1 and S_2 with linear kernels.

We also experimented with different values of the regularisation constant C, but found that this had marginal impact on AROC in the above settings.

In addition, our experiments with different label noise settings (LN_+ and LN_-) show that the pattern of decreasing performance with increasing amounts of negative class instances persists with different levels of label noise.

5   Discussion

It is interesting that in our extensive experiments, while positive one-class classifiers using $\approx 3\%$ of the AHR-data provide best models, learning with negative class examples only provides very poor models (eg., mean AROC $= 62.4 \pm 3.4\%$

vs. $38.3 \pm 3.3\%$ for centroid models and $hSVM^1$, $hSVM^2$ and hRN^2 classifiers at the low C limit). Further, the unbalanced 2-class SVMs perform consistently below the level of the trivial random classifier $AROC < 50\%$. The reasons for this behaviour need to be investigated more thoroughly.

Our experiments indicate that performance of one class learning method is dependent on learning machine, though the dominant tendency of performance deterioration with increased influence of the majority (the negative) class was visible for our range of classifiers. In particular, the popular support vector machine with linear penalty (SVM^1) performed poorly for minority class dominated learning while performing much better for the majority class learning setting. The minor modification of this algorithm ($hSVM^1$) has demonstrated quite opposite properties, while the support vector machine with quadratic penalty ($hSVM^2$) performed relatively well over the whole range of settings. This appraisal of the positive one-class $hSVM^2$ holds true in particular, for its performance on AHR-data, where learning is implemented with 85 minority class instances in the full 18330 dimensional feature space.

The degradation of learning performance in the presence of abundant negative examples has been noted in [9]. Their solution of focusing on the best positive region works in low dimensional input space when there is a single region to be labelled as positive and the minority class is around 10% of the data. For our situation of very high dimensional input space with around 3% minority class data, the more drastic solution of totally ignoring negative class examples seems to work better for all machines.

Further, even when the sparse high dimensional space is reduced to a more dense representation via aggressive feature reduction methods, the advantage of one-class learners persist. This indicates that there is a combination of factors involved in this phenomenon, and a more thorough investigation with synthetic data is warranted.

Experiments with the polynomial kernels seem to indicate that interactions between the 19 noisy attributes in the set S_3 are equivalent to explicit addition of hundreds of extra noise attributes in the datasets S_1 and S_2. The higher the degree of the kernel, the more such 'noisy' virtual attributes are added (on the level of the feature space) and the more pronounced is the difference between one-class and two-class learning. Note that in this case, in contrast to the case of AHR-data case, the range of AROC values is around 60-90%.

6   Conclusion

We have shown that learning from positive examples only can be advantageous for real life data such as AHR-data used in KDD Cup 2002 in term of classifier accuracy but is not restricted to this data set. A few synthetic data sets tested in this paper show that favourable conditions for such learning method can naturally arise in many other situations, in particular when popular support vector machines with non-linear kernels are used. More research is required to study these conditions.

Our experiments demonstrate that one-class learning from positive class examples can be a very robust classification technique when dealing with very unbalanced data and high dimensional noisy feature space. It can be used as an alternative to aggressive feature selection usually used in such situations and can be very attractive for learning with non-linear kernels, when direct feature selection on the feature space level cannot be implemented.

Acknowledgements

The permission of the Managing Director, Telstra Research Laboratories, to publish this paper is gratefully acknowledged.

References

1. D. Bamber. The area above the ordinal dominance graph and the area below the receiver operating characteristic graph. *J. Math. Psych.*, 12:387 – 415, 1975.
2. R. Centor. The use of ROC curves and their analysis. *Med. Decis. Making*, 11:102 – 106, 1991.
3. Y. Chen, X. Zhou, and T. Huang. One-class svm for learning in image retrieval. In *Proceedings of IEEE International Conference on Image Processing (ICIP'01 Oral)*, 2001.
4. M. Craven. The Genomics of a Signaling Pathway: A KDD Cup Challenge Task. *SIGKDD Explorations*, **4(2)**, 2002.
5. N. Cristianini and J. Shawe-Taylor. *An Introduction to Support Vector Machines and other kernel-based learning methods.* Cambridge University Press, Cambridge, 2000.
6. F. Girosi, M. Jones, and T. Poggio. Regularization theory and neural networks architectures. *Neural Computation*, 7(2):219–269, 1995.
7. N. Japkowicz and S. Stephen. The class imbalance problem: A systematic study. *Intelligent Data Analysis Journal*, 6(5), 2002.
8. A. Kowalczyk and B. Raskutti. One Class SVM for Yeast Regulation Prediction. *SIGKDD Explorations*, **4(2)**, 2002.
9. M. Kubat, H. R., and S. Matwin. Learning when negative examples abound. In *Proceedings of the Ninth European Conference on Machine Learning ECML97*, 1997.
10. D. Lewis and J. Catlett. Training Text Classifiers by Uncertainty Sampling. In *Proceedings of the Seventeenth International ACM SIGIR Conference on Research and Development in Information Retrieval*, 1994.
11. L. M. Maneivitz and M. Yousef. One-class SVMs for Document Classification. *Journal of Machine Learning Research*, 2:139–154, 2002.
12. J. R. Quinlan. Induction of Decision Trees. *Machine Learning*, **1**(1), (1986).
13. G. Salton and M. J. McGill. *Introduction to Modern Information Retrieval.* McGraw Hill, 1983.
14. B. Schölkopf, J. Platt, J. Shawe-Taylor, A. Smola, and R. Williamson. Estimating the support of a high-dimensional distribution, 1999.
15. B. Schölkopf and A. J. Smola. *Learning with Kernels: Support Vector Machines, Regularization, Optimization and Beyond.* MIT Press, 2001.

16. V. Vapnik. *Statistical Learning Theory*. Wiley, New York, 1998.
17. G. Whaba. Support vector machines, reproducing Hilbert spaces and the randomised GACV. In B. Schölkopf, C. Burges, and A. J. Smola, editors, *Advances in Kernel Methods*, pages 69–88, Cambridge, Ma., 1999. MIT Press.
18. Y. Yang and J. O. Pedersen. A Comparative Study on Feature Selection in Text Categorization. In *Proceedings of the Fourteenth International Conference on Machine Learning*, 1997.

Rule Discovery and Probabilistic Modeling for Onomastic Data

Antti Leino[1,2], Heikki Mannila[1,3], and Ritva Liisa Pitkänen[2,4]

[1] Helsinki Institute for Information Technology, Basic Research Unit
Department of Computer Science
P.O. Box 26, FIN-00014 University of Helsinki, Finland
[2] Research Institute for the Languages of Finland
Sörnäisten rantatie 25, FIN-00500 Helsinki, Finland
[3] Helsinki University of Technology
Laboratory of Computer and Information Science
P.O. Box 5400, FIN-02015 HUT, Finland
[4] University of Helsinki, Department of Finnish
P.O. Box 3, FIN-00014 University of Helsinki, Finland

Abstract. The naming of natural features, such as hills, lakes, springs, meadows etc., provides a wealth of linguistic information; the study of the names and naming systems is called onomastics. We consider a data set containing all names and locations of about 58,000 lakes in Finland. Using computational techniques, we address two major onomastic themes. First, we address the existence of local dependencies or repulsion between occurrences of names. For this, we derive a simple form of spatial association rules. The results partially validate and partially contradict results obtained by traditional onomastic techniques. Second, we consider the existence of relatively homogeneous spatial regions with respect to the distributions of place names. Using mixture modeling, we conduct a global analysis of the data set. The clusterings of regions are spatially connected, and correspond quite well with the results obtained by other techniques; there are, however, interesting differences with previous hypotheses.

1 Introduction

In spatial statistics, a *point process* is a random process that produces points in the Euclidean plane. A realization of such a process, i.e., a set of points, is called a *point pattern*, or *spatial point data* [1,2]. A *marked point process* consists of several point processes producing different types of points. The points are often also called *events*.

Marked point processes arise in many applications, such as linguistics (in the study of dialects or place names, each word, grammatical construct, name, etc. corresponds to a different type of event), biodiversity studies (different types of events correspond to, e.g., different types of plants, and the locations are the places in which the plant has been observed), business applications (locations of customers etc.). There are some fundamental differences in the point data

N. Lavrač et al. (Eds.): PKDD 2003, LNAI 2838, pp. 291–302, 2003.

in these applications. The most relevant here is that in some of these cases the point data represents an underlying phenomenon that is contionuous (e.g. the occurrence area of a species, or the area in which a particular word is used), while in others the underlying phenomenon is itself discrete. In the current study we discuss the latter type of point processes.

The analysis of high-dimensional point processes can be quite demanding. The data is often sparse, i.e., we have only fragmentary information of the underlying phenomenon. When there are several different types of events, modeling their interaction can be complex. In many cases the observed quantities are results of several unobserved processes. The granularity and accuracy of the locations of the points can vary: sometimes the event can be localized perfectly, sometimes not.

Spatial statistics (see, e.g., the books [1,2]) has developed several strong methods for analyzing a single point process. However, marked point processes with a high number of different types of events have received less attention.

This paper is a case study in the use of pattern discovery and mixture modeling for the analysis of a high-dimensional marked point processes.

Our application is in the area of linguistics, especially *onomastics* (the study of names), particularly place names. The naming of natural features, such as hills, lakes, springs, meadows etc., provides a wealth of information. Our example data consists of full information about place names in Finland. The names tend to be fairly old, and they provide information about the population history and linguistic conditions at the time when the names where given.

Research in onomastics has traditionally been conducted by selecting a single name, or a group or related names, drawing maps of their occurrences, and doing qualitative analysis of the patterns of occurrences. Global analyses of the spatial distributions of different names are non-existent.

Our case study concerns two major themes in onomastics. The first is dependence between occurrences of names. It has long been assumed that the name of a nearby location has an influence on the naming of a location. For example, if a lake is called "Black Lake" (usually because the water is sufficiently clear that one can see the dark bottom of the lake), then a nearby lake might be named "White Lake". No quantitative evidence for this phenomenon is known, however. A special case of the local influence of names is *repulsion*: if a location is called B, then it makes sense to assume that other similar locations near this will *not* be called B: after all, the purpose of naming is to assign identifiers to locations. Our first goal is to study the local interactions between names.

The second theme we want to verify is the existence of relatively homogeneous spatial regions with respect to the distribution of place names. It is typically assumed that the naming conventions in nearby areas should be more or less similar, i.e., that there are clear regional trends in the style of names. The occurrence maps of individual names support this hypothesis, but virtually no global analyses exist.

In this paper we address both these themes. We first show how one can modify the basic ideas of association rule techniques to obtain local descriptions of the dependencies between the occurrences of names. The results show that indeed there are statistically significant associations between the occurrences of names. As for repulsion effects, we show that they are far less noticeable than expected.

For the second theme, we demonstrate the use of mixture modeling for the data at the granularity of municipalities, and show that the resulting clusters of municipalities are spatially extremely coherent. Thus the results verify the basic hypothesis that spatial homogeneity exists and provide new data for further onomastic research into the naming processes that cause the phenomenon.

The rest of this paper is organized as follows. The data set is described in Section 2. In Section 3 we show how the basic ideas of association rules can be generalized to the case of spatial point patterns, and give a sample of the results. Section 4 describes how mixture modeling applies to this data set, and discusses the results briefly. Section 5 is a brief conclusion.

2   The Data Set

Our example data set is a subset of the Finnish names occurring in the National Place Name Register, a part of the Geographic Names Register kept by the National Land Survey of Finland. The register contains all place names that appear on the 1:20 000 Basic Map and is maintained for the purposes of creating these maps. The size of the register, as well as that of our subsets, can be found in table 1, which shows the total number of Finnish names (or name instances), the number of different names, and the number of different municipalities in which these names are found.

Table 1. National Place Name Register data

	Name instances	Different names	Municipalities
Entire Register	717 747	303 626	447
Lakes	58 267	25 178	408
Common lake names	9 008	54	315
Name endings	55 538	45	407

The full data model of the register is explained in [3], but for the present study it is sufficient to note that the register includes a *language* field, a *feature type* field and the spatial information in different formats, including two co-ordinate systems and several administrative divisions. The *feature type* categorizes geographical features into such classes as *lake or pond*, or *river*, or *stretch of river*, or *forest*. For lakes, the location is fixed to be a selected point inside of the lake.

For our study we selected first all lake names in Finnish. This selection we pruned further along two different lines. For our primary data set we chose the names that have at least 90 instances. While our aim was to concentrate on the most common names, the limit of 90 instances is somewhat arbitrary. To supplement the primary data set we selected for clustering purposes a second data set, consisting not of complete place names but of derivational suffixes and final parts of compound names.

The two different subsets were selected mainly for onomastic reasons. Our working hypothesis was that spatial associations are in a large part related to the phenomenon of contrastive names — that is, pairs of names that refer to similar geographical features and differ only by the first part of the name in some sort of contrastive manner. To study this we needed to search for spatial associations for full names. Similarly, both intuition and onomastic consensus would say that there is a repulsion effect between two instances of the same name which is closely related to the use of place names to identify a place: a name cannot normally be used by the same group of people to denote two different places of the same type[1]. Again, this means we have to study the full names. In either case it seems appropriate to restrict ourselves to relatively common names, to make sure there are enough instances of each of them to get valid results.

With clustering the situation is somewhat different. The obvious way to start is to use full names, like we do with the association rules, and there is no reason to doubt that this approach works. However, it is also reasonable to postulate that by studying word endings — both derivative suffixes and end-parts of compound names — we can get insight into differences in naming practices. Using name endings is thus an attempt to do cluster analysis based on the distribution of various name types, not just names as such.

3   Spatial Association Rules

In this section we consider the first theme: finding local effects between the occurrences of different names. As an example, consider Figures 1—3 showing the occurrences of certain pairs of names. How do the occurrences of one name affect the probability of occurrence of another name? It is fairly clear that the maps alone cannot answer the question.

In spatial statistics questions such as this have been addressed by using, e.g., nearest neighbor distances or the K function and its derivatives [4,2]. Here we describe a similar approach, but using the terminology of association rules.

Given a set of observations over 0-1 attributes $A_1, \ldots, A_n$, an *association rule* is an expression $X \Rightarrow Y$, where $X, Y \subseteq \{A_1, \ldots, A_n\}$. Given a set X of attributes, the frequency $f(X)$ of X is the fraction of observations that have a 1 in all attributes of X. The frequency of the rule is defined to be $f(X \cup Y)$, and the accuracy (confidence) of the rule is $f(X \cup Y)/f(X)$.

We consider spatial association rules of the form $A \Rightarrow_r B$. The interpretation of such a rule is that given a location (x, y) in which event of type A occurs, one is likely to see at least one event of type B within distance r from (x, y). This definition is close to the ones used by [5,6,7,8,9]. From an onomastic point of

---

[1] It is, however, relatively common to name e.g. a farm after a nearby lake.

view it seems prudent to start with restricting ourselves to associations between two names.

To test the significance of a rule $A \Rightarrow_r B$ we start with a set of places named A and another set of places named B. We want to evaluate whether the occurrence of a B is more likely in the context of a nearby A than in general. Note, however, that a B can only occur if there is a suitable natural feature present: we cannot observe a "Pike Lake" at position (x, y) unless there is a lake at (x, y). To take this into account we consider as a set of reference points all points belonging to the the same type of feature as B; call this set C_B. In our case we used all Finnish lakes as C_B.

The probability that a given place that belongs to C_B is named B is $P(B) = \frac{N(B)}{N(C_B)}$, where $N(B)$ is the total number of places named B and $N(C_B)$ is the total number of all the places of the same type. We now select the places belonging to set C_B which are within the given radius r of a place named A. We denote the size of this selection by $n(C_B)$ and the number of B places in it by $n(B)$. As null hypothesis we can now assume that the occurrences of A and B are independent. Under this hypothesis our selection can be viewed as a random sample, which can be approximated by the Poisson distribution, $X \sim \text{Poisson}(\lambda)$, where $\lambda = n(C_B)\frac{N(B)}{N(C_B)}$. To correct for multiple testing, we use the Bonferroni correction.

Repulsion. Repulsion is essentially a special case of a spatial association rule $A \Rightarrow_r B$, where $A = B$. However, in this situation we select points based on the spatial distribution of A; it is not immediately obvious that this can be considered a random sample with regard to A. We have therefore used another method to confirm the results on repulsion.

In the general case we again start with two kinds of points, A and B, the latter of which belong to set C_B. The overall number of points B and C_B is $N(B)$ and $N(C_B)$, respectively; the probability of a given C_B point being a B point is $p = \frac{N(B)}{N(C_B)}$.

Within a given radius of the ith point with name A there are $n(C_{B_i})$ points of set C_B. We use random variable X_i to denote the number of points named B in this set. If the B points are distributed independently of each other, $X_i \sim \text{Bin}(n(C_{B_i}), p)$, so $E(X_i) = n(C_{B_i})$ and $D^2(X_i) = n(C_{B_i})p(1-p)$. Summing, we obtain a variable $S_m = \sum_{i=1}^{m} X_i$, and by assuming independence of the variables X_i, we have $E(S_m) = \sum_{i=1}^{m} E(X_i)$ and $D^2(S_m) = \sum_{i=1}^{m} D^2(X_i)$. Applying the central limit theorem we can obtain confidence estimates.

Results. Applying the method presented above to the common names data set gave both expected and unexpected results. As expected, most of the pairs of names had no significant associations either way. Also to be expected was that there were pairs that had significant repulsion between the names: the spatial distributions of these names just don't overlap, for various reasons related to such things as geography or variation in dialects.

One interesting sub-category of the association rules was what can be called contrasting names. These have traditionally considered only for such pairs as *Mustalampi* "Black Lake" — *Valkealampi* "White Lake" where the contrasting element in at least one of the names refers to a notable property of the lake and there is a clear antonymic relation between the two names. Our study indicates that this kind of variation is used in the naming process more widely and with far less strict semantic constraints for the elements than onomasticians have thought. For instance, there was a group of three names, *Ahvenlampi* "Perch Lake", *Haukilampi* "Pike Lake" and *Särkilampi* "Roach Lake", all of which had significant associations with each other even over small distances. Figure 1 shows the spatial distribution for *Ahvenlampi* and *Haukilampi* on a map with main dialectal regions, along with Poisson-approximated probabilities before and after the Bonferroni correction.

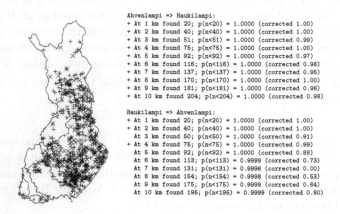

```
Ahvenlampi => Haukilampi:
+ At 1 km found 20; p(n<20) = 1.0000 (corrected 1.00)
+ At 2 km found 40; p(n<40) = 1.0000 (corrected 1.00)
+ At 3 km found 51; p(n<51) = 1.0000 (corrected 0.99)
+ At 4 km found 75; p(n<75) = 1.0000 (corrected 1.00)
+ At 5 km found 92; p(n<92) = 1.0000 (corrected 0.97)
+ At 6 km found 116; p(n<116) = 1.0000 (corrected 0.98)
+ At 7 km found 137; p(n<137) = 1.0000 (corrected 0.95)
+ At 8 km found 170; p(n<170) = 1.0000 (corrected 1.00)
+ At 9 km found 181; p(n<181) = 1.0000 (corrected 0.96)
+ At 10 km found 204; p(n<204) = 1.0000 (corrected 0.98)


Haukilampi => Ahvenlampi:
+ At 1 km found 20; p(n<20) = 1.0000 (corrected 1.00)
+ At 2 km found 40; p(n<40) = 1.0000 (corrected 1.00)
  At 3 km found 50; p(n<50) = 1.0000 (corrected 0.91)
+ At 4 km found 75; p(n<75) = 1.0000 (corrected 0.99)
  At 5 km found 92; p(n<92) = 1.0000 (corrected 0.88)
  At 6 km found 113; p(n<113) = 0.9999 (corrected 0.73)
  At 7 km found 131; p(n<131) = 0.9996 (corrected 0.00)
  At 8 km found 154; p(n<154) = 0.9998 (corrected 0.53)
  At 9 km found 175; p(n<175) = 0.9999 (corrected 0.64)
  At 10 km found 195; p(n<195) = 0.9999 (corrected 0.80)
```

Fig. 1. Spatial distribution of *Haukilampi* (x) and *Ahvenlampi* (+)

There were, however, other pairs that would at first glance appear to be similarly contrasting, but whose associations are somewhat weaker and start to show at significantly longer distances. In fact, the question arises whether there is a connection in the naming process or whether the names just have a similar distribution. One such case is the pair of *Joutenlampi* "Swan Lake" and *Hanhilampi* "Goose Lake", as shown in Figure 2. The reasons for the difference between this pair and that of *Ahvenlampi* — *Haukilampi* are not very obvious, and further onomastic study of these phenomena is needed.

Then there are pairs of names that have a significant association but are not contrasting, like *Lehmilampi* "Cow Lake" and *Likolampi* "Retting Lake"[2], as shown in Figure 3. In some cases another reason for the association can be seen; here, for instance, both names have similar agricultural origins. Although one can make such guesses about the reasons for the association, the phenomenon itself is a new discovery, and again further study would be strongly indicated.

---

[2] The name refers to a step in the processing of flax into linen.

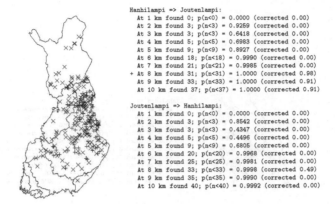

```
Hanhilampi => Joutenlampi:
  At 1 km found 0; p(n<0) = 0.0000 (corrected 0.00)
  At 2 km found 3; p(n<3) = 0.9259 (corrected 0.00)
  At 3 km found 3; p(n<3) = 0.6418 (corrected 0.00)
  At 4 km found 5; p(n<5) = 0.6983 (corrected 0.00)
  At 5 km found 9; p(n<9) = 0.8927 (corrected 0.00)
  At 6 km found 18; p(n<18) = 0.9990 (corrected 0.00)
  At 7 km found 21; p(n<21) = 0.9985 (corrected 0.00)
+ At 8 km found 31; p(n<31) = 1.0000 (corrected 0.98)
  At 9 km found 33; p(n<33) = 1.0000 (corrected 0.91)
  At 10 km found 37; p(n<37) = 1.0000 (corrected 0.91)

Joutenlampi => Hanhilampi:
  At 1 km found 0; p(n<0) = 0.0000 (corrected 0.00)
  At 2 km found 3; p(n<3) = 0.8542 (corrected 0.00)
  At 3 km found 3; p(n<3) = 0.4347 (corrected 0.00)
  At 4 km found 5; p(n<5) = 0.4496 (corrected 0.00)
  At 5 km found 9; p(n<9) = 0.6805 (corrected 0.00)
  At 6 km found 20; p(n<20) = 0.9968 (corrected 0.00)
  At 7 km found 25; p(n<25) = 0.9981 (corrected 0.00)
  At 8 km found 33; p(n<33) = 0.9998 (corrected 0.49)
  At 9 km found 35; p(n<35) = 0.9990 (corrected 0.00)
  At 10 km found 40; p(n<40) = 0.9992 (corrected 0.00)
```

Fig. 2. Spatial distribution of *Hanhilampi* (x) and *Joutenlampi* (+)

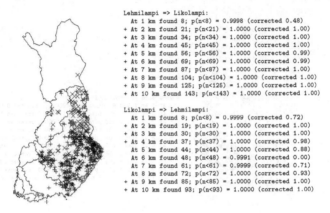

```
Lehmilampi => Likolampi:
  At 1 km found 8; p(n<8) = 0.9998 (corrected 0.48)
+ At 2 km found 21; p(n<21) = 1.0000 (corrected 1.00)
+ At 3 km found 34; p(n<34) = 1.0000 (corrected 1.00)
+ At 4 km found 45; p(n<45) = 1.0000 (corrected 1.00)
+ At 5 km found 56; p(n<56) = 1.0000 (corrected 0.99)
+ At 6 km found 69; p(n<69) = 1.0000 (corrected 0.99)
+ At 7 km found 87; p(n<87) = 1.0000 (corrected 1.00)
+ At 8 km found 104; p(n<104) = 1.0000 (corrected 1.00)
+ At 9 km found 125; p(n<125) = 1.0000 (corrected 1.00)
+ At 10 km found 143; p(n<143) = 1.0000 (corrected 1.00)

Likolampi => Lehmilampi:
  At 1 km found 8; p(n<8) = 0.9999 (corrected 0.72)
+ At 2 km found 19; p(n<19) = 1.0000 (corrected 1.00)
+ At 3 km found 30; p(n<30) = 1.0000 (corrected 1.00)
+ At 4 km found 37; p(n<37) = 1.0000 (corrected 0.98)
  At 5 km found 44; p(n<44) = 1.0000 (corrected 0.88)
  At 6 km found 48; p(n<48) = 0.9991 (corrected 0.00)
  At 7 km found 61; p(n<61) = 0.9999 (corrected 0.71)
  At 8 km found 72; p(n<72) = 1.0000 (corrected 0.93)
+ At 9 km found 85; p(n<85) = 1.0000 (corrected 1.00)
+ At 10 km found 93; p(n<93) = 1.0000 (corrected 1.00)
```

Fig. 3. Spatial distribution of *Lehmilampi* (x) and *Likolampi* (+)

The repulsion between different instances of the same name does not seem to be a very common phenomenon. Onomastically, this is rather surprising. It is true that our data set contains such names as *Pahalampi* "Evil Lake"[3] (shown in Figure 4) or *Palolampi* "Burnt Lake"[4], where there are no instances within 2 km of each other. However, the area covered by such selections is rather small, and most of these findings cannot be considered significant. The repulsion effects are for the most part insignificant even without the Bonferroni correction. One possible explanation for the scarcity of significant repulsion is that the body of Finnish lake names is relatively large and the distance a name needs to retain

---

[3] Some of these — possibly even a large amount — are euphemisms for a vulgar name that the locals considered too offensive to tell outsiders they perceived as being of a higher social standing, such as visiting onomasticians or geographers.

[4] These names are related to the agricultural method of burn-beating, practiced in some places in Finland until the early 20th century.

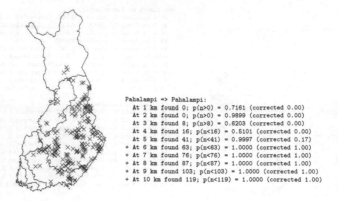

Fig. 4. Spatial distribution of *Pahalampi*

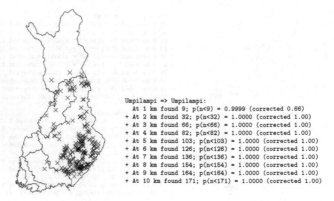

Fig. 5. Spatial distribution of *Umpilampi*

its usefulness as an identifier quite small: the name of a typical small lake is only used within a single village. The latter of these two factors may be sufficient to keep the repulsion small enough to disappear into the random variation caused by the former.

With all this in mind, it is still somewhat surprising to find that there are cases like *Umpilampi* "Closed Lake"[5] (shown in Figure 5) where there is a visible association even at distances of 1 km or less. Again, one can guess for the reasons why this is possible — these are mostly small ponds, and in many cases the need to refer to one of them exists only within one farmer family — but nevertheless this would appear to contradict the onomastic consensus that the basic unit for name use in rural areas is one village.

---

[5] That is, a small lake overgrown with weeds.

4   Probabilistic Modeling

We now turn to the second onomastic theme, the existence or nonexistence of homogeneous regions with respect to place names. We tested this hypothesis by considering the municipalities as observations, and using mixture modeling and the EM algorithm to obtain a clustering of the municipalities.

In more detail, we took the 315 municipalities, and created 54 variables, one for each of the names in the common names data set. This gives us 54-dimensional data set, where each column indicates the number of occurrences of the name in the municipality. We then took the 407 municipalities and 45 name endings, and conducted a similar test on that set.

We use mixture modeling to this data set [10,11]. A (finite) mixture model assigns a probability $P(\mathbf{x}|\Theta)$ to an observation $\mathbf{x}$ as weighted sum $\sum_j P(\mathbf{x}|\theta_j)$ of component distributions $P(\mathbf{x}|\theta_j)$ for $j = 1, \ldots, K$, where the weights (or mixing proportions) π_j satisfy $\pi_j \geq 0$ and $\sum \pi_j = 1$.

For each single component of the model for an observation $\mathbf{x} = (x_1, \ldots, x_d)$ we assume independence between variables and use the multinomial Bernoulli distribution

$$P(\mathbf{x}|\theta) = \prod_{i=1}^{d} \theta_i^{x_i}$$

with the constraint $\sum_{i=1}^{d} \theta_i = 1$. A finite mixture of multivariate Bernoulli probability distributions is thus specified by the equation

$$P(\mathbf{x}|\Theta) = \sum_{j=1}^{K} \pi_j P(\mathbf{x}|\theta_j) = \sum_{j=1}^{K} \pi_j \prod_{i=1}^{d} \theta_{ji}^{x_i}$$

with the parameterization $\theta = \{\pi_1, \ldots, \pi_K, (\theta_{ji})\}$ containing $K(d+1)$ parameters for data with d dimensions.

Given a data set R with d binary variables and the number K of mixture components, the parameter values of the mixture model can be estimated using the Expectation Maximization (EM) algorithm [12,13,14]. The EM algorithm has two steps which are applied alternately in an iterative fashion. Each step is guaranteed to increase the likelihood of the observed data, and the algorithm converges to a local maximum of the likelihood function [12,15]. The method gives for each component and each observation a probability of the observation stemming from that component.

We applied mixture modeling to the data described above; for each municipality $\mathbf{x}$ and component j we can compute the probability of the observation $\mathbf{x}$ stemming from component j by

$$P(\mathbf{x}|j) = \frac{P(\mathbf{x}|\theta_j)}{\sum_i P(\mathbf{x}|\theta_i)}.$$

For most municipalities there is clearly one component j which gives the municipality the highest probability. Example results are shown in Figures 6 and 7.

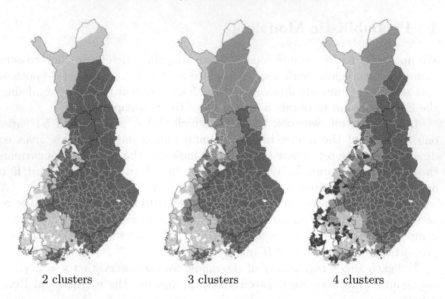

2 clusters   3 clusters   4 clusters

Fig. 6. Clustering based on the most common lake names

The different clusters are shown in shades of grey; white municipalities have no lakes in the data set[6].

Several features are of interest. First of all, the clusters of municipalities obtained in this way are spatially very well connected. Note that the method in itself has no information about the locations of the municipalities, and hence the spatial connectedness of the clusters is interesting. Second, as the number of clusters increases, the existing cluster boundaries tend not to change very much, but rather existing clusters split. Third, the clusters obtained correspond fairly well with the previous onomastic information about the distribution of names.

Specifically, in roughly the southernmost third of the map the boundary seen in the two-cluster maps corresponds rather well with the division between the eastern and western dialectal groups of Finnish. There is a small but noticeable deviation in Tavastland, and this is in line with our knowledge of the history of the settlement of Finland. Likewise, the western cluster continues north along the coast, and this too is in line with what we know from history. However, the middle third looks rather interesting: large regions that were designated and used as hunting grounds for the dialectally western Tavastland communities as late as the 16th century are not associated with the parent province but instead with the eastern regions, from where they were to a large extent populated in the 17th century. This would appear to imply that there is far less old influence in the names of that region than has been commonly believed, and this in turn opens up a variety of interesting onomastic questions.

---

[6] This is mostly because the common names data set contains only 15% of the lakes, but also because Finland is a bilingual country, and there are some municipalities that are uniformly Swedish.

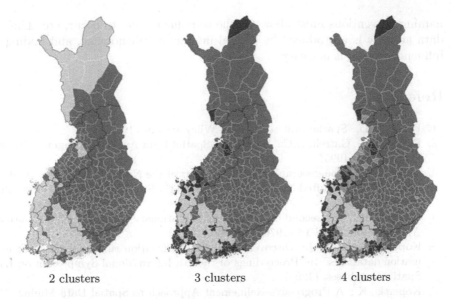

2 clusters   3 clusters   4 clusters

Fig. 7. Clustering based on the name ends

5   Conclusions

We have described a case study in the area of high-dimensional spatial point processes. We showed how one can use the basic principles of rule discovery and mixture modeling to analyze an onomastic data set about place names. The discovered rules of association and repulsion between names show fascinating local effects between the occurrences. The global analysis of name distribution by using mixture modeling demonstrated that homogeneous onomastic regions do exist. The methods lead to novel onomastic results. While the computational techniques we used are fairly standard, their application was not trivial. The global and local analysis of names has been shown to be very useful, and the study is continuing in several directions.

The existing techniques can be used to answer many onomastic questions. While computational methods of this type have not been applied to onomastic data, the reactions of various researchers in that field have been promising. However, there are also computational open problems. Finding more complex local interactions between names is a particularly interesting one. If A and B occur close to each other, then C is likely to occur close, too. While straightforward generalizations of association rules of the type $AB \Rightarrow_r C$ are possible, it might be more useful to investigate rules of the form $\Gamma \Rightarrow_r C$, where Γ is a derived predicate of position, e.g., of the type "there are names of type α in the neighborhood".

A deeper issue is separating the different layers in the process leading to a particular name occurring in a particular location. In order for a lake at location (x, y) to be called "Black Pond", there has to be a lake at that location, the people who named it must use words "black" and "pond" in their dialect, their naming conventions must allow for the combined name to occur, etc. Thus the data actually is a produced by several interacting phenomena, and finding the influence of each is not easy.

References

1. Ripley, B.D.: Spatial Statistics. John Wiley & Sons (1981)
2. Bailey, T.C., Gatrell, A.C.: Interactive Spatial Data Analysis. Longman Scientific & Technical (1995)
3. Leskinen, T.: The geographic names register of the National Land Survey of Finland. In: Eighth United Nations Conference on the Standardization of Geographical Names. (2002)
4. Ripley, B.D.: The second-order analysis of stationary point processes. Journal of Applied Probability **13** (1976) 255–266
5. Koperski, K., Han, J.: Discovery of spatial association rules in geographic information databases. In: Proceedings of the 4th International Symposium on Large Spatial Databases. (1995)
6. Koperski, K.: A Progressive Refinement Approach to Spatial Data Mining. PhD thesis, Simon Fraser University (1999)
7. Estivill-Castro, V., Lee, I.: Data mining techniques for autonomous exploration of large volumes of geo-referenced crime data. In: 6th International Conference on Geocomputation. (2001)
8. Huang, Y., Shekhar, S., Xiong, H.: Discovering co-location patterns from spatial datasets: A general approach. Submitted to IEEE Transactions on Knowledge and Data Engineering (TKDE), under second round review (2002)
9. Huang, Y., Xiong, H., Shekhar, S., Pei, J.: Mining confident co-location rules without a support threshold. To appear in Proceedings of the 18th ACM Symposium on Applied Computing (ACM SAC) (2003)
10. McLachlan, G., Peel, D.: Finite Mixture Models. Wiley Series in Probability and Statistics. John Wiley & Sons (2000)
11. Everitt, B., Hand, D.: Finite Mixture Distributions. Monographs on Applied Probability and Statistics. Chapman and Hall (1981)
12. Dempster, A., Laird, N., Rubin, D.: Maximum likelihood from incomplete data via the EM algorithm. Journal of the Royal Statistical Society, Series B **39** (1977) 1–38
13. Redner, R., Walker, H.: Mixture densities, maximum likelihood and the EM algorithm. SIAM Review **26** (1984) 195–234
14. McLachlan, G.J.: The EM Algorithm and Extensions. Wiley & Sons (1996)
15. Wu, C.J.: On the convergence properties of the EM algorithm. The Annals of Statistics **11** (1983) 95–103

Constraint-Based Mining of Sequential Patterns over Datasets with Consecutive Repetitions*

Marion Leleu[1,2], Christophe Rigotti[1],
Jean-François Boulicaut[1], and Guillaume Euvrard[2]

[1] LIRIS CNRS FRE 2672
Bâtiment Blaise Pascal, INSA Lyon, 69621 Villeurbanne Cedex, France
{crigotti,jfboulic}@lisisun1.insa-lyon.fr
[2] Direction de la Stratégie - Informatique CDC, 113 rue Jean-Marin Naudin, F-92220
Bagneux, France
{marion.leleu,guillaume.euvrard}@caissedesdepots.fr

Abstract. Constraint-based mining of sequential patterns is an active research area motivated by many application domains. In practice, the real sequence datasets can present consecutive repetitions of symbols (e.g., DNA sequences, discretized stock market data) that can lead to a very important consumption of resources during the extraction of patterns that can turn even efficient algorithms to become unusable. We propose a constraint-based mining algorithm using an approach that enables to compact these consecutive repetitions, reducing drastically the amount of data to process and speeding-up the extraction time. The technique introduced in this paper allows to retain the advantages of existing state-of-the-art algorithms based on the notion of occurrence lists, while permitting to extend their application fields to datasets containing consecutive repetitions. We analyze the benefits obtained using synthetic datasets, and show that the approach is of practical interest on real datasets.

Keywords: constraint-based mining, sequential pattern, generalized occurrence

1 Introduction

Sequential pattern mining has been introduced in 1995 [1]. It concerns pattern discovery (e.g., regularities) from ordered data, typically sequence databases. It has many applications, e.g., customer purchase analysis, Web Usage Mining, DNA sequence analysis. Looking for efficient algorithms has received a lot of attention (e.g., [8,11,9,5,10,12,14,13]). Each of these algorithms has its own pros and cons. Their efficiency depends on the characteristics of the data and on the kind of user-defined selection criteria, i.e., the constraints that must be satisfied by the extracted patterns. Several available algorithms are based on the so-called *occurrence lists*, i.e., lists that contain the location of the patterns in the

---

* This research is partially funded by the European Commission IST Programme - Future and Emergent Technologies, cInQ project (IST-2000-26469).

N. Lavrač et al. (Eds.): PKDD 2003, LNAI 2838, pp. 303–314, 2003.

data. This technique has been proved very useful for frequent pattern extraction (e.g., [8,12,14,3,13]).

Independently, the use of user-defined constraints to reduce the search space during sequential pattern extraction has been developed (e.g., [11,9,4,2]). Indeed, it has also been integrated in the occurrence list approach in the *cSpade* algorithm [13], resulting in one of the most efficient algorithms proposed for constraint-based mining of sequential patterns.

We have two main application domains for which we need efficient sequential pattern algorithms: financial data (stock market data) analysis for CDC (a major financial company in France) and DNA sequence database analysis. When considering the *cSpade* approach on these data, we understood that the benefits of the use of occurrence lists are lost when mining sequences containing consecutive repetitions of symbols. It comes from an explosion of the number of occurrences due to the repetition of the symbols. We recently proposed to handle efficiently the repetitions in the occurrence lists [7] when considering only a minimal frequency constraint. In this paper, we present how to generalize the notion of occurrence to perform efficient constraint-based mining on collections of sequences that contain repetitions. From a practical point of view, this leads to a technique that retains the advantages of the *cSpade* approach, while being able to address efficiently a broader scope of applications. The key idea is to use a single generalized occurrence to represent several occurrences while keeping enough information for the mining process.

This paper is organized as follows. Section 2 recalls the constraint-based sequential pattern mining problem and gives an abstract formulation of an algorithm for sequential pattern mining using occurrence lists. The notion of generalized occurrence is introduced in Section 3, and the corresponding modifications of the mining algorithm is presented. The practical impact of the use of generalized occurrences is demonstrated by means of experiments in Section 4. We conclude in Section 5.

2   Problem Statement and Abstract Algorithm

2.1   Constrained Sequential Pattern

The problem is to mine all frequent sequential patterns, verifying some user-defined constraints, that can be found in a sequence database. The constraints considered in this paper are the so-called *minimum* and *maximum gap* constraints, that enable to specify the minimum or maximum time interval between the occurrences of two events inside a pattern. Another similar constraint considered is the *time window* constraint, that enables to limit the maximum time between the first event and the last event of a pattern. Basically, the problem can be presented as follows: Let $I = \{i_1, i_2, \ldots, i_m\}$ be a set of m distinct items. An *event* (also called *itemset*) of size l is a non empty set of l items from $I : (i_1 i_2 ... i_l)$. A *sequence* α of *length* L is an ordered list of L events $\alpha_1, \ldots, \alpha_L$, denoted as $\alpha_1 \rightarrow \alpha_2 \rightarrow ... \rightarrow \alpha_L$. A database is composed of sequences, where each sequence has a unique sequence identifier (*sid*) and each event of each sequence has a temporal event identifier (*eid*) called timestamp. For a sequence in the database, each *eid* associated to an event is unique and if an event e_i precedes event e_j in a sequence, then the *eid* of e_j must be strictly greater than the *eid* of e_i. A *sequential pattern* (or *pattern*) is a sequence. Due to the lack of space, we considered only single-item events in patterns, that is patterns composed of events of size 1. The extension to pattern composed of events of size greater than 1 is straightforward and can be found in an extended version of the paper [6].

We are interested in the so-called constrained sequential patterns defined as follows. A sequence $s_a = \alpha_1 \rightarrow \alpha_2 \rightarrow \ldots \rightarrow \alpha_n$ is called a *subsequence* of another sequence $s_b = \alpha'_1 \rightarrow \alpha'_2 \rightarrow \ldots \rightarrow \alpha'_m$ if and only if there exists integers $1 \leq i_1 < i_2 < \ldots < i_n \leq m$ such that $\alpha_1 \subseteq \alpha'_{i_1}$, $\alpha_2 \subseteq \alpha'_{i_2}$, ..., $\alpha_n \subseteq \alpha'_{i_n}$. Let *supMin* be a positive integer called *absolute support threshold*, a pattern p verifies the minimum frequency constraint in a database D if p is a subsequence of at least *supMin* sequences of D. In this paper, we also use interchangeably the *relative support threshold* expressed in the percentage of the number of sequences of D. Let *gapMin* be the fixed value of the *minimum gap* constraint. A pattern $p = \alpha_1 \rightarrow \alpha_2 \rightarrow \ldots \rightarrow \alpha_n$ verifies the minimum gap constraint if and only if, for all α_i, $i = 1 \ldots n-1$, $eid(\alpha_{i+1}) - eid(\alpha_i) \geq gapMin$. Similarly, let *gapMax* be the fixed value of the *maximum gap* constraint. Pattern p verifies the maximum gap constraint if and only if, for all α_i, $i = 1 \ldots n-1$, $eid(\alpha_{i+1}) - eid(\alpha_i) \leq gapMax$. Now, let *winMax* be the fixed value of the *time window* constraint. Pattern p verifies this constraint, if and only if $eid(\alpha_n) - eid(\alpha_1) \leq winMax$.

2.2   Abstract Mining Algorithm

We present in this section an abstract algorithm corresponding to the general principle used in algorithms based on the use of occurrence lists for mining sequential patterns (e.g., [8,12,14,3,13]). The algorithm repeats two operations: a generation of candidate patterns and a support counting step. Let us introduce some needed concepts. A pattern with k items is called a *k-pattern*. A *prefix* of a k-pattern z is a subpattern of z constituted by the $k-1$ first items of z and its *suffix* corresponds to its last item. We extend the notion of *prefix* and *suffix* to occurrence. Let $y = e_1 \rightarrow e_2 \rightarrow \ldots \rightarrow e_{k-1} \rightarrow e_k$ be an occurrence of a k-pattern z, then $prefix(y) = e_1 \rightarrow e_2 \rightarrow \ldots \rightarrow e_{k-1}$ and $suffix(y) = e_k$.

The algorithm uses two frequent k-patterns z_1 and z_2 having the same $(k-1)$-pattern as prefix to generate a $(k+1)$-pattern z. This operation is denoted as $merge(z_1, z_2)$ and generates a single k-pattern: $z = z_1 \rightarrow suffix(z_2)$. The support counting for the newly generated pattern is not made by scanning the whole database. Instead, the algorithm has stored in specific lists, called *occLists*, the positions where z_1 and z_2 occur in the database. It then uses these two lists denoted $occList(z_1)$ and $occList(z_2)$ to determine where z occurs. Then $occList(z)$ allows to compute directly the support of z, by counting the number of distinct *sids* present in this list. The computation of $occList(z)$ is a kind of *join* and is denoted $join(z_1, z_2)$. The abstract algorithm is presented as Algorithm 1.

Algorithm 1 (Abstract Mining Algorithm)
Input: *a database of sequences and a support threshold.*
Output: *the frequent sequential patterns contained in the database.*

Use the database to compute:
  - F_1 the set of all frequent items
  - $occList(z)$ for all element z of F_1
let $i := 1$
while $F_i \neq \emptyset$ **do**
  let $F_{i+1} := \emptyset$
  for all $z_1 \in F_i$ **do**
    for all $z_2 \in F_i$ **do**
      if z_1 *and* z_2 *have the same prefix* **then**
        let $z := merge(z_1, z_2)$
        let $occList(z) := join(occList(z_1), occList(z_2))$
        Use $occList(z)$ *to determine if* z *is frequent*
        if z *is frequent* **then**
          $F_{i+1} := F_{i+1} \cup \{z\}$
        fi
    fi
    od
  od
  $i := i + 1$
od
output $\bigcup_{1 \leq j < i} F_j$

Fig. 1. Abstract mining algorithm using occurrence lists.

3 Generalized Occurrences and *GoSpec* Algorithm

3.1 Constrained Generalized Occurrences

The structure of a constrained generalized occurrence list is designed to reduce the size of the occurrence lists by representing several occurrences with a single more general one. In case of data presenting consecutive repetitions of items, this leads to an important gain in term of memory space used, and since the lists proceeded by the *join* operation are shorter, it results also in the reduction of the overall execution time.

For example, let us consider the following toy database containing three sequences. In these sequences the events are located at consecutive timestamps (i.e., 1,2,3, ...) and each sequence begin at timestamp 1.
Sequence 1:
   $\{A\}, \{A\}, \{A\}, \{A\}, \{A\}, \{B\}, \{B\}, \{B\}, \{B, C\}, \{B, C\}, \{B, C\}, \{B, C\},$
$\{B\}, \{B\}, \{B\}$
Sequence 2:

$\{B\}, \{A, B\}, \{A, B\}, \{B\}, \{B, C\}, \{B, C\}, \{B, C\}, \{B, C\}, \{B, C\}, \{B, C\},$
$\{C\}, \{C\}, \{C\}, \{C\}$

Sequence 3:

$\{\}, \{A\}, \{\}, \{B\}, \{B\}, \{B\}, \{B\}, \{B, C\}, \{B, C\}, \{C\}, \{C\}, \{C\}, \{C\}, \{C\}$

A classical representation of occurrence lists like the one used by *cSpade* [13] is depicted in Figure 2, in the left tables of each three areas. These tables represent the occurrence lists of *cSpade* for patterns A, B, C, A → B, A → C and A → B → C, with supMin = 2, gapMin = 2, gapMax = 5 and winMax = 10. In the tables, the column *sid* corresponds to the identifier of the sequence in which the pattern occurs, *eid* corresponds to the timestamp of the last event of this occurrence, and *diff* corresponds to the difference between the timestamps of the first and the last event of the occurrence (used by *cSpade* to check the *time window* constraint).

We propose a notion of *constrained generalized occurrence* (generalized occurrence for short) to compact such consecutive occurrences. This notion is straightforward for pattern of size 1, but not so trivial for longer patterns since it has to enable the handling of the various constraints. For a pattern z, the form of a generalized occurrence is $\langle sid, tBeg, [min, max], gmax \rangle$, and contains:

- An identifier sid that corresponds to identifier of a sequence where pattern z occurs.
- A timestamp $tBeg$ that corresponds to the timestamp of an occurrence of the first event of the pattern z (the detailed construction of $tBeg$ will be given in Algorithm 2).
- An interval $[min, max]$ corresponding to *eids* of consecutive occurrences of the last event of pattern z.
- A value $gmax$ that indicates the timestamp of the last occurrence of the last event of pattern z respecting the *gapMax* constraint. If no such occurrence exists then $gmax$ is set to -1.

Examples of generalized occurrences for the toy database are given in Figure 2, in the right tables of each three areas. In the case of pattern B, it is possible to reduce its 10 consecutive occurrences in the first sequence to a single generalized occurrence $\langle 1, 6, [6, 15], 15 \rangle$, where the interval [6,15] compacts all 10 *eids*. It should be noticed that for patterns of size 1 the fields $tBeg$ and $gmax$ are useless. However this is not the case for longer patterns. For example, let us consider the last generalized occurrence of the constrained generalized occurrence list of pattern A → B. This generalized occurrence is $\langle 3, 2, [4, 9], 7 \rangle$, indicating that it appears in sequence 3 and starts at timestamp 2. The interval [4,9] means that it represents several occurrences ending from 4 to 9. The *gmax* value of 7 notifies that occurrences ending from 4 to 7 satisfy the *maxGap* constraint, while for occurrences ending strictly after timestamp 7 only the prefix of the occurrence satisfies *maxGap*.

In the case of a generalized occurrence that does not represent any occurrence that satisfy the *maxGap* constraint for all its events, but that represents only occurrences satisfying this constraint up to this its last event, then the *gmax*

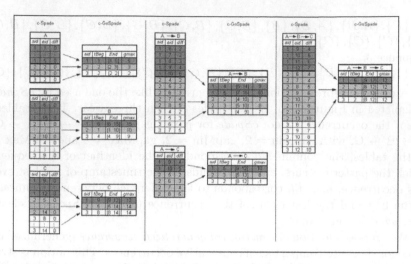

Fig. 2. Occurrence lists vs. Generalized occurrence lists for patterns A, B, C, A → B, A → C and A → B → C, with supMin = 2, gapMin = 2, gapMax = 5 and winMax = 10.

value is set to -1 (as for example in the generalized occurrence $\langle 3, 2, [8, 12], -1 \rangle$ of pattern A → C in Figure 2).

3.2 Dedicated Join Algorithm

The *GoSpec* Algorithm is an instance of the abstract algorithm 1 using a join designed for the generalized occurrence lists.

The join process is called when the merge operation has been done. It computes the constrained generalized occurrence list of a candidate pattern z, from the *occLists* of two generator patterns z_1 and z_2 having the same prefix.

Two different procedures are called depending on the level of the extraction process, *JoinLevel*$_2$ (Algorithm 4) and *Join* (Algorithm 3). The first one is a specific algorithm dedicated to the particular case of a 2-pattern candidate and the second one to the general case of a k-pattern candidate with $k > 2$. These two algorithms use a common function, *LocalJoin* (Algorithm 2), that computes a generalized occurrence $v = \langle sid, tBeg, [min, max], gmax \rangle$ of z from a single generalized occurrence of z_1 and a single generalized occurrence of z_2.

The *LocalJoin*(Algorithm 2), first verifies that the input generalized occurrences satisfy necessary conditions to be joined, performing the tests of line 3 and that the two generalized occurrences are from a same sequence, that is $sid_1 = sid_2$ (line 4). One line 3, the first comparison verifies that there exists at least one suffix of an instance of $\langle sid_2, tBeg_2, [min_2, max_2], gmax_2 \rangle$ that follows the first suffix of an instance of $\langle sid_1, tBeg_1, [min_1, max_1], gmax_1 \rangle$ and that satisfies the *gapMin* constraint. The second comparison checks that there exists at least one suffix of an instance of $\langle sid_2, tBeg_2, [min_2, max_2], gmax_2 \rangle$ that satisfies the *winMax* constraint wrt. $tBeg_1$. The last comparison ensures that

---

Algorithm 2 (*LocalJoin*)
Input: *Two generalized occurrences*
$\langle sid_1, tBeg_1, [min_1, max_1], gmax_1 \rangle$
and $\langle sid_2, tBeg_2, [min_2, max_2], gmax_2 \rangle$
Output: $\langle v, add \rangle$, *where* $v = \langle sid, tBeg, [min, max], gmax \rangle$ *and* add *is
a boolean value that is false if v cannot be created.*

1. let $add := false$
2. let $v := null$
3. if $(min_1 + gapMin \le max_2)$ *and* $(tBeg_1 + winMax \ge min_2)$
      and $(min_1 \le gmax_1)$**then**
4.   **if** $(sid_1 = sid_2)$ **then**
5.   let $sid := sid_1$
6.   let $tBeg := tBeg_1$
7.   **find** min *the minimum element x of* $[min_2, max_2]$
          such that $x \ge min_1 + gapMin$
8.   **find** max *the maximum element x of* $[min_2, max_2]$
          such that $x \le tBeg_1 + winMax$
9.   **find** $gmax$ *the maximum element x of* $[min_2, max_2]$
          such that $x \le gmax_1 + gapMax$
10.   **fi**
11.   **if** *(min and max exist)* *and* $(min \le max)$ **then**
12.   **if** *(gmax not exists)* **then let** $gmax := -1$
13.   **else if** $(gmax > max)$ **then**
14.   **let** $gmax := max$ **fi**
15.   **fi**
16.   let $v := \langle sid, tBeg, [min, max], gmax \rangle$
17.   let $add := true$
18.   **fi**
19. **fi**
20. **output** $\langle v, add \rangle$

Algorithm 3 (*Join*)
Input: $occList(z_1)$ *and* $occList(z_2)$, *generalized occurrence lists
of two patterns that share a same prefix.*
Used subprograms: *Algorithm 2*
Output: *a new occList*

Initialize GoIdList to the empty list.
1. **for all** $occ_1 \in GoIdList(z_1)$ **do**
2.   **for all** $occ_2 \in GoIdList(z_2)$ **do**
3.   **let** $\langle v, add \rangle := LocalTemporalJoin(occ_1, occ_2)$
4.   **if** add **then**
5.   *Insert v in occList*
6.   **fi**
7.   **od**
8. **od**
9. **output** $occList$

Fig. 3. *LocalJoin* and *Join* algorithms.

$\langle sid_1, tBeg_1, [min_1, max_1], gmax_1 \rangle$ has at least one instance that satisfies the *gapMax* constraint.

Lines 5 to 9 generate a new generalized occurrence. *min* is the timestamp of the earliest suffix of an instance of $\langle sid_2, tBeg_2, [min_2, max_2], gmax_2 \rangle$ that follows the earliest suffix of an instance of $\langle sid_1, tBeg_1, [min_1, max_1], gmax_1 \rangle$ and that verifies the *minimum gap* constraint. In a same way, *max* is the timestamp of the latest suffix of an instance of $\langle sid_2, tBeg_2, [min_2, max_2], gmax_2 \rangle$ that verifies the *time window* constraint wrt $tBeg_1$. *gmax* indicates the timestamp of the latest suffix of an instance of $\langle sid_2, tBeg_2, [min_2, max_2], gmax_2 \rangle$ that can form an occurrence of v that verifies *gapMax*.

This *LocalJoin* algorithm is called by *Join* (Algorithm 3) that generates a new *occList* from the *occLists* of two generator patterns z_1 and z_2. The Algorithm 3 iterates on the elements of $occList(z_1)$ and $occList(z_2)$. For each pair (occ_1, occ_2) a new constrained generalized occurrence is generated when possible using *LocalJoin*. Algorithm 3 is the general join operation used for k-patterns when $k > 2$. A dedicated join is needed to generate the occurrence lists of 2-patterns (i.e., z_1 and z_2 contain a single item. It is called $JoinLevel_2$ and is presented as Algorithm 4. Contrarly to the general *Join*, $JoinLevel_2$ performs several calls to the *LocalJoin* procedure. Indeed, the instances of the generalized occurrence $\langle sid_1, tBeg_1, [min_1, max_1], gmax_1 \rangle$ must be proceeded separately because they correspond, in the data, to different starting timestamps of the 1-pattern z_1. Thus, several calls are made on all generalized occurrences $\langle sid_1, p, [p, p], p \rangle$ with p varying between the values min_1 and max_1.

Proofs of the correctness of the representation using generalized occurrences (and the corresponding join process) can be found in [6].

Algorithm 4 (*JoinLevel₂*)
Input: *occList(z₁)*, *occList(z₂)*
Used subprograms: *Algorithm 2*
Output: *a new occList*

Initialize occList to the empty list.
1. **for all** $\langle sid_1, tBeg_1, [min_1, max_1], gmax_1 \rangle \in occList(z_1)$ **do**
2.   **for all** $\langle sid_2, tBeg_2, [min_2, max_2], gmax_2 \rangle \in occList(z_2)$ **do**
3.   **for all** $p \in [min_1 , max_1]$ **do**
4.   let $\langle v, add \rangle := LocalTemporalJoin(\langle sid_1, p, [p, max_1], p \rangle,$
                $\langle sid_2, tBeg_2, [min_2, max_2], gmax_2 \rangle)$
5.   **if** *add* **then**
6.   *Insert v in occList*
7.   **fi**
8.   **od**
9.   **od**
10. **od**
11. **output** *occList*

Fig. 4. *JoinLevel₂* algorithm.

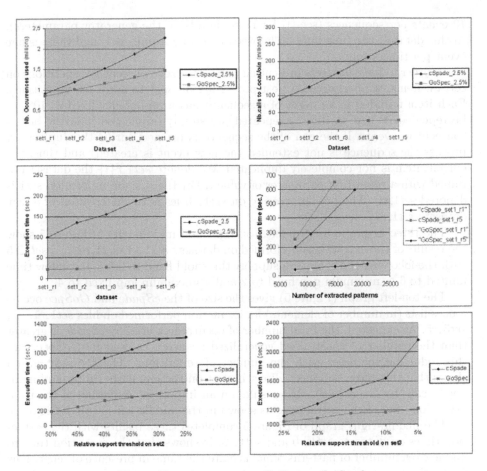

Fig. 5. Experiments using *GoSpec* and *cSpade*.

4   Experimental Results

In this section, we present experimental results and compare the behaviors of *GoSpec* and of *cSpade* [13] (one of the most efficient algorithm proposed in the literature and based on occurrence lists).

Both algorithms have been implemented using Microsoft Visual C++ 6.0, with the same kind of low level optimization to allow a fair comparison. All experiments have been performed on a PC with 196 MB of memory and a 500 MHz Pentium III processor under Microsoft Windows 2000.

4.1   Experiments on Synthetic Datasets

The synthetic dataset has been generated using the Dataquest generator of IBM [1] and the following parameters: C10-T2.5-S4-I1.25-D1K over an alphabet of 100 items (called *set1*). It contains 1000 sequences with an average size of

10 events per sequences (see [1] for more details on the generator parameters). In this dataset, the time interval between two time stamps is 1, and there is one event per time stamp.

In order to have datasets presenting parameterized consecutive repetitions on certain items, we performed a post-processing on $set1$ to add such repetitions. Each item founded in an event of a sequence has a probability fixed to 10% to be repeated. When an item is repeated, we simply duplicated it in the next i consecutive events. If the end of the sequence is reached during the duplication process the sequence is not extended (no new event is created) and thus, the current item is not completely duplicated. We denote $set1\_r\{i\}$ the dataset obtained with a repetition parameter of value i. For the sake of uniformity, $set1$ is denoted $set1\_r0$. The post-processing on $set1\_r0$ leads to the creation of 5 new datasets $set1\_r1,\ldots, set1\_r5$.

The three first graphs (top-left, top-right and middle-left) of Figure 5 show the results of the extractions performed on datasets $set1\_r1, set1\_r2, \ldots, set1\_r5$ with the following constraints: a support threshold fixed to 2.5%, a window time limited to 6, a minimum gap fixed to 2 and a maximum gap fixed to 4.

The top-left graph (Figure 5) gives the size of the $cSpade$ and $GoSpec$ occurrences lists (in number of elements) for extraction performed on files $set1\_r1, \ldots, set5\_r5$. As expected, the total number of occurrences used by $cSpade$ is greater than the number of constrained generalized occurrences used by $GoSpec$, and this reduction increases with the number of repetitions. The top-right graph shows that this reduction has a direct impact on the join costs (in term of number of calls to $LocalJoin$), that results on an important reduction of the total execution time of the extractions as shown in the middle-left graph of Figure 5.

The middle-right graph of Figure 5 completes these results with the extraction times on datasets $set1\_r0$ and $set1\_r5$. It shows that the execution time to find a given number of patterns remains quite the same in presence of repetitions for $GoSpec$.

4.2   Experiments on Real Datasets

The first real dataset is a financial dataset provided by the CDC financial company (Caisse des Dépôts et Consignations) and contains the variations of stock prices over one year. The discretized data results in a set (called $set2$) of 2830 sequences with an average length of 15 events per sequence. These sequences have been built from an alphabet of 17 items. The extractions have been performed using the extended version of the algorithm ([6]), that is without any limitation on the number of item per event composing the generated patterns. The following constraints have been used: winMax = 10, maxGap = 4 and minGap = 2. The bottom-left graph of Figure 5 represents the total execution time of both $cSpade$ and $GoSpec$ for minimum support thresholds varying from 25% to 50% and shows that $GoSpec$ offers a significant gain wrt. $cSpade$.

The second real dataset corresponds to a dataset of DNA sequences called $set3$. It contains 1778 sequences with an average length of 102 events composed by only one item per event over the nucleic alphabet {A,T,G,C}. The extractions have been performed using a window time constraint sets to 6, a maximum gap constraint of 3, no minimum gap constraint, and a minimum support threshold varying from 5% to 25%. The bottom-left graph of Figure 5 illustrates the total execution time used by the extractions and shows the advantages of *GoSpec* in practice on this second kind of data.

5 Conclusion

In this paper we presented an algorithm that enables to manage efficiently the constraint-based mining task when the sequential databases contain consecutive repetitions of their items. Such a situation can appear in several domains (e.g., discretized quantitative time series and DNA sequences). This can cause an explosion of the number of pattern occurrences and thus to an important loss of efficiency for algorithms based on an occurrence list approach (e.g., [8,12,14,3,13], while this algorithm family has shown its interest in many situations (e.g., low support mining and active constraint handling)). The algorithm presented in this paper, extends this family to tackle with these domains. It is based on the notion of constrained generalized occurrences, that have the particularity to compact several consecutive occurrences of patterns while keeping enough information for a constraint-based mining process. We showed by means of experiments , that the gain in term of memory space and execution time is important and that it increases with the number of consecutive repetitions contained in the input sequences.

References

1. R. Agrawal and R. Srikant. Mining sequential patterns. In *Proc. of the 11th International Conference on Data Engineering (ICDE'95)*, pages 3–14, Taipei, Taiwan, March 1995. IEEE Computer Society.
2. H. Albert-Lorincz and J.-F. Boulicaut. Mining frequent sequential patterns under regular expressions: a highly adaptive strategy for pushing constraints. In *Proceedings of the Third SIAM International Conference on Data Mining SDM 2003*, San Francisco, USA, May 2003.
3. J. Ayres, J. Flannick, J. Gehrke, and T. Yiu. Sequential pattern mining using bitmap representation. In *Proc. of the 8th ACM SIGKDD International Conference on Knowledge Discovery and Data Mining*, Edmonton, Alberta, Canada, July 2002.
4. M. Garofalakis, R. Rastogi, and K. Shim. SPIRIT: Sequential pattern mining with regular expression constraints. In *Proc. of the 25th International Conference on Very Large Databases (VLDB'99)*, pages 223–234, Edinburgh, United Kingdom, September 1999.
5. J. Han, J. Pei, B. Han Mortazavi-Asl, Q. Chen, U. Dayal, and M.-C. Hsu. Freespan: Frequent pattern-projected sequential pattern mining. In *Proc. 2000 Int. Conf. Knowledge Discovery and Data Mining (KDD'00)*, pages 355–359, Boston, MA, USA, August 2000.
6. M. Leleu, C. Rigotti, J.-F. Boulicaut, and G. Euvrard. Constrained-based mining of sequential patterns over datasets with consecutive repetitions. Technical report, LIRIS, INSA Lyon, Bat. Blaise Pascal, 69621 Villeurbanne Cedex, France, 2003.

7. M. Leleu, C. Rigotti, J.-F. Boulicaut, and G. Euvrard. Go-spade: Mining sequential patterns over datasets with consecutive repetitions. In *Proc.2003 Int. Conf. Machine Learning and Data Mining (MLDM'03)*, Leipsig, Germany, July 2003.

8. H. Mannila, H. Toivonen, and A. Verkamo. Discovery of frequent episodes in event sequences. *Data Mining and Knowledge Discovery*, 1(3):259–298, November 1997.

9. F. Masseglia, F. Cathalat, and P. Poncelet. The PSP approach for mining sequential patterns. In *Proc. of the 2nd European Symposium on Principles of Data Mining and Knowledge Discovery in Databases (PKDD'98)*, pages 176–184, Nantes, France, September 1998. Lecture Notes in Artificial Intelligence, Springer Verlag.

10. J. Pei, B. Han, B. Mortazavi-Asl, and H. Pinto. Prefixspan: Mining sequential patterns efficiently by prefix-projected pattern growth. In *Proc. of the 17th International Conference on Data Engineering (ICDE'01)*, 2001.

11. R. Srikant and R. Agrawal. Mining sequential patterns: Generalizations and performance improvements. In *Proc. of the 5th International Conference on Extending Database Technology (EDBT'96)*, pages 3–17, Avignon, France, September 1996.

12. M. Zaki. Efficient enumeration of frequent sequences. In *Proc. of the 7th International Conference on Information and Knowledge Management (CIKM'98)*, pages 68–75, November 1998.

13. M. Zaki. Sequence mining in categorical domains: incorporating constraints. In *Proc. of the 9th International Conference on Information and Knowledge Management (CIKM'00)*, pages 422–429, Washington, DC, USA, November 2000.

14. M. Zaki. Spade: an efficient algorithm for mining frequent sequences. *Machine Learning, Special issue on Unsupervised Learning*, 42(1/2):31–60, Jan/Feb 2001.

Symbolic Distance Measurements
Based on Characteristic Subspaces

Marcus-Christopher Ludl

Austrian Research Institute for Artificial Intelligence, Vienna

Abstract. We introduce the *subspace difference metric*, a novel heterogeneous distance metric for calculating distances between points with
both continuous and (unordered) categorical attributes. Our approach
is based on the computation and comparison of *characteristic subspaces*
(i.e. contexts) for each of the symbols and can be viewed as a generalization of the well-known *value difference metric*.
Subsequently, as one possible extension, we propose a linearization of
the computed symbolic distances by *multidimensional scaling*, thereby
mapping a set of symbols onto the interval $[0, 1]$. Thus, even algorithms,
which have originally been designed for usage with continuous attributes
(e.g. clustering algorithms like k-means), may be applied to datasets
containing discrete attributes, without having to adapt the algorithm
itself.
Finally, we evaluate the proposed metric and the linearization in quantitative and qualitative settings and exemplify the applicability in clustering domains.

1 Introduction and Motivation

Many inductive algorithms in machine learning and data mining make strict assumptions on the attribute types of the database. On the one hand, algorithms
for dealing with continuous data may naturally utilise nearness measurements
and exploit the metric properties of the instance space. On the other hand,
however, restricting a knowledge structure to categorical data allows for more
"exact" induction methods (like association rule or functional dependency mining), because, intuitively, categorical data does not contain that kind of inherent
"fuzziness" which continuous data usually exhibits.

Many learning algorithms, most notably instance based techniques, neural
networks and some of the most widely used clustering methods, necessitate all
attribute types to be continuous – in these cases categorical or nominal data or
missing values can often not be handled appropriately. To overcome these difficulties, in the context of classication learning, the *value difference metric* (VDM,
refer to [SW86] or [CS93]) or one of its generalizations [WM97] has been used
to good effect. Furthermore some interesting approaches for clustering heterogeneous data (i.e. datasets with mixed continuous and categorical attributes) have
been published in the recent years.

N. Lavrač et al. (Eds.): PKDD 2003, LNAI 2838, pp. 315–326, 2003.

Some of these novel approaches can be regarded as self-contained methods for clustering: E.g., [GKR00] and [ZFCH00] describe iterative clustering approaches based on dynamical systems, [SCC00] use a generalized notion of entropy, [GRS00] propose a concept of links to measure similarity and [GGR99] introduce a summarization-based algorithm. Additionally, extensions for some well-known clustering algorithms have been developed: E.g., [NH98] describe incremental and sparse variants of the EM algorithm and [GRB99] develop a discrete KMeans algorithm.

Whereas the problem of transforming continuous attributes to discrete types for the application of symbolic algorithms is well-known and usually termed *discretization*[1], none of the aforementioned methods solves the inverse problem: transforming discrete symbols to ordered (continuous or nominal) types. They either represent complete self-contained clustering procedures with little or no affinity to any continuous clustering scheme or require a rewriting of the distance metric used by the clusterer.

In this paper we propose the *heterogeneous subspace difference metric* (HSDM), a novel heterogeneous distance metric for computing distances between points with both continuous and categorical attributes. The basic idea is the computation and comparison of *characteristic subspaces* for each of the symbols, thus, the HSDM can be viewed as a generalization of the HVDM (*heterogeneous value difference metric*).

Furthermore, for usage as a pre-processing step, we propose a linearized extension of the SDM component, the *linearized subspace difference metric* (lSDM), which induces a strict ordering of the involved symbols. For this task we make use of *multidimensional scaling* (MDS) to map a set of symbols onto a one-dimensional scale, given high-dimensional distance measurements calculated by the SDM. Thus, even algorithms, which have originally been designed for usage with continuous attributes, may be applied to datasets containing discrete attributes, without having to adapt the algorithm itself.

2   Categorical Metrics

As [GGR99] note, distance functions on categorical attributes are not naturally defined, because it is difficult to reason that, e.g., "one color is 'like' or 'unlike' another color in a way similar to real numbers." This is due to the fact that (unordered) categorical attributes typically do not contain any information other than the symbols themselves. Whereas with continuous numbers various calculations and pairwise comparisons can be performed, usually all we can say about two colors, is whether they are equal or not.

[.] Some induction algorithms can be viewed as being able to discretize "on the fly" (e.g. C4.5), however also several methods for automatic pre-discretization have appeared in print: [DKS95] give an excellent overview, while [Lud00] present an unsupervised multivariate approach.

2.1   Overlap and VDM

This, then, is also the most widely used (e.g. [AKA91]) method for comparing two symbols: For calculating the distance between two instances with mixed continuous and categorical values, the following *heterogeneous euclidean overlap metric* (HEOM, refer to [WM97]) is used:

$$HEOM(\boldsymbol{x}, \boldsymbol{y}) = \sqrt{\sum_{i=1}^{m} d_i(x_i, y_i)^2}$$

Here, $\boldsymbol{x}$ and $\boldsymbol{y}$ are instance vectors, m is the number of attributes and d_i is the following function:

$$d_i(x, y) = \begin{cases} 1 & \text{if } x \text{ or } y \text{ is missing} \\ overlap(x, y) & \text{if attribute } A_i \text{ is categorical} \\ \frac{|x-y|}{range_i} & \text{otherwise} \end{cases}$$

The following simple *overlap function* is used:

$$overlap(x, y) = \begin{cases} 0 & \text{if } x = y \\ 1 & \text{otherwise} \end{cases}$$

Clearly, the HEOM is overly simplistic in handling categorical attributes and although it may be appropriate in some cases, its use can lead to poor performance [CS93].

A more sophisticated alternative, the *value difference metric* (VDM), introduced by [SW86], in most cases provides a better distance measurement for categorical attributes: Basically, it consists of considering two symbols to be similar, if they make similar predictions. The following definition is a simplified version without weighting terms [Dom96]:

$$SVDM(x, y) = \sum_{i=1}^{c} |p(c_i|x) - p(c_i|y)|^q$$

where c is the number of classes, $p(c_i|x)$ is the conditional probability that the output class is c_i given the input value x and q is a constant (usually 1 or 2). This categorical distance function can then be used as a replacement for the overlap function, yielding the *heterogeneous value difference metric* (HVDM, refer to [WM97]).

As [Dom96] note, the SVDM "attenuates [the problem of sensitivity to irrelevant attributes] for symbolic attributes, as long as a large number of examples is available, [...] due to the fact that by definition $p(c_i|x_j)$ will be roughly the same for all values x_j of an irrelevant attribute, leading to zero distance between them." However, the SVDM is obviously only applicable in cases, where class values are available for all instances (e.g. classification tasks).

2.2   The Subspace Difference Metric (SDM)

In situations, where a class attribute is not readily available (e.g. unsupervised learning), the VDM obviously cannot be used by definition. In such cases one might be tempted to fall back on the simpler *overlap function* to compute heterogeneous distances. With datasets, where the majority of attributes is continuous, this procedure might be appropriate – or at least not too harmful –, but when many or all of the attributes are of categorical type, much information about the distribution of the symbols will be lost thereby.

Rethinking the first paragraph in section 2, we realize that it is not exactly true. Actually, we do have more information about the symbols of a categorical attribute: we know (or can compute) the distribution of each symbol within the instance space, i.e. we know what values in other attributes each symbol co-occurs with. And this is exactly the information that we need to be able to argue that, say, the color red is more similar to orange than to blue.

Intuitively, to be able to argue about the pairwise similarity of two symbols, we have to make the assumption that the symbols do not exactly partition the domain space of another attribute. E.g. to be able to say that red is more similar to orange than to blue, we could argue that there are more instances in the dataset which can be red or orange than there are instances which can be red or blue. I.e. the context, in which the color red occurs overlaps more with the context of the color orange than with that of the color blue.

Characteristic Subspaces. The ideal tool for computing the overlaps between such contexts would be *characteristic rules* or the *characteristic subspaces* induced by them. Formally:

Definition 1. *Let A_i represent the i-th attribute ($i \in \{1, \ldots, m\}$) and V_i be a set of values from the domain of attribute i. A characteristic rule is an implication rule of the form*

$$A_i = x \rightarrow A_1 \in V_1 \wedge \cdots \wedge A_m \in V_m$$

where the attribute A_i does not occur on the right hand side. A characteristic rule is said to hold with confidence *p, formally*

$$A_i = x \rightarrow_p A_1 \in V_1 \wedge \cdots \wedge A_m \in V_m$$

if, with probability at least p, whenever condition $A_i = x$ holds for an instance, the right hand side holds as well.

A characteristic rule is basically an *association rule*, where the left hand side is restricted to only one condition. V_i is a discrete set of values, if A_i is categorical, and a continuous interval, if A_i is continuous.

Usually characteristic rules are constructed for values of a designated class attribute, we may, however, induce such rules for symbols from any categorical attribute. Thereby we obtain information about the context in which a certain symbol occurs – the subspaces induced by these rules will in the following be called *characteristic subspaces*. Intuitively, a *characteristic subspace* gives an extensional description of a symbol and by comparing these spaces we could argue about the similarity of the symbols.

Projected Characteristic Subspaces. Unfortunately, computing such *characteristic subspaces* with high *confidences* would necessitate running a characteristic rule induction algorithm for each single symbol, which is obviously computationally infeasible. For rules with high confidence it would, e.g., not suffice to compute the projections of a symbol onto each of the other attributes and combine them, because the cartesian product of two or more highly probable regions in different attributes need not necessarily be highly probable as well, formally:

$$A_i = x \to_p (A_j \in V_j), A_i = x \to_p (A_k \in V_k) \not\Rightarrow A_i = x \to_p (A_j \in V_j \wedge A_k \in V_k)$$

However, in our case, we may relax the requirements, in that we actually need not construct full *characteristic subspaces*. For the comparison of two symbols it does suffice to compare the projections onto each of the other attributes. We call the resulting spaces *projected characteristic subspaces (pc-subspaces)*.

Definition 2. *With D_i representing the domain of attribute A_i, let the discrete domain $\overline{D}_i$ be defined as follows:*

$$\overline{D}_i = \begin{cases} D_i & \text{if attribute } A_i \text{ is categorical} \\ \{1, \dots, d\} & \text{if attribute } A_i \text{ is continuous, for } d \in \mathbb{N} \end{cases}$$

To construct the *discrete domains* of the attributes, we have to discretize the continuous attributes by some simple unsupervised discretization method. Preferably we should use *equal-width* discretization for this task, because this method visualizes the distribution of the values within the interval (in the experiments we chose $d = 6$ as standard setting).

Definition 3. *Let p_j be the discrete probability density of the values from the discrete domain $\overline{D}_j$ of attribute j and let $p_{j,A_i=x}$ be this discrete probability density under the condition $A_i = x$ (i.e. $p_{j,A_i=x}(y) = p_j(y|A_i = x)$). Then the probabilistic projected characteristic subspace (ppc-subspace) of the symbol x in attribute A_i is defined as the collection of all densities $p_{j,A_i=x}$ for $j \neq i$.*

Definition 4. *With m being the number of attributes, x and y symbols from a categorical attribute A_i and $q \in \mathbb{N}$, the subspace difference metric (SDM) is defined as follows:*

$$SDM_i(x,y) = \sum_{\substack{j=1 \\ j \neq i}}^{m} \sum_{v \in \overline{D}_j} |p_{j,A_i=x}(v) - p_{j,A_i=y}(v)|^q$$

For convenience and ease of computation, the *ppc-subspace* of a symbol can be written as a matrix of dimensions $m \times s$, with m being the number of attributes and $s := \max\{|\overline{D}_1|, \ldots, |\overline{D}_m|\}$. The positions in this matrix are the respective conditional probabilities; columns having less than s symbols are filled with zeros up to s elements. That way, the calculation of a symbolic distance can be seen as a matrix operation.

Finally, plugging the SDM into the heterogeneous metric, where previously the (S)VDM or the overlap function was used, yields the HSDM:

Definition 5. *With m being the number of attributes and $q \in \mathbb{N}$ the heterogeneous subspace difference metric (HSDM) of two instances x and y is defined as follows:*

$$HSDM(\boldsymbol{x}, \boldsymbol{y}) = \sum_{i=1}^{m} d_i(x_i, y_i)^q$$

with the distance function d_i defined heterogeneously as follows:

$$d_i(x, y) = \begin{cases} 1 & \text{if } x \text{ or } y \text{ is missing} \\ SDM_i(x, y) & \text{if attribute } A_i \text{ is categorical} \\ \frac{|x-y|}{range_i} & \text{otherwise} \end{cases}$$

Complexity. Due to the fact that the *ppc-subspace* of a symbol can be written as a matrix of dimensions $m \times s$ (section 2.2) the comparison of two symbols is linear in the number of attributes m and in the maximum number of unique symbols s in any categorical attribute. Thus, the computation of a heterogeneous distance between two instances is, in the worst case, $O(m^2 s)$. Unfortunately, however, the explicit pre-computation of the *ppc-matrices* would take time $O(m^2 n^2)$ (n being the number of instances), yielding $O(n^2)$ complexity for the computation of a dissimilarity matrix as well.

By means of a hashtable, however, many of the computations can be simplified. If, e.g., a complete dissimilarity matrix is to be computed, the *ppc-subspaces* of all symbols can be pre-computed by scanning through all instances and recording each co-occurrence of two symbols x and y in two attributes a_1 and a_2. That way, the *ppc-subspace* of a symbol is only implicitly represented by all associations (along with their counts) for the according symbol stored in the hashtable. This pre-computation step basically consists of linearly scanning the database and thus can be realized in time $O(m^2 n)$.

3   Linearization by Multidimensional Scaling

One possibility for modularly implementing the SDM in a learning algorithm is the HSDM as defined above. However, this necessitates (at least partly) a rewriting of the involved algorithm. Unfortunately, this is not always possible. Also, it may prove advantageous for subsequent quantitative data mining techniques to have available nearness information between the symbolic values.

A transformation of the categorical attributes to continuous types in a pre-processing step could thus be a possible solution. However, we have to be aware of the fact that this task can not always be performed well and that in some cases the simple overlap metric might indeed be more appropriate.

We now extend the SDM to a linearized version, the *linearized subspace difference metric* (lSDM). We accomplish this by applying *multidimensional scaling* (MDS) methods to the distance measurements induced by the SDM.

MDS is typically used for computing representative data points for high-dimensional data (which should, e.g., be visualized) or proximity data (sometimes even incomplete) in a suitable low-dimensional space, such that the distances between the projected data points match the original distance values as faithfully as possible. The basic idea consists of minimizing a cost function, usually *stress, raw stress, strain* or something similar. The original algorithm for minimizing stress can be found in [Kru64], Sammon mapping, also one of the oldest approaches, can be found in [Sam69]. [dLJ77] describe the widely used majorization method for MDS and [KB97] present an application of deterministic annealing to the problem.

In our case the task can be defined as projecting data points of $m \times s$ dimensions (see section 2.2) onto a continuous scale, i.e. one dimension. For this projection we applied a simple gradient descent approach, minimizing *raw stress*.

The usual problem of finding good initial configurations applies here as well: Applying a strict gradient descent algorithm to a one-dimensional configuration can only move the points around a bit, but cannot change their relative order. However, because the target space is only one-dimensional, we have available a few canonical configurations, which should work reasonably well as starting points:

For a categorical attribute A_i we construct $|\overline{D}_i|$ different initial configurations by using one of the symbols x_j as the point of origin and placing the other symbols at positions which conform to their distances from x_j. We then run the gradient descent algorithm on each of the $|\overline{D}_i|$ configurations and accept the final configuration with the lowest *raw stress* as the resulting projection.

4   Experimental Evaluation

4.1   Quantitative Results

In a first set of experiments we exemplify the applicability of our approach in a clustering setting: We chose several datasets with varying characteristics from the UCI Machine Learning Repository and from the Esprit Project StatLog.

Apart from `hayes-roth` and `postoperative` (`patient`), all datasets had both continuous and categorical source attributes. As reference clusterings (for the calculation of *recall*) we used the class labels[2]; in case of the `servo` dataset,

---

* It is reasonable to assume that the existing class labels correlate with well-defined subspaces within the instance space. There is, however, no theoretical argument corroborating this assumption.

where the target variable is continuous, we chose to pre-discretize the target by equal-width into 6 classes (i.e. clusters).

The synthetic dataset had five attributes: The attributes 1, 3 and 4 were continuous (ranges $[1,100]$, $[1,50]$ and $[1,100]$, respectively), while the attributes 2 and 5 were symbolic: $\{A, B\}$ and $\{w, y, o, r\}$. The following rules held for the class attribute: $(a_2 = A \wedge a_3 \geq 20 \rightarrow class = 1)$, $(a_2 = B \wedge a_3 \leq 15 \rightarrow class = 2)$, $(a_5 = w \rightarrow class = 3)$ and $(else \rightarrow class = 4)$.

As a clusterer we used the k-means implementation pam [KR90], which relies on the pre-computation of a dissimilarity matrix before clustering, from the $cluster$ package of the freely available and widely used statistics software R. We compared three different methods: In the first two runs we used $daisy$[3] and the HSDM[4], respectively, for the computation of the dissimilarity matrix before applying pam. In a third run we used the lSDM to transform the symbols into real numbers and applied pam to the continuous dataset.

For each clustering thus generated we computed two different quality measures: On the one hand we used the $silhouette\ coefficient$ from [KR90] as an internal measure. On the other hand we used $recall$ (refer to [LW02]) to compare the resulting clustering to the reference clustering. Table 1 shows the results.

Table 1. Comparative clustering results for various datasets. Shown are the $silhouette$ $widths$ ("silh") and $recall$ values (tolerance $= 0.5$) of k-means clusterings. The highest numbers are printed in bold (no significance test applied).

dataset	rows	cols	classes	daisy silh.	daisy recall	hsdm silh.	hsdm recall	lsdm silh.	lsdm recall
postoperative	90	8	3	0.18	0.24	0.18	0.24	**0.25**	0.24
hayes-roth	132	4	3	0.23	0.00	0.29	0.00	**0.33**	0.00
servo	167	4	6	0.14	0.12	**0.54**	**0.16**	0.38	0.13
synthetic	200	5	4	0.25	0.10	0.40	**0.19**	**0.44**	0.05
heart	270	13	2	0.26	**0.68**	0.24	0.56	**0.38**	0.40
crx	690	15	2	0.17	0.64	0.18	0.63	**0.19**	**0.66**

As can be seen, the HSDM and/or the lSDM in most cases allow for better clusterings: In all cases the $silhouette\ width$ is higher, which is not really surprising, given that the HSDM naturally allows for more accurate distance measurements than the HEOM – a fact that typically has a positive impact on internal quality measures. We have illustrated this consideration by 2-dimensional cluster plots in figure 1.

---

[3] Basically, $daisy$ relies on a slightly more sophisticated HEOM to calculate heterogeneous distances by using weighting factors for each attribute. $daisy$ is described in detail in [KR90].

[4] To be able to use the same format for the dissimilarity matrix we implemented the HSDM in pure R-code (no Fortran), which did not use hashtables (refer to section 2.2). Running times are therefore not comparable.

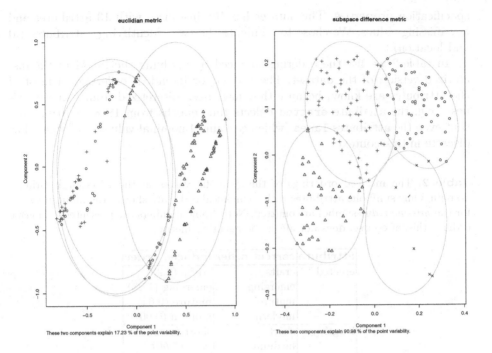

Fig. 1. Cluster plots (mapped to two dimensions) for the synthetic dataset (see text), clustered by k-means (4 centers) in conjunction with the HEOM (daisy) and the HSDM, respectively.

However, in most cases also the *recall* values of SDM-clusterings increase. From these results it is reasonable to assume that a lower clustering quality is largely due to an inappropriate handling of symbolic attributes.

It has to be noted that the HEOM achieved a better *recall* rating in one case and did not do much worse in two others. As mentioned in section 3, this might be due to the fact that in some cases – especially when the symbols have no inherent canonical order – inducing an artificial order instead of using 0-1-equality might actually be detrimental to clustering quality.

4.2 Qualitative Results

In a second experiment we wanted to investigate whether the linearized SDM *lSDM* is able to capture canonical orderings within symbolic attributes, i.e. whether the concept of characteristic subspaces can indeed reflect intuitively obvious orderings and whether these orderings are preserved by scaling the multidimensional information down to only one dimension.

For this test we chose a more complex, but rather small dataset: *Pittsburgh Bridges* (from the UCI repository) is a "design domain" rather than a classification domain, where 5 design descriptions need to be predicted based on 7 specification properties. The dataset has 108 instances with 13 attributes and few missing values. We chose to eliminate the two "identifying" attributes (id and location).

In table 2 we list the orderings induced by applying the lSDM to the discretized version of the dataset. The correct (or intended) orderings of most of the attributes are know, because they have been discretized from numeric values – e.g. the attribute **erected** reflects time epochs from 1818 to 1986. Note that even the attributes **lanes** (which contains 4 nominal values) was treated as discrete in this context.

Table 2. The induced orderings of the symbolic values in the Pittsburgh Bridges domain. Only symbolic attributes with a canonical order are shown, the induced values for the *missing value*-symbol are omitted. Note: Some numbers are presented in reverse order – this, of course, does not affect the resulting metric.

attribute	correct order	induced order
erected	crafts	crafts (2.93)
	emerging	emerging (1.28)
	mature	mature (0.64)
	modern	modern (0.00)
length	short	short (1.83)
	medium	*long (1.00)*
	long	*medium (0.49)*
lanes	1	1 (4.22)
	2	2 (2.16)
	4	4 (1.09)
	6	6 (0.00)
span	short	short (0.00)
	medium	medium (1.96)
	long	long (2.72)

As can be seen, linearizing the symbols by the lSDM can re-construct the former orderings in all but one attribute, where the order of *medium* and *long* is inverted. Note that we only list the results for attributes with a known canonical ordering here and omit results for attributes like, e.g., **purpose** (with the values *walk, aqueduct, rr* and *highway*).

5   Summary and Discussion

We have presented the novel heterogeneous metric HSDM for computing distances between points with categorical or mixed-continuous-categorical attribute values. The approach is based on the concept of *(probabilistic projected) characteristic subspaces*, which can also be viewed as a generalization of the well-known *value difference metric*. Additionally, we have introduced the idea of linearizing the high-dimensional distance matrices by *multidimensional scaling*, thereby yielding the lSDM, a reverse transformation to discretization: encoding symbols by continuous numbers.

We have exemplified in a clustering setting that the HSDM most often yields better results in terms of internal (*silhouette width*) and external (*recall*) clustering quality measures than the widely used HEOM. Finally, in a qualitative setting, we have shown that the lSDM yields good transformations of the involved symbols into real numbers, especially when an obvious canonical order among the symbols exists.

It has to be noted that the transformation does not always produce superior (quantitative or qualitative) results. Especially in cases, when no intuitively obvious order among the symbols can be found, the naive *overlap* function yields slightly better clusterings. This is obviously due to the fact that in such cases, the computation of 0-1-equality is "as good as it gets" and any artificial ordering of the symbols might deteriorate the results. A combination of both approaches in a single metric is one of our future research topics.

Furthermore, the mapping of the high-dimensional subspace data to one continuous dimension is obviously a crucial step in our algorithm. Using a more sophisticated method for MDS might further improve the final transformation and thus is also one of our topics for future research.

Finally, as one reviewer pointed out, it may be expected that the SDM should lend itself well for application in instance based learning tasks (nearest neighbor classification). We have chosen to evaluate the proposed metric in a clustering setting mainly because – unlike the VDM – it was intended to be used without a designated class attribute and will, eventually, be part of a novel clustering algorithm. Evaluating the SDM in IBL settings is a valuable suggestion, which we plan to tackle in the near future.

Acknowledgements

This research is supported by the Austrian *Fonds zur Förderung der Wissenschaftlichen Forschung (FWF)* under grant no. P12645-INF. The Austrian Research Institute for Artificial Intelligence acknowledges basic financial support by the Austrian Federal Ministry for Education, Science, and Culture.

We would like to thank Peter Filzmoser for invaluable discussions and ideas. We would also like to thank the anonymous reviewers for their helpful comments and suggestions.

References

AKA91.   D.W. Aha, D. Kibler, and M.K. Albert. Instance-based learning algorithms. *Machine Learning*, 6:37–66, 1991.

CS93.   S. Cost and S. Salzberg. A weighted nearest neighbor algorithm for learning with symbolic features. *Machine Learning*, 10:57–78, 1993.

DKS95.   J. Dougherty, R. Kohavi, and M. Sahami. Supervised and unsupervised discretization of continuous features. In Prieditis A. and Russell S., editors, *Proceedings of the 12^{t}h International Conference on Machine Learning (ML'95)*, San Francisco, CA, 1995. Morgan Kaufmann.

dLJ77.   de Leeuw. J. Applications of convex analysis to multidimensional scaling. In J. Barra, F. Brodeau, G. Romier, and B. van Cutsem, editors, *Recent Developments in Statistics*, pages 133–145. North Holland Publishing Company, The Netherlands, 1977.

Dom96.   P. Domingos. Unifying instance-based and rule-based induction. *Machine Learning*, 24(2):141–168, August 1996.

GGR99.   V. Ganti, J. Gehrke, and R. Ramakrishnan. Cactus - clustering categorical data using summaries. In *Knowledge Discovery and Data Mining*, pages 73–83, 1999.

GKR00.   D. Gibson, J. Kleinberg, and P. Raghavan. Clustering categorical data: An approach based on dynamical systems. *VLDB Journal (Very Large Databases)*, 8(3–4):222–236, 2000.

GRB99.   S.K. Gupta, K.S. Rao, and V. Bhatnagar. K-means clustering algorithm for categorical attributes. In *Data Warehousing and Knowledge Discovery*, pages 203–208, 1999.

GRS00.   S. Guha, R. Rastogi, and K. Shim. ROCK: A robust clustering algorithm for categorical attributes. *Information Systems*, 25(5):345–366, 2000.

KB97.   H. Klock and J.M. Buhmann. Multidimensional scaling by deterministic annealing. In *Energy Minimization Methods in Computer Vision and Pattern Recognition*, pages 245–260, 1997.

KR90.   L. Kaufman and P.J. Rousseeuw. *Finding Groups in Data: An Introduction to Cluster Analysis*. Wiley, NY, USA, 1990.

Kru64.   J. Kruskal. Multidimensional scaling by optimizing goodness of fit to a nonmetric hypothesis. *Psychometrika*, 29(1):1–27, March 1964.

Lud00.   M.-C. Ludl. Relative unsupervised discretization of continuous attributes. Master's thesis, Institut für Medizinische Kybernetik und Artificial Intelligence, University of Vienna, Austria, 2000.

LW02.   M.-C. Ludl and G. Widmer. Towards a simple clustering criterion based on minimum length encoding. In *Proceedings of the 13th European Conference on Machine Learning (ECML-02)*, pages 258–269, Helsinki, Finland, 2002. Springer Verlag.

NH98.   R.M. Neal and G.E. Hinton. A view of the EM algorithm that justifies incremental, sparse and other variants. In M.I. Jordan, editor, *Learning in Graphical Models*. Kluwer, 1998.

Sam69.   J.W. Sammon. A nonlinear mapping for data structure analysis. *IEEE Trans. Comp.*, C-18(5):401–409, 1969.

SCC00.   D.A. Simovici, D. Cristofor, and L. Cristofor. Generalized entropy and projection clustering of categorical data. In *Proceedings of the 4th European Conference on Principles and Practice of Knowledge Discovery in Databases (PKDD2000)*. Springer, 2000.

SW86.   C. Stanfill and D. Waltz. Toward memory-based reasoning. *Communications of the ACM*, 29:1213–1228, 1986.

WM97.   D.R. Wilson and T.R. Martinez. Improved heterogeneous distance functions. *Journal of Artificial Intelligence Research*, 6:1–34, 1997.

ZFCH00.   Y. Zhang, A.W. Fu, C.H. Cai, and P.-A. Heng. Clustering categorical data. In *ICDE*, page 305, 2000.

The Pattern Ordering Problem

Taneli Mielikäinen and Heikki Mannila

HIIT Basic Research Unit, Department of Computer Science, University of Helsinki
{Taneli.Mielikainen,Heikki.Mannila}@cs.helsinki.fi

Abstract. Many pattern discovery methods provide fast tools for finding the frequently occurring patterns in large data sets. Such pattern collections can also be used to approximate the underlying joint distribution, and they summarize the data set well. However, a large set of patterns is unintuitive and not necessarily easy to use. In this paper we consider the problem of ordering a collection of patterns so that each prefix of the ordering gives as good a summary of the data as possible. We formulate this problem for general loss functions, show that the problem has an efficient solution, and prove that its natural variant is NP-complete but the greedy approximation algorithm gives an $e/(e-1) \approx 1.58$ approximation quality. We apply the general technique to approximation of frequencies of frequent sets, and show that the method gives good empirical results.

1   Introduction

Many pattern discovery methods provide fast tools for finding the frequently occurring patterns in large data sets. However, many methods also result in large collections of patterns which are difficult to use. There has been lots of work on techniques for pruning the pattern collections without losing too much information (see, e.g., [1,2,3,4,5]).

A collection S of patterns whose frequencies are known can be used to estimate the frequencies of other patterns in several ways. For example, in frequent set mining with threshold σ, if we know the frequencies of AB, AC, and BC we can estimate the frequency of ABC by at least three methods: by $\sigma/2$, by the minimum of the frequencies of AB, AC, and BC, or by maximum entropy methods [6]. Other techniques exist, too, see e.g. [7,8,9].

In this paper we consider the following simple problem: given a collection of patterns and an estimation method for the frequencies of unknown patterns, how should we sort the known patterns in order of decreasing informativeness for the estimation? The solution of this problem gives an ordering such that each prefix is as informative as possible with respect to the following patterns.

We formulate this problem for general pattern classes, estimation methods, and loss functions. We show that the problem can be solved efficiently, and prove that its natural variant is NP-complete but the greedy method yields an $e/(e-1)$ approximation algorithm for certain loss functions and estimation

N. Lavrač et al. (Eds.): PKDD 2003, LNAI 2838, pp. 327–338, 2003.
© Springer-Verlag Berlin Heidelberg 2003

methods, where e is the base of natural logarithms. We apply the general technique to approximation of frequencies of frequent sets, and show that the method gives good empirical results.

The rest of this paper is organized as follows. Section 2 gives background on the general framework of pattern discovery and on condensed representations. Section 3 describes the pattern discovery problem shows that an optimal solution for the problem can be used as a good approximation of the pattern collection. In Section 4 the approximation technique is illustrated with concrete estimation methods and loss functions. The technique is experimentally evaluated in Section 5. Section 6 is a short conclusion.

2   Background and Related Work

Pattern discovery, i.e., finding interesting patterns from a data set, is the central task in data mining [10,11]. The pattern discovery problem can be formulated as follows: given a pattern collection $\mathcal{P}$ and a *quality predicate* (or an *interestingness predicate*) $q : \mathcal{P} \to \{0, 1\}$, find all interesting patterns, i.e., the patterns $p \in \mathcal{P}$ such that $q(p) = 1$. The predicate is usually defined by using a *quality measure* $\phi : \mathcal{P} \to [0, 1]$ and a threshold value $\sigma \in [0, 1]$:

$$q(p) = \begin{cases} 1 & \text{if } \phi(p) \geq \sigma, \text{ and} \\ 0 & \text{otherwise.} \end{cases}$$

Several measures of quality have been proposed [12]. The most prominent measure of quality is the frequency of the pattern w.r.t. the data set. Especially, frequent set mining has received considerable attention [13]. In frequent set mining, the data set d is a finite sequence $d = d_1 \ldots d_n$ of subsets of a set R, the pattern collection $\mathcal{P}$ is the collection of subsets of R, the frequency of a set $X \subseteq R$ (w.r.t. the data set d) is

$$fr(X) = fr(X, d) = \frac{|\{i : X \subseteq d_i, 1 \leq i \leq n\}|}{n}.$$

The set $X \subseteq R$ is considered interesting if and only if $fr(X) \geq \sigma$.

Finding a good measure of quality and an adequate threshold value is not easy. To avoid missing interesting patterns, very low quality threshold values might be needed. This implies that the number of patterns deemed to be interesting can be quite large. This is not necessarily a problem if the true objective of the pattern discovery was to find all the interesting patterns w.r.t. the quality predicate.

Several methods have been developed for finding *condensed representations* of the pattern collections (see e.g. [14,15,1,16,17,18,19,20,21,22,23,5]). The condensed representations are small descriptions of the pattern collections such that it is possible to infer the original collection of interesting patterns and the quality values (approximately) using some inference method. They depend on some structural properties of the pattern collection and the quality measure. Usually condensed representations choose a subset of all interesting patterns and infer the quality values of the interesting patterns from that subset.

The most popular condensed representation of pattern collections is the concept of *closed patterns*. The representation depends only on the partial order of the pattern collection and the antimonotonicity of the quality measure. Let $\prec$ be a partial order for the pattern collection $\mathcal{P}$. A pattern $p \in \mathcal{P}$ is *closed* if and only if its quality value is greater than any of its superpattern's quality value, i.e., if and only if

$$\forall q \in \mathcal{P} : p \prec q \Rightarrow \phi(p) > \phi(q).$$

The number of closed patterns can be much smaller than the number of all patterns.

Unfortunately even the condensed representations of the pattern collection can be very large. Thus we suggest ordering the patterns w.r.t. their informativeness. From the ordered collection of patterns, the user can interactively choose the appropriate trade-off between the number of chosen patterns and the accuracy of the approximation.

For brevity, for the rest of the paper we shall consider frequencies instead of arbitrary quality measures.

3 Pattern Ordering and Frequency Estimation

Most condensed representations of a pattern collection consist of a subset of the pattern collection. Thus a simple approach to simplify the condensed representation is to order the patterns in the condensed representation so that that the next pattern increases the knowledge about the pattern collection most.

Given a collection of patterns, we are interested in finding an ordering such that for each $i = 1, \ldots, n$ the prefix $p_1, \ldots, p_i$ of the ordering $p_1, \ldots, p_n$ gives as much information about $p_{i+1}, \ldots, p_n$ as possible. To formulate this we need to define *estimation methods* and *loss functions*. An estimation method ψ takes a subcollection $\mathcal{S}$ of all patterns $\mathcal{P}$ with known frequencies, and provides approximations of the frequencies of all patterns. I.e., an estimation method is a function

$$\psi : \mathcal{P} \times [0, 1]^{\mathcal{S}} \to [0, 1].$$

A trivial example is the estimation method which gives the known frequencies for patterns in $\mathcal{S}$ and 0 for everything else.

The loss function ℓ tells what penalty is to be paid for errors in estimating the frequencies. The loss function takes as inputs the true frequencies of the patterns and the estimated frequencies, and returns a score for the estimation:

$$\ell : [0, 1]^{\mathcal{P}} \times [0, 1]^{\mathcal{P}} \to \mathbb{R}.$$

A typical example of a loss function would be the L_2 metric

$$\ell(x, y) = \sqrt{\sum_{q \in \mathcal{P}} \left(x\left(q\right) - y\left(q\right) \right)^2}.$$

The task of ordering the patterns can be formulated as a computational problem as follows:

Input: A pattern collection $\mathcal{P}, |\mathcal{P}| = n$, a quality measure $\phi : \mathcal{P} \to [0, 1]$, an estimation method $\psi : \mathcal{P} \times [0, 1]^{\mathcal{S}} \to [0, 1], \mathcal{S} \subseteq \mathcal{P}$, and a loss function $\ell : [0, 1]^{\mathcal{P}} \times [0, 1]^{\mathcal{P}} \to \mathbb{Q}$.

Output: The pattern collection $\mathcal{P}$ as an ordered sequence $p_1, p_2, \ldots, p_n$ such that

$$\ell\left(\phi\left(\mathcal{P}\right), \psi\left(\mathcal{P}, \phi|\left\{p_1, \ldots, p_{i-1}, p_i\right\}\right)\right) \leq \ell\left(\phi\left(\mathcal{P}\right), \psi\left(\mathcal{P}, \phi|\left\{p_1, \ldots, p_{i-1}, p_j\right\}\right)\right)$$

for each $1 \leq i < j \leq n$, where $\phi|\mathcal{S}$ is the restriction of the mapping ϕ to the set $\mathcal{S}$.

We call this problem the *pattern ordering problem*. The problem can be solved by a greedy algorithm as follows:

ORDER-PATTERNS$(\mathcal{P}, \phi, \psi, \ell)$
1 $\mathcal{P}_0 = \emptyset$
2 **for** $i \leftarrow 0$ **to** $n - 1$
3   **do** $p_{i+1} \leftarrow \arg_p \min \left\{\ell\left(\phi\left(\mathcal{P}\right), \psi\left(\mathcal{P}, \phi|\mathcal{P}_i \cup \{p\}\right)\right) : p \in \mathcal{P} \setminus \mathcal{P}_i\right\}$
4          $\mathcal{P}_{i+1} \leftarrow \mathcal{P}_i \cup \{p_{i+1}\}$
5 **return** $p_1, \ldots, p_n$

The running time of the algorithm depends on the efficiency of finding in each iteration i the pattern p_{i+1} that decreases the error most. The time complexity is the combined time complexity of finding the minimums. Let $M(\mathcal{P})$ be the maximum time complexity of finding a pattern p_{i+1} such that the loss for $\mathcal{P}_i \cup \{p_{i+1}\}$ is as small as possible. Then the time complexity of the algorithm is bounded by $O(nM(\mathcal{P}))$. For example, using the trivial estimation method which gives the known frequencies for the chosen patterns and 0 for the others, the minimum in each iteration can be found in logarithmic time in n using a heap [24] (assuming the loss depends only on the differences $fr(p) - \psi(p, fr|S)$).

The ordering $p_1, \ldots, p_n$ of the pattern collection $\mathcal{P}$ found by the algorithm ORDER-PATTERNS can be interpreted as a refining approximation of the pattern collection: each prefix $\mathcal{P}_k = \{p_1, \ldots, p_k\}$ approximates the whole pattern collection $\mathcal{P}$. The ordering might itself shed some light to the relationships between the patterns. In addition, for several combinations of estimation methods and loss functions it can be shown that each prefix of the ordering gives a frequency approximation that is guaranteed to be at most a constant factor worse than the frequency approximation from any subset of $\mathcal{P}$ of same size.

The greedy approach, in general, offers an efficient approach to find solutions for a wide variety of problems and several exact and approximate algorithms have been successfully derived by this approach [25,26,27,28]. Also in the case of the pattern ordering problem it is possible to show for certain estimation methods and loss functions that any prefix $\mathcal{P}_k = \{p_1, \ldots, p_k\}$ of the optimal solution $p_1, \ldots, p_n$ for the pattern ordering problem is at most $e/(e - 1)$ worse and has at most $(e - 1)/e$ times smaller decrease in loss than any size k subset $\mathcal{S}$ of $\mathcal{P}$. (For more in-depth introduction to approximability, see e.g. [29].) On the other hand, the problem of finding the k patterns that describe the collection best can be shown to be NP-hard. Thus finding the size k optimal subset of $\mathcal{P}$ seems to be infeasible all but very small k.

Let us define some notation. The decrease of loss w.r.t. frequency estimation without any known frequencies is denoted by

$$\Delta\left(\mathcal{S}\right) = \ell\left(\phi\left(\mathcal{P}\right), \psi\left(\mathcal{P}, \phi|\emptyset\right)\right) - \ell\left(\phi\left(\mathcal{P}\right), \psi\left(\mathcal{P}, \phi|\mathcal{S}\right)\right).$$

Let $\mathcal{P}_k^*$ be a size k subset of $\mathcal{P}$ with the smallest loss. The prefix of length k of the optimal solution for the pattern ordering problem is denoted by $\mathcal{P}_k$. The problem of finding the best k subset of the pattern collection resembles a lot the *minimum set cover problem*, i.e., given a collection $\mathcal{T}$ of subsets of a finite set R, find the smallest subset $\mathcal{S}$ of $\mathcal{T}$ such that $\bigcup \mathcal{S} = R$ [30]. Thus the approximation quality of the algorithm ORDER-PATTERNS can be proven similarly to the approximability of certain variants of the minimum set cover problem [25].

First we prove Lemma 1 below which can be used to show that certain combinations of estimation methods and loss functions guarantee that $\Delta\left(\mathcal{P}_k\right) \geq \frac{e-1}{e}\Delta\left(\mathcal{P}_k^*\right)$ holds for all $1 \leq i \leq k$. I.e., the decrease of the error in the frequency estimation (w.r.t. the frequency estimation with no patterns) from the length k prefix of the pattern ordering found by the algorithm ORDER-PATTERNS is at least a fraction $(e-1)/e$ of the best decrease of the error over the size k subsets of $\mathcal{P}$. All one has to show is that the error decreases sufficiently in each iteration. Lemma 1 will be used in Section 4.

Lemma 1. *If*

$$\Delta\left(\mathcal{P}_i\right) - \Delta\left(\mathcal{P}_{i-1}\right) \geq \frac{\Delta\left(\mathcal{P}_k^*\right) - \Delta\left(\mathcal{P}_{i-1}\right)}{k} \tag{1}$$

holds for all $1 \leq i \leq k$ then

$$\Delta\left(\mathcal{P}_k\right) \geq \frac{e-1}{e}\Delta\left(\mathcal{P}_k^*\right).$$

holds for all $1 \leq k \leq n$.

Proof. From Equation 1 we get

$$\Delta\left(\mathcal{P}_i\right) \geq \frac{1}{k}\Delta\left(\mathcal{S}\right) + \left(1 - \frac{1}{k}\right)\Delta\left(\mathcal{P}_{i-1}\right) \geq \frac{1}{k}\Delta\left(\mathcal{P}_k^*\right) + \left(1 - \frac{1}{k}\right)\Delta\left(\mathcal{P}_{i-1}\right)$$

$$\geq \frac{1}{k}\Delta\left(\mathcal{P}_k^*\right)\sum_{j=0}^{i-1}\left(1 - \frac{1}{k}\right)^j = \frac{1}{k}\Delta\left(\mathcal{P}_k^*\right)\frac{\left(1 - \frac{1}{k}\right)^i - 1}{\left(1 - \frac{1}{k}\right) - 1}$$

$$= \left(1 - \left(1 - \frac{1}{k}\right)^i\right)\Delta\left(\mathcal{P}_k^*\right).$$

Thus

$$\Delta\left(\mathcal{P}_k\right) \geq \left(1 - \left(1 - \frac{1}{k}\right)^k\right)\Delta\left(\mathcal{P}_k^*\right) \geq \left(1 - \frac{1}{e}\right)\Delta\left(\mathcal{P}_k^*\right)$$

as claimed. $\qquad\square$

It is possible to show the similar result for the loss instead of the decrease of the loss:

Lemma 2. *If*

$$\ell\left(\phi\left(\mathcal{P}\right),\psi\left(\mathcal{P},\phi|\mathcal{P}_{i-1}\right)\right) - \ell\left(\phi\left(\mathcal{P}\right),\psi\left(\mathcal{P},\phi|\mathcal{P}_i\right)\right) \geq$$

$$\frac{1}{k}\left(\ell\left(\phi\left(\mathcal{P}\right),\psi\left(\mathcal{P},\phi|\mathcal{P}_{i-1}\right)\right) - \ell\left(\phi\left(\mathcal{P}\right),\psi\left(\mathcal{P},\phi|\mathcal{P}_k^*\right)\right)\right) \qquad (2)$$

holds for all $1 \leq i \leq k$ *then*

$$\ell\left(\phi\left(\mathcal{P}\right),\psi\left(\mathcal{P},\phi|\mathcal{P}_k\right)\right) \leq \frac{e}{e-1}\ell\left(\phi\left(\mathcal{P}\right),\psi\left(\mathcal{P},\phi|\mathcal{P}_k^*\right)\right)$$

holds for all $1 \leq k \leq n$.

Proof. The proof is essentially identical to the proof of the Lemma 1.   □

4   Case Study: Approximating by Maximums of Superpattern Frequencies

In this section we consider approximating the frequencies of the frequent patterns using the maximums of known superpattern frequencies. To define what superpattern is, we need a partial order $\prec$ for the pattern collection $\mathcal{P}$. A partial order $\prec$ for the collection $\mathcal{P}$ is transitive ($p \prec q \wedge q \prec r \Rightarrow p \prec r$) and irreflexive ($p \prec q \Rightarrow p \neq q$) binary relation on $\mathcal{P}$. We denote $(p,q) \in \prec$ by $p \prec q$. We further assume that the partial order is antimonotone w.r.t. the frequencies, i.e., $p \prec q \Rightarrow fr\left(p\right) \geq fr\left(q\right)$. For example, the set inclusion relation is such a partial order. A pattern q is a superpattern of p if and only if $p \prec q$. The estimation method of maximum of superpattern frequencies is

$$\psi\left(p, fr|\mathcal{S}\right) = \max_{q \in \mathcal{S}}\left(\{fr\left(p\right) : p = q\} \cup \{fr\left(q\right) : p \prec q\} \cup \{0\}\right).$$

The smallest subset of frequent patterns that is sufficient to describe the frequencies $fr\left(p\right)$ of the frequent patterns p in $\mathcal{P}$ correctly is called a collection of *closed frequent patterns*. More precisely, a pattern $p \in \mathcal{P}$ is closed if and only if

$$fr(p) > \max_{q \in \mathcal{P}}\{fr(q) : p \prec q\}.$$

A closure of a pattern $p \in \mathcal{P}$, denoted by $cl(p)$, is the largest superpattern $q \in \mathcal{P}$ of p such that $fr\left(p\right) = fr\left(q\right)$. The set of closures of a pattern collection $\mathcal{P}$ is denoted by $cl\left(\mathcal{P}\right)$.

Theorem 1. *The collection* $cl\left(\mathcal{P}\right)$ *of closed frequent patterns is the smallest collection such that for all frequent patterns* $p \in \mathcal{P}$ *we have*

$$fr\left(p\right) = \psi\left(p, fr|\,cl\left(\mathcal{P}\right)\right).$$

Proof. By definition, for each $p \in P$ there is $q = cl(p) \in cl(P)$ such that $fr(p) = fr(q)$. Also, no closed pattern can be left out from the collection. $\square$

It follows from the definition of closed frequent patterns that they can be chosen from the collection of the frequent patterns by simply checking for each pattern whether any its superpatterns (subpatterns) have equal frequency and pruning the pattern (subpattern) if that holds. The efficiency of the method depends strongly on the pattern collection, the partial order and their representations. For example, the closed patterns from the collection P of frequent sets, i.e., the closed frequent sets (or frequent closed sets), can be found in time

$$\sum_{X \in P, X \neq \emptyset} \binom{|X|}{|X|-1} (|X|-1) = O\left(|R|^2 |P|\right)$$

by applying the fact that $X \in cl(P)$ if and only if $fr(X) \neq fr(Y)$ for all $Y \in P, Y \supset X, |Y| = |X| + 1$.

The problem turns out to be NP-hard if we allow errors. Let us first consider the maximum of absolute errors, i.e.,

$$\ell(fr(P), \psi(P, fr|S)) = \max_{X \in P} |fr(X) - \psi(X, fr|S)|$$

$$= \max_{X \in P} \left| fr(X) - \max_{Y \in S} \{fr(Y) : X \subseteq Y\} \right|$$

$$= \max_{X \in P} \left(fr(X) - \max_{Y \in S} \{fr(Y) : X \subseteq Y\} \right).$$

Then the problem is NP-hard even for the pattern class of frequent sets:

Theorem 2. *Given a collection P of frequent sets and a rational number ϵ, it is NP-hard to find a smallest subset S of P such that*

$$\max_{X \in P} \left(fr(P) - \max_{Y \in S} \{fr(Y) : X \subseteq Y\} \right) \leq \epsilon.$$

Proof. We show the NP-hardness by a reduction from the decision version of the minimum set cover problem, where the objective is, instead of finding the smallest subset $S \subseteq T$ such that $\bigcup S = R$, to decide whether there is a subset $S \subseteq T$ of size at most k such that $\bigcup S = R$. We can assume that each element in R occurs in some set T, the cardinality of each set in T is greater than one, and no set in T is contained in another set in T.

Let us construct the data set d of subsets of R as follows: d consists of T and appropriate number of one-element subsets of R such that $fr(\{x\}, d) = fr(\{y\}, d)$ for all $x, y \in R$. Let $\epsilon = fr(\{x\}, d) - 1/n, x \in R$.

Then for each $S \subseteq T, |S| \leq k$, holds:

$$\bigcup S = R \Leftrightarrow \max_{X \in P} \left(fr(P) - \max_{Y \in S} \{fr(Y) : X \subseteq Y\} \right) \leq \epsilon.$$

$\square$

Corollary 1. *Given a pattern collection* $\mathcal{P}$ *and a rational number* ϵ, *it is NP-hard to find a smallest subset* $\mathcal{S}$ *of* $\mathcal{P}$ *such that*

$$\ell\left(\phi\left(\mathcal{P}\right),\psi\left(\mathcal{P},\phi|\mathcal{S}\right)\right) \leq \epsilon.$$

On the positive side, it can be shown for the estimation method of choosing the maximums of known superpattern frequencies that the problem of choosing size k subset of patterns such that the maximum absolute error is minimized is a special case of the minimum weight set cover, which is, given a collection $\mathcal{T}$ of subsets of a finite set R and a weight function $w : \mathcal{T} \to [0,1]$, to find a subset $\mathcal{S} \subseteq \mathcal{T}$ of smallest weight

$$w\left(\mathcal{S}\right) = \sum_{p \in \mathcal{S}} w\left(p\right)$$

such that $\bigcup \mathcal{S} = R$ [29].

If the loss function is, e.g., the average error instead of the maximum error, the connection to set cover is not so obvious. Also in that case the approximability guarantees can be established:

Theorem 3. *For the prefix* $\mathcal{P}_k$ *of length* k *of an optimal solution for the pattern ordering problem and the size* k *subset of* $\mathcal{P}$ *with the smallest loss we have*

$$\Delta\left(\mathcal{P}_k\right) \geq \frac{e-1}{e}\Delta\left(\mathcal{P}_k^*\right)$$

with respect to any loss function

$$\ell\left(fr\left(\mathcal{P}\right),\psi\left(\mathcal{P},fr|\mathcal{S}\right)\right) = \sum_{p \in \mathcal{P}} f\left(|fr(p) - \psi\left(p,fr|\mathcal{S}\right)|\right)$$

where f *is a convex strictly increasing function.*

Proof. It suffices to show that Equation 1 holds. We have

$$\Delta\left(\mathcal{P}_i\right) - \Delta\left(\mathcal{P}_{i-1}\right)$$
$$= \sum_{p \in \mathcal{P}} f\left(|fr(p) - \psi\left(p,fr|\mathcal{P}_{i-1}\right)|\right) - \sum_{p \in \mathcal{P}} f\left(|fr(p) - \psi\left(p,fr|\mathcal{P}_i\right)|\right)$$
$$\geq \frac{1}{k}\left(\sum_{p \in \mathcal{P}} f\left(|fr(p) - \psi\left(p,fr|\mathcal{P}_{i-1}\right)|\right) - \sum_{p \in \mathcal{P}} f\left(|fr(p) - \psi\left(p,fr|\mathcal{P}_k^*\right)|\right)\right)$$
$$= \frac{\Delta\left(\mathcal{P}_k^*\right) - \Delta\left(\mathcal{P}_{i-1}\right)}{k}$$

because $\{p_i\} = \mathcal{P}_i \setminus \mathcal{P}_{i-1}$ is the pattern that decreases the error most and

$$\sum_{p \in \mathcal{P}} f\left(|fr\left(p\right) - \psi\left(p,fr|\mathcal{S}\right)|\right)$$
$$= \sum_{p \in \mathcal{P}} \min\left\{f\left(|fr\left(p\right) - \psi\left(p,fr|\mathcal{S}\setminus\mathcal{T}\right)|\right), f\left(|fr\left(p\right) - \psi\left(p,fr|\mathcal{T}\right)|\right)\right\}.$$

holds for all $\mathcal{T} \subseteq S$.   □

The computation of the approximation can be made more efficient by observing the following fact that all but the closed patterns in $\mathcal{P}$ can be neglected.

Theorem 4. *For all ℓ and $\mathcal{S} \subseteq \mathcal{P}$ we have*

$$\ell\left(fr\left(\mathcal{P}\right), \psi\left(\mathcal{P}, fr|\mathcal{S}\right)\right) = \ell\left(fr\left(\mathcal{P}\right), \psi\left(\mathcal{P}, fr|\, cl\left(\mathcal{S}\right)\right)\right).$$

Proof. Any pattern $p \in \mathcal{S}$ can be replaced by $cl\left(p\right)$ as $fr\left(p\right) = fr\left(cl\left(p\right)\right)$, and if $\psi\left(p, fr|\mathcal{S}\right) = fr\left(p\right)$ then $\psi\left(p, fr|\mathcal{S}\right) = fr\left(cl\left(p\right)\right)$. $\square$

5   Experiments: Approximating Frequent Sets

We implemented the ORDER-PATTERNS algorithm to evaluate the practical usefulness of the method. In the experiments we computed frequent sets with different minimum frequency thresholds for two data sets from UCI KDD Repository[1]: Internet Usage data set consisting 10104 rows and 10674 attributes, and IPUMS Census data set consisting of 88443 rows and 39954 attributes.

The estimation method was the maximum of chosen superset frequencies, i.e., $\psi\left(X, fr|\mathcal{S}\right) = \max_{Y \in \mathcal{S}}\left\{fr\left(Y\right) : X \subseteq Y\right\}$, and the loss function was the mean of absolute errors

$$\ell\left(fr\left(\mathcal{P}\right), \psi\left(\mathcal{P}, fr|\mathcal{S}\right)\right) = \frac{1}{|\mathcal{P}|}\left(\sum_{X \in \mathcal{P}} fr\left(X\right) - \max_{Y \in \mathcal{S}}\left\{fr\left(Y\right) : X \subseteq Y\right\}\right).$$

The results are shown in Figure 1, and in Tables 1 and 2. The results show that relatively short prefixes can be used to obtain a good accuracy in estimating the frequencies. The inversion of the order of the error curves in Figure 1 is due to the combination of the estimation method and the loss function: As the initial frequency estimates are all zero, the average absolute error is smaller for lower minimum frequency thresholds. On the other hand the frequencies can be estimated exactly from the closed frequent sets and the number of closed frequent sets is smaller for higher minimum frequency thresholds.

6   Conclusions

We have considered the problem of ordering a pattern collection in such a way that each prefix of the ordered sequence of patterns would be as good a summary of the pattern collection as possible. A general algorithm was given, the problem complexity and the algorithm were analyzed and the approach was justified experimentally. It seems that the problem of finding good orderings of pattern collections is useful and interesting. Several open problems remain. One specially interesting one is combining pattern discovery and ordering steps: could we somehow discover patterns in an order that approximates the most informative pattern ordering. To do this exactly is impossible, but there might be some possibilities of obtaining approximate results in the style of competitive analysis.

---

* http://kdd.ics.uci.edu

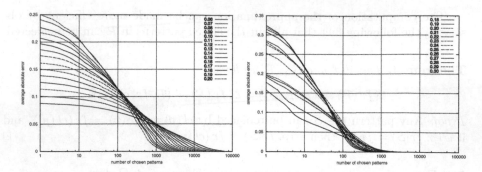

Fig. 1. Internet Usage data (left) and IPUMS Census data (right). The axes are the length of the prefix of the pattern ordering and the average absolute error of the frequency estimation from the prefix. Each curve corresponds to the minimum frequency threshold given as its label.

Table 1. Internet Usage data. The column σ corresponds to the minimum frequency threshold. Columns $|\mathcal{P}|$ and $|cl(\mathcal{P})|$ correspond to the cardinalities of the frequent sets and the closed frequent sets, respectively. Each column $|\tau(x)|$ corresponds to the length of the shortest prefix found by the algorithm ORDER-PATTERNS such that the average absolute error is at most x.

σ	$|\mathcal{P}|$	$|cl(\mathcal{P})|$	$|\tau(0.001)|$	$|\tau(0.005)|$	$|\tau(0.01)|$	$|\tau(0.02)|$	$|\tau(0.04)|$	$|\tau(0.08)|$
0.17	3246	3246	2672	1925	1421	970	597	231
0.16	4013	4013	3254	2295	1671	1132	655	242
0.15	4983	4983	3994	2764	1995	1377	775	270
0.14	6291	6290	4955	3339	2362	1602	860	261
0.13	8000	7998	6208	4093	2881	1972	1034	281
0.12	10476	10472	7970	5118	3562	2414	1189	289
0.11	13813	13802	10267	6352	4305	2804	1284	264
0.10	18615	18594	13468	8068	5409	3395	1423	245
0.09	25729	25686	18035	10399	6920	4094	1587	203
0.08	36812	36714	24870	13681	9032	5008	1708	153
0.07	54793	54550	35441	18477	12147	6276	1803	95

Table 2. IPUMS Census data. The columns are as in Table 1.

σ	$|\mathcal{P}|$	$|cl(\mathcal{P})|$	$|\tau(0.001)|$	$|\tau(0.005)|$	$|\tau(0.01)|$	$|\tau(0.02)|$	$|\tau(0.04)|$	$|\tau(0.08)|$
0.28	11443	1696	551	351	260	184	120	66
0.27	13843	1948	624	395	292	203	128	68
0.26	17503	2293	725	456	338	233	147	71
0.25	20023	2577	810	502	369	256	161	77
0.24	23903	3006	944	583	427	293	185	92
0.23	31791	3590	1093	661	477	328	196	85
0.22	53203	4271	1194	678	481	316	171	57
0.21	64731	5246	1454	813	573	372	189	62
0.20	86879	6689	1771	949	661	424	218	67
0.19	151909	8524	1974	953	628	363	151	27
0.18	250441	10899	2212	992	625	312	99	10

References

1. Boros, E., Gurvich, V., Khachiyan, L., Makino, K.: On the complexity of generating maximal frequent and minimal infrequent sets. In Alt, H., Ferreira, A., eds.: STACS 2002. Volume 2285 of Lecture Notes in Computer Science., Springer-Verlag (2002) 133–141
2. Gouda, K., Zaki, M.J.: Efficiently mining maximal frequent itemsets. In Cercone, N., Lin, T.Y., Wu, X., eds.: Proceedings of the 2001 IEEE International Conference on Data Mining. IEEE Computer Society (2001) 163–170
3. Gunopulos, D., Khardon, R., Mannila, H., Saluja, S., Toivonen, H., Sharma, R.S.: Discovering all most specific sentences. ACM Transactions on Database Systems **28** (2003) 140–174
4. Kryszkiewicz, M.: Concise representation of frequent patterns based on disjunction-free generators. In Cercone, N., Lin, T.Y., Wu, X., eds.: Proceedings of the 2001 IEEE International Conference on Data Mining, IEEE Computer Society (2001) 305–312
5. Zaki, M.J., Hsiao, C.J.: CHARM: An efficient algorithms for closed itemset mining. In Grossman, R., Han, J., Kumar, V., Mannila, H., Motwani, R., eds.: Proceedings of the Second SIAM International Conference on Data Mining, SIAM (2002)
6. Mannila, H., Pavlov, D., Smyth, P.: Prediction with local patterns using cross-entropy. In: Proceedings of the Fifth ACM SIGKDD International Conference on Knowledge Discovery and Data Mining, ACM (1999) 357–361
7. Kessler, D., Schiff, J.: Inclusion-exclusion redux. Electronic Communications in Probability **7** (2002) 85 – 96
8. Mannila, H., Toivonen, H.: Multiple uses of frequent sets and condensed representations. In Simoudis, E., Han, J., Fayyad, U.M., eds.: Proceedings of the Second International Conference on Knowledge Discovery and Data Mining (KDD-96), AAAI Press (1996) 189–194
9. Pavlov, D., Mannila, H., Smyth, P.: Beyond independence: probabilistic methods for query approximation on binary transaction data. IEEE Transactions on Data and Knowledge Engineering (2003) To appear.
10. Hand, D.J.: Pattern detection and discovery. In Hand, D., Adams, N., Bolton, R., eds.: Pattern Detection and Discovery. Volume 2447 of Lecture Notes in Artificial Intelligence., Springer-Verlag (2002) 1–12
11. Mannila, H.: Local and global methods in data mining: Basic techniques and open problems. In Widmayer, P., Triguero, F., Morales, R., Hennessy, M., Eidenbenz, S., Conejo, R., eds.: Automata, Languages and Programming. Volume 2380 of Lecture Notes in Computer Science., Springer-Verlag (2002) 57–68
12. Tan, P.N., Kumar, V., Srivastava, J.: Selecting the right interestingness measure for association patterns. In Hand, D., Keim, D., Ng, R., eds.: Proceedings of the Eight International Conference on Knowledge Discovery and Data Mining (KDD-2002), ACM (2002)
13. Hipp, J., Güntzer, U., Nakhaeizadeh, G.: Algorithms for association rule mining – a general survey and comparison. SIGKDD Explorations **1** (2000) 58–64
14. Boulicaut, J.F., Bykowski, A.: Frequent closures as a concise representation for binary data mining. In Terano, T., Liu, H., Chen, A.L.P., eds.: Knowledge Discovery and Data Mining. Volume 1805 of Lecture Notes in Artificial Intelligence., Springer-Verlag (2000) 62–73
15. Boulicaut, J.F., Bykowski, A., Rigotti, C.: Free-sets: a condensed representation of Boolean data for the approximation of frequency queries. Data Mining and Knowledge Discovery **7** (2003) 5–22

16. Bykowski, A., Rigotti, C.: A condensed representation to find frequent patterns. In: Proceedings of the Twentieth ACM SIGACT-SIGMOD-SIGART Symposium on Principles of Database Systems, ACM (2001)

17. Calders, T., Goethals, B.: Mining all non-derivable frequent itemsets. In Elomaa, T., Mannila, H., Toivonen, H., eds.: Principles of Data Mining and Knowledge Discovery. Volume 2431 of Lecture Notes in Artificial Intelligence., Springer-Verlag (2002) 74–865

18. Kryszkiewicz, M., Gajek, M.: Concise representation of frequent patterns based on generalized disjunction-free generators. In Chen, M.S., Yu, P., Liu, B., eds.: Advances in Knowledge Discovery and Data Mining. Volume 2336 of Lecture Notes in Artificial Intelligence., Springer-Verlag (2002) 159 – 171

19. Mielikäinen, T.: Frequency-based views to pattern collections. In: IFIP/SIAM Workshop on Discrete Mathematics and Data Mining. (2003)

20. Pasquier, N., Bastide, Y., Taouil, R., Lakhal, L.: Discovering frequent closed itemsets for association rules. In Beeri, C., Buneman, P., eds.: Database Theory - ICDT'99. Volume 1540 of Lecture Notes in Computer Science., Springer-Verlag (1999) 398–416

21. Pei, J., Dong, G., Zou, W., Han, J.: On computing condensed pattern bases. In: Proceedings of the 2002 IEEE International Conference on Data Mining (ICDM 2002), 9-12 December 2002, Maebashi City, Japan, IEEE Computer Society (2002) 378–385

22. Pei, J., Han, J., Mao, T.: CLOSET: An efficient algorithm for mining frequent closed itemsets. In Gunopulos, D., Rastogi, R., eds.: ACM SIGMOD Workshop on Research Issues in Data Mining and Knowledge Discovery. (2000) 21–30

23. Stumme, G., Taouil, R., Bastide, Y., Pasquier, N., Lakhal, L.: Computing iceberg concept lattices with TITANIC. Data & Knowledge Engineering 42 (2002) 189–222

24. Knuth, D.E.: Sorting and Seaching. second edn. Volume 3 of The Art of Computer Programming. Addison-Wesley (1998)

25. Feige, U.: A threshold of $\ln n$ for approximating set cover. Journal of the ACM 45 (1998) 634 – 652

26. Guha, S., Khuller, S.: Greedy strikes back: Improved facility location algorithms. Journal of Algorithms 31 (1999) 228 – 248

27. Helman, P., Moret, B.M.E., Shapiro, H.D.: An exact characterization of greedy structures. SIAM Journal on Discrete Mathematics 6 (1993) 274 – 283

28. Kempe, D., Kleinberg, J., Éva Tardos: Maximizing the spread of influence through a social network. In: Proceedings of the Ninth ACM SIGKDD International Conference on Knowledge Discovery and Data Mining, ACM (2003)

29. Ausiello, G., Crescenzi, P., Kann, V., Marchetti-Spaccamela, A., Protasi, M.: Complexity and Approximation: Combinatorial Optimization Problems and Their Approximability Properties. Springer-Verlag (1999)

30. Garey, M.R., Johnson, D.S.: Computers and Intractability: A Guide to the Theory of NP-Completeness. W.H. Freeman and Company (1979)

Collaborative Filtering
Using Restoration Operators

Atsuyoshi Nakamura, Mineichi Kudo, and Akira Tanaka

Division of Systems and Information Engineering
Graduate School of Engineering
Hokkaido University, Sapporo 060-8628 Japan
{atsu,mine,takira}@main.eng.hokudai.ac.jp

Abstract. We propose a new collaborative filtering method that uses restoration operators. The problem of restoration by operators was originally studied in the field of digital image restoration [9]. We also consider the problem of selecting items that users should be asked to rate in order to achieve a small expected squared error, and we propose a greedy method as a solution of this problem. According to our experimental results, prediction performance of restoration operators is good when the number of observed ratings is small, and our greedy method outperforms random query item selection.

1 Introduction

Information filtering has become an important technology in recent years due to wide spread of Internet. This technology enables systems to learn a user's personal preference, and recommend items (such as news articles, music and books) that are preferred by the user. Collaborative filtering, an information filtering technique, has been studied extensively in recent years [10, 11, 2, 8]. This filtering technique enables a system to recommend items to a user that are preferred by similar users. Generally, a user's preference for an item is represented by a rating, and calculation of similarities between users and judgement on whether similar users prefer or not are based on the ratings. Collaborative filtering problem can be also considered as the problem of predicting unknown entry values from known entry values of a partially known user-item rating matrix.

The performance of a system in giving initial recommendations for a new user is important because a user will not continue to use the system if it takes long time for the system to learn the user's preference. In this paper, we propose a collaborative filtering method using restoration operators that can make good predictions based on a small number of ratings. Furthermore, we consider the problem of selecting items that a new user is asked to rate in order to achieve optimal prediction performance. This kind of learning, in which a learner obtains information needed to learn actively, is called *active learning* [4] and it is one of the current hot research topics.

There have been a few studies on collaborative filtering using active learning. Goldberg et al. [7] developed a system that asks new users to rate the same set of

N. Lavrač et al. (Eds.): PKDD 2003, LNAI 2838, pp. 339–349, 2003.

items called the *gauge set*. The system uses the ratings for the items in the gauge set to classify the user into one of the clusters and then recommends the items that are popular among the users belonging to that cluster. However, Goldberg et al. did not describe how the gauge set is selected. Boutiler and Zemel [3] proposed a system that asks new users to rate an item with the maximum *expected value of information* with respect to the current probabilistic model. Dasgupta et al. [5] considered that there are a number of typical users and that each user's ratings are close to those of one typical user. They proposed a query selection algorithm and analyzed the number of queries to an arbitrary user needed for finding a typical user with similar ratings. Here, a query means asking a user to rate a given item. Query selection, therefore, means determination of which item is given.

Restoration operators have been extensively studied in the field of digital image restoration [1, 9]. For a known degration operator P, they have studied the problem of how to restore a digital image $x \in \Re^m$ from a given degraded image $Px \in \Re^k (k < m)$. Collaborative filtering problem can be also seen as a restoration problem in which a vector x of user's preference for all items is restored from a partial vector Px of x, where P is an operator that restricts the components to those having observed values. Considering characteristics of collaborative filtering problem, we propose a collaborative filtering method using an unbiased Wiener filter without additive noise. Our experimental results showed that our method outperforms the correlation-based method with the best performance when the number of observed ratings is small.

In our problem setting, the active learning problem, the problem of selecting the best items that a new user is asked to rate, is that of finding a restriction operator P that minimizes the expected squared error. However, to the best of our knowledge, there are no efficient methods to find the best operator P among $O(m^k)$ operators, where m is the number of items and k is the number of queries. Instead of finding the best operator, we propose a greedy method that finds an operator with small expected error by greedy search which involves comparison of $O(km)$ operators. Our experimental results showed that our greedy method is better than random selection in terms of prediction performance.

This paper is organized as follows. In Section 2, we formalize a collaborative filtering problem as a restoration problem and show its solution. In Section 3, we describe an optimal query item selection problem and our greedy selection method. Results of experiments using the EachMovie data set are shown in Section 4, and future work is discussed in Section 5.

2 Collaborative Filtering Problem as a Restoration Problem

Let m denote the number of items and $x \in \Re^m$ denote a vector of the user's preference for all items. Assume that x is generated according to an arbitrary distribution over $\Re^m$. The collaborative filtering task can be regarded as a restoration of vector x from a given partial vector of x. Such a restoration problem has been studied in the area of digital image restoration. In this section, we formalize a collaborative filtering problem like the formalization of a digital image restoration problem presented in [9].

For $J = \{i_1, i_2, ..., i_k\} \subseteq \{1, 2, ..., m\}$, let P_J denote a *restriction transformation (matrix)* that restricts components of a vector to those in J, namely, P_J is a transformation such that $P_J x = (x_{i_1}, x_{i_2}, ..., x_{i_k})^T$ for $x = (x_1, x_2, ..., x_m)^T$. Observing $P_J x$ for some J, we want to find an estimation $\hat{x}$ of x which is close to x, that is, the mean squared error (MSE) $||x - \hat{x}||^2$ is as small as possible, where $|| \cdot ||$ denotes Euclidean norm. Since $\hat{x}$ is calculated by applying a restoration transformation to $P_J x$, this problem is reduced to that of finding an optimal transformation. For fixed J, this problem can be formalized as follows if restoration transformations are restricted to affine transformations.

Problem 1. For given J with[1] $|J| = k$, find $(B_{\mathrm{opt}}, w_{\mathrm{opt}})$ satisfying

$$(B_{\mathrm{opt}}, w_{\mathrm{opt}}) = \arg \min_{(B,w)} E_x ||x - BP_J x - w||^2, \qquad (1)$$

where the minimization is with respect to all possible linear transformations $B : \Re^k \to \Re^m$ and all possible vectors $w \in \Re^m$.

This problem setting can be regarded as that of an unbiased Wiener filter without additive noise. (See [9].) The solution for this problem is as follows. Note that $\cdot^T$ and $tr(\cdot)$ denote the transpose and trace of a matrix, respectively.

Solution 1.
$$B_{\mathrm{opt}} = RP_J^T(P_J RP_J^T)^+,$$
$$w_{\mathrm{opt}} = (I - B_{\mathrm{opt}} P_J) E_x x \text{ and}$$
$$min_{(B,w)} E_x ||x - BP_J x - w||^2 = tr(R - B_{\mathrm{opt}} P_J R),$$
where $R = E_x(x - E_x x)(x - E_x x)^T$ and $(P_J RP_J^T)^+$ is the Moore-Penrose inverse of $P_J RP_J^T$.

Why do not we take noise into account? In digital image restoration, we assume that an original image x contains no noise and that as training data we observe $Px + n$, an image which is degraded by a degradation operator P and to which noise n is added, as well as x. If we take noise into account in our collaborative filtering problem, it should be assumed that we observe a user's preference x that contains noise and a partial vector $P_J x$ that does not contain additional noise. Assume that the observed vector x contains noise n. Then, what we want to estimate is $\hat{x}$ that minimizes $||x - n - \hat{x}||^2$. Thus, instead of Equation (1), we should use the equation

$$(B_{\mathrm{opt}}, w_{\mathrm{opt}}) = \arg \min_{(B,w)} E_x E_n ||x - n - BP_J x - w||^2. \qquad (2)$$

If we assume that x and n are mutually independent and that $E(n) = 0$,

$$E_x E_n ||x - n - BP_J x - w||^2 = E_x ||x - BP_J x - w||^2 + E_n ||n||^2.$$

---

[1] Here, $|J|$ denotes the number of elements in J.

Since the term $E_{\boldsymbol{n}}||\boldsymbol{n}||^2$ does not affect the minimization, Equation (2) coincides with Equation (1). Therefore, our problem setting deals with the case in which noise is contained in a user's preference, which can be inevitable because user's evaluation values are quantized.

When Solution 1 is used, consideration must be given to the method used for estimating the covariance matrix $R = E_{\boldsymbol{x}}(\boldsymbol{x} - E_{\boldsymbol{x}}\boldsymbol{x})(\boldsymbol{x} - E_{\boldsymbol{x}}\boldsymbol{x})^T$. What makes this difficult is that this estimation must be done from the user's preference dataset X, the elements of which have many missing values. Here, we estimate R as follows. Let n_i denote the number of elements in X whose ith component is not missing. Let $\hat{\bar{\boldsymbol{x}}} = \{\hat{\bar{x}}_1, \hat{\bar{x}}_2, ..., \hat{\bar{x}}_m\}$ denote an estimation of $E_{\boldsymbol{x}}\boldsymbol{x}$. Then, in our estimation,

$$\hat{\bar{x}}_i = \sum_{\boldsymbol{x} \in X, x_i \neq *} x_i/n_i, \tag{3}$$

where $*$ denotes the missing value. For $\boldsymbol{x} \in X$, let $\boldsymbol{x}'$ denote the vector that is made from $\boldsymbol{x}$ by replacing all missing values x_i with $\hat{\bar{x}}_i$. Our estimation $\hat{r}_{i,j}$ of the (i, j) entry value of R is calculated as

$$\hat{r}_{i,j} = \frac{1}{n} \sum_{\boldsymbol{x} \in X} (x'_i - \hat{\bar{x}}_i)(x'_j - \hat{\bar{x}}_j),$$

where n is the number of elements in X.

3   Optimal Query Item Selection

Another concern is for which items we should know the user's preference in order to estimate the whole vector of the user's preference as precisely as possible. The problem of optimal query item selection is that of finding the best J, which is fixed in Problem 1.

Problem 2. For given $1 \leq k \leq n$, find $J_{\mathrm{opt}} \subseteq \{1, 2, ..., m\}$ satisfying

$$J_{\mathrm{opt}} = \arg \min_{J:|J|=k} \min_{(B, \boldsymbol{w})} E_{\boldsymbol{x}}||\boldsymbol{x} - BP_J\boldsymbol{x} - \boldsymbol{w}||^2, \tag{4}$$

where the second minimization is with respect to all possible linear transformations $B : \Re^k \to \Re^m$ and all possible vectors $\boldsymbol{w} \in \Re^m$.

By plugging Solution 1 into Equation (4), we obtain the following equation:

$$J_{\mathrm{opt}} = \arg \max_{J:|J|=k} tr(RP_J^T(P_JRP_J^T)^+P_JR). \tag{5}$$

Finding J_{opt} requires calculation of $tr(RP_J^T(P_JRP_J^T)^+P_JR)$ for $m(m - 1)\cdots(m - k + 1)/k!$ combinations J of items, so it is not practical for a large value of m. In that case, we use a greedy method by which optimal items are added to J one by one until the number of items in J becomes k. Our greedy method uses J_k instead of J_{opt}. J_k is defined as follows:

$$J_k = \arg \max\{tr(RP_J^T(P_JRP_J^T)^+P_JR) : J \subset J_{k-1}, |J| = k\}, \quad J_0 = \emptyset. \tag{6}$$

Note that the number of combinations J in calculation of J_k is $O(km)$, while that in calculation of J_{opt} is $O(m^k)$.

4   Experiments

We conducted two experiments: one for evaluation of prediction performance based on a small number of rated items, and the other for evaluation of query item selection by our greedy method.

4.1   Correlation-Based Methods

We compared the performance of our method to the performances of correlation-based methods [10, 11]. By correlation-based methods, predictions are made using correlation coefficients between each pair of users that are calculated from ratings for the commonly rated items. The correlation coefficient between users x and y is calculated as follows:

$$w_{x,y} = \frac{\sum\limits_{x_j \neq *, y_j \neq *} (x_j - c_{x,j})(y_j - c_{y,j})}{\sqrt{\sum\limits_{x_j \neq *, y_j \neq *} (x_j - c_{x,j})^2 \sum\limits_{x_j \neq *, y_j \neq *} (y_j - c_{y,j})^2}},$$

where $c_{x,j}$ is a center value for user[2] x and item j. Prediction of x_i for user x with $x_i = *$ are made using correlation coefficients $w_{x,y}$ between user x and other users y with $y_i \neq *$ as follows:

$$x_i = c_{x,i} + \frac{\sum\limits_{y_i \neq *} w_{x,y}(y_i - c_{y,i})}{\sum\limits_{y_i \neq *} |w_{x,y}|}.$$

With respect to $c_{x,i}$, we consider the following three variations.

1. Fixed center
   In this variation, $c_{x,i}$ is the same constant for all x. The method using this correlation coefficient is called *constrained Pearson r* algorithm in [11].
2. User mean
   In this variation, $c_{x,i}$ is the average of x_j for all item j with $x_j \neq *$. The method using this correlation coefficient is called *Pearson r* algorithm in [11]. In the case of new users having no ratings, we use the average of all scores rated by all users.
3. Item mean
   In this variation, $c_{x,i}$ is the average of x_i for all users x with $x_i \neq *$, namely, $c_{x,i} = \hat{\bar{x}}$, where $\hat{\bar{x}}$ is defined in Equation (3).

---

[2] We are somewhat loose in our notation here, using a vector of user's preference as if it were the user itself.

4.2   Data

In our experiment, we used the EachMovie collaborative filtering data set [6]. The data set consists of 2,811,983 numeric ratings for 1,628 movies evaluated by 72,916 users. The numeric rating for a pair of user and movie represents how much the user likes the movie on a six-point scale (0.0, 0.2, 0.4, 0.6, 0.8, 1.0). Note that only 2.37% of the user-movie matrix is filled.

Table 1. Statistical information on the data used in our experiments

#Users	#Movies	#Ratings							Filled %
		0.0	0.2	0.4	0.6	0.8	1.0	total	
2,000	1,618	93,980 (17.0%)	32,895 (5.9%)	73,277 (13.2%)	138,337 (24.9%)	135,406 (24.4%)	80,950 (14.6%)	554,845	17.1

	max	min	average			max	min	average
#(Ratings by a user)	1455	191	277.4		#(Ratings for an item)	1796	1	342.9

In our experiments, We used the ratings of 2,000 users with the largest number of ratings. Statistical information on our data is shown in Table 1. Note that 17.1% of the user-movie matrix of the data is filled.

4.3   Performance Measures

We evaluate the performance by two measures: mean squared error and recall-precision curve. The former is suitable to our problem setting, but the latter is more appropriate from a practical point of view. The following is an explanation of the method used for drawing recall-precision curves.

Rating values are divided into two classes, hot (0.8 or 1.0) and cold (0.0, 0.2, 0.4 and 0.6), as in [2]. Precision and recall are used to evaluate what degree a method correctly predicts a set of *hot* movies for each user. Let T denote the set of hot movies for an arbitrary user and S denote the set of movies predicted to be hot for that user. Then, precision and recall for this prediction is defined as follows:

$$\text{precision} = \frac{|T \cap S|}{|S|} \qquad \text{recall} = \frac{|T \cap S|}{|T|}.$$

For methods in which prediction values are given by real numbers, the set of movies predicted to be hot can be obtained by determining a threshold and selecting movies whose predicted values are greater than the threshold. In such cases, a recall-precision curve can be drawn by moving the threshold. We want to know the performance averaged over all users, but the problem is the difference in user's thresholds for the same recall. Thus, for recall r and user u, we find the set $S_{r,u}$ of movies predicted to be hot by lowering the threshold until recall r is achieved, and we calculate averaged precision for recall r as follows:

$$\text{averaged precision for recall } r = \frac{\sum_{u \in U} |T_u \cap S_{r,u}|}{\sum_{u \in U} |S_{r,u}|},$$

where T_u is the set of hot movies for user u and U is the set of users. We draw the recall-precision curve by moving recall r.

4.4   Experimental Methodology

We randomly divided the 2,000 users into 10 groups and carried out a 10-fold cross validation. We first estimated the covariance matrix R using only training data. Then, for each user in a test data, we selected a set of movies for queries from the movies whose scores are known. Based on the ratings for the selected movies, predictions for the other movies whose scores are known were made. We evaluated methods by prediction performance averaged over all users. In the case of random query item selection, prediction performance was also averaged over five runs.

4.5   Results

Prediction Performance Based on a Small Number of Rated Items. We compared prediction performance of restoration operators with prediction performances of correlation-based methods when a small number of items are selected randomly as a set of items that are assumed to be rated.

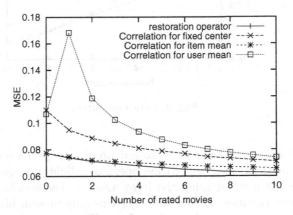

Fig. 1. Learning curves

Fig. 1 shows the relation between the number of rated movies and MSE. Here, predictions by restoration operators using no rating are those by the estimated mean vector $\hat{\bar{x}}$ defined by Equation (3). Among the correlation-based methods, the method using the item mean performed best. As can be seen in Fig. 1, prediction performance of restoration operators was better than prediction performances of correlation-based methods.

Fig. 2 shows recall-precision curves when the number of rated movies are five and ten. The results are consistent with the results in terms of MSE. Prediction performances of two correlation-based methods using the fixed center

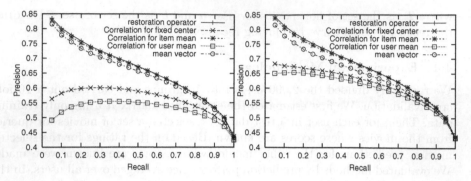

Fig. 2. Recall-precision curves (Left: prediction using 5 rated movies, Right: prediction using 10 rated movies)

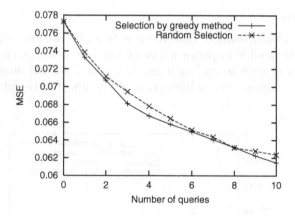

Fig. 3. Learning curves

and the user mean could not even exceed prediction performance of the mean verctor. The method using restoration operators slightly outperformed the best correlation-based method using the item mean at low recalls, which are more important parts because recommender systems only present high-ranking items.

Effectiveness of Query Item Selection. We investigated the effectiveness of our greedy query item selection.

Fig. 3 shows the relation between the number of queries and MSE. AS can be seen in the figure, MSE decreases as the number of queries increases. We compared prediction performance of our greedy methhod with that of random selection. The graph shows the greedy method outperformed random selection.

The left panel of Fig. 4 shows recall-precision curves for predictions made by the greedy method and the mean vector. It can be seen that the perfomance improves as the number of queries increases, particularly at low recalls. Improvement in performance at low recalls with respect to number of queries is shown in

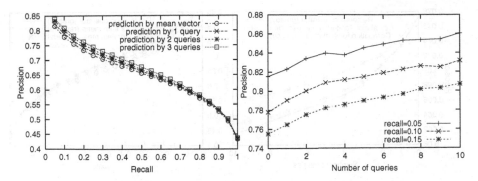

Fig. 4. Left: Recall-precision curves, Right: #Queries-precision curves

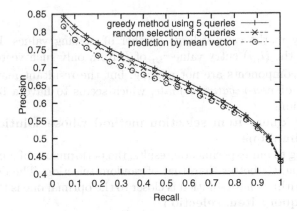

Fig. 5. Comparison with random selection

the right panel of Fig. 4. The rate of improvement decreases when the number of queries is larger than three.

Fig. 5 shows a comparison of the performances of our greedy method and random selection in terms of recall-precision curves. The graph shows the greedy method also outperformed radom selection with respect to this measure.

Are the query items selected by our greedy method effective for the predictions of correlation-based methods? According to the results of our experiments, this is true for the correlation-based method using the item mean. (See Fig. 6.)

5 Future Work

The following issues must be considered in future works.

1. **Better estimation of the covariance matrix R**
   In order to estimate R from a user's preference dataset X, we must deal with the problem of missing values. In our experiments, we replaced all missing values with the item mean, but this does not seem to be the best method,

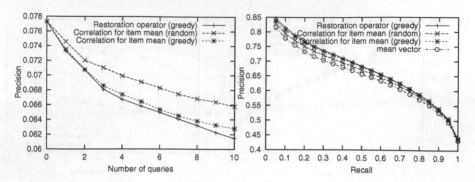

Fig. 6. Left: Learning curves, Right: Recall-precision curves for predictions using 5 queries

especially in the case of large number of missing values. It is possible to estimate the (i, j) entry value $r_{i,j}$ of R from only such vectors x whose ith and jth components are not missing, but the resultant matrix $\hat{R}$ loses the property of *non-negative definite*, which seems to have a bad influence on the solution.

2. **Efficient query item selection method whose solution is closer to the optimal one**
   According to our experimental results, the performance of our greedy method is only slightly better than that of random selection. The development of a fast algorithm whose solution is closer to the optimal one is therefore needed.

3. **Online query item selection**
   The methods proposed in [3, 5] are *online*, that is, the next query item is decided on the basis of answers to the previous queries. Development of such methods in our framework is preferable because there is a possibility that a set of query items more appropriate for the predictions of a user's preference can be selected by using information obtained by the previous queries.

Acknowledgement

This work was partly supported by a Grant-in-Aid for Scientific Research (B), No. 14380151, from the Japan Society for the Promotion of Science.

References

1. Andrews, H., Hunt, B.: Digital Image Restoration. Prentice-Hall, Englewood Cliffs, N.J., 1977.
2. Billsus, D., Pazzani, M.: Learning Collaborative Information Filters. Proc. of the 15th International Conference (ICML'98) (1998) 46–54.
3. Boutilier, C. and Zemel, R. S.: Online queries for collaborative filtering. Ninth International Workshop on Artificial Intelligence and Statistics (2002).

4. Cohn, D., Ghahramani, Z., Jordan, M.: Active Learning with Statistical Models. Journal of Artificial Intelligence Research 4 (1996) 129–145.
5. Dasgupta,S., Lee, W., Long, P.: A Theoretical Analysis of Query Selection for Collaborative Filtering. Machine Learning 51 (2003) 283–298.
6. EachMovie collaborative filtering data set, 1997. research.compaq.com/SRC/eachmovie/.
7. Goldberg, K., Roeder, T., Gupta, D., Perkins, C.: Eigentaste: A Constant Time Collaborative Filtering Algorithm. Information Retrieval Journal 4(2) (2000) 133–151.
8. Nakamura, A., Abe, N.: Collaborative Filtering using Weighted Majority Prediction Algorithms. Proc. of 15th International Conference on Machine Learning (1998) 395–403.
9. Ogawa, H., Oja, E.: Projection Filter, Wiener Filter, and Karhunen-Loève Subspaces in Digital Image Restoration. Journal of Mathematical analysis and applications 114 (1986) 37–51.
10. Resnick, P., Iacovou, N., Suchak, M., Bergstom P., Riedl, J.: GroupLens: An Open Architecture for Collaborative Filtering of Netnews. Proc. of CSCW (1994) 175–186.
11. Shardanand, U., Maes, P.: Social Information Filtering: Algorithms and Automating "Word of Mouth". Proc. of CHI95 (1995) 210–217.

Efficient Frequent Query Discovery in FARMER

Siegfried Nijssen and Joost N. Kok

Leiden Institute of Advanced Computer Science
Niels Bohrweg 1, 2333 CA, Leiden, The Netherlands
snijssen@liacs.nl

Abstract. The upgrade of frequent item set mining to a setup with multiple relations – frequent query mining – poses many efficiency problems. Taking Object Identity as starting point, we present several optimization techniques for frequent query mining algorithms. The resulting algorithm has a better performance than a previous ILP algorithm and competes with more specialized graph mining algorithms in performance.

1 Introduction

Recently, multi-relational or structured data mining has gained much interest. Especially frequent structure mining similar to APRIORI [1] was discussed in a number of recent publications, such as the gSpan algorithm by Yan et al. [10] and FSG by Kuramochi et al. [7]. Given a database of complex structures —in the case of gSpan and FSG a collection of graphs— the task of these algorithms is to find those substructures that occur in many of the complex structures. Already several years ago, Dehaspe et al. [3] introduced an algorithm called WARMR for frequent pattern mining in relational databases. WARMR was built on the solid theoretical foundations of Inductive Logic Programming (ILP). It accomplished similar tasks as the more recent algorithms. When comparing WARMR to graph mining algorithms such as gSpan, we note the following points:

- the greater expressiveness of WARMR: specialized mining algorithms often concentrate on one type of database, for example databases of labeled undirected graphs. For different kinds of structures, modified algorithms are required. In ILP algorithms, such as WARMR, any structure can be expressed easily. The incorporation of background knowledge is also straightforward.
- the choice for traditional clause based query evaluation in WARMR: in this case, two variables may have the same value during evaluation. In the subgraph mining algorithms two nodes in a subgraph cannot be mapped to one node in a database graph.
- in publications of subgraph mining algorithms [6,7,10], much attention is given to efficiency issues. The WARMR algorithm can be considered as a proof-of-concept of a framework; efficiency issues have not been given too much attention.

In this paper, we will introduce a new algorithm for frequent query mining. While it is largely comparable to WARMR from an expressive point of view,

N. Lavrač et al. (Eds.): PKDD 2003, LNAI 2838, pp. 350–362, 2003.
© Springer-Verlag Berlin Heidelberg 2003

it uses techniques introduced by subgraph mining algorithms as well as new techniques. The main contributions of the paper are:

- We show how the query discovery task can be changed to query discovery under *Object Identity*. Although the focus of this paper is not on the semantic consequences of this choice, we will argument that this approach closely matches that of subgraph mining, is very natural and does not pose restrictions in many data mining situations.
- Building upon this evaluation under Object Identity, and using a tree data structure, we will define an order on queries that allows for more efficient search space traversals than the approach used by WARMR. To some extent this order is equivalent to that of gSpan; it is however more flexible and allows for some new optimizations.
- We will show how this order can be exploited in both breadth-first and depth-first algorithms. For the latter case we will introduce optimizations that are allowed by the query ordering, including hash structures and sorting to reduce the cost of query evaluations that result in false.
- We will present experimental results showing large speed-ups in comparison with a recent implementation of WARMR. We will also compare our results to those obtained by gSpan and FSG. In some cases, our algorithm obtains similar run times as FSG, but it does not equal the efficiency of gSpan. We will give some arguments for this difference in performance.

Our aim is to use ILP formalisms that are very close to WARMR and to reach the efficiency of algorithms like gSpan and FSG.

Our depth-first and breadth-first algorithms are major revisions of our previous FARMER algorithm for mining multiple relations [8]. The algorithm in [8] was restricted to some variants of labeled, unordered trees and did not use Object Identity for query evaluation. In the breadth-first algorithm presented here, only the tree-like notation is reused. Restrictions that were present in the previous version of FARMER, do no longer exist in our new algorithm. We will however still use the name FARMER to denote our class of algorithms.

2   Search Space Specification and Object Identity

We will introduce some notation. Any capital A denotes an atom. An *atom set* S is an unordered set of atoms. An ordered atom set is called a *query* and is denoted by a capital Q. With (Q, A) we denote the query Q to which atom A is concatenated. With $last(Q)$ we denote the last atom of Q. The variables in $A = last(Q)$ that do not occur in $Q \backslash A$ are called the *new variables* of A in Q.

Every predicate p is considered to be *typed*: each argument has a type. A variable or constant that is used as an argument of a predicate, has the same type as the argument. Types are frequently used in ILP systems to allow the definition of more narrow search spaces. With $var(S, T)$ (or $var(Q, T)$) we denote the set of all variables of type T in an atom set S.

We will first introduce a mechanism that defines the search space of queries that our algorithm will investigate. It uses a similar mode mechanism as WARMR.

Definition 1 (Bias). *A mode declaration $p(c_1, \dots, c_n)$ consists of a predicate with arguments c_i, each of which is either '+' (input), '-' (output) or '#' (constant). The bias $\mathcal{B}$ of a search space consists of: 1) the type definitions of predicates, 2) a set of modes $\mathcal{M}$, 3) an operator $const(T)$ which defines for each type T a set of constants, 4) a function $\max$ which assigns an integer to each predicate in $\mathcal{M}$, and 5) one atom $k(X)$ (this atom is called the key of the search).*

Definition 2 (Search space). *Given a bias $\mathcal{B}$, a query Q and an atom $A = p(t_1, \dots, t_n)$, atom A is a (mode) refinement of Q iff there is a mode $M = p(c_1, \dots, c_n) \in \mathcal{M}$ such that for every $1 \leq i \leq n$ either:*

- *t_i is a variable in $var(Q, T_i)$ and $c_i = '+'$, or*
- *t_i is a variable not in $\cup_j var(Q, T_j)$ and $c_i = '-'$, or*
- *t_i is a constant in $const(T_i)$ and $c_i = '\#'$.*

Here, T_i is the type of argument position i. The search space $\mathcal{S}(\mathcal{B})$ defined by a bias $\mathcal{B}$ consists of all queries Q that can be built iteratively starting from the key atom $k(X)$ using valid refinements. Each atom A that is added to a query Q should satisfy the following restrictions to be a valid refinement: 1) A is a mode refinement; 2) A does not already occur literally in Q; 3) the predicate p used in A does not occur more than $\max(p)$ times in the new query.

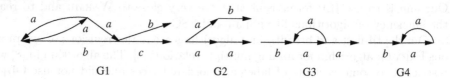

Fig. 1. Directed, edge labeled graphs.

Example 1. As an example we will use the representation of a directed, edge labeled graph using a predicate $e(G, N, N, L)$. Graph G1 in Fig. 1 can be represented using the following facts:

$$K = \{k(g_1), e(g_1, n_1, n_2, a), e(g_1, n_2, n_1, a), e(g_1, n_2, n_3, a), e(g_1, n_3, n_1, b),$$
$$e(g_1, n_3, n_4, b), e(g_1, n_3, n_5, c)\}.$$

Of course, the choice of constants n_i is arbitrary here. Using the set of modes $\mathcal{M} = \{e(+, -, -, \#), e(+, +, -, \#), e(+, +, +, \#)\}$ the following queries can be constructed:

$$Q2 = k(G), e(G, N_1, N_2, a), e(G, N_2, N_3, a), e(G, N_1, N_4, a), e(G, N_4, N_5, b),$$
$$Q3 = k(G), e(G, N_1, N_2, b), e(G, N_2, N_3, a), e(G, N_3, N_2, a), e(G, N_3, N_4, a),$$
$$Q4 = k(G), e(G, N_1, N_2, b), e(G, N_2, N_3, a), e(G, N_3, N_2, a);$$

they correspond to graphs G2, G3 and G4 in Fig. 1.

Given a knowledge base K the support of a query Q can be defined as:

$$support_K(Q) = \#\{\theta \mid K \models Q\theta\},$$

where θ is a substitution to constants of all variables in the key of Q; $Q\theta$ denotes the application of this substitution to Q.

In WARMR, to compute the $\models$ relation, a Prolog engine based on θ-subsumption is used. For a knowledge base containing only facts, this evaluation comes down to the discovery of a substitution such that $Q\theta \subseteq K$. On the other hand, we will use an evaluation technique based on subsumption under *Object Identity (OI-subsumption)*. Under Object Identity, the satisfying substitution θ is constrained in two ways: no two variables in Q may be mapped to the same constant, and no variable may be mapped to a constant already occurring in Q.

We will briefly illustrate some consequences of this choice. Under usual θ-subsumption, example query Q_2 is a consequence of the knowledge base K, as Q_2 can be satisfied by mapping $N_1 \rightarrow n_2, N_3 \rightarrow n_2, N_2 \rightarrow n_1, N_5 \rightarrow n_1, N_4 \rightarrow n_3$. In the graph notation of Fig. 1, some nodes in G2 are mapped to the same nodes in G1. Under Object Identity, this is not allowed: the mapping must be injective. Such an injective mapping is also used in gSpan for labeled undirected graphs. Similar arguments show that G3 is not included in G1 under Object Identity, while it would be included under traditional θ-subsumption.

An important issue is that of query *equivalency*. In general, two queries Q_1 and Q_2 are equivalent iff for every possible knowledge base K: $K \models Q_1 \Leftrightarrow K \models Q_2$. For evaluation without Object Identity, one can prove that Q_1 and Q_2 can only be equivalent when Q_1 and Q_2 mutually subsume each other. Without OI, G3 and G4 in our example are equivalent. Every graph which contains G4 also contains G3, as node N_4 can always be mapped to the same node as N_2. The first reason for choosing Object Identity is that these counterintuitive situations are prevented under OI. Under OI queries are equivalent iff they are *alphabetic variants* [4,5]. We will define this equivalency relation more precisely. Given a query Q, let $vars(Q)$ denote the set of all variables occurring in Q and let $varlf(Q)$ denote the list of all variables Q in order of first occurrence.

Definition 3 (Equivalency of queries). *Given a query Q, the normally named query $n(Q)$ is the query Q to which the following renaming substitution is applied: $\theta = \{V/V_i \mid V \in vars(Q), i = ord(V, Q)\}$. Here $ord(V, Q)$ is the position of V in $varlf(Q)$. Two queries Q_1 and Q_2 are equivalent (denoted by $Q_1 \equiv Q_2$) if there exists a permutation π of the atoms in Q_1 such that $n(\pi(Q_1)) = n(Q_2)$.*

To determine whether two queries are equivalent, is therefore 'only' a problem of finding a permutation which transforms the one query into the other. This is still a difficult problem; it can be shown that to compute whether two queries are equivalent, one has to solve a graph isomorphism problem, and vice versa. The complexity of graph isomorphism is currently unknown: no polynomial algorithm is known, and a proof of NP completeness does not exist either. In comparison with full θ-subsumption, however, OI makes the computation of equivalency slightly easier. This property is the second reason for choosing OI.

We will now present our pattern mining task.

Definition 4. *Given a bias $\mathcal{B}$, a knowledge base K and a threshold minsup,* FARMER *should discover a set of queries $\mathcal{Q}$ such that for every $Q \in \mathcal{S}(\mathcal{B})$ with* $support_K(Q) \geq minsup$, *there is exactly one $Q' \in \mathcal{Q}$ such that $Q' \equiv Q$.*

A query for which $support_K(Q) \geq minsup$ is said to be *frequent*. The single query in $\mathcal{Q}$ to which a query Q is equivalent is considered to be its *normal form* or its *canonical label*.

The third advantage of OI can be understood by considering Q_2 in conjunction with the following modes, which define a search space of edge labeled *trees*: $\{e(+, -, -, \#), e(+, +, -, \#)\}$. Query Q_2 is not equivalent with any smaller query. Every subquery $Q_2' \in \mathcal{S}(\mathcal{B})$ of Q_2 with $|Q_2| = |Q_2'| + 1$ is however equivalent with a query smaller than $|Q_2'|$. An algorithm which relies on refinement with building blocks of one atom, will not construct Q_2 if it removes equivalent queries immediately. Such difficulties with refinement are avoided under OI.

The choice for Object Identity has many consequences on the types of patterns that can be discovered. As an illustration consider a situation in which one also allows *wildcards* as labels. A possible query in this case would be:

$$k(G), e(G, N_1, N_2, L_1), L_1 = a, e(G, N_2, N_3, L_2), e(G, N_4, N_5, L_3), L_3 = b.$$

Under full OI, all labels L_1, L_2 and L_3 must be different. Although for clear objects (such as nodes), an inequality constraint is a natural choice, for properties (such as the label of an edge) inequality can be undesirable. An elegant solution could be to use a variant of Object Identity which does not force OI on variables for such properties; in this weaker OI, one can sometimes (and also in this example) still guarantee the three properties of Object Identity that we exploit. Due to lack of space, we refer to [9] for more details about OI related issues.

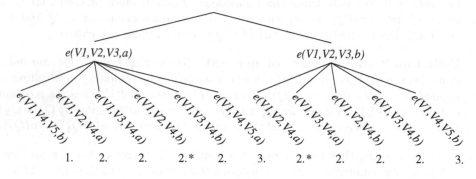

Fig. 2. A query tree.

3   A Tree Based Normal Form

In our algorithm, all queries are stored in an ordered tree as given in Fig. 2. Every node in this tree is labeled with an atom. Every path starting starting in the root represents a query. Every node has therefore an associated query. Once a query is counted, its support is stored in the associated node. The query tree is similar to the *query pack* tree used by WARMR for efficient query evaluation [2]. By introducing an order on nodes in the tree, FARMER adds as main application of the tree the efficient determination of candidate queries. The order of queries is determined by their order in the tree:

Definition 5 (Order of queries). *Let* $Q_1 = (Q_p, A_p, A_1, Q_1')$ *and* $Q_2 = (Q_p, A_p, A_2, Q_2')$, $A_1 \neq A_2$ *be two queries (where* Q_p, Q_1' *and* Q_2' *may be empty), then* $Q_1 <_T Q_2$ *iff* $A_1 < A_2$ *in the child list of* (Q_p, A_p).

If $Q_1 <_T Q_2$, then Q_1 is called an *earlier* query than Q_2 or Q_2 is called a *later* query than Q_1.

An outline of the FARMER algorithm is given in Algorithm 1 and 2. In line (5) of Algorithm 1, the order in which nodes are expanded is intentionally left unspecified. The order is only restricted by the precondition of FARMER-EXPAND. In line (5) of FARMER, and line (3) of Algorithm 2 this observation is used: $\forall Q_2 : (\exists Q_1 \subseteq Q_2 : support(Q_1) < minsup) \Rightarrow support(Q_2) < minsup$.

Algorithm 1: FARMER
Input: A bias $\mathcal{B}$, a knowledge base K and a threshold *minsup*.
Output: A tree T with all queries according to Definition 4.
(1)   Read K and determine $const(T)$ for each type T
(2) $T :=$ a tree with only the key atom in the root
(3)   **repeat**
(4)   Count the frequency of all uncounted queries.
(5)   **for** one or more uncounted, unmarked, frequent leafs **do**
(6)   Expand that leaf
(7)   **until** T contains no uncounted queries
(8)   Remove all marked nodes

Algorithm 2: FARMER-EXPAND
Input: A query Q in a tree T with counts for (1) all ancestor queries of Q, (2) all earlier queries Q', $|Q'| \leq |Q|$; (3) all later queries Q' which are a brother of an ancestor of Q.
Output: A query tree with uncounted expansions Q' of Q, $|Q'| = |Q| + 1$.
(1)   Let A be $last(Q)$; let A_p be the parent of A and Q_p the query associated with A_p.
(2)   Add as child of A all valid refinements $A'' = last(n(Q, A'))$, where A' is either:
(3)   1. a frequent atom occurring after A in A_p's child list, where new variables in A' are renamed such that they are also new in (Q, A').
(4)   2. a *dependent* atom, which is any atom that uses at least one variable that was new in A.
(5)   3. a copy of A if A has new variables; those new variables are given new names in the copy.
(6)   Remove the new child query A'' if it is equivalent to an earlier query, unless the child is only equivalent to a brother. In that case A is marked but kept in the tree.

It is this property that has led to the popularity of APRIORI-like algorithms: this property restricts the search space in such a way that is possible to compute all frequent patterns if the threshold is not too low.

We will first consider the resulting tree T when all queries are frequent and FARMER-EXPAND line (6) is absent. We will show that for every query in the search space, at least one equivalent query can be found in this tree.

Example 2. Under these assumptions, and given modes $e(+, -, -, \#)$ and $e(+, +, -, \#)$, Fig. 2 shows for each query of length 2 how they are obtained from queries of length 1 by applying FARMER-EXPAND. Each number indicates which of the three possibilities is applied to generate a new atom.

Lemma 1. *Given is a query Q which occurs in a Query Tree T generated by* FARMER, *and an atom $A \notin Q$ which is a valid refinement of Q. Then a query $Q' = n(Q_1, A, Q_2)$ exists in the tree T, for some subdivision of Q into Q_1 and Q_2, $Q = (Q_1, Q_2)$. Furthermore, Q is either a prefix of Q' or $Q' <_T Q$.*

Proof. As A is a valid refinement of Q, there is a prefix (Q_p, A_p) of Q such that the normalized atom $A' = last(n(Q_p, A_p, A))$ is a dependent atom of A_p. This dependent atom is generated in line 4 of the FARMER-EXPAND algorithm. If A_p is the last atom of Q, our statement is clear. Therefore assume that A_p has a different successor A_{p+1} in Q. This atom A_{p+1} is also a child of A_p in T. Consider the order of A' and A_{p+1} in the list of children:

- if A' occurs before A_{p+1}, A_{p+1} is a right-hand child of A'. The copying mechanism in line 3 will copy A_{p+1} as a child of A'; all steps which created Q are applicable subsequently and result in a query Q'.
- if A' equals A_{p+1}, both have output variables. In line 5 a self-duplicate A_{p+1} of A' is generated. All steps which created Q are applicable subsequently.
- if A' occurs after A_{p+1}, A' is copied as a child of A_{p+1}. This child of A_{p+1} may be left or right from A_{p+2} (the next atom in the original query). We can recursively apply our arguments on the situation for $p + 1$ until one of the above conditions holds.

Also the order of the old and new query follows from these arguments.   □

Theorem 1 (Completeness of search). *For every query Q_1 in the search space, there is at least one equivalent query Q_2 in the tree T.*

Proof (Sketch). This can be shown by induction on the length of the query. A query with only the key occurs in T. By inductive assumption, an equivalent query for $Q_1 \backslash last(Q_1)$ exists in the tree, and a corresponding variable renaming. When $last(Q_1)$ is renamed accordingly, this renamed atom is a valid refinement of the equivalent query, and one can apply Lemma 1.   □

Two equivalent queries that still coexist without line 6 in Algorithm 2 are indicated with a (*) in Fig. 2. We will now consider the algorithm with this line added. We have to prove that by removing an atom from the tree, we do not remove an atom that otherwise would have been used to create a query for which no equivalent query exists.

Lemma 2. *Let T be the tree obtained after iterative application of Algorithm 2 without line 6. Assume that a query Q_2 is equivalent with a query $Q_1 <_T Q_2$. Then every query Q_2' which has Q_2 as prefix must have an equivalent query more left in the tree.*

Proof. As Q_2 is equivalent with Q_1, there is a permutation of atoms of Q_2 followed by a renaming θ that makes Q_2 equal to Q_1. This substitution θ can be applied to all atoms in $Q_s = Q_2' \backslash Q_2$. Some of these atoms are now valid refinements of Q_1. According to Lemma 1 one by one these atoms can be added to Q_1, yielding queries Q_1' that are either extensions of Q_1 or occur $Q_1' < Q_1$.   □

Theorem 2. *For every query defined by the bias, Algorithm 1 generates exactly one normal form if all queries are frequent.*

Proof. It is clear that no two normal forms can occur: in line 6 of Algorithm 2 and line 8 of Algorithm 1, any query which has an equivalent lower query is removed. Theorem 1 showed that if equivalents were not removed, the search is complete. According to Lemma 2, if a query Q is equivalent to an earlier query, all of its descendents must also be equivalent to an earlier query. Q should therefore not be expanded further. The only remaining function of atom $last(Q)$ is its function as an expansion for earlier brothers in line 3 of Algorithm 2. In case Q is equivalent to an earlier query Q' which is not a brother, $last(Q)$ is not required as a building block for earlier brothers: the brother atom can be added to Q to yield a query $Q'' < Q'$ and every expansion of Q' can also be added to Q'' (similar to the construction of Lemma 2). By the marking mechanism only those atoms are kept as building block that are equivalent to an earlier brother.   □

A consequence of the monotonicity constraint is that every building block of a query Q must also be frequent. From this observation it follows that our algorithm performs exactly the task that was defined in Definition 4.

4 Depth First and Breadth First Algorithms

In the algorithm discussed in the previous section, many elements have been kept unspecified. In this section, we give an overview of some details.

Equivalency Check. To determine whether an earlier equivalent query exists, we essentially use an exhaustive search algorithm. Given a query Q, the mode mechanism is used recursively to build queries Q' that contain atoms in Q. After an atom is added, the tree T is consulted to determine whether Q' is later (in which case Q' is not further expanded) or infrequent (in which case Q cannot be frequent and is pruned). Once a query $Q' < Q$ is found which contains all atoms of Q, Q is pruned. Especially the combination of frequency pruning with equivalency pruning is a distinctive feature of our algorithm. Although it requires infrequent nodes to be stored in the tree, it could give the exhaustive exponential search an additional value and could reduce the number of queries that should be counted later significantly.

Order of Query Expansion. We distinguish two query expansion orders: *breadth first* and *depth first*. In the breadth first approach, all nodes at the lowest level of the tree are expanded. This yields a tree in which all nodes at the new lowest level are uncounted. The nodes are counted next, and the process is repeated until no new level can be added.

In the depth first approach, only one node is expanded; the new children are counted immediately. Starting with the first child, the process is recursively repeated. Only after the complete subtree of the first child has been constructed, the next child is recursively expanded.

In both approaches, the precondition of Algorithm 2 is satisfied. Breadth first is the traditional approach and corresponds to the evaluation order of APRIORI [1], WARMR [3] and FSG [6]. The depth first order matches that of gSpan.

Query Counting. To determine whether a query is OI-subsumed by a knowledge base of facts, an exponential search is required (one can easily see that this problem is equivalent to the subgraph isomorphism problem, which is known to be NP hard). Especially those queries which can *not* be satisfied for a given key substitution are computationally very expensive as many variable assignments have to be checked before this can be concluded. The task of the algorithm is to reduce the number of key substitutions which result in *false* as much as possible, and to reduce the cost of such an evaluation if the computation is required.

One strategy to reduce the computational cost, is to overlap the computation of queries. Consider a query Q with several child expansions. One can backtrack over all possible assignments of Q as long as one of the child expansions is not satisfied. This is more efficient than to evaluate each child expansion separately.

The advantage of the breadth-first approach is that the number of queries that should be evaluated at a certain level is maximal. For a given substitution of key variables, the evaluation of many queries can be combined. Our breadth first implementation uses this evaluation technique, which is similar to *query packs* as discussed in [2] for WARMR and [8] for our previous FARMER algorithm.

To reduce the number of false evaluations, a substitution ID list approach can be used. For each query that is evaluated, one can store the list of all key substitutions for which the query can be satisfied. One can easily see that a query which is constructed from a query Q (either by copying $last(Q)$ or by expanding Q) can never be *true* for key substitutions for which Q is *false*. Therefore only substitutions in Q's SID list need to be evaluated.

To reduce the cost of evaluation, with each key substitution θ one can also store the variable assignments that satisfy each query Q. If the backtracking over variables is performed in a deterministic order from left to right, one can continue the evaluation of each expansion of Q starting from the assignment that satisfied Q without having to recompute that assignment. Some assignments are skipped in this way, but one can show that this can safely be done. To reduce the memory demand of the approach, for each query Q and key substitution θ we only store the difference $\Delta(\theta, Q)$ between the first variable assignment that satisfies Q and the first assignment that satisfies the parent in T of Q.

Order of Children. There are many possible child orders:

- The order in which children are generated in Algorithm 2. This is the order that we used in [8] and yields queries that are very well readable.
- A lexicographical order. To determine query equivalency, one repeatedly has to search for a given atom in a set of children. With a lexicographical order, in combination with binary search and hashing, we speed up this search.
- Sorted by support. Atoms with a lower support occur earlier in a query in this case, which results in a quicker evaluation of queries that cannot be satisfied (the most selective atoms occur earliest).
- Sorted by *backtracking progression*. Consider a query Q, a key substitution θ and a set $\Delta(\theta, Q)$ of variable assignment changes. The position of the leftmost variable affected by $\Delta(\theta, Q)$ in Q is the backtracking progression of Q for θ. By averaging $\Delta(\theta, Q)$ over all θ one can compute the average backtracking progression of each query. When a candidate query (Q, A) is generated by copying an atom A below a query Q, both $\Delta(\theta, Q)$ and $\Delta(\theta, A)$ could be used as starting point for the evaluation of (Q, A); best would be to always use the assignment which has backtracked most. However, when the evaluation of several queries is overlapped, much additional bookkeeping would be required. As tradeoff we always use $\Delta(\theta, Q)$ as starting point, but sort to make sure that the parent has backtracked most on average.

Note that in the last two orders, some special care has to be taken in the equivalency procedure, as the order of children is only known *after* they are counted.

5   Experimental Results

From the possibilities discussed in the previous section, we implemented and tested several (see [9]). We implemented a breadth-first algorithm with naive sorting order and evaluation without substitution ID lists as a reference algorithm. Furthermore we implemented a depth-first algorithm which incorporated overlapping evaluation and a complex sorting order: given a query Q that is going to be expanded, all children of nodes that are not an ancestor of Q are stored in lexicographical order to allow for quick equivalency checks; nodes on the path corresponding to Q are also sorted first on backtracking cost, then on support and finally lexicographically. These two orders can be combined in an efficient way. From our experiments, we concluded that it is most beneficial.

Bongard Dataset[1]. The Bongard dataset [2] was used to compare WARMR, depth-first and breadth-first FARMER (Fig. 3). In the experiments, FARMER was

---

[1] Experiments were performed on a Linux Pentium II 350Mhz with 192MB RAM, using the GNU C++ compiler, version 2.96 with O3 code optimization setting.

clearly several orders of magnitude faster than WARMR. One should however realize that in these experiments, WARMR was provided with a bias that forced Object Identity by adding inequality atoms. WARMR was not optimized for this. Part of the efficiency difference may also be due to the different programming language that was used (Prolog).

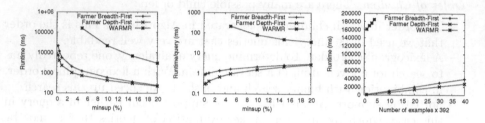

Fig. 3. Results on the Bongard dataset. Default dataset size is 392, $minsup = 5\%$.

Predictive Toxicology Evaluation Challenge (PTE). Execution times for PTE were published in [10], [6] and [7]; in these publications, labeled, *undirected* graphs were constructed from the atom and bond information; one searches for *connected* frequent subgraphs. To emulate the injective setup of gSpan and FSG, Object Identity is a necessity. To deal efficiently with connected, undirected graphs, the mode mechanism that was described in this article is not powerful enough. Therefore, we incorporated a more powerful declarative formalism based on *mode trees* in FARMER. Due to space limitations, we omit the details.

Table 1 and Fig. 4 display some execution times. We also show some execution times of other publications to set these into a perspective. Note that our algorithm runs on computers with relatively few memory, even though the ID lists augmented with variable assignments have to be stored in main memory.

Table 1. Comparison of execution times on the PTE dataset for $minsup \in \{6\%, 7\%\}$.

Machine	Algorithm	6% (s)	7% (s)
Intel Pentium III 500Mhz 448MB	gSpan [10]	5s	
AMD Dual Athlon MP1800+ 2GB	FSG Iterative Partitioning [7]	11s	7s
AMD Athlon XP1600+ 265MB	FARMER	72s	48s
Intel Pentium II 350Mhz 192MB	FARMER	224s	148s
Intel Pentium III 500Mhz 448MB	FSG [10]	248s	
AMD Dual Athlon MP1800+ 2GB	FSG Inverted index [7]	675s	23s
Intel Pentium III 650Mhz 2GB	FSG [6]		600s

We may conclude that our algorithm does not reach the state-of-the-art performance of gSpan. Compared to other graph mining algorithms, its performance is reasonable. We could easily compute all frequent subgraphs down to a support of 3%. The performance of gSpan is hard to obtain with the more general setup

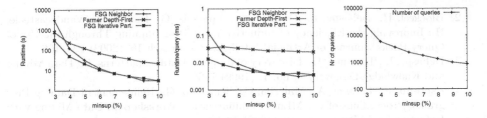

Fig. 4. Results on the PTE dataset. FARMER was run on an AMD Athlon XP1600+. For FSG results published [7] for an AMD Dual Athlon MP1800+ are used.

that we are dealing with. For example, gSpan orders the labels on the vertices first and performs a depth-first search to discover the graphs with the first label first. Next, all vertices with the first label are removed from the database, and the process is repeated for the remaining graphs. In general, this optimization is harder to apply. Therefore, one could better use gSpan if one is exactly searching for the kind of patterns that gSpan is optimized for.

Mutagenesis. The Mutagenesis dataset is very similar to the PTE dataset and was also used in [2]. We use it to compare FARMER with WARMR without Object Identity. Using a minimum support of 20%, WARMR discovers 91 frequent queries in 207s (of which 205s are spent while generating candidates). On the same Intel Pentium II FARMER discovers 1075 frequent queries in 73s. The different number of queries is due to the fact that WARMR does not discover graphs like $C - C - C$, as these are equivalent to $C - C$ without Object Identity. The set of queries found by FARMER is a proper superset of those found by WARMR.

6 Conclusion

In this article we presented an efficient algorithm for discovering frequent queries. We used Object Identity and a tree data structure to introduce several optimizations. Experiments showed that the algorithm outperforms WARMR and is comparable with some more specialized algorithms, but is not as efficient as recently published graph mining algorithms.

Acknowledgements

We are grateful to the Artificial Intelligence research group of the KU Leuven and to Xifeng Yan for their help with some experiments.

References

1. Agrawal, R., Manilla, H., Srikant, R., Toivonen, H., Verkamo, A.: Fast Discovery of Association Rules. In: U.M. Fayyad et al. (eds). Advances in Knowledge Discovery and Datamining. AAAI/MIT Press (1996) 307–328.

2. Blockeel, H., Dehaspe, L., Demoen, B., Janssens, G., Ramon, J., Vandecasteele, H.: Improving the Efficiency of Inductive Logic Programming Through the Use of Query Packs. Journal of Artificial Intelligence Research 16. (2002) 135–166.
3. Dehaspe, L., Toivonen, H.: Discovery of frequent Datalog patterns. In: Data Mining and Knowledge Discovery 3, no. 1. (1999) 7–36.
4. Esposito, F., Laterza, A., Malerba, D., Semeraro, G.: Refinement of Datalog Programs. Proceedings of the MLnet Familiarization Workshop on Data Mining with Inductive Logic Programming. (1996) 73–94
5. Ferilli, S., Fanizzi, N., Mauro, N., Basile, T.: Efficient θ-subsumption under Object Identity. In: Proceedings of the ICML'02. (2002)
6. Kuramochi, M., Karypsis, G.: Frequent Subgraph Discovery. In: Proceedings of the ICDM'01. (2001) 313–320
7. Kuramochi, M., Karypsis, G.: An Efficient Algorithm for Discovering Frequent Subgraphs. Technical Report 02-026, University of Minesota. (2002).
8. Nijssen, S., Kok, J.N.: Faster Association Rules for Multiple Relations. Proceedings of the IJCAI'01. (2001) 891–897
9. http://www.liacs.nl/home/snijssen/farmer
10. Yan, X., Han, J.: gSpan: Graph-Based Substructure Pattern Mining. In: Proceedings of the ICDM'02. (2002)

Towards Behaviometric Security Systems: Learning to Identify a Typist

Mordechai Nisenson[2], Ido Yariv[1], Ran El-Yaniv[1], and Ron Meir[2]

[1] Department of Computer Science
{yariv@vipe,rani@cs}.technion.ac.il
[2] Department of Electrical Engineering
Technion - Israel Institute of Technology
{sm0ti@t2,rmeir@ee}.technion.ac.il

Abstract. We consider the problem of identifying a user typing on a computer keyboard based on patterns in the time series consisting of keyboard events. We develop a learning algorithm, which can rather accurately learn to authenticate and protect users. Our solution is based on a simple extension of the well known Lempel-Ziv (78) universal compression algorithm. A novel application of our results is a second-layer *behaviometric security system*, which continually examines the current user without interfering with this user's work while attempting to identify unauthorized users pretending to be the user. We study the utility of our methods over a real dataset consisting of 5 users and 30 'attackers'.

1 Introduction

Many security systems rely on a single log-on entry, typically a password, for access. Such systems can be compromised if the password is discovered, or is easy to attack. Greater security is achieved by relying on a physical means of identification, most often an access card (which may also include a One-Time-Pad to generate secure passwords). But if the card is lost it too could become a security risk. In general, all of these systems are vulnerable to an attacker co-opting a user's session; either by physically taking the place of the user, or by some exploitable weakness in the system. Recently, biometric methods have begun to appear in widespread use (see e.g. [1]). These typically rely on fingerprints or retinal structure. While some biometric security methods are considered rather safe, by and large these systems are only used for single log-on, and require additional hardware.

A different class of identification methods can rely on patterns appearing in a user's behavior when interacting with a machine. Possible examples could be driving a car, or interacting with a computer through typing, mouse control, navigation patterns, and so on. Such *behaviometric identification* is different from biometric identification in two respects[1]. On the one hand, behaviometric measurements can be intentionally biased (or corrupted) to some extent by

---

[1] The field dealing with measurements, theories and analysis of patterns in all aspects of human behavior is called *behaviometrics*.

N. Lavrač et al. (Eds.): PKDD 2003, LNAI 2838, pp. 363–374, 2003.

users who can control their behavior. On the other hand, unlike biometric measurements, behaviometric readings can be done on a continuous basis without interrupting or interfering with users' activities. This possibility allows for creating a secondary security system, which is continually operated after log-on. Such a system need not only be applicable to computers but to a wider array of other devices as well; conceivably any device with a sufficiently complex input system.

The aim of this study is to examine the question of whether a behaviometric security method can be automatically learned by a machine. In particular, we focus on the problem of *typist identification*. While being a particular instance of behavior, we believe that typing can represent some essential and general issues in behaviometric identification. Like other types of interactions with machines it is suggested that every person types differently, not only having to deal with typing method (e.g. touch typing), but more importantly with a person's physical and mental attributes. The size of one's hands, length of one's fingers, fine motor skills, language skills, and knowledge of keyboard layout could all come into play to affect how one types. Thus, identifying a typist is an interesting and challenging problem worthy of behavior analysis.

Our solution to the learning of typist classifiers is based on a number of simple ideas, which combine well into an effective method. We represent sequences of typing events as discrete sequences over finite (and rather small) alphabets, and then use universal prediction machines (based on known universal compression algorithms) to generate probabilistic behaviometric models for users. Using these models we then solve instances of *single-class* classification problems. We describe the method and evaluate its performance over a real dataset collected from various typists. Our examination provides a proof of concept indicating that automatic learning of behaviometric identification of typists is a feasible task.

2   Problem Setup and Preliminaries

With a security application in view (as mentioned above), we model the typist identification problem as a *single-class* classification problem where we have a training set of typing samples from one user u and we would like to construct a classifier capable of distinguishing new typing sequences generated by u from sequences generated by other users. Usually, in this single-class setting the other samples (not from u) are referred to as *outliers*.

In general, a single-class classification formulation is required whenever it is possible to acquire training examples of the target class (e.g. typing sequences of the user u) but hard or impossible to collect examples of the outliers (e.g. sequences of intruders). Thus, while the desired classifier is still binary and should discriminate between the target and the outliers, only one side of the boundary is supported by the data. Therefore single-class classification problems are harder (and much less studied) than standard binary classification problems (see also Section 6). The performance of a single-class classifier is best measured using standard statistical distinctions between error of the first type δ_1, giving the proportion of target samples which are classified as outliers, and error of the second type δ_2 measuring the proportion of outlier samples classified as target samples. A plausible requirement, in the context of security systems, is that the tradeoff between δ_1 and δ_2 is controlled by the user.

We now characterize more formally the typing sequences we consider. The output generated when a user operates a keyboard is a sequence of events which can be described as follows. Standard keyboards usually have 104 keys and the keyboard outputs *events* when keys are pressed and released. Let K be the set of keys on the keyboard (that is, $|K| = 104$). Each key can be in one of two states: *pressed* or *released*. Let $A = \{\text{press}, \text{release}\}$ be this set of states. A *keyboard event*, $e = (k, a)$, where $k \in K, a \in A$, occurs whenever a key is pressed or released. Let E be the set of all keyboard events. Clearly, $|E| = |K||A| = 2|K| = 208$. A sequence $e_1, e_2, \ldots, e_n$ of keyboard events is viewed as a *time series* $x_1, x_2, \ldots, x_n$ where $x_i = (e_i, t_i)$ and t_i is the time recorded for the event e_i. Any such finite time series of keyboard events is called a *sentence*.

3 Typist Identification via Universal Prediction

The proposed solution to typist identification is based on universal prediction algorithms for discrete sequences. In this section we first describe a transformation of input sentences into a suitable representation for the use of such prediction algorithms. We then describe the prediction algorithm, which is obtained by extending a standard Lempel-Ziv compression algorithm.

3.1 Representation via Quantized Time Differentials

As described above each input sample is a time sequence $(e_1, t_1), (e_2, t_2), \ldots, (e_n, t_n)$ of keyboard events. The exact times at which events take place are of little value. Of much greater interest is the time differential between two events. Not surprisingly (and as noted by others, e.g. [2]), these differentials contain much of the discriminative information between typists. Setting $\Delta_i = t_{i+1} - t_i$, we transform the sentence into a sequence of its differentials so that $(e_i, t_i) \rightarrow (e_i, \Delta_i, e_{i+1})$. The resulting sequence of triplets consisting of events and differentials faithfully represents the time transitions between events which are relevant to typist discrimination. However, we choose to use the following slightly different differential representation which can be uniquely determined from a triplet sequence.

$$(e_1, \Delta_1, e_2), \ldots, (e_{n-1}, \Delta_{n-1}, e_n) \leftrightarrow e_1, \Delta_1, e_2, \Delta_2, \ldots, e_{n-1}, \Delta_{n-1}, e_n.$$

This last representation (on the right-hand side) is simpler in the sense that it has a smaller "alphabet" size. However, while the number of events is finite the number of time differentials is not. First, unbounded differentials can be avoided by specifying that all values larger than a specific Δ_{max} represent the start of a new sentence (keystrokes that are minutes apart are unlikely to be related in any fashion). Second, for a given a set of sentences D_u, emitted by the typist u, which are to be learned, an additional transformation on the time differentials is performed with the goal of limiting the number of time differentials and smoothing over them. Fewer symbols make the data easier to learn, by reducing statistical sparseness (and thus reducing variance). To accomplish this, vector quantization [3] is used to cluster the time differentials into Q clusters, with Q centroids $c_1, c_2, ...c_Q$. The time differentials then undergo the following transformation:

$$\Delta \Rightarrow c^* \quad \text{where} \quad c^* = \underset{c_i}{\operatorname{argmin}} |\Delta - c_i|.$$

This transformation is used on all sentences that are to be learned or ranked by the user's model. Thus, the final makeup of a sentence is $\{e_1, q_1, e_2, q_2, ..., q_{n-1}, e_n\}$, where $q_i \in \{c_1, \ldots, c_Q\}$ represents some time differential Δ. Therefore, the number of symbols in our alphabet is $|E| + Q$. Note that the number Q of centroids becomes a parameter of the algorithm.

3.2 Lempel-Ziv Universal Prediction

Having represented a keyboard event sequence as a sequence of discrete symbols over a finite alphabet, we can now use any universal prediction algorithm for discrete sequences to generate conditional likelihood estimates of unseen sequences. Specifically, given a set D_u of training sentences for user u, we use a universal prediction algorithm to train a model M_u which is then capable of estimating $\Pr(x|D_u)$, the conditional probability distribution of an unseen sentence x. Using such conditional estimates we then solve the single-class classification problem.

There are a number of universal prediction algorithms whose empirical performance for lossless text compressions is considered state-of-the-art. Notable examples are the context tree weighting method (CTW) [4], the Burrows-Wheeler Transform (BWT) (see e.g. [5]) and variants of Prediction by Partial Matching (PPM) [6]. For simplicity and for computational efficiency we compromise likelihood estimation accuracy and rely on the Lempel-Ziv algorithm (lz78 [7]). In particular, we use the prediction component of the lz78 algorithm as described in [8]. Besides being very simple and fast this algorithm enjoys performance guarantees of various types (see e.g. [9]). We also propose two improvements to the algorithm, which appear to increase its prediction accuracy.

The lz78 Universal Prediction is a one-pass algorithm. It builds a weighted tree from sequences over a finite alphabet, and can assign probability estimates to new sentences given such a tree. The lz78 phrase tree holds a "dictionary" of phrases parsed from the training sequence and is constructed by parsing input sequences as follows. At each stage the algorithm parses the smallest prefix which is not yet in the tree. For example, the string "ababbac" is parsed into: a, b, ab, ba, c. This set of phrases can be viewed as a phrase tree such that each parsed phrase is a path from the root to a leaf (see [8] for a detailed exposition).

As described in [8] the phrase tree can be extended to provide count statistics by adding a counter to each node. These count statistics can be used to calculate a probability estimate for traversing from a parent node to one of its children.

Given a set of sequences, $s_1, \ldots, s_k$, emitted by some source (say the user u), a parse tree with appropriate counter statistics can be constructed for all s_i (e.g. by concatenating the s_i into one long sequence). The resulting statistical model is denoted by M_u. M_u can be used to compute the conditional probability $\Pr(x|M_u) \approx \Pr(x|s_1, \ldots, s_k)$ of a new sequence x. This is done by traversing down from the root according to the letters of x, and multiplying the probability estimates of the traversals, until a leaf is reached. Then the traversal resumes from the root. In practice, the normalized (negative) log-likelihood is used,

$$V(x, M_u) = \frac{-\log_2 \Pr(x|M_u)}{|x|} . \tag{1}$$

This value is non-negative for all x and is 0 (for finite length strings) only when $Pr(x|M_u) = 1$, which is the ideal prediction for any sentence emitted by u.

3.3 Improvements to Standard LZ Prediction

A major advantage of the lz78 parsing technique is its speed. This speed is possible by compromising a systematic consideration of all substrings. While for very large training sets this compromise will not affect the results significantly, for small training sets (and short test sequences) this results in sparser and noisier statistics. We propose two simple modifications to the algorithm which increase the number of phrases extracted and improve performance of the lz78 estimation. The two modifications are termed *input shifting*, and *back-shift parsing*.

Input shifting is used during the learning process to extract more phrases from a sentence. Considering a sentence $x = x_1 x_2 \cdots x_n$, the sentence is parsed once as described above. Then it is parsed s more times, where in the ith additional parsing we parse the suffix $x_{i+1} x_{i+2} \cdots x_n$ in the usual way (but starting with the aggregated model constructed by previous parsings). The effect of input shifting is to increase the number of phrases thus making the phrase tree larger. As s grows so does the height of the tree as longer and longer phrases are parsed. Note that by taking $s = 0$ we leave the lz78 algorithm intact.

Another deficiency of the lz78 algorithm is the loss of context when parsing a sequence (and when calculating the likelihood of a sequence). Specifically, each time the algorithm returns to the root (see description in Section 3.2) after parsing a phrase in a sequence, the entire context consisting of previous symbols is lost. In order to remedy this, we propose a method which utilizes the last m letters parsed to provide a prior context for the next phrase (taking $m = 0$ leaves lz78 intact). This method, which we term *Back-shift parsing (BSP)* seeks to achieve this by back-shifting m letters after parsing each phrase. This approach is problematic for $m > 1$, however, since more letters may be back-shifted than parsed (which occurs often in practice). This seriously impedes progress and compromises speed, which is one of the advantages of lz78.

We prevent this by requiring that the
m letters come from the last phrase
parsed. This slight change is implemented by utilizing a "marker" as
described in Figure 1. The resulting
Back-Shift Parsing with a marker prevents back-tracking beyond the marker
thus guaranteeing rapid progress. The
overall effect is to quickly build a tree
with no path shorter than $m + 1$ in
length, or to make the tree deeper
while minimally affecting its width.

```
Initialization: marker = start of sentence

Repeat until no more phrases to be parsed:
    phrase = next phrase parsed
            (starting at marker)
    add phrase to dictionary
    if (length(phrase) > m)
        marker = marker + length(phrase) - m
```

Fig. 1. Pseudo-code for Back-Shift Parsing (BSP) with a marker.

BSP also affects the calculation of a probability estimate for an unseen sentence. Instead of returning to the root after traversing to a leaf, the last m letters
traversed are first traced down from the root to some node v, and then the new
traversal begins from v (if v does not exist, then the new traversal continues
from the root instead).

The modified algorithm now has two parameters and is denoted by $\mathtt{lz78}(s, m)$
where s determines the number of input shifts and m determines the context
length for back-shifting. The following example shows the parsed phrases generated by some $\mathtt{lz78}(s, m)$ algorithms for the sequence "ababbac". Note that the
phrases appear in the order of their parsing.

Algorithm	Phrases Parsed from "ababbac"
lz78(0,0)	{a,b,ab,ba,c}
lz78(1,1)	{a,ab,b,ba,abb,bac,c,bab,bb}
lz78(2,2)	{a,ab,aba,b,ba,bab,abb,bb,bba,bac,ac,babb,bbac,abba}

3.4 Single-Class Classification and Model Selection

Let $D_u = \{S_1, \ldots, S_n\}$ be a training set of sentences emitted by u. Given a fixed
choice of the parameters s and m we use the $\mathtt{lz78}(s, m)$ algorithm to build a
model $M_u = M(D_u, s, m)$ for the user u (thus, for a particular user, a model
corresponds to a choice of s and m). This model can provide likelihood estimates for unseen sentences. Given an unseen sequence x we should determine
whether $\Pr(x|M)$ is sufficiently large to "accept" x (alternatively, that $V(x, M_u)$
is sufficiently small; see Eq. (1)). To this end, a cutoff point, or threshold t, is
necessary. We determine a threshold using the following leave-one-out methodology. For each training sentence S_i in D_u we calculate the likelihood of S_i given
a model trained on D_u excluding S_i. More formally, for each $S_i \in D_u$ let

$$V_i = V\left(S_i, M(D_u \setminus \{S_i\}, s, m)\right) . \tag{2}$$

Let μ_M and σ_M be the empirical average and standard deviation of the V_i,
$i = 1, \ldots, n$. An ideal (but perhaps not achievable) threshold t places all (future) user's sentences below the cutoff and attackers' sentences above. Given
the evidence we have (the training sentences for u) we attempt to guarantee
results for the user by setting the threshold to $t(M) = \mu_M + k_\sigma \sigma_M$ where k_σ is

sufficiently large. Using Chebyshev's inequality, for any k_σ we can provide for u a *confidence level* as follows. For any random variable X whose mean and standard deviation are μ and σ, respectively, a one-tailed version of Chebyshev's inequality [10] states that for any $k \geq 0$, $\Pr\{X - \mu \geq k\sigma\} \leq \frac{1}{1+k^2}$. Considering future sentences emitted by u as observations of a random variable S and taking μ_M and σ_M as estimates of the true mean and standard deviation of the random variable $V(S, M(s, m, D_u))$, we have for any choice k_σ (and using $t(M) = \mu_M + k_\sigma \sigma_M$), $\Pr\{V(S, M(s, m, D_u)) > t(M)\} \leq \frac{1}{1+k_\sigma^2}$. Thus the confidence level is $1 - \delta = 1 - 1/(1 + k_\sigma^2) = k_\sigma^2/(1 + k_\sigma^2)$.

To summarize, our typist identification algorithm has four parameters: Q, the quantization level; m and s, the parameters of the improved 1z78 algorithm; and k_σ, which determines acceptance threshold. Our goal is to set values to these parameters based only on the training set D_u.

Within a minimax setting, we choose the best model which maximizes the likelihood of the "hardest" training sentence. Specifically, we take

$$M_u^* = \underset{M}{\arg\min} \left\{ \max_{S_i \in D_u} V_i \right\} \tag{3}$$

This optimization determines values for the parameters Q, m and s. The parameter k_σ is set such that the maximum V_m value in (3) is just below the threshold $t(M_u^*)$ and will be accepted by the model. Specifically, $\max_{S \in D_u} V_M(S, D_u) = \mu_M + k_\sigma \sigma_M$ and solving for k_σ we get

$$k_\sigma = \frac{\max_{S \in D_u} V_{M_u^*}(S, D_u) - \mu_{M_u^*}}{\sigma_{M_u^*}}.$$

In addition to the above single-class setting we also consider a setting where a (small) set of "attacker" sentences is available for training. Clearly, if such a set of "outliers" is not very large, it is not likely to faithfully represent the general statistics of outliers. However, it is interesting to investigate whether this additional piece of information can be exploited to improve performance. Although this problem is typically not a standard two-class problem, for the rest of the paper we call this setting the 'two-class' setting. Denote by D_a the set of attacker sentences available for training. For each model M, let $t_M(k_\sigma) = \mu_M + k_\sigma \sigma_M$ where μ_M and σ_M are estimated as described above. The *accuracy* of the model M with respect to the decision threshold is given by the ratio of the number of correctly classified strings to the total number of strings,

$$A(M, k_\sigma) = \frac{|\{V(x, M) \mid v < t_M(k_\sigma), x \in D_u\}| + |\{a \mid V(x, M) \geq t_M(k_\sigma), x \in D_a\}|}{|D_u| + |D_a|}$$

Let ε be any limit on the desired accuracy ($0 \leq \varepsilon \leq 1$), The *robustness* $R_\varepsilon(M)$ of the model M is defined as

$$R_\varepsilon(M) = \int_{k>0 \,:\, A(M,k) \geq 1-\varepsilon} A(M, k)\, dk.$$

That is, the robustness is the area below the accuracy curve viewed as a function of threshold magnitude. Note that in practice the robustness can be rather accurately estimated using the average accuracy of the model over a number of suitable k_σ representatives. Figure 2 depicts the accuracy curves of various $\mathtt{lz78}(s, m)$ models. The areas enclosed by these curves and the 90% asymptote are the ε-robustness values of these models (with $\varepsilon = 0.1$).

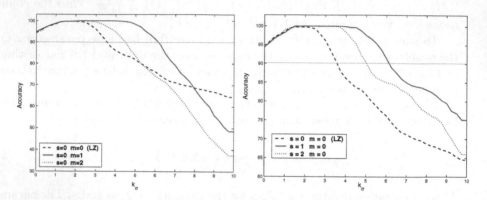

Fig. 2. Accuracy as a function of threshold magnitude for various $\mathtt{lz78}(s, m)$ models. Areas above the 90% asymptote are robustness values. For example, $R_{0.1}(\mathtt{lz78}(0, 1))$ and $R_{0.1}(\mathtt{lz78}(1, 0))$ are the largest robustness values in the left and right panels, respectively.

The model M_u^* which maximizes robustness is selected in this two-class setting and k_σ is set to maximize accuracy as measured over the training set $D_u \cup D_a$. That is, $k_\sigma = \mathrm{argmax}_k A(M_u^*, k)$. Note that there may be more than a single value which gives the maximum accuracy. In this case, there may be several peaks in the accuracy curve. We note however that in practice a single broad peak is typically observed. Whenever there is more than one maximum, we heuristically choose the threshold as the midpoint of the widest peak.

4   Dataset and Experimental Setup

For evaluating the proposed algorithms a dataset of keyboard event sentences was collected from 5 users and 30 attackers. We note that the recording of keyboard events including their precise time stamps is not straightforward using user-level programs on most standard operating systems. Thus, a suitably adapted system was constructed including a modified keyboard interrupt service routine[2]. Each of the users and attackers typed several sentences. The user input sequences were on average longer than the attackers' input. The text typed by

---

[2] In particular, a Linux system was used with all non-essential modules and services removed or disabled. System calls were used to request **unbuffered** keyboard events.

users corresponded to answers to open ended questions (e.g. "What did you do today?") and to a specific sentence ("To be or not to be. That is the question."). Additionally a (completely) free text section was also allowed. On average, each user recorded 2551 ± 1866 keystrokes. Each of the thirty "attackers" was asked two open ended questions, and was required to type the specific sentence "To be or not to be. That is the question." They were also allowed to type in free text. On average 660 ± 597 keystrokes were logged for an attacker[3]. To maximize the utility of this dataset, the sentences, both before learning and before testing, were split into segments of 100 keystrokes (arbitrarily set). Additionally, all of the attackers' sentences (120 in total) were used to attack each model selected.

We selected a set of "feasible" parameter values for the models[4]. To maximize evaluation accuracy we used the following leave-one-out protocol: For each user u, each of the sentences in D_u was in turn selected to be in the test set and the rest of the sentences remained in the training set. Once a model was selected, it was tested if the model can identify and accept the left out sentence.

For the two-class problem, where we wish to see if providing attacker data can improve performance, the attackers were partitioned into two groups: a group of 10 attackers to be used for training (40 sentences), and a group of 20 attackers for testing (80 sentences). Other than the partitioning of the attackers, testing was identical to that of the single-class case, although the 10 attackers used for training were not used for testing. One hundred cross-validation folds were made.

5   Experimental Results

We begin by considering the results obtained for the single-class setting. Table 1 specifies the results in the single-class setting. As can be seen, impressive performance can be achieved by the system. The system performs well even when limited information is available (for example, user 5), though performance, particularly in self identification, does slightly suffer. Table 2 shows the results for the two-class experiments. Performance, on average, was similar to the single-class results, though user 5 did have a marked decline in self identification success. This does not seem to be dependent on the amount of data available, as user 1's performance also dropped, though less significantly. Performance in terms of successfully defending did improve, however, achieving perfect scores for nearly all of the users, which resulted in a higher break-even point.

In addition, we examined the performance of the algorithm when models were restricted to use the "pure" `lz78` algorithm (i.e. the `lz78(0, 0)` model with Q and k_σ still variable was trained with the same methodology), both for the single-class and two-class problems. Due to space limitations we only report on the estimated break-even points for these experiments which were 93.57 and 96.42 for the single-class and two-class problems, respectively. These results indicate

---

[3] The complete dataset will be available at
  http://www.cs.technion.ac.il/~rani/typist.

[4] The particular values we tested are $Q = 80, 90, 100, 110, 120$; $m = 0, 1, 2, 3, 4$; $s = 0, 1, 2, 3, 4$; and $k_\sigma = 0, 0.25, 0.5, 0.75, \ldots, 10$.

Table 1. Single-Class Results: Individual users, averages and estimated break-even point (defined to be the harmonic mean of the averages).

· ···	· · ·· ··· ····	· · ··· ··· ···	· · ··· ··· ·· ····	· ··· ·· ·· ·	· · ··· ·	· ··· ·· ···· ·
1	5344	13	114	97.37 ± 2.79	1560	98.33 ± 1.08
2	4156	16	90	97.78 ± 8.53	1920	100.0 ± 0.63
3	1630	5	36	94.44 ± 11.65	600	99.67 ± 0.41
4	1076	5	23	91.3 ± 10.85	600	99.33 ± 0.97
5	548	5	14	92.86 ± 9.58	600	97.0 ± 3.82
Averages				· ··· · $\pm$ · ··· ·		· ··· · $\pm$ · ··· ·
Estimated Break-Even Point				· · ··· ·		

Table 2. Two-Class Results: Individual user, averages and break-even point. Results are across all 100 cross-validation folds.

· ···	· · ·· ··· ····	· · ··· ··· ···	· · ··· ··· ·· ····	· ··· ·· ·· ·	· · ··· ·	· ··· ·· ···· ·
1	5344	13	11400	93.86 ± 7.01	104000	100.0 ± 0.0
2	4156	16	9000	100.00 ± 0.0	128000	100.0 ± 0.0
3	1630	5	3600	97.22 ± 11.45	40000	100.0 ± 0.0
4	1076	5	2300	95.65 ± 11.23	40000	99.75 ± 0.5
5	548	5	1400	85.71 ± 17.2	40000	100.0 ± 0.0
Averages				· ··· · $\pm$ · ··· ·		· ··· · $\pm$ · ··
Estimated Break-Even Point				· · ··· ·		

that the lz78(s,m) modifications have a significant advantage in the single-class setting, particularly when there is little data available for training. For example, for users 4 and 5, the estimated break-even points for the "pure" lz78 algorithm are 90.2 and 80.9, respectively. With our improvements the values obtained for these users are 95.1 and 94.9, respectively.

6   Related Work

There is quite extensive literature on "keystroke dynamics" by attempting to identify characterizing features in keystroke sequences. One of the earliest works is [11], which introduce the use of "digraph times" in this context. For each pair of keys typed, its *digraph time* is the interval between the pressing of the first key and the pressing of the second. Many other works later use this basic idea or its extensions to "trigraphs", etc. Due to space limitations we limit the discussion here to two of the most recent papers, which present the most impressive results to-date. The work presented in [2] uses a combination of digraph times and keystroke latencies to generate feature vectors. Factor analysis is then used to select discriminative features. Using a nearest neighbor approach together with clustering, the authors examine the classification success rate of a number of distance functions. On a dataset consisting of 63 users, the best results are obtained using a Bayesian distance function. The stated results are approximately 92%. These results were obtained over a dataset where all users typed fixed text selections from "a list of phrases". There was also a free text component in this study though results are not presented and are stated to be inferior. The recent results of [12] consider again identifying typists of a fixed phrase. This phrase consists of 683 characters (which form 125 words). Using a fixed trigraph vector representation the authors obtain very high accuracy using a heuristic distance measure between trigraph vectors. Their best results for the single-class problem are 1.8% false alarm rate and 0.01% for "imposter pass" rate. The authors also test higher order "graphs" and experiment with subsets of the fixed phrase. While higher order graphs (e.g. 6-graphs) do not improve results, the use of sub-phrases can drastically increase the false alarm rate (e.g. by taking 1/4 of the phrase the false alarm rate increases to more than 12%). While these two works indicate that very high precision can be obtained in recognizing keystroke "signatures" over a fixed text, these methods fall short in handling free text, particularly when little data is available. The main contribution of the present work is in showing for the first time a new representation and algorithms that can attain very high accuracy also for *free text*. The results we obtain (e.g. over 96% break-even for the single-class authentication problem) enable a practical behaviometric security system for continual non-intrusive authentication, which can handle any text. These results are not directly comparable to the above results. However, when considering sample sizes and accuracy, it appears that our results may be significantly better than the results of [12]. Nevertheless, these other results are obtained with an impressive database consisting of typed sentences from 44 users and 110 attackers whereas our primarily free text dataset consists of 5 users and 30 attackers.

As noted previously, the more challenging and perhaps common setting for a security system as described here, is that of a single-class problem. This variant of binary classification has various other jargon names, such as: novelty detection outlier detection, one-class classification. For other approaches for setting the boundary in single-class problems see e.g. [13–15].

7   Conclusions and Future Work

We have introduced an approach to modeling keystroke dynamics of users based on using the universal Lempel-Ziv compression algorithm as a generator of the predictive distribution of future strings, based on statistics collected from an individual user. We use this predictive distribution in the context of single-class learning, where particular values for the augmented Lempel-Ziv algorithm are selected based on cross-validation. While previous work tended to focus on fixed representations based on N-graphs which can be considered to be fixed order Markov models, our representation allows for variable length contextual information. As a result of this, our statistical model is capable of retaining more robust statistics, possibly at the cost of increased space requirements.

Our method can be potentially improved in several ways. First, other universal prediction algorithms (such as CTW; see Section 3.2) could perhaps improve prediction accuracy, at the expense of speed. Such a compromise may be unacceptable for the particular application of continual non-intrusive authentication.

It may be interesting to investigate whether taking *relative* time-differentials (rather than absolute time-differentials) can improve performance, perhaps by a reduction of the variance caused by the variability in typing speeds of users. This direction is particularly promising when considering the successful technique of [12], which achieved impressive performance on a fixed text by ignoring absolute differential times (but utilizing the relative sizes of trigraph times).

While our results are impressive, they can only be viewed as a proof-of-concept due to the limited sample size, and the use of a single session for data acquisition. Finally, an advantage of our techniques is that they are not specifically targeted to the keyboard, and can be easily extended to other devices.

References

1. V. Matyas Jr. and Z. Riha. Biometric authentication systems. Technical report, ECOM-MONITOR, 2000.
2. F. Monrose and A.D. Rubin. Keystroke dynamics as a biometric for authentication. *Future Generation Computer Systems*, 16(4):351–359, 2000.
3. A. Gersho and R.M. Gray. *Vector Quantization and Signal Compression*. Kluwer Academic Publishers, Boston, 1992.
4. F. M. J. Willems, Y. M. Shtarkov, and Tj. J. Tjalkens. The context-tree weighting method: basic properties. *IEEE Trans. Info. Theory*, pages 653–664, 1995.
5. G. Manzini. The burrows-wheeler transform: Theory and practice. In *Symposium on Mathematical Foundations of Computer Science (MFCS '99)*, volume 1672, pages 34–47. Springer Verlag Lecture Notes in Computer Science, 1999.
6. J. G. Cleary and W. J. Teahan. Unbounded length contexts for PPM. *The Computer Journal*, 40(2/3):67–75, 1997.
7. J. Ziv and A. Lempel. Compression of individual sequences via variable rate coding. *IEEE Transactions on Information Theory*, 24:530–536, 1978.
8. G.G. Langdon. A note on the lempel-ziv model for compressing individual sequences. *IEEE Transactions on Information Theory*, 29:284–287, 1983.
9. M. Feder. Gambling using a finite state machine. *IEEE Transactions on Information Theory*, 37:1459–1465, 1991.
10. D.R. Stirzaker G.R. Grimmett. *Probability and Random Processes*. Oxford University Press, third edition, 2002.
11. R. Gaines, W. Lisowski, , S. Press, and W. Shapiro. Authentication by keystroke timing: Some preliminary results. Report R-256-NSF, Rand Corp., 1980.
12. F. Bergadano, D. Gunetti, and C. Picardi. User authentification through keystroke dynamics. *ACM Transactions on Information and System Security*, 5(4):367–397, 2002.
13. C. Bishop. Novelty detection and neural network validation. *IEEE Proceedings on Vision, Image and Signal Processing*, 141(4):217–222, 1994.
14. N. Japkowicz. *Concept-Learning in the absence of counterexamples: an autoassociation-based approach to classification*. PhD thesis, Rutgers, New Brunswick, 1999.
15. D.M.J. Tax. *One-Class Classification*. PhD thesis, The Delft University of Technology, 2001.

Efficient Density Clustering Method for Spatial Data

Fei Pan, Baoying Wang, Yi Zhang, Dongmei Ren, Xin Hu, and William Perrizo

Computer Science Department
North Dakota State University
Fargo, ND 58105
{fei.pan,baoying.wang,yi.zhang,dongmei.ren,xin.hu,
William.perrizo}@ndsu.nodak.edu
Tel: (701) 231-6403/6257
Fax: (701) 231-8255

Abstract. Data mining for spatial data has become increasingly important as more and more organizations are exposed to spatial data from sources such as remote sensing, geographical information systems, astronomy, computer cartography, environmental assessment and planning, etc. Recently, density based clustering methods, such as DENCLUE, DBSCAN, OPTICS, have been published and recognized as powerful clustering methods for data mining. These approaches have run time complexity of $O(n \log n)$ when using spatial index techniques, R^+ tree and grid cell. However, these methods are known to lack scalability with respect to dimensionality. In this paper, a unique approach to efficient neighborhood search and a new efficient density based clustering algorithm using EIN-rings are developed. Our approach exploits compressed vertical data structures, Peano Trees (P-trees[1]), and fast P-tree logical operations to accelerate the calculation of the density function within EIN-rings. This approach stands in contrast to the ubiquitous approach of vertically scanning horizontal data structures (records). The average run time complexity of our algorithm for spatial data in d-dimension is $O(dn\sqrt{n})$. Our proposed method has comparable cardinality scalability with other density methods for small and medium size of data, but superior speed and dimensional scalability.

1   Introduction

With the rapid growth of large quantities of spatial data collected in various application areas, such as remote sensing, geographical information systems, astronomy, computer cartography, environmental assessment and planning, efficient spatial data mining methods are in great demand. Density based cluster algorithms have been widely used in the mining of large spatial data. Density based cluster algorithms group the attribute objects into a set of connected dense components separated by regions of low density. A cluster is regarded as a connected dense region of objects, which grows in any direction that density leads. Density based cluster algorithms have

---

[1] Patents are pending on the P-tree technology. This work is partially supported by GSA Grant ACT#: K96130308.

N. Lavrač et al. (Eds.): PKDD 2003, LNAI 2838, pp. 375–386, 2003.

been recognized as a powerful clustering approach capable of discovering arbitrary shape of clusters as well as dealing with noise and outliers for spatial data mining.

There are two major approaches for density-based methods. The first approach is represented by DENCLUE [3]. It exploits a density function, e.g., step function or Gaussian function to measure the density in attribute metric space. Clusters are identified by determining corresponding density attractors. Thus, clusters of arbitrary shape can be easily determined by overall density functions. This algorithm scales well with run time complexity $O(n \log n)$ by means of grid cells techniques. However, it requires careful selection of the density parameter σ and noise threshold ξ, which may significantly influence the quality of the clustering results [10].

The second approach calculates the density of all data points and groups them based on density connectivity. Typical algorithms in this approach include DBSCAN [6] and OPTICS [8]. DBSCAN first defines a core object as a set of neighbor points consisting of more than a specified number of data points. All the data points reachable within a chain of overlapping core objects define a cluster. The run time complexity of DBSCAN is $O(n \log n)$ for spatial data when using a spatial index. Otherwise, it is $O(n^2)$ [10]. OPTICS can be considered as an extension of DBSCAN without providing global density. It assumes each cluster has its own density parameter and uses a random variable to learn its probability distribution. It has the same run time complexity as DBSCAN, that is, $O(n \log n)$ if a spatial index is used and $O(n^2)$ otherwise.

However, the spatial index techniques, such as R tree, R^+ tree, and grid cell, are known to be suitable for low dimensional data sets. They perform well in 2-3 dimensions. In high dimensional spaces they exhibit poor behavior in the worst case and in typical cases as well [0]. The reason is that the data space becomes sparse at high dimensionalities causing the bounding regions to become large. In this paper, a unique approach to efficient neighborhood search using EIN-rings, and a new efficient density based clustering algorithm are developed. The center idea is to make use of P-trees and EIN-rings to calculate the density function in $O(\sqrt{n})$ time, on the average. Our approach exploits compressed vertical data structures, Peano Trees (P-trees), and fast P-tree logical operations to accelerate the calculation of the density function within EIN-rings. This approach stands in contrast to the ubiquitous approach of vertically scanning horizontal data structures (records). Furthermore, we adopt a *look around* pruning method to combine the density calculation and a hill climbing technique. The overall run time complexity is $O(dn\sqrt{n})$ for a d-dimensional data set, on the average. Experimental results show that the algorithm works efficiently on large-scale, high-dimensional spatial data, outperforming other density methods significantly.

This paper is organized as follows. In section 2, we first briefly review the basic P-trees, and then present a variation of P-tree, range predicate tree. In section 3, we define a unique equal interval neighborhood rings, EIN-rings, and then present the new efficient density clustering method using EIN-rings. Finally, we compare our method with other density methods experimentally in section 4 and conclude the paper in section 5.

2   Extended Peano Trees

A new tree structure, the Peano tree (P-tree), was developed to facilitate efficient data mining [1][2]. In this section, we first briefly review the basic P-trees, and then develop a new calculation method of a variation of P-tree, range predicate trees. In this paper, we use $\wedge$, $\vee$ and prime (') to denote P-tree operations AND, OR and NOT, respectively.

2.1   Review of Basic Peano Trees

A basic P-tree is a lossless, bitwise, vertical quadrant-based compressed tree, which can be 1-dimensional, 2-dimensional, 3-dimensional, etc. For a data set with d feature attributes, $X = (A_1, A_2 \ldots A_d)$, and the binary representation of jth feature attribute A_j as $b_{j,m} b_{j,m-1} \ldots b_{j,i} \ldots b_{j,1} b_{j,0}$, we strip each feature attribute into several files, one file for each bit position. Such files are called bit files. A bit file is then recursively partitioned into quadrants and each quadrant into sub-quadrants until the sub-quadrant is pure (entirely 1-bits or entirely 0-bits). The recursive raster ordering is called the Peano or Z-ordering in the literature – therefore, the name Peano tree.

We illustrate the detailed construction of P-trees using an example shown in Fig.1. The spatial data is the red reflective value of a 2-dimensional spatial data, which is shown in a). We represent the reflectance as binary values, e.g., $(7)_{10} = (111)_2$. Then strip them into three separate bit files, one file for each bit, as shown in b), c), and d). The corresponding basic P-trees, P_1, P_2 and P_3, are constructed by recursive partition, which are shown in e), f) and g).

As shown in e) of Fig.1, the root of P_1 tree is 36, which is the 1-bit count of the entire bit file-1. The second level of P_1 contains the 1-bit counts of the four quadrants, 16, 7, 13, and 0. Since quadrant 0 and quadrant 3 are pure, there is no need to partition these quadrants. Quadrant 1 and 2 are further partitioned recursively. We note here that we identify quadrants using a Quadrant identifier, Qid - the string of successive sub-quadrant numbers (01,2 or 3 in Z or Peano order, separated by "." (as in IP addresses). Thus, the Qid of the bolded and underlined quadrant in Fig.1 is 2.2.

AND, OR and NOT logic operations are the most frequently used P-tree operations. The P-tree logical operations are performed level-by-level starting from the root level. They are commutative and distributive, since they are simply pruned bit-by-bit operations. For instance, ANDing a pure-0 node with anything results in a pure-0 node, ORing a pure-1 node with anything results in a pure-1 node.

2.2   Range Predicate Trees

Range predicate tree, $P_{x \prec y}$, is a basic P-tree that satisfies predicate $x \prec y$, where y is a boundary value, and $\prec$ is the comparison operator, i.e., $<, >, \geq$, and $\leq$. Without loss of generality, we only present the calculation of range predicate $P_{A > c}$, $P_{A \leq c}$ and their proof as follows.

Lemma 1. Complement Rule of P-tree Let P_1, P_2 be basic P-trees, and P_1' is the complement P-tree of P_1, then $P_1 \vee (P_1' \wedge P_2) = P_1 \vee P_2$.

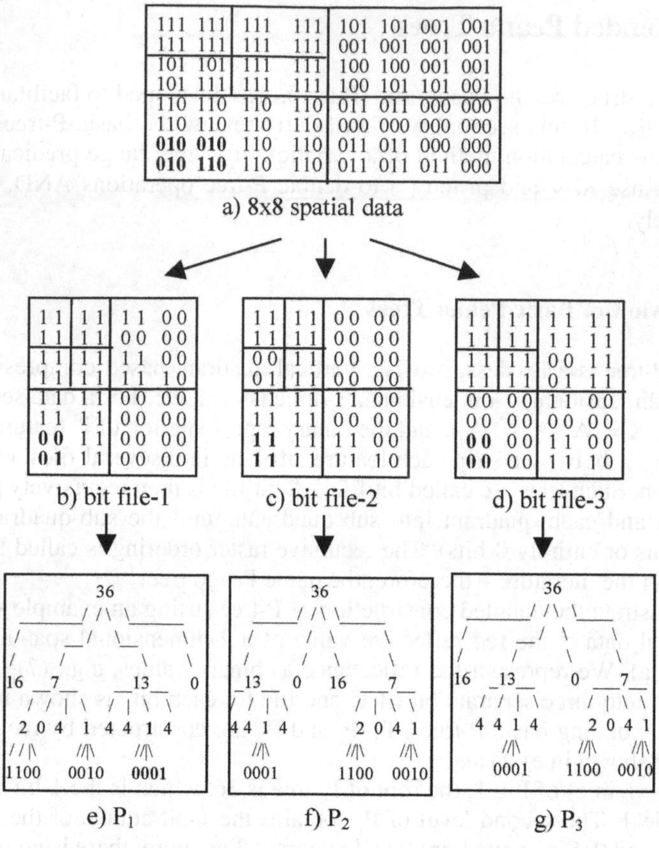

```
111 111 111  111 101 101 001 001
111 111 111  111 001 001 001 001
101 101 111  111 100 100 001 001
101 111 111  111 100 101 101 001
110 110 110  110 011 011 000 000
110 110 110  110 000 000 000 000
010 010 110  110 011 011 000 000
010 110 110  110 011 011 011 000
```

a) 8x8 spatial data

```
11 11 11 00        11 11 00 00        11 11 11 11
11 11 00 00        11 11 00 00        11 11 11 11
11 11 11 00        00 11 00 00        11 11 00 11
11 11 11 10        01 11 00 00        11 11 01 11
11 11 00 00        11 11 11 00        00 00 11 00
11 11 00 00        11 11 00 00        00 00 00 00
00 11 00 00        11 11 11 00        00 00 11 00
01 11 00 00        11 11 11 10        00 00 11 10
```

b) bit file-1   c) bit file-2   d) bit file-3

```
         36                  36                  36
      / /\ \               / /\ \               / /\ \
     /  /  \  \           /   /  \   \         /   / \   \
   16  7   13   0       13   0  16    7      16  13   0   7
   //\  //\ //\         //\    //\ //\      //\   //\  //\
  2 0 4 1 4 4 1 4      4 4 1 4 2 0 4 1     4 4 1 4 2 0 4 1
  //\   //\  //\        //\    //\ //\      //\   //\  //\
 1100  0010 0001       0001   1100 0010    0001  1100 0010
```

e) P_1 f) P_2 g) P_3

Fig. 1. Construction of 2-D Basic P-trees for Spatial Data.

Proof:

$P_1 \vee (P_1' \wedge P_2)$

(according to the distribution property of P-tree operations)

$= (P_1 \vee P_1') \wedge (P_1 \vee P_2)$

$=$ True $\wedge (P_1 \vee P_2)$

$= P_1 \vee P_2$

Proposition 1. Let A be j^{th} attribute of data set X, m be its bit-width, and $P_m, P_{m-1}, \dots P_0$ be the basic P-trees for the vertical bit files of A. Let $c = b_m \dots b_i \dots b_0$, where b_i is i^{th} binary bit value of c, and $P_{A>c}$ be the predicate tree for the predicate A>c, then

$$P_{A>c} = P_m \, op_m \, \dots \, P_i \, op_i \, P_{i-1} \, \dots \, op_{k+1} \, P_k, \quad k \leq i \leq m, \qquad (1)$$

where 1) op_i is $\wedge$ if $b_i=1$, op_i is $\vee$ otherwise, 2) k is the rightmost bit position with value of "0", i.e., $b_k=0$, $b_j=1$, $\forall j<k$, and 3) the operators are right binding. Here the

right binding means operators are associated from right to left, e.g., P_2 op$_2$ P_1 op$_1$ P_0 is equivalent to $(P_2$ op$_2$ $(P_1$ op$_1$ $P_0))$.

Proof (by induction on number of bits):

Base case: without loss of generality, assume $b_0=1$, then need show $P_{A>c} = P_1$ op$_1$ P_0 holds. If $b_1=1$, obviously the predicate tree for $A>(11)_2$ is $P_{A>c} = P_1 \wedge P_0$. If $b_1=0$, the predicate tree for $A>(01)_2$ is $P_{A>c} = P_1 \vee (P_1' \wedge P_0)$. According to Lemma 1, we get $P_{A>c} = P_1 \vee P_0$ holds.

Inductive step: assume $P_{A>c} = P_n$ op$_n$ ... P_k, we need to show $P_{A>c} = P_{n+1}$op$_{n+1}$$P_n$ op$_n$ ...P_k holds. Let $P_{right} = P_n$ op$_n$ ... P_k, if $b_{n+1}=1$, then obviously $P_{A>c} = P_{n+1} \wedge P_{right}$. If $b_{n+1}= 0$, then $P_{A>c} = P_{n+1} \vee (P'_{n+1} \wedge P_{right})$. According to Lemma 1, we get $P_{A>c} = P_{n+1} \vee P_{right}$ holds.

Proposition 2. Let A be j^{th} attribute of data set X, m be its bit-width, and P_m, P_{m-1}, ... P_0 be the basic P-trees for the vertical bit files of A. Let $c=b_m...b_i...b_0$, where b_i is i^{th} binary bit value of c, and $P_{A \le r}$ be the predicate tree for $A \le c$, then

$$P_{A \le c} = P'_m op_m ... P'_i op_i P'_{i-1} ... op_{k+1} P'_k, \quad k \le i \le m, \tag{2}$$

where 1). op$_i$ is $\wedge$ if $b_i=0$, op$_i$ is $\vee$ otherwise, 2) k is the rightmost bit position with value of "0", i.e., $b_k=0$, $b_j=1$, $\forall$ j<k, and 3) the operators are right binding.

Proof (by induction on number of bits):

Base case: without loss of generality, assume $b_0=0$, then need show $P_{A \le c} = P'_1$ op$_1$ P'_0 holds. If $b_1=0$, obviously the predicate tree for $A \le (00)_2$ is $P_{A \le c} = P'_1 \wedge P'_0$. If $b_1=1$, the predicate tree for $A \le (10)_2$ is $P_{A \le c} = P'_1 \vee (P_1 \wedge P'_0)$. According to Lemma 1, we get $P_{A \le c} = P'_1 \vee P'_0$ holds.

Inductive step: assume $P_{A \le c} = P'_n$ op$_n$ ... P'_k, we need to show $P_{A \le c} = P'_{n+1}$op$_{n+1}$$P'_n$ op$_n$ ...P'_k holds. Let $P_{right} = P'_n$ op$_n$ ... P'_k, if $b_{n+1}=0$, then obviously $P_{A \le c} = P'_{n+1} \wedge P_{right}$. If $b_{n+1}= 1$, then $P_{A \le c} = P'_{n+1} \vee (P_{n+1} \wedge P_{right})$. According to Lemma 1, we get $P_{A \le c} = P'_{n+1} \vee P_{right}$ holds.

Theorem 1. Complement Rule Let A be j^{th} attribute of data set X, $P_{A \le c}$ and $P_{A>c}$ are the predicate tree for $A \le c$ and $A>c$, where c is a boundary value, then $P_{A \le c} = P'_{A>c}$.

Proof: It is obvious. This theorem can be exploited to reduce computation time of predicate trees.

3   The EIN-Ring Based Density Clustering Algorithm

In this section, we present an EIN-ring based Density Clustering approach (EDC). We first define neighborhood rings and equal interval neighborhood ring (EIN-ring), and then describe the approach of calculation of EIN-ring using P-trees. In section 3.2, we describe calculation of the density function using EIN-rings. In section 3.3, the algorithm for finding density attractors is developed. Finally, the efficiency of our algorithm is analyzed in terms of time complexity.

3.1   EIN-Ring Based Neighborhood Search

Definition 1. The Neighborhood Ring of data point c with radii r_1 and r_2 is defined as the set $R(c, r_1, r_2) = \{x \in X \mid r_1 < |c-x| \le r_2\}$, where $|c-x|$ is the distance between x and c.

Definition 2. The Equal Interval Neighborhood Ring of data point c with radii r and fixed interval λ is defined as the neighborhood ring $R(c, r, r+\lambda) = \{x \in X \mid r < |c-x| \le r+\lambda\}$, where $|c-x|$ is the distance between x and c.

The interval λ is a user-defined parameter based on accuracy requirements. The higher the accuracy requirement, the smaller the interval. For $r = k\lambda$, k=1,2,…, the rings called the kth *EIN-rings*. Fig.2 shows 2-D EIN-rings with k = 1, 2, and 3.

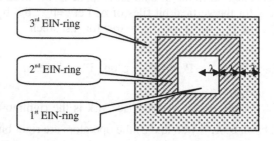

Fig. 2. Diagram of EIN-rings.

The calculation of neighbors within EIN-ring $R(x, r, r+\lambda)$ is as follows. Let $P_{r,\lambda}$ be the P-tree representing data points within EIN-ring $R(x, r, r+\lambda)$. We note $P_{r,\lambda}$ is just the predicate tree corresponding to the predicate $x-r-\lambda < X \le x-r$ or $x+r < X \le x+r+\lambda$. We first calculate the data points within neighborhood ring $R(x, 0, r)$ and $R(x, 0, r+\lambda)$ by $P_{x-r \le X \le x+r}$ and $P'_{x-r-\lambda \le X \le x+r+\lambda}$ respectively. $P_{x-r < X \le x+r}$ is shown as the shadow area of a) and $P'_{x-r-\lambda < X \le x+r+\lambda}$ is the shadow area of b) in Fig.3. The data points within the EIN-ring $R(x, r, r+\lambda)$ are those that are in $R(x, 0, r+\lambda)$ but not in $R(x, 0, r)$. Therefore $P_{r,\lambda}$ is calculated by the following formula, which is the shadow area shown in c) of Fig.3.

$$P_{r,\lambda} = P_{x-r-\lambda < X \le x+r+\lambda} \wedge P'_{x-r < X \le x+r} \qquad (3)$$

3.2   Calculation of the Density Function Using EIN-Ring

Density based clustering algorithm is a clustering method based on a set of density distribute function, called an influence function, which describes the impact of a data point within its neighborhood. Our algorithm employs a special EIN-ring based influence function. The overall density of the data space is then modeled as the sum of the influence functions of all data points. Clusters are determined by identifying density attractors, where density attractors are local maxima of the overall density function.

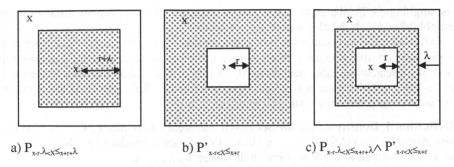

a) $P_{x-r-\lambda < x \le x+r+\lambda}$ b) $P'_{x-r < x \le x+r}$ c) $P_{x-r-\lambda < x \le x+r+\lambda} \wedge P'_{x-r < x \le x+r}$

Fig. 3. Calculation of Data Points within EIN-ring $R(x, r, r+\lambda)$.

Let x and y be data points in F^d, a d-dimensional feature space. The influence function of the data point y on x is a function $f_H^y: F^d -> R_0^+$, which is defined based on EIN-ring:

$$f_{r1,r2}^{y}(x)=1 \qquad \text{if } y \in R(c, r_1, r_2)$$

$$=0 \qquad \text{if } y \notin R(c, r_1, r_2) .$$

(4)

The EIN-ring based density function of x is defined as the weighted summation of $RC(x,r)$, which is calculated as follows

$$f_h^D(x) = \sum_{r=1}^{m} w_r * f_{r1,r2}^{y}(x)$$

(5)

$$= \sum_{r=1}^{m} w_r * RC(x,r) .$$

where $f_h^D(x)$ denotes the EIN-ring based density of data point x, with respect to weights, w_r. The selection of this weight is based on a RBF kernel function of the radius of EIN-ring, such as Gaussian function, step function, etc.

3.3   Finding Density Attractors Using the *Look Around* Pruning Technique

Once the density of each data point is defined, the next step is to define *density attractors*, i.e., local maxima of the overall density function. Having a high density doesn't necessarily make a point a density attractor - it must have the highest density among its neighbors. Instead of using formal hill climbing as is done in DENCLUE [3], we adopt a simpler heuristic *look around* technique.

Algorithm 1. Look Around Pruning We first define a neighborhood as a ball of some chosen radius r. The number r can range from 0 to the maximal bit length of the attributes. After finding the density function D_x of a point, x, we compare that density with that of data points within its neighborhood. If it is greater than the density of all its neighbors, it is labeled as a new density attractor. Any old density attractor in that neighborhood is de-labeled as a density attractor.

After all the data points have gone through the process above, we have a set of *intermediate* density attractors. We compare each intermediate attractor's density with that of its nearest neighbor data point. If the former is less than the latter, the attractor is de-labeled. Otherwise, it is a final density attractor. This step finds attractors that are isolated and therefore should be removed as noise.

Definition 3. Density Attractor Set Given a sequence of points $x_1, x_2, \ldots x_n$, the Density Attractor Set DAS $(x_1, x_2, \ldots x_n)$ is a set of attractors produced by the look around algorithm applied to the data points in the order, $x_1, x_2, \ldots x_n$.

Definition 4. A data point x is *reachable* from data point y if $x \in R$ (y, 0, r), where r is the user-defined radius for the density clustering. If x is reachable from y, y is also reachable from x.

The *look around* pruning algorithm is robust, which means the clustering results are independent of data point treating order. The proof is given as follows.

Lemma 2. (Density Characterization Lemma) Data point y is a density attractor iff $Dy \geq Dz$, $\forall z \in R(y,0,r)$. If y is not a density attractor, $\exists z \in R(y, 0, r) \ni: Dz > Dy$.

Lemma 3. Given a data point y, and $Dy \geq Dz$, $\forall z \in R(y, 0, r)$, y is the density attractor independent of the order in which y and z are treated in the *look around* process.

Proof (Proof by contradiction):

Assume the statement is not true, i.e. y is an attractor, but $\exists z \in R(y, 0, r)$, $\ni: Dz > Dy$. If z is treated first and z is an attractor, then when y gets treated, y would not be an attractor (Lemma 3.2.1). If y is treated before z, y could be designated an attractor at that time. But when z gets treated, y will be de-labeled according to *look around* pruning algorithm 3.2.1. Therefore y is not an attractor. Contradiction!

Theorem 2. Given data set X in two different sequences: $\{x_{i1}, x_{i2}, \ldots x_{in}\}$ and $\{x_{j1}, x_{j2}, \ldots, x_{jn}\}$, then $DAS(x_{i1}, x_{i2}, \ldots x_{in}) = DAS(x_{j1}, x_{j2}, \ldots x_{jn})$.

Proof (Proof by contradiction):

Assume the statement is not true, i.e. $DAS(x_{i1}, x_{i2}, \ldots x_{in}) \neq DAS(x_{j1}, x_{j2}, \ldots x_{jn})$. That means $\exists x \in DAS(x_{i1}, x_{i2}, \ldots x_{in})$ but $x \notin DAS(x_{j1}, x_{j2}, \ldots x_{jn})$. According to $x \in DAS(x_{i1}, x_{i2}, \ldots x_{in})$ and Lemma 3.2.1, $\forall z \in R(x,0,r)$ $Dx \geq Dz$. Also according to $x \notin DAS(x_{j1}, x_{j2}, \ldots x_{jn})$ and Lemma 3.2.1, $\exists z \in R(x,0,r)$ $\ni: Dz > Dx$. Contradiction!

We illustrate the finding of density attractors using look around pruning algorithm as follows. Suppose Qid of data point X is 0.3.2 and $D_x = 250$. We need compare D_x with the neighbor's density. From the Px,σ, x has four neighbors with Qids of 0.0.2, 0.3.1, 2.3.0 and 2.3.3. If densities of these points are respectively 300, 0, 220 and 0, and 0.0.2 and 2.3.0 are labeled as density attractors. By comparing Dx with the maximal density of 0.0.2 and 2.3.0, 250 < max(300, 220), therefore we determine that x is not a density attractor. Otherwise if $D_x = 350$, 350 > max (300, 220), x is labeled as the new density attractor. The old density attractors 0.0.2 and 2.3.0 are de-labeled and will not be considered later. Finally, clusters are determined by the density attractors. The pseudo code of overall algorithm is shown in Fig. 4.

```
INPUT: P-tree Set P_{ij} for bit j and attribute i, HOBBit ring R(i, 0, σ)
OUTPUT: Density attractors
// P_{ij} – P-tree for attribute i and bit j;
// P'_j – Neighborhood P-tree;
// N - # of data points; n - # of attributes;
// flag[i] – label array of cluster center of data point i.
BEGIN
        FOR i=1 to N DO
            flag[i] ← 0
            P_i ← Pure1 P-tree, DENS[i] ← 0, PrevRC ← 0
            FOR h = 1 TO m - 1 DO
                P_v ← Pure1 P-tree
                FOR j = 1 TO n DO
                    GET b_{jh}[i]
                    IF b_{jh}[i] = 1
                                PX_{jh} ← P_{j,h}
                    ELSE
                                PX_{jh} ← P'_{jh}
                        P_v[h] ← P_v [h] & PX_{jh}

            END FOR
            P_i ← P_i & P_v [h]
            w [i] = h * 2^{h*n}
            DENS[i]← DENS[i]+ w * (RootCount(P_i )- PrevRC);

            PrevRC ← RootCount(P_i);

            IF h = m - σ
                    P_σ ← P_i
            END FOR
            IF DENS[i] > the density of attractors within neighborhood ,
                    flag[i] ← 1, clear the flags of its neighbors.

        END FOR
        // Final look around pruning to intermediate attractors
        FOR i = 1 to N DO
        IF flag[i] = 1DENS[i] < The density of the closest neighbor
                Clear its flag
        END FOR
```

Fig. 4. EIN-ring base density Clustering algorithm

3.4   Time Complexity Analysis

Let f be the fan-out of a P-tree and let n be the number of data points it represents. We first present some Lemmas on P-trees, and then derive the average run time complexity to be $O(n\sqrt{n})$.

Lemma 4. The number of level of P-tree k = log(f) n

Proof: The numbers of nodes in each level of P-trees are: 1, f, f^2, f^3, ... f^k. Obviously the leaf level k is n bits long, i.e. f^k = n. Thus k = log(f) n.

Lemma 5. The maximum number of nodes in P-tree in the worst case $\eta = (n - 1) / (f - 1)$.

Proof: Without compression, the total number of nodes is $\eta = 1 + f + f^2 + f^3 + \dots f^{k-1}$ $= (f^k - 1) / (f - 1)$. According to Lemma 3.3.1, $f^k = n$, we get
$\eta = (n - 1) / (f - 1)$

Lemma 6. Total number of nodes in a P-tree with a compression ratio of ρ ($\rho < 1$) is $\eta = 1 + (\rho^k * n - f) / (f * \rho - 1)$, where k is the number of levels of P-tree.

Proof: The numbers of nodes in each level of a P-tree with compression ratio ρ at level i is $f^i * \rho^{i-1}$., where i ranges from 1 to k.. For example, at level 2, there are $(f * \rho) * f = f^2 * \rho$ nodes. We get the total number of nodes in the case that the P-tree has a compression ratio of ρ as

$$
\begin{aligned}
\eta \quad &= 1 + f + f^2 * \rho + f^3 * \rho^2 + \dots + f^{k-1} * \rho^{k-2} \\
&= 1 + f * (f^{k-1} * \rho^{k-1} - 1) / (f*\rho - 1) \\
&= 1 + (f^k * \rho^k - f) / (f * \rho - 1) \\
&= 1 + (\rho^k * n - f) / (f * \rho - 1)
\end{aligned}
$$

Corollary 1. When $\rho = 0$, the total number of nodes in the P-tree is 1; when $\rho = 1$, the total number of nodes in the P-tree is $(n - f)(f - 1) + 1$. When $\rho = 0.5$ and $f = 4$, the total number of nodes in a P-tree with compression ratio η is

$$
\begin{aligned}
\eta \quad &= 1 + (4^k/2^k - 2 *4) / (4 - 2) \\
&= 1 + (4^{k/2} - 8) /2 \\
&= 1 + (\sqrt{n} - 8) /2
\end{aligned}
$$

Theorem 3. The average run time complexity of EDC with compression ratio 0.5 and fan-out 4 is $O(d*n * \sqrt{n})$, where d is the number of dimensions.

Proof: The P-tree ANDing operation is executed node by node when calculating the density. Each node ANDing is counted as one operation. For n data points in d-dimension, there are $d*m$ basic P-trees, here m is the maximal bit size of each dimension. The total run time to get density P-trees is $d*m*n*\eta$, where η is the total number of nodes of a P-tree.

For data sets with fan-out $f = 4$ and average compress rate $\rho = 0.5$, according to Corollary 3.4.1, the total number of nodes of a P-tree $\eta = 1 + (\sqrt{n} - 8) /2$. Therefore, the total time to get the density for n data points in d-dimension is $d*m*n * (1 + (\sqrt{n} - 8) /2)$. Thus, the average time complexity of density based clustering using P-tree with compression ratio 0.5 and fan-out of 4 is $O(d*n * \sqrt{n})$.

4   Experiment Evaluation

Our experiments were implemented in the C++ language on a 1GHz Pentium PC machine with 1GB main memory, running on Debian Linux 4.0. The test data includes the aerial TIFF image (with Red, Green and Blue band reflectance values), moisture, and nitrate map of the Oakes Irrigation Test Area in North Dakota. The data is prepared in five sizes, that is, 128x128, 128x256, 256x256, 256x512, 512x512. The data sets are available at [4]. We evaluate our proposed EIN-ring base density Clustering algorithm (EDC) with respect to scalability, which is tested by increasing number of data records and number of attributes.

In this experiment, we compare our proposed EDC with Density Function based Clustering method using Euclidian distance (DFC). The experiment was performed on the five different sizes of data sets. The average CPU run time of 30 runs is shown in Fig.5.

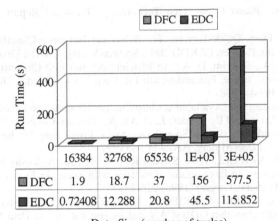

	16384	32768	65536	1E+05	3E+05
DFC	1.9	18.7	37	156	577.5
EDC	0.72408	12.288	20.8	45.5	115.852

Data Size (number of tuples)

Fig. 5. Running Time Comparison of EDC with other Density Clustering

From Fig. 5, we see that EDC method is much faster than all of them on these five data sets. Especially when the data set size increases, the time of EDC method increases at a much lower rate than other methods. The experiment results show that EDC method is fast and scalable for large spatial data set.

5   Conclusion

In this paper, a unique approach to efficient neighborhood search using EIN-rings, and a new efficient density based clustering algorithm are developed. Our approach exploits compressed vertical data structures, Peano Trees (P-trees), and fast P-tree logical operations to accelerate the calculation of the density function within EIN-rings. This approach stands in contrast to the ubiquitous approach of vertically scanning horizontal data structures (records). The overall run time complexity is $O(dn\sqrt{n})$

for a d-dimensional data set, on the average. Experimental results show that the algorithm works efficiently on large-scale, high-dimensional spatial data, outperforming other density methods significantly.

Our method is particularly useful for data streams. In data streams, such as large sets of transactions, remotely sensed images, multimedia video, etc., new data keeps on arrival continually. Therefore both speed and accuracy are critical issues. Achieving high speed using P-tree, and high accuracy using the weighted EIN-rings provides a density based clustering method that is well suited to the clustering of steam data. Besides spatial data, our method also has potential applications in other areas, such as DNA micro array and medical image analysis.

Reference

1. Perrizo, W.: Peano Count Tree Technology. Technical Report NDSU-CSOR-TR-01-1 (2001)
2. Khan, M., Ding, Q., & Perrizo, W.: K-Nearest Neighbor Classification on Spatial Data Streams Using P-Trees. PAKDD 2002, Spriger-Verlag, LNAI 2336 (2002) 517-528
3. Hinneburg, A., & Keim, D. A.: An Efficient Approach to Clustering in Large Multimedia Databases with Noise. Proceeding 4th Int. Conf. on Knowledge Discovery and Data Mining, AAAI Press (1998)
4. TIFF image data sets. Available at <http://midas-10cs.ndsu.nodak.edu/data/images/>.
5. Ester, M., Kriegel, H.P., Sander, J., & Xu, X.: Density-Connected Sets and their Application for Trend Detection in Spatial Databases. Proceeding 3rd Int. Conf. On Knowledge Discovery and Data Mining, AAAI Press (1997)
6. ESTER, M., KRIEGEL, H-P., SANDER, J. & XU, X.: A density-based algorithm for discovering clusters in large spatial databases with noise. In Proceedings of the 2nd ACM SIGKDD, Portland, Oregon (1996) 226-231
7. SANDER, J., ESTER, M., KRIEGEL, H.-P., & XU, X.: Density-based clustering in spatial databases: the algorithm GDBSCAN and its applications. In Data Mining and Knowledge Discovery, 2 (1998) 169-194.
8. ANKERST, M., BREUNIG, M., KRIEGEL, H.-P., & SANDER, J.: OPTICS: Ordering points to identify clustering structure. In Proceedings of the ACM SIGMOD Conference, Philadelphia, PA (1999) 49-60
9. XU, X., ESTER, M., KRIEGEL, H.-P., & SANDER, J.: A distribution-based clustering algorithm for mining in large spatial databases. In Proceedings of the 14th ICDE, Orlando, FL (1998) 324-331
10. HAN, J. & KAMBER, M.: Data Mining. Morgan Kaufmann Publishers. San Francisco, CA (2001)
11. HAN, J., KAMBER, M., & TUNG, A. K. H.: Spatial clustering methods in data mining: A survey. In Miller, H. and Han, J. (Eds.) Geographic Data Mining and Knowledge Discovery, Taylor and Francis (2001)
12. Arya, S., Mount, D. M. & Narayan, O.: Accounting for boundary effects in nearest-neighbor searching. Discrete and Computational Gemetry (1996) 155-176

Statistical σ-Partition Clustering over Data Streams

Nam Hun Park and Won Suk Lee

Department of Computer Science, Yonsei University
134 Shinchon-dong Seodaemun-gu Seoul, 120-749, Korea
{zyonix,leewo}@amadeus.yonsei.ac.kr

Abstract. This paper proposes a grid-based clustering method that dynamically partitions the range of a grid-cell based on its distribution statistics of data elements in a data stream. Initially the multi-dimensional space of a data domain is partitioned into a set of mutually exclusive equal-size *initial cells*. As a new data element is generated continuously, each cell monitors the distribution statistics of data elements within its range. When the support of data elements in a cell becomes high enough, the cell is dynamically divided into two mutually exclusive smaller cells called *intermediate cells* by assuming the distribution of data elements is a normal distribution. Eventually, the dense sub-range of an initial cell is recursively partitioned until it becomes the smallest cell called *a unit cell*. In order to minimize the number of cells, a sparse intermediate or unit cell can be pruned if its support becomes much less than a minimum support. The performance of the proposed method is comparatively analyzed through a series of experiments.

1   Introduction

Recently, several data mining methods[1,2,3] for a data stream are actively proposed. A data stream is a massive unbounded sequence of data elements continuously generated at a rapid rate. Due to this reason, it is impossible to maintain all elements of a data stream. Consequently, data stream processing should satisfy the following requirements[4]. First, each data element should be examined at most once to analyze a data stream. Second, memory usage for data stream analysis should be restricted finitely although new data elements are continuously generated in a data stream. Third, newly generated data elements should be processed as fast as possible to produce the up-to-date analysis result of a data stream, so that it can be instantly utilized upon request. To satisfy these requirements, data stream processing sacrifices the correctness of its analysis result by allowing some errors.

   This paper proposes a grid-based clustering method that dynamically partitions the range of a grid-cell based on its distribution statistics of data elements in a data stream. Initially the multi-dimensional space of a data domain is partitioned into a set of mutually exclusive equal-size initial cells. As a new data element is generated continuously, each cell monitors the distribution statistics of data elements within its range. When the support of a cell becomes high enough, the cell is dynamically di-

N. Lavrač et al. (Eds.): PKDD 2003, LNAI 2838, pp. 387–398, 2003.

vided into two mutually exclusive smaller cells, called *intermediate cells*, based on its distribution statistics. Similarly, a dense intermediate cell itself can be partitioned but it is replaced by its two-divided cells. Eventually, the dense sub-range of an initial cell is recursively partitioned until it becomes the smallest cell called a *unit cell*. A cluster of a data stream is a group of adjacent dense unit cells. As the size of a unit cell is set to be smaller, the resulting set of clusters is more accurately identified. In order to minimize the number of cells, a sparse intermediate or unit cell is pruned if its support becomes much less than a minimum support.

The rest of this paper is organized as follows. Section 2 presents related works. Section 3 presents the proposed statistical σ-partition clustering algorithm in detail. In Section 4, several experimental results are comparatively analyzed to illustrate the various characteristics of the proposed method. Finally, Section 5 presents conclusions.

2   Related Works

Clustering is a process of finding groups of similar data elements which are defined by a given similarity measure. Clustering techniques are categorized into several methods: partitioning, hierarchical, density-based and grid-based. The partitioning method such as k-means[5] and k-medoid[6] divides the data space of a data set into *k* mutually disjoint regions called clusters. The number of clusters should be predefined in advance. The k-medoid algorithm selects *k* data elements as the centers of *k* clusters initially, and repeatedly replaces one of the selected centers until it finds the best set of *k* centers. In this method, noise data elements can substantially influence the generation of a cluster, so that it may be difficult to produce a correct result in some cases. The hierarchical method such as BIRCH[7] and CURE[8] decomposes a data set into a tree-like structure. In BIRCH, a CF(Clustering Feature) tree which is used to summarize cluster representations is generated dynamically. After the CF tree is built, any clustering algorithm such as a typical partitioning algorithm is then used. In CURE, instead of using a single centroid to represent a cluster, a fixed number of well-scattered data objects is selected to represent a cluster. The selected representative data objects are shrunk towards the centroid of their cluster by a specified shrinking factor in the process of clustering. Among the clusters, two adjacent clusters whose representative data objects are the closest can be merged into one cluster until a predefined number of clusters is left. A typical density-based clustering algorithm[9] which regards a cluster as a region in a data space with a high density of data elements. Its strong points are that it can discover an arbitrarily shaped cluster, and control noise data easily. In the grid-based clustering method, the data space of a problem is divided statistically into a set of equal-size cells. A cluster is generated by merging adjacent cells that have more than a predefined number of data elements. Its time complexity is very efficient but the accuracy of a cluster is affected by the size of a cell. STING[10] uses a grid-based multi-resolution data structure in which a data space is divided into rectangular cells. There are several levels of such rectangular cells corresponding to the different levels of resolution.

Most conventional clustering algorithms assume a data set is fixed and focuses on how to minimize processing time or memory usage algorithmically. When a data set is enlarged incrementally, it is more efficient to use incremental clustering algorithms[7,11] which mainly focus on how to utilize the previous clustering result of an original data set in clustering its enlarged data set efficiently. In other words, the set of old data elements is scanned only when a new possible cluster may be found by the set of newly added data elements. Therefore, all the old data elements should be maintained physically.

In [13], a k-median algorithm is proposed to find the clusters of data elements generated in a data stream. It regards a data stream as a sequence of stream chunks. A stream chunk is a set of consecutive data elements generated in a data stream. Whenever a new stream chunk containing a set of newly generated data elements is formed, the LSEARCH routine which is an O(1)-approximate k-medoid algorithm is performed to select k data elements from the data elements of the stream chunk as the local centers of the chunk. The algorithm confines its memory space to holding a fixed number of local centers for previous stream chunks. Therefore, if retaining ik centers is impossible at the ith stream chunk, the LSEARCH routine is performed again to cluster the weighted ik points to retain k centers.

3   σ-Partition Clustering

Given a data stream D of d-dimensional data space $N=N_1 \times N_2 \times ... \times N_d$, a data element generated at the jth turn is denoted by $e^j=<e_1^j, e_2^j, ..., e_d^j>$, $e_i^j \in N_i$ $1 \le i \le d$. When a new data element e^t is generated at the tth turn in a data stream D, the current data stream D^t is composed of all the data elements that have ever been generated so far i.e. $D^t=\{e^1, e^2, ..., e^t\}$. The total number of data elements generated in the current data stream D^t is denoted by $|D^t|$.

Finding a cluster of similar data elements in the current data stream D^t is identifying a region whose current density of data elements is dense enough. A unit cell whose length in each dimension is less than λ is used to define the similarity between data elements. The current support of a cell is the ratio of the number of those data elements in D^t that are inside the cell over the total number of data elements in D^t. Therefore, a cluster at D^t is a group of adjacent dense unit cells whose current supports are greater than or equal to a predefined minimum support S_{min}.

The range of each dimension N_i is initially partitioned by p number of mutually exclusive equal-size intervals $I_i^j = [s_i^j, f_i^j)$ $1 \le j \le p$ where s_i^j and f_i^j denote the start and end values in the jth interval of the ith dimension. Consequently, p^d number of initial cells are formed in N and each initial cell g is defined by a set of d intervals $\{I_1, I_2, ..., I_d\}$ $I_i \subseteq N_i$ $1 \le i \le d$. The range R(g) of an initial cell g is a rectangular space $rs = I_1 \times ... \times I_d$. However, the initial rectangular space of an initial cell becomes a set of rectangular spaces $RS=\{rs_1, rs_2, ..., rs_q\}$ as a series of cell partitioning and pruning operations are performed subsequently. When these rectangular spaces are projected to the ith dimension, the intervals of the ith dimension of a cell g can be found and they are denoted by

$IS_i(g)=\{I_i^1,I_i^2,...,I_i^q\}$. The sum of these intervals is defined as the interval size of the ith dimension of the cell g. The range of the cell g is the united spaces of all the rectangular spaces $rs_1,...,rs_q$, $R(g)= \bigcup\limits_{i=1}^{q} rs_i$. Each cell keeps the current distribution statistics of those data elements in the current data stream D^t that are within its range as defined in Definition 1.

[Definition 1] Distribution Statistics of a grid-cell $g(RS,c,\mu,\sigma)$
For the current data stream D^t, a term $g(RS, c^t, \mu^t, \sigma^t)$ is used to denote the distribution statistics of a cell g which is defined by a set of its rectangular spaces RS. Let D_g^t denote those elements in D^t that are in the range of the cell g, i.e., $D_g^t=\{ e|\ e \in D^t$ and $e \in R(g) \}$. The distribution statistics of the cell g are defined as follows:

i) c^t : the number of data elements in D_g^t
ii) $\mu^t=<\mu_1^t,...,\mu_d^t>$: μ_i^t denotes the average of the ith dimensional values of the data elements in D_g^t.

$$\mu_i^t= \sum_{j=1}^{c^t} e_i^j / c^t \ , 1\leq i \leq d$$

iii) $\sigma^t=<\sigma_1^t,...,\sigma_d^t>$: σ_i^t denotes the standard deviation of the ith dimensional values of the data elements in D_g^t.

$$\sigma_i^t=\sqrt{\sum_{j=i}^{c^t}(e_i^j - \mu_i^t)^2/c^t} \ , 1\leq i \leq d$$

When a new data element e^t is generated in the current data stream D^t, its corresponding initial cell among the p^d initial cells is identified based on the initial partitions of the data space N. If the data element is in the range of the initial cell g and the distribution statistics of the cell g was updated most recently at the insert of the vth data element ($v\leq t$), its statistics remain the same as $g(RS, c^v,\mu^v,\sigma^v)$ and they are updated to $g(RS, c^t,\mu^t,\sigma^t)$ as follow: for $\forall i$, $1\leq i \leq d$

$$c^t=c^v+1, \ \mu_i^t= \frac{\mu_i^v \times c^v + e_i^t}{c^t}, \ \sigma_i^t=\sqrt{\frac{c^v}{c^t}\times(\sigma_i^v)^2 + \frac{(\mu_i^v)^2 +(e_i^t)^2}{c^t} -(\mu_i^t)^2}$$

For the current data stream D^t, the current support of an initial cell $g(RS, c^t, \mu^t, \sigma^t)$ is defined by the ratio of its count over the total number of data elements generated so far, i.e. $c^t/|D^t|$. When the current support of the cell becomes the same as a *predefined split support* $S_{split}(S_{split}<S_{min})$, two intermediate cells g_1 and g_2 are created as the children of the initial cell. To split the range of the cell g, a dividing dimension is selected based on the distribution deviation of data elements in the cell g. Among the dimensions whose interval sizes for the cell g are larger than λ, the one with the smallest standard deviation, say σ_k^t, is chosen as a dividing dimension. Based on the standard deviation σ_k^t in the dividing dimension, the set of intervals in the dividing dimension k is partitioned into two sets of intervals. One contains those intervals that are within

the interval $[\mu_k - \sigma_k, \mu_k + \sigma_k)$ in which the 68 percentage of data elements in the cell g is assumed to be distributed according to a normal distribution. The other includes the remaining intervals. The rectangular spaces of the cell g are divided into two mutually exclusive sets by a statistical σ-partition method with respect to the two sets of intervals in the dividing dimension. These two sets of the rectangular spaces are assigned to the ranges of the two divided cells g_1 and g_2 respectively. If a rectangular space of the cell g includes μ_k^i, the corresponding interval of the dividing dimension is actually divided. Figure 1 illustrates how to divide the rectangular spaces of a cell g (RS, c, μ, σ) in a two-dimensional data space.

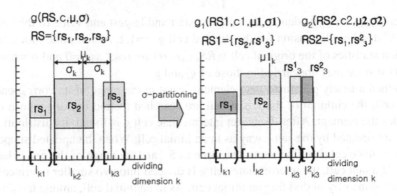

Fig. 1. σ-partition on a cell g

When a cell $g(RS, c^i, \mu^i, \sigma^i)$ is partitioned by the above σ-partition method into cells $g_1(RS_1, c1^i, \mu1^i, \sigma1^i)$ and $g_2(RS_2, c2^i, \mu2^i, \sigma2^i)$, the distribution statistics of g_1 and g_2 are initialized as follows. Let $\varphi(x) = \dfrac{1}{\sqrt{2\pi}\sigma^i} e^{-\frac{(x-\mu^i)^2}{2(\sigma^i)^2}}$ be the normal distribution function of the data elements in the dividing dimension k for the cell g.

$$c1^i = c^i \times \int_{\mu_k^i - \sigma_k^i}^{\mu_k^i + \sigma_k^i} \varphi(x)dx, \ c2^i = c^i - c1^i \tag{1}$$

$$\mu1^i = \mu2^i = \mu^i \text{ except } \mu1_k^i \text{ and } \mu2_k^i, \ \mu1_k^i = \int_{s_k(g_1)}^{f_k(g_1)} x\varphi(x)dx$$

$$\text{if } s_k(g_2) < \mu1_k^i < f_k(g_2), \ \mu2_k^i = \int_{s_k(g_2)}^{f_k(g_2)} x\varphi(x)dx - \mu1_k^i$$

$$\text{else } \mu2_k^i = \int_{s_k(g_2)}^{f_k(g_2)} x\varphi(x)dx$$

$$\sigma 1' = \sigma 2' = \sigma^t \text{ except } \sigma 1_k' \text{ and } \sigma 2_k', \ \ \sigma 1_k' = \sqrt{\int_{s_k(g_1)}^{f_k(g_1)} x^2 \varphi(x) dx - \left(\mu 1_k^t\right)^2} \quad \text{and}$$

$$\text{if } s_k(g_2) < \mu 1_k' < f_k(g_2), \ \ \sigma 2_k' = \sqrt{\int_{s_k(g_2)}^{f_k(g_2)} x^2 \varphi(x) dx - \int_{s_k(g_1)}^{f_k(g_1)} x^2 \varphi(x) dx - \left(\mu 2_k^t\right)^2}$$

$$\text{else } \sigma 2_k' = \sqrt{\int_{s_k(g_2)}^{f_k(g_2)} x^2 \varphi(x) dx - \left(\mu 2_k^t\right)^2}$$

where $s_k(g_i)$ and $f_k(g_i)$ denote the smallest start and largest end value of the intervals of the kth dividing dimension for the divided cell g_i, $i=1,2$. At the same times, the distribution statistics of the original cell $g(RS, c^t, \mu^t, \sigma^t)$ are reset as $c^t=0$ and $\mu_i^t=\sigma_i^t=0$ for $\forall i$, $1 \leq i \leq d$ since they are carried to those of g_1 and g_2.

When a newly generated data element is not in the range of its corresponding initial cell, the children of the initial cell are searched to find the one whose range includes the element. After the target intermediate cell g is found, its distribution statistics are updated by the same way as in an initial cell. When the updated support of the intermediate cell itself becomes the same as S_{splt} and the range of the cell is larger than that of a unit cell, the intermediate cell g is divided into two smaller intermediate cells by the same way of dividing an initial cell. As in an initial cell, among the dimensions whose interval sizes are larger than λ, the one with the smallest standard deviation is chosen as a dividing dimension. However, unlike an initial cell, the original intermediate cell is replaced by the two divided cells. Consequently, the parent initial cell of the original cell becomes the parent of each divided cell.

On the other hand, when the current support of an intermediate cell g becomes less than a *predefined pruning support* S_{prn}, i.e., $c^t/|D|^t < S_{prn}$, the probability of finding a cluster in the range of the cell in the near future is very low. Consequently, the cell is removed and its distribution statistics $g(RS, c^t, \mu^t, \sigma^t)$ are returned back to its parent initial cell g_p. Suppose the distribution statistics of the parent cell g_p were updated lastly at the vth element($v \leq t$) and they are denoted by $g_p(RSp, cp^v, \mu p^v, \sigma p^v)$ where μp^v $=<\mu p_1^v, \mu p_2^v, ..., \mu p_d^v>$ and $\sigma p^v =<\sigma p_1^v, \sigma p_2^v, ..., \sigma p_d^v>$. Its new statistics $g_p(RSp, cp^t, \mu p^t, \sigma p^t)$ at D^t is updated as follows:

For all dimensions $i(1 \leq i \leq d)$, $cp^t = cp^v + c^t$ and

$$\mu p_i^t = \frac{\mu p_i^v \times c^v + \mu_i^t \times c^t}{cp^t} \quad \text{and}$$

$$\sigma p_i^t = \sqrt{\frac{cp^v \times (\sigma p_i^v)^2 + c^t \times (\sigma_i^t)^2}{cp^t} + \frac{(\mu p_i^v)^2 + (\mu_i^t)^2}{cp^t} - (\mu p_i^t)^2}$$

When a cell is divided, the 68% of its count is assigned to one divided cell and the rest, i.e., 32% is assigned to the other in Equation (1). Therefore, the value of a pruning support S_{prn} should be less than the 32% of S_{splt} in order to avoid pruning a newly divided cell too soon.

As mentioned, a sparse intermediate or unit cell can be pruned when a data element in the range of the cell is generated. However, a considerable number of such sparse cells may not be pruned since the possibility of encountering a data element in the range of a sparse cell is very low. All sparse intermediate or unit cells can be forced to be pruned together by examining their current supports. This mechanism is called as a *force-pruning operation*. Since the distribution statistics of all intermediate or unit cells should be examined, the processing time of a force-pruning operation takes relatively long. Due to this reason, it can be performed periodically or when the current number of cells reaches a predefined threshold value.

---

divide $N_1,...,N_d$ into p intervals and create p^d initial cells;
/* $S(g)$: the support of cell g, $S(g)=c^t/|D^t|$ */

for a data stream $D^t=\{ e^1, e^2, ..., e^t \}$ **do** /* t is enlarging */
   read current data element e^t;
   search the cell g which includes e^t;
   update μ^i, σ^i, c^t of the cell g;
   if g is a unit cell or an intermediate cell{
      if $S(g) >= S_{splt}$ {
         if $|f_i(g) - s_i(g)| > \lambda$ in any i dimension {
            /* dividing an intermediate cell g*/
            find the largest σ_k^t among σ^t where $|f_k(g) - s_k(g)| > \lambda$;
            generate g_1 and g_2;
            set the statistics of g_1 and g_2; eliminate cell g;
         }
      }
      else if $S(g) <= S_{prn}$ {
         /* pruning a cell g */
         find the parent initial cell g_p where g is included to g_p;
         update g_p with statistics of g;
         eliminate cell g;
      }
   }
   else if g is an initial cell {
      if $S(g) >= S_{splt}$ {
         /* dividing an initial cell g */
         find the largest σ_k^t among σ^t where $|f_k(g) - s_k(g)| > \lambda$;
         generate g_1 and g_2;
         set the statistics of g_1 and g_2;
         set $c^t=0$ and $\mu_i^t=\sigma_i^t=0$ for $\forall i$ dimension;
      }
   }
end

Fig. 2. The statistic σ-partition clustering

Figure 2 shows the detailed steps of the proposed algorithm. When a cell is split, the counts of two divided cells are initialized by assuming the actual distribution of data elements in the dividing dimension of the cell is a normal distribution. However, if the actual distribution of data elements in the original cell is not the normal distribution, there is a certain estimation error. In other words, the count of each divided cell may be incorrectly initialized. This estimation error of a certain range in a data space is accumulated until the range is represented by a unit cell through a series of cell partitioning. Once the range becomes a unit cell, there is no additional estimation error since the number of data elements in its range is actually counted. Therefore, this accumulated error count of a unit cell is constant. However the support error of the accumulated error count in a dense unit cell is continuously decreased due to Property 1. For a new unit cell $g(RS, c^t, \mu^t, \sigma^t)$ created in the current data stream D^t, let $|D_g^t|$ denote the actual count of data elements in its range $R(g)$ up to D^t. The estimation error $E(g)$ in the count c^t of the cell g is constant and is defined by the difference between $|D_g^t|$ and its estimated count c^t, i.e., $E(g) = | |D_g^t| - c^t|$.

Property 1. (Support error decreasing property) When a unit cell g is newly created in the current data stream D^t, its support error is $E(g)/|D|^t$ and the estimation error $E(g)$ is constant. After m new additional data elements are processed subsequently, the total number of data elements is increased to $|D^{t+m}|$ in D^{t+m} and $|D^{t+m}| > |D^t|$ is satisfied. Consequently, the support error of the cell g in $|D^{t+m}|$ becomes $\dfrac{E(g)}{|D^{t+m}|}$, and

$\dfrac{E(g)}{|D^{t+m}|} < \dfrac{E(g)}{|D^t|}$ is satisfied. As m is increased infinitely, $\dfrac{E(g)}{|D^{t+m}|}$ converges to 0,

i.e. $\lim\limits_{m \to \infty} \dfrac{E(g)}{|D^{t+m}|} \approx 0$. Therefore, it can be ignorable.

A unit cell in the current data stream D^t is dense if its current support is greater than or equal to a predefined minimum support S_{min}. A cluster in the current data stream is a set of adjacent dense unit cells. As the size of a unit cell λ is defined to be smaller, the range of a cluster is more precisely identified. On the other hand, a possible dense cell is split earlier as the value of a split support S_{splt} is lower. Due to this reason, a dense unit cell is found earlier, which enables a unit cell to monitor its actual count more accurately. Furthermore, as the gap between S_{prn} and S_{splt} is increased, more number of intermediate cells are maintained while the support of identified clusters are more precisely found.

4   Experimental Results

In order to analyze the performance of the proposed method, a data set containing one million 4-dimensional data elements is generated by the data generator used in ENCLUS [14]. Most of data elements are concentrated on randomly chosen 10 distinct data regions whose sizes in each dimension are also randomly varied from 10 to 20 respectively. The result of the proposed method is compared with that of the grid-based clustering algorithm STING. The values of a pruning support S_{prn} and a split support S_{splt} are assigned relatively to a predefined minimum support S_{min}. The entire data space of the data set is divided into 4 initial cells. In all experiments, data elements are looked up one by one in sequence to simulate the environment of a data stream.

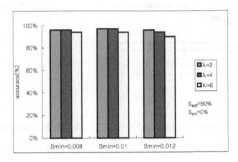

Fig. 4. Accuracy variations to S_{min} and λ

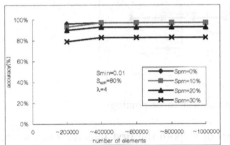

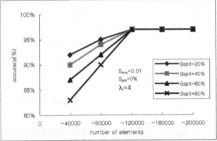

Fig. 5. Accuracy variations to S_{prn} **Fig. 6.** Accuracy variations to S_{splt}

Figure 4 shows the accuracy of the proposed method by varying the values of λ and S_{min}. The accuracy of the proposed method is measured relatively to that of STING. In other words, it is the ratio of the number of correctly clustered elements by the proposed method over the total number of data elements clustered by STING.

Figure 5 shows the accuracy of the proposed method by varying the value of S_{prn}. The sequence of generated data elements is divided into 5 intervals each of which consists of 200 thousand elements. The average accuracy in each interval is shown. As noticed in this figure, the accuracy of the first interval is relatively lower than those of the other intervals. This is because the support of an intermediate cell is too sensitively varied in the first interval. As a result, a lot of cell partitioning operations are performed in the first interval to produce a set of meaningful unit cells. However, it becomes stabilized as the total number of data elements is increased. As the value of S_{prn} is increased, the accuracy becomes lower since a considerable number of possible dense intermediate cells are pruned before they become unit cells. Figure 6 shows the effect of S_{splt} on the accuracy in the first interval. As the value of S_{splt} is set to be lower, unit cells are generated more quickly, so that the accuracy is improved in the early stage of clustering. However, regardless of the values of S_{splt}, the stabilized accuracy is the same.

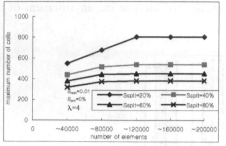

Fig. 7. Memory usage variations to S_{splt} **Fig. 8.** Memory usage variations to S_{prn}

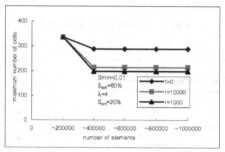

Fig. 9. Memory usage with force-pruning

Figure 7 shows the maximum number of cells in the first interval when no cell is pruned. The maximum number of cells is stabilized after the first interval. As the value of S_{splt} is set to be low, the maximum number of cells is increased since cell partitioning operations are performed frequently to generate meaningful unit cells. As the number of dense unit cells is increased, the maximum number of cells is stabilized. In Figure 8, the variation of the maximum number of cells is shown when sparse cells are pruned. After most of dense unit cells are generated, the maximum number of cells can be decreased by setting the value of S_{prn} adequately. When S_{prn} is set to the 30% of S_{splt}, the memory usage is not decreased. The reason is that most of divided intermediate cells are pruned too quickly and their initial cells are repeatedly partitioned again. On the contrary, when the value of S_{prn} is set to 10%, the memory usage is minimized since dense intermediate cells are successfully divided into its dense unit cells while sparse ones are pruned properly.

A force-pruning operation is usually performed periodically or when it is needed. Figure 9 shows the memory usage of the proposed method by varying the period of a force-pruning operation. In this experiment, two force-pruning periods f=1,000 and f=10,000 are compared. A force-pruning period f=1,000 means that a force-pruning operation is performed whenever 1,000 new data elements are processed. The memory usage of each interval is represented by the maximum number of cells. The number of cells is decreased as the period is shortened. As noticed by this experiment, a force pruning operation does not have to be performed frequently. Instead, it only needs to be performed when lots of intermediate cells are partitioned.

5   Conclusion

In this paper, a grid-based statistical σ-partition clustering method for a data stream is proposed. The multi-dimensional data space is dynamically divided into a set of cells with different sizes. By maintaining only the distribution statistics of data elements in each cell, its current support is precisely monitored. A dense sub-range of a data space is partitioned repeatedly until it becomes a set of dense unit cells. Two thresholds S_{split} and S_{prn} are proposed to control the performance of the proposed method in a data stream. A split support S_{split} is used to determine how fast dense unit cells are identified. A pruning support S_{prn} is used to remove meaningless sparse intermediate or unit cells. Therefore, it can be used to minimize the usage of main memory. However, if it is too high, a less accurate clustering result can be obtained. By controlling these two thresholds properly, the performance of the proposed algorithm can be flexibly controlled.

References

1. M. Datar, A. Gionis, P. Indyk and R. Motwani. Maintaining stream statistics over sliding windows. In *Proc. Of the 13th Annual ACM-SIAM Symp. on Discrete Algorithms, Jan.* 2002
2. M. Charikar, K. Chen and M. Farach-Colton. Finding Frequent Items In Data Streams. In *Proc. Of the 29th Int'l Colloq. on Automata, Language and Programming,* 2002
3. G. S. Manku and R. Motwani. Approximate frequency counts over data streams. In *Proc. Of the 28th Int'l Conference on Very Large Databases*, Hong Kong, China, Aug. 2002.
4. M. Garofalakis, J. Gehrke and R. Rastogi. Querying and mining data streams: you only get one look. In *the tutorial notes of the 28th Int'l Conference on Very Large Databases*, Hong Kong, China, Aug. 2002.
5. R. O. Duda and P. E. Hart. *Pattern Classification and Scene Analysis*. Wiley, 1972.
6. L. Kaufman and P.J. Rousseeuw. *Finding Groups in Data. An Introduction to Cluster Analysis*. Wiley, New York, 1990.
7. T. Zhang, R. Ramakrishnan, and M. Livny. BIRCH: an efficient data clustering method for very large databases. In *Proc. SIGMOD*, pages 103-114, 1996
8. S. Guha, R.Rastogi, and K. Shim. CURE: An efficient clustering algorithm for large databases. In *Proc. SIGMOD*, pages 73-84, 1998
9. M. Ester, H. Kriegel, J. Sander, and X. Xu. A density-based algorithm for discovering clusters in large spatial databases, 1996.
10. W. Wang, J. Yang, and R. Muntz. Sting: A statistical information grid approach to spatial data mining, 1997.
11. M. Ester, H. Kriegel, J. Sander, M. Wimmer, and X. Xu. Incremental clustering for mining in a data warehousing environment, In Proc. VLDB 24th, New York, 1998

12. G. S. Manku and R. Motwani. Approximate frequency counts over data streams. In *Proc. of the 28ʰ Int'l Conference on Very Large Databases,* Hong Kong, China, Aug. 2002.
13. Liadan O'Callaghan, Nina Mishra, Adam Meyerson, Sudipto Guha, and Rajeev Motwani. STREAM-data algorithms for high-quality clustering. In *Proc. of IEEE International Conference on Data Engineering*, March 2002.
14. Cheng, C., Fu, A., and Zhang, Y. *Entropy-based subspace clustering for mining numerical data*. KDD-99, 84-93, San Diego, August 1999.

Enriching Relational Learning
with Fuzzy Predicates

Henri Prade, Gilles Richard, and Mathieu Serrurier

IRIT - Université Paul Sabatier
118 route de Narbonne 31062 Toulouse France
{henri.prade,serrurier}@irit.fr, grichard@ifi.edu.vn

Abstract. The interest of introducing fuzzy predicates when learning rules is twofold. When dealing with numerical data, it enables us to avoid arbitrary discretization. Moreover, it enlarges the expressive power of what is learned by considering different types of fuzzy rules, which may describe gradual behaviors of related attributes or uncertainty pervading conclusions. This paper describes different types of first-order fuzzy rules and a method for learning each type. Finally, we discuss the interest of each type of rules on a benchmark example.

Keywords: Inductive Logic Programming, relational learning, fuzzy rule, confidence degree

1 Introduction

Inductive Logic Programming (ILP) [9] provides a general framework for learning classical first-order logic rules, for which reasonably efficient algorithms have been developed (Progol [6], FOIL [13],...). Relational learning can be presented as a subfield of ILP that concerns the induction process on relational databases compiled in first-order logic. In this scope, we have only to consider function-free Horn clauses. But first-order logic cannot directly handle rules with exceptions, which are common in practice. This has been a motivation for introducing probabilities in ILP [7]. In fact, probabilities, implicitly appear in the FOIL control procedure. Indeed, during the gain computation, the value associated to a rule can be viewed as a confidence degree expressed in terms of "domain probabilities". Such probabilities, together with "world probabilities", are the basic notions of Halpern's first-order probabilistic logic [4]. Domain probabilities are used to capture statistical information for a fixed first-order logic interpretation. These probabilities are obtained by applying a probability measure to the set of valuations making rules true in the interpretation. So, there is no longer any genuine quantifier in a rule when the probability to encounter exceptions is non-zero.

One of the difficulties of the induction of rules from examples is to manage real numbers and imprecision when attributes are non-binary. Classical method for handling real-valued attributes is to turn them into (symbolic) qualitative labels by discretization. Fuzzy sets are known to provide a gradual interface

N. Lavrač et al. (Eds.): PKDD 2003, LNAI 2838, pp. 399–410, 2003.

with numerical data, by escaping the problem of sharp transitions between categories. In the propositional framework, confidence degrees have been integrated in learning methods, together with the handling of fuzzy properties. At least three main trends of works can be distinguished w.r.t. this latter concern. First, neuro-fuzzy learning techniques have been developed for tuning fuzzy membership functions in fuzzy rules; see [8] for a survey. The fuzzy rules, which are produced in that way, are used for functions approximation in automatic control problems. Another research line has been investigated with a greater concern for the descriptive power of the fuzzy rules from the user's point of view, by extending Quinlan's [12] ID3 algorithm to fuzzy decision trees, involving a fuzzy descriptions of classes and making use of entropy measures (extended to fuzzy sets) for building the fuzzy rules; see [1] for a survey. More recently, the use of fuzzy membership functions has been advocated by several researcher for providing association rules in data mining with a better representation power, e.g. [5].

Presently, the majority of the methods for learning fuzzy rules are propositional. A version of FOIL that handles membership degrees has already been developed [15] but the rules induced still keep a classical meaning. In this paper, we propose a method for inducing first-order rules that may include fuzzy predicates. We first explain how a classical database is read in terms of fuzzy predicates, and we further discussed different types of fuzzy rules recently introduced in a learning perspective [11]. For each type of rules, the FOIL algorithm is adapted by defining the corresponding confidence degree. The paper is organized as follows. Sections 2 provides a brief background on ILP. Section 3 presents different types of fuzzy rules and the fuzzy database. Section 4 describes our algorithm and section 5 illustrates the approach on an toy example and a benchmark.

2   Background

We first briefly recall the standard definitions and notations. Given a first-order language $\mathcal{L}$ with a set of variables Var, we build the set of terms $Term$, atoms $Atom$ and formulas as usual. The set of ground terms is the Herbrand universe $\mathcal{H}$ and the set of ground atoms or facts is the Herbrand base $\mathcal{B} \subset Atom$. A *literal* l is just an atom a (positive literal) or its negation $\neg a$ (negative literal). A (resp. ground) substitution σ is an application from Var to (resp. $\mathcal{H}$) $Term$ with inductive extension to $Atom$. We denote $Subst$ the set of ground substitutions. A *clause* is a finite disjunction of literals $l_1 \vee \ldots \vee l_n$ also denoted $\{l_1, \ldots, l_n\}$. A Horn clause is a clause with at most one positive literal. A Herbrand interpretation I is just a subset of $\mathcal{B}$: I is the set of true ground atomic formulas and its complementary denotes the set of false ground atomic formulas. Let us denote $\mathcal{I} = 2^{\mathcal{B}}$, the power set of $\mathcal{B}$ i.e. the set of all Herbrand interpretations. We can now proceed with the notion of logical consequence.

Definition 1. *Given A an atomic formula, $I, \sigma \models A$ means that $\sigma(A) \in I$. As usual, the extension to general formulas F uses compositionality.*

$I \models F$ means $\forall \sigma$, $I, \sigma \models F$ (we say I is a model of F).
$\models F$ means $\forall I \in \mathcal{I}$, $I \models F$.
$F \models G$ means that all models of F are models of G.

Stated in the general context of first-order logic, the task of *induction* is to find a set of formulas H such that:

$$B \cup H \models E \qquad\qquad (1)$$

given a background theory B and a set of observations E (training set), where E, B and H here denote sets of clauses. A set of formulas is here, as usual, considered as the conjunction of its elements.

Of course, one may add two natural restrictions:

- $B \not\models E$ since, in such a case, H would not be necessary to explain E.
- $B \cup H \not\models \bot$: this means $B \cup H$ is a consistent theory.

In ILP, there are two ways for describing examples. The first describes the set of positives examples E^+ and the set of negative examples E^-. The other describes only positive examples in E and make the closed world assumption. It is this hypothesis that we will uses along this paper. Each element of E is called an example and we call a counter-example a fact on the target concept which is not in E. In the setting of relational databases, inductive logic programming is often restricted to Horn clauses and function-free formulas, E is just a set of ground facts. Moreover, the set E itself satisfies the previous requirement but it is generally not considered as an acceptable solution since it has no predictive ability. Usually, rules extraction fits with the idea of providing a compression of the information content of E.

There are two general types of algorithms, *top down* and *bottom up* algorithms. *Top down* ones start from the most general clause and specialize it step by step. *Bottom up* procedures start from a fact and generalize it. In our case, we will use the FOIL algorithm [13] which is a *top down* process. The goal of FOIL is to produce rules until all the examples are covered. Rules with conclusion part C, the target predicate, are found in the following way:

1. take $A \rightarrow C$ as the most general clause with $A = \top$
2. choose the literal l such as the clause $l \wedge A \rightarrow C$ maximizes the gain function
3. $A = l \wedge A$
4. if confidence($A \rightarrow C$)< threshold goto 2
5. return $A \rightarrow C$

The gain function is computed by the formula:

$$gain(l \wedge A \rightarrow C, A \rightarrow C) = n * (log_2(cf(l \wedge A \rightarrow C)) - log_2(cf(A \rightarrow C)))$$

where n is the number of distinct examples covered by $l \wedge A \rightarrow C$. Given a Horn clause $A \rightarrow C$, the confidence $cf(A \rightarrow C) = \frac{P(A \wedge C)}{P(A)}$. Confidence degrees are computed according to the definition of domain probabilities [4]. ILP data are supposed to describe one interpretation under Closed World Assumption. We call I_{ILP} this interpretation. So, given a fact f:

$$I_{ILP} \models f \quad \text{iff} \quad B \wedge E \models f.$$

The domain $\mathcal{H}$ is the Herbrand domain described by B and E. We take P as a uniform probability on $\mathcal{H}$. So we deduce that the confidence in a clause $A \rightarrow C$, with $\overrightarrow{t}$ as vector on the n free variables, is:

$$cf(A(\overrightarrow{t}) \rightarrow C(\overrightarrow{t}))_{I_{ILP}} = \frac{|\{\overrightarrow{x} \in \mathcal{H}^n \mid I_{ILP} \models \sigma[\overrightarrow{t}/\overrightarrow{x}](A(\overrightarrow{t}) \wedge C(\overrightarrow{t}))\}|}{|\{\overrightarrow{x} \in \mathcal{H}^n \mid I_{ILP} \models \sigma[\overrightarrow{t}/\overrightarrow{x}](A(\overrightarrow{t}))\}|} \quad (2)$$

where $|\;|$ denotes cardinality. Another possible definition of a confidence degree might be taken here as the proportion of the number of positive examples covered by the rule w.r.t. the number of total examples (positive and negative) covered by the rule. This confidence degree would represent the probability that a fact deduced from the rule is true. But this definition would not take into account the number of situations covered in the condition part of the rule, which is not always the total number of examples covered since we are in a first-order setting.

In ILP, the goal is to learn a concept represented by a predicate. E is the set of all facts pertaining to the target predicate. B is the set of facts pertaining to predicates other than the target one. So the learned rules are (in the non-recursive case) composed by predicates that appear in B for the condition part and by the target predicate in the consequence part.

3   Induction in Fuzzy Database

3.1   Fuzzy Databases and Fuzzy Rules

We consider a first-order logic database K with fuzzy predicates (e.g., heavy, old ...) as a set of positive facts labeled by real numbers in $[0, 1]$. For instance in Section 5 we shall deal with a database containing facts such as $(weight(a, heavy), 0.9)$ which means that the car a is very representative of heavy cars. Thus, K is made of pairs of the form $(A(\overrightarrow{x}), \mu(A(\overrightarrow{x})))$ for $\overrightarrow{x} \in \mathcal{H}^n$, where $A(\overrightarrow{x})$ is a fact, and $\mu(A(\overrightarrow{x}))$ is the satisfaction degree associated with the fuzzy property A for $\overrightarrow{x}$.

There exist at least two reasons for introducing fuzzy predicates in universally quantified rules. This may be for making them either more flexible or more expressive. Indeed, a fuzzy predicate can be viewed as a family of ordinary predicates whose characteristic functions are the level cut functions μ_{F_α} associated to the fuzzy set membership function μ_F, namely $\mu_{F_\alpha}(\overrightarrow{x}) = 1$ iff $\mu(F(\overrightarrow{x})) \geq \alpha$ and $\mu_{F_\alpha}(\overrightarrow{x}) = 0$ otherwise. Thus a rule "$A(\overrightarrow{t}) \rightarrow C(\overrightarrow{t})$" is naturally associated with the crisp rules "$A_\alpha(\overrightarrow{t}) \rightarrow C_\beta(\overrightarrow{t})$". Note that, if $A_\beta(\overrightarrow{t})$ holds then $A_\alpha(\overrightarrow{t})$ also holds for $\alpha \leq \beta$. So we may only consider the crisp approximations "$A_\alpha(\overrightarrow{t}) \rightarrow C_\alpha(\overrightarrow{t})$".

Then, if we are concerned with *flexibility*, a possible understanding of the fuzzy rule "$A(\overrightarrow{t}) \to C(\overrightarrow{t})$" can be

$$\forall \overrightarrow{x}, \exists \alpha \ \ A_\alpha(\overrightarrow{x}) \to C_\alpha(\overrightarrow{x}), \tag{3}$$

i.e. there exists a crisp understanding of the fuzzy rule which covers each example (but it is not necessary the same for each example since α depends on $\overrightarrow{x}$). This is a kind of rule yet considered in [10]. By flexible rules, we mean here rules which are robust since their predicates can be adapted to borderline situations.

If we are concerned with *expressivity*, we may look for fuzzy rules such that the rule holds for *each of* its level cut counterpart. This means that we have

$$\forall \overrightarrow{x}, \forall \alpha \ \ A_\alpha(\overrightarrow{x}) \to C_\alpha(\overrightarrow{x}). \tag{4}$$

This is clearly more restrictive than (3) since the fuzzy rule is equivalent to a set of ordinary rules with nested predicates and summarizes it into a unique fuzzy rule. In fact (4) is nothing but a gradual rule [3] expressing "The more $\overrightarrow{x}$ satisfies A, the more $\overrightarrow{x}$ satisfies C" (since they are modeled by a constraint of the form $\mu(A(\overrightarrow{x})) \geq \mu(C(\overrightarrow{x}))$).

Gradual rules are one of the four basic kinds of fuzzy rules [3]. Two of them, namely gradual rules and certainty rules, are based on implication connectives and express constraints on the possible models of the world. The two other types, named possibility rules and antigradual rules, rather express that some values are guaranteed to be possible (i.e. that they exist in the base of examples). For instance, let us take possibility rules of the form "The more $\overrightarrow{x}$ is A, the more all the interpretations which makes C true (truth becomes a matter of degree when C is fuzzy) are guaranteed to be possible ". This means in practice that "The more $\overrightarrow{x}$ is A, the more there are examples for any possible interpretation of C". Note that this rule cannot have any "classical" counter-example since we are interested in the distribution of the membership degrees in the database.

In the following, we only consider gradual and certainty rules. Certainty rules contrast with possibility rules, and express that "the more $\overrightarrow{x}$ is A, the more certain $\overrightarrow{x}$ is C". Let us first consider the case where "A" is a fuzzy predicate and "C" is an ordinary predicate. This expresses that "the more $\overrightarrow{x}$ is A, i.e. the greater α such that $A(\overrightarrow{x}) \geq \alpha$, the smaller the number of exceptions of the rule $A_\alpha(\overrightarrow{t}) \to C(\overrightarrow{t})$". Indeed when α decreases, the number of exceptions cannot but increase since the scope of A_α is then enlarged. When C is also a fuzzy predicate, in order to preserve this understanding of the rule, we are led to look for rules of the form

$$\forall \overrightarrow{x}, \forall \alpha \ \ A_\alpha(\overrightarrow{x}) \to C_{1-\alpha}(\overrightarrow{x}), \tag{5}$$

since when α increases $C_{1-\alpha}$ cover more cases.

3.2   Application to ILP

It is well known that algorithms for learning rules have difficulties for handling real-valued attributes. In fact, numerical values may lead to an infinite hypothesis space. In relational learning, this problem is deeper since the hypothesis space is already large. The difficulties grow up when real numbers appears in the concept we want to learn. Real numbers are essentially treated in two way: either by introducing constraints or by discretization.

Introducing constraints in first order logic consists in the use of several operators such as inequalities or mathematical functions (average, ...) [14]. Rules induced by this method may suffer of a lack of expressivity and generality. Furthermore, algorithms dealing with constraints go out of the scope of the standard resolution process in first-order logic.

The second way is to use discretization and clusterization for transforming continuous information into qualitative information. Then, information can be directly treated in the classical logic setting. This method is the most currently used since it allows to cope with numerical values and to improves the readability. Since the clusters are usually defined in an arbitrary way before the induction process, the rules which are produced depend on the quality of the clusters. These clusters are often represented by predicates having an imprecise meaning. For example in the auto-mpg data in UCI, the mpg (city fuel consumption in miles per gallons) value can be represented in terms of the predicates "low consumption", "medium consumption" and " high consumption". In this case, fuzzy labels, represented by fuzzy sets, are more appropriate for describing the mpg values since they avoid arbitrary thresholds between low and medium (see Fig. 1 for description).

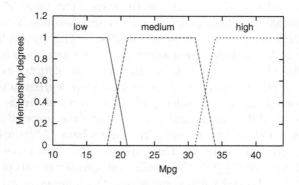

Fig. 1. Fuzzy cluster of mpg

Finally, using fuzzy predicates allows to relax the rigidity of crisp clustering and keeps the readability of the induced rules. Moreover, the different types of fuzzy rules we have described allow a better description and provide new types of summarization of the data. *Flexible rules* can be viewed as a fuzzy adaptation of a crisp rules in the sense that there exists a reading of the fuzzy predicates, corresponding to a high level-cut of its fuzzy representation, which leads to a meaningful rule w.r.t. data. *Gradual rules* and *Certainty rules* are new type of rules, which describe new implicative properties of the data. In the case of

ILP, the goal is to find an hypothesis which is sound and complete with respect to the examples. The hypothesis is sound if it does not cover facts on the target concept which are false in the interpretation defined by the background and the examples.It is complete if it covers all facts on the target concept which are true in the latter interpretation. In the case of fuzzy ILP, the definition of an example covered by a rule will depend on the type of the fuzzy rule and of the membership degrees of facts validating the rules.

4   Algorithm

In the FOIL algorithm, the guidelines for the process are: the confidence degree, the halting condition and the number of distinct examples covered by the rule. We consider that an example is covered by a fuzzy rule if it is itself covered by the classical counter-part of the rule.So we describe these guidelines for each kind of rules (see [10] [11] for details).

Flexible Rules. This first type of meaning for a fuzzy rule "$A(\overrightarrow{t}) \rightarrow C(\overrightarrow{t})$" is close to the one of a classical rule. Of course, we are now expecting that the satisfaction degrees of $A(\overrightarrow{x})$ and $C(\overrightarrow{x})$ are as high as possible. So we can introduce classical interpretations associated with each α-cut.

Definition 2. *An α-interpretation I_α, given a fact f, is defined by:*

$$I_\alpha \models f \ \text{ iff } \ B \wedge E \models f \text{ and } \mu(f) \geq \alpha$$

In this type of interpretations, only facts having a satisfaction degree greater than α are true. Now we have to compute the confidence degree of the rule in the classical way (using (2)) for each α-interpretation. According to the intended meaning of the fuzzy rule, we must favor the confidence degrees of the rule computed in high α-interpretations. Indeed, we prefer the examples be covered with a high degree of satisfaction. The following definition, which is an adaptation in term of first-order logic of the one proposed by [2], takes this into account:

$$cf_{flex}(A(\overrightarrow{t}) \rightarrow C(\overrightarrow{t})) = \sum_{\alpha_i}(\alpha_i - \alpha_{i+1}) * cf(A(\overrightarrow{t}) \rightarrow C(\overrightarrow{t}))_{I_{\alpha_i}}$$

where $\alpha_1 = 1, ..., \alpha_t = 0$ is the decreasing list of the satisfaction degrees that appear in the database. This confidence degree corresponds to the discretization of a Choquet integral of the confidence degrees on α-interpretations. We deduce the number of distinct examples covered:

$$n_{flex}(A(\overrightarrow{t}) \rightarrow C(\overrightarrow{t})) =$$
$$\sum_{\alpha_i}(\alpha_i - \alpha_{i+1}) * |\{\overrightarrow{x_1} \in \mathcal{H}^q, \exists \overrightarrow{x_2} \in \mathcal{H}^r \mid I_{\alpha_i} \models \sigma[\overrightarrow{t_1}, \overrightarrow{t_2}/\overrightarrow{x_1}, \overrightarrow{x_2}](A \wedge C)\}|$$

Gradual Rules. In this case, the values of the satisfaction degrees are only useful for comparing satisfaction degrees in condition and conclusion parts. So, we do not privilege the confidence degree in high α-interpretation as previously.
$$cf_{grad}(A(\overrightarrow{t}) \rightarrow C(\overrightarrow{t})) =$$
$$\frac{|\{\overrightarrow{x} \in \mathcal{H}^n \mid I_{ILP} \models \sigma[\overrightarrow{t}/\overrightarrow{x}](A \wedge C), \mu(\sigma[\overrightarrow{t}/\overrightarrow{x}]C) \geq \mu(\sigma[\overrightarrow{t}/\overrightarrow{x}]A)\}|}{|\{\overrightarrow{x} \in \mathcal{H}^n \mid I_{ILP} \models \sigma[\overrightarrow{t}/\overrightarrow{x}](A)\}|}$$

When the valuation of the condition part of the rule is a conjunction of grounds literals, the satisfaction degree of this conjunction is the minimum of the degree of each literal. We deduce the number of distinct examples covered:

$$n_{grad}(A(\overrightarrow{t}) \to C(\overrightarrow{t})) = |\{\overrightarrow{x_1} \in \mathcal{H}^q, \exists \overrightarrow{x_2} \in \mathcal{H}^r \ |$$
$$I_{ILP} \models \sigma[\overrightarrow{t_1}, \overrightarrow{t_2}/\overrightarrow{x_1}, \overrightarrow{x_2}](A \wedge C, \mu(\sigma[\overrightarrow{t_1}/\overrightarrow{x_1}]C) \geq \mu(\sigma[\overrightarrow{t_1}, \overrightarrow{t_2}/\overrightarrow{x_1}, \overrightarrow{x_2}]A)\}|$$

Type 1 Certainty Rules. The meaning of the fuzzy rule "$A(\overrightarrow{t}) \to C(\overrightarrow{t})$" is then "the more $\overrightarrow{x}$ is A, the more certain $\overrightarrow{x}$ is C". For these rules we are not interested in the satisfaction degrees of the consequence parts. This type of rule will be referred to as type 1 certainty rules in the following. The α-cut for these rules correspond to the following type of classical interpretation:

Definition 3. *An α-certainty interpretation, given a fact f, is defined by:*

$$I_{\alpha-cert} \models f \quad iff \quad (B \models f \ and \ \mu(f) \geq \alpha) \ or \ E \models f$$

With this kind of rules, confidence degrees are expected to be high for high α-certainty interpretation. The idea is that we can be more permissive with respect to exceptions for the classical counterparts of the rule "$A(\overrightarrow{t}) \to C(\overrightarrow{t})$" corresponding small values of α. So, we are led to use the following Choquet integral.

$$cf_{cert1}(A(\overrightarrow{t}) \to C(\overrightarrow{t})) = \sum_{\alpha_i}^{t}(\alpha_i - \alpha_{i+1}) * cf(A(\overrightarrow{t}) \to C(\overrightarrow{t}))_{I_{\alpha_i - cert}}$$

We deduce the number of distinct examples covered:

$$n_{cert1}(A(\overrightarrow{t}) \to C(\overrightarrow{t})) =$$
$$\sum_{\alpha_i}(\alpha_i - \alpha_{i+1}) * |\{\overrightarrow{x_1} \in \mathcal{H}^q, \exists \overrightarrow{x_2} \in \mathcal{H}^r \ | \ I_{\alpha \ cert} \models \sigma[\overrightarrow{t_1}, \overrightarrow{t_2}/\overrightarrow{x_1}, \overrightarrow{x_2}](A \wedge C)\}|$$

Type 2 Certainty Rules. The above definition is modified in the following way for taking care of the satisfaction degree of the consequence of the rules. This type of rule will be referred to as type 2 certainty rules in the following

$$cf_{cert2}(A(\overrightarrow{t}) \to C(\overrightarrow{t})) =$$
$$\frac{|\{\overrightarrow{x} \in \mathcal{H}^n \ | \ I_{ILP} \models \sigma[\overrightarrow{t}/\overrightarrow{x}](A \wedge C), \mu(\sigma[\overrightarrow{t}/\overrightarrow{x}]C) > 1 - \mu(\sigma[\overrightarrow{t}/\overrightarrow{x}]A)\}|}{|\{\overrightarrow{x} \in \mathcal{H}^n \ | \ I_{ILP} \models \sigma[\overrightarrow{t}/\overrightarrow{x}](A)\}|}$$

We deduce the number of distinct examples covered:
$$n_{cert2}(A(\overrightarrow{t}) \to C(\overrightarrow{t})) = |\{\overrightarrow{x_1} \in \mathcal{H}^q, \exists \overrightarrow{x_2} \in \mathcal{H}^r \ |$$
$$I_{ILP} \models \sigma[\overrightarrow{t_1}, \overrightarrow{t_2}/\overrightarrow{x_1}, \overrightarrow{x_2}](A \wedge C), \mu(\sigma[\overrightarrow{t_1}/\overrightarrow{x_1}]C) > 1 - \mu(\sigma[\overrightarrow{t_1}, \overrightarrow{t_2}/\overrightarrow{x_1}, \overrightarrow{x_2}]A)\}|$$

Thus, we can use the FOIL algorithm for inducing various kinds of first-order fuzzy rules by adapting confidence degree and cardinality with the type of rules we want to learn.

5 Results

5.1 Illustrative Example

Let us consider a database that describes 21 houses in a town. First we have some fuzzy relational predicates such as $(close(x, y), \alpha)$ which means that the house x is close to the house y with a membership degree α, or $(know(x, y), \alpha')$ which means that the owner of house x knows the owner of the house y at a degree α' (from 0 for unknown to 1 for friends). The houses are also described with some nearly propositional fuzzy predicates such as $price(x, expensive)$ or $size(x, small)$.

So, in this context, we can find fuzzy rules of each type with a good confidence degree. For example, we find the *flexible* rule

$$close(x, y), price(y, expensive) \rightarrow price(x, expensive),$$

with 0.81 of confidence degree, because we can reasonably expect that a house which is close to an expensive one, is expensive as well (since expensive houses are often located in the same area). A typical *gradual* rule is

$$size(x, large) \rightarrow price(x, expensive),$$

i.e. "the larger the house, the more expensive", which describes the fact that price grows up with size. Its confidence degree is 0.80. A good example of *certainty rules* is

$$close(x, y) \rightarrow know(x, y)$$

with 0.95 of confidence degree if the rule is viewed as type 1 certainty rule and 0.88 of confidence degree if the rule is viewed as type 2 certainty rule. This rule means "the closer the houses, the more we are sure that the owners know together". The fact that owners of very close houses have a high probability to know each other is realistic. This probability can decrease when the distance between the houses grow up. As ending remark, we may observe that all these rules are obviously subject to exceptions and, despite their interest, they cannot be obtained in any way by a classical ILP machine.

5.2   Benchmark

As a benchmark, we use the "auto-mpg" database from UCI [1]. This database is constituted with informations about cars and the concept we want to learn is the city-cycle fuel consumption in miles per gallon. There are 398 instances of cars described by 9 attributes of which 5 are continuous, including the concept to be learn. This database con be represented in propositional logic but is sufficient to illustrate the interest of the approach. First, the database has been "discretized" with fuzzy sets. Moreover, we also built three crisp discretizations corresponding to i) the crisp partition of the attribute domain which is the closest to the fuzzy partition, ii) the support of the fuzzy sets, iii) the core of the fuzzy sets. Then, we learn the class of city-cycle fuel consumption according to the crisp and fuzzy set for all the types of fuzzy rules.

---

[1] http://www.ics.uci.edu/ mlearn/MLRepository.html

Types of rules	nbr of rules	coverage	avg cf
classical rules	8	0.77	0.84
classical rules with the core	11	0.59	0.87
classical rules with the support	10	0.85	0.80
flexible rules	3	0.51	0.84
gradual rules	4	0.47	0.91
type 1 certainty rules	2	0.62	0.75
type 2 certainty rules	2	0.59	0.76

Here are examples of rule induced by the algorithm for each type of them (classical rules are one induced on the discretisation corresponding to the crisp partition of the attribute domain which is the closest to the fuzzy partition).

Classical rules

$cylinders(A, 8) \rightarrow mpg(A, low)$

flexible rules

$displacement(A, low), weight(A, medium) \rightarrow mpg(A, medium)$

gradual rules

$weight(A, high) \rightarrow mpg(A, low)$

type 1 certainty rules

$cylinders(A, 6), weight(A, high) \rightarrow mpg(A, low)$

type 2 certainty rules

$cylinders(A, 6), weight(A, high), origin(A, 1), acceleration(A, low),$
$horsepower(A, low) \rightarrow mpg(A, low)$

As expected, the coverage score of classical rules is between the score of classical rules with the core of fuzzy sets and classical rules with the support of fuzzy sets. It is due to the fact that, with the core of the fuzzy rules, the examples that are in the boundary of the crisp classes are not treated. On the contrary, with the support of fuzzy sets, the example of that are in the boundary of the crisp classes can belong to two classes. The smaller score of fuzzy rules w.r.t. coverage is due to the fact that fuzzy rules are harder to find than classical ones. This result is expected because fuzzy rules are more constrained since they take into account the membership degree of the valuations of each predicate. In fact, confidence degrees of classical rules do not rely on the distance of the data to the boundaries of the discretized sets. For example, let us consider a classical rule F with a good confidence degree, and its fuzzy flexible counterpart F'. If many example of F are borderline w.r.t. fuzzy sets, the confidence degree of F' will be lower than the one of F. On the contrary, if many counter-examples of F are borderline, the confidence degree of F' will be greater than the one of F. So, the confidence degree of fuzzy flexible counterpart of a rule is a good indicator of the robustness of the classical rule w.r.t. small variation of the boundaries of the sets.

Type 1 certainty rules focus on membership degrees of the conditional parts of the rules. Gradual and certainty rules show how conditions and conclusions parts evolve together. These rules have a meaning far from the classical one and the rules that we find have not necessarily a crisp counterpart or approximation. The fuzzy rules that handle certainty tend to favor the non-fuzzy predicates in condition part because they leave more freedom with respect to the satisfaction degree of covered examples. Note that some rules could be described in propositional logic, but here the instantiations are automatically generated by the algorithm. As shown in some rules, the algorithm can mix fuzzy predicates and non-fuzzy predicates.

6   Conclusion

In this paper, we have provided a formal framework and a procedure for dealing with fuzzy predicates and learning fuzzy first-order rules of different kinds in the case of relational databases. Since the confidence degree computation is a weighted version of FOIL's one, it is easy to deduce that the complexity of our algorithm is the same as the FOIL's one. The definition of confidence degrees for each kind of rules allows us to take into account the fuzzy predicates in the algorithms that use confidence degrees for guiding the learning process. It is obvious that using fuzzy predicates for managing real-valued data instead of using crisp discretization or constraint-based induction is a good compromise between the readability of the rules and the flexibility of the discretization. Moreover, fuzzy predicates allow to extract new kinds of relations.

Through the example, we see that fuzzy rules are often too constrained for covering all the examples of the target concept, but they convey information on the robustness of the rules w.r.t. borderline examples. So, it can be useful to learn fuzzy rules together with classical ones.

In this paper, we focus on the search of different kinds of fuzzy rules and the definition of confidence degrees associated to each of them. In further works, it will be interesting to show how much fuzzy discretization is efficient in a learning point of view. More generally, a formal definition of ILP that handles all the types of rules must be defined. In this context, automatic deduction mechanisms may be developed for testing the efficiency of fuzzy rules in terms of classification.

References

1. B. Bouchon-Meunier and C. Marsala. Learning fuzzy decision rules, in. *Fuzzy Sets in Approximate Reasoning and Information Systems*,(J.C. Bezdek, D. Dubois, H. Prade, eds.), The Handbooks of Fuzzy Sets Series. Kluwer Academic Publishers, 1999, 279-304.
2. M. Delgado, D. Sanchez, and M.A. Vila. Fuzzy cardinality based evaluation of quantified sentences. *Inter. J. of Approximate Reasoning*, pages 23:23–66, 2000.
3. D. Dubois and H. Prade. What are fuzzy rules and how to use them. *Fuzzy Sets and Systems*, 84(2):169–189, 1996.
4. J. Halpern. An analysis of first-order logics of probability. *Artificial Intelligence*, 46:310–355, 1990.
5. E. Hüllermeier. Implication-based fuzzy association rules. In L. De Raedt and A. Siebes, editors, *Proc. PKDD-01, 5th Conf. on Principles and Pratice of Knowledge Discovery in Databases*, number 2168 in LNAI, pages 241–252, 2001.

6. S.H. Muggleton. Inverse entailment and Progol. *New Generation Computing*, 13:245–286, 1995.
7. S.H. Muggleton. Learning stochastic logic programs. *Electronic Transactions in Artificial Intelligence*, 5(041), 2000.
8. D. Nauck and R. Kruse. Neuro-fuzzy methods in fuzzy rule generation, in. *Fuzzy Sets in Approximate Reasoning and Information Systems*,(J.C. Bezdek, D. Dubois, H. Prade, eds.), The Handbooks of Fuzzy Sets Series. Kluwer Acad. Pub., 1999, 305-334.
9. S-H Nienhuys-Cheng and R. de Wolf. *Foundations of Inductive Logic Programming*. Number 1228 in LNAI series. Springer, 1997.
10. H. Prade, G. Richard, and M. Serrurier. Learning first order fuzzy rules. In *Proc. of 10 th Int. Fuzzy Systems Association (IFSA-03)*, Istanbul, 2003.
11. H. Prade, G. Richard, and M. Serrurier. On the induction of different kinds of first-order fuzzy rules. In *Proc7 th European Conference on Symbolic and Qualitative Approaches to Reasoning with Uncertainty (ECSQARU-03)*, Aalborg, 2003.
12. J. R. Quinlan. Induction of decision trees. *Machine Learning*, 1(1):81–106, 1986.
13. J. R. Quinlan. Learning logical definitions from relations. *Machine Learning*, 5:239–266, 1990.
14. J. R. Quinlan. Learning with continuous classes. In *Proc. Artificial Intelligence (AI'92)*, 343–348, Singapore, 1992.
15. D. Shibata, N. Inuzuka, S. Kato, T. Matsui and H. Itoh In Ning Zhong and Lizhu Zhou, editors, *Proceedings of The Third Pacific-Asia Conference on Knowledge Discovery and Data Mining (PAKDD-99): Methodologies for Knowledge Discovery and Data Mining*, 268–273, LNAI 1574, Beijing, China, April 1999.

Text Categorisation Using Document Profiling

Maximilien Sauban and Bernhard Pfahringer

Department of Computer Science
University of Waikato
Hamilton, New Zealand
{m.sauban,bernhard}@cs.waikato.ac.nz

Abstract. This paper presents an extension of prior work by Michael D. Lee on psychologically plausible text categorisation. Our approach utilises Lee's model as a pre-processing filter to generate a dense representation for a given text document (a document profile) and passes that on to an arbitrary standard propositional learning algorithm. Similarly to standard feature selection for text classification, the dimensionality of instances is drastically reduced this way, which in turn greatly lowers the computational load for the subsequent learning algorithm. The filter itself is very fast as well, as it basically is just an interesting variant of *Naive Bayes*. We present different variations of the filter and conduct an evaluation against the Reuters-21578 collection that shows performance comparable to previously published results on that collection, but at a lower computational cost.

1 Introduction

In the last decade the amount of textual information in digital form has grown exponentially, mainly due to the forever-increasing accessibility of the Internet. It is crucial to create tools to organise the amount of information available. Text categorisation is one such tool. It aims at classifying textual documents into pre-defined categories. Text categorisation applications are manifold and are ranging from *automated meta-data extraction* for indexing to *document organisation* for databases or web pages (see Yang et al., [1]). Other interesting uses of text categorisation include *text filtering*, generally as part of a producer-consumer relationship, or *word sense disambiguation* when dealing with natural languages processing (see Roth, [2]).

1.1 Existing Text Categorisation Methods

It is difficult to be exhaustive when listing the existing text categorisation methods. Amongst the main approaches, decision tree methods ("divide-and-conquer" approach) have the advantage of being "human readable" in the sense that they deal with symbolic entities and not numeric values [3]. Investigations using probabilistic models usually focus on *Naive Bayes* and its variants [4]. Joachims [5] introduced the support vector machine method to text categorisation. Also

N. Lavrač et al. (Eds.): PKDD 2003, LNAI 2838, pp. 411–422, 2003.

worth mentioning is the Rocchio method [6]; this method creates a prototype document for each class from the training set. A test document will be assigned to the class of the closest prototype found. Yang [7] invented a mapping approach using a multivariate regression model and investigated, together with Pedersen, lazy learning for text categorisation [8]. Frank et al. [9] investigated text categorisation using compression models, and Wiener et al. [10] were using neural networks. Yet other approaches have tried to improve predictive performance by incorporating semantic information like WordNet hypernyms [11].

1.2   David Lee's Method

Lee [12] came up with a psychologically plausible approach considering three different insights. Firstly, Lee noted that people are able not only to state that a given document is about a given topic but also that a document is not about a topic. Take for example "middle east conflict" as a topic; the occurrence of the word "rugby" in a document would give a strong hint about the document not being about the topic. Secondly, humans are able to make non-compensatory decisions: one can decide if a document is about a topic or not without necessarily having to read the whole document. Using our previous middle-east conflict example, if the document starts with something like "The south African rugby team just arrived in Auckland..." most people would not need to read any further to reach a conclusive, in this case negative, decision. Thirdly, people have the capacity to give answers with a level of confidence and so they are able to state if a document is either definitely about a topic or alternatively just remotely related to a topic.

Lee's model's formal definition is based on a Bayesian analysis, which states that it is possible to compute the posterior odds of a document being about a topic or not by multiplying the prior odds—chances of a document to be about a topic before looking at it—and the evidence—probability that a document would have been generated under the assumption that it is about a topic (or not). Lee considers the document as a *sequence* of words. The evidence then becomes the product of the probability of each word being in a document about the topic (or not). Note that this approach follows the *Naive Bayes* assumption that all words are independent of one another, also called the independence assumption. The evidences' probabilities are quantified using the number of occurrences of the given word over the total number of words and are calculated for both categories, for and against the topic. The independence assumption allows analysing words sequentially, which permits monitoring the evolution of the posterior odds word by word in the same order as they appear in the document. Considering the logarithm of the posterior odds and using evidences for and against the topic leads to the following equation:

$$\ln \frac{\Pr(c_j|d_i)}{\Pr(\neg c_j|d_i)} = \ln \frac{\Pr(c_j)}{\Pr(\neg c_j)} + \sum_{k=1}^{n} \ln \frac{\Pr(w_{ni}|c_j)}{\Pr(w_{ni}|\neg c_j)}$$

Figure 1 shows the evolution of the posterior odds of two documents processed by the text classifier. The graph on the left depicts the partial log-odds sums for a document about the topic, while the graph on the right depicts those sums for a document that is not about the topic. Note that the partial log-odds sums are computed over larger and larger initial subsequences of the document, which causes the order of the words in the document to become significant. This example also shows the possibility of non-compensatory decision making by setting two thresholds, one for the document being about the topic and the other for the document not being about the topic. The decision is taken when one of the thresholds is reached. Let's assume the thresholds in Figure 1 are 100 for a document about the category and -20 for a document not about the category. In the left hand side case, the decision is taken after reading the 120th word (when the curve meets with $y = 100$). On the right hand side example, the decision can be taken after reading the 45th word (the curve meets with $y = -20$).

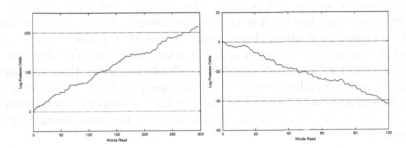

Fig. 1. Illustration of document profiles, the left hand side one is about the topic while the right hand side one is not.

1.3 Our Approach

The investigation presented here is an extension of Lee's work [12]. An interesting aspect of Lee's method is that the document is processed sequentially and the odds of the document with respect to a given category can be tracked as the words are fed to the system. We call this sequence of the partial sums of the log-odds of the words of a document a *document profile*. Usually those profiles are not as clear-cut and easy to classify as the ones shown in Figure 1. We have therefore decided to investigate a two-step process, where a first step generates document profiles according to Lee's method, and a second step extracts propositional information from these profiles that then can be fed into any arbitrary propositional learner. Thus, Lee's system is used as a dimensionality-reducing pre-processing step.

The next section will explain this process in more detail, discuss issues with the dictionary, and basically describe two different ways of extracting attributes from document profiles. Sections 3 and 4 explain the experimental setup and give and discuss experimental results. In Section 5 we present conclusions and discuss further work.

2   Generating and Manipulating Document Profiles

To construct a model, each word in the vocabulary is assigned two probabilities, the probability of the word being about the topic $\Pr(w_k|c_j)$ and the probability of the word not being about the topic $\Pr(w_k|\neg c_j)$ where w_k is the word, and c_j the topic. The word's *influence* $(\mathcal{I}_{w_k})$ is then calculated as follow:

$$\mathcal{I}_{w_k} = \ln \frac{\Pr(w_k|c_j)}{\Pr(w_k|\neg c_j)}$$

The probabilities are based on the rate at which the word has occurred in the training documents about and not about the category. Figure 2 (on the left hand side) portraits one such dictionary. The higher the magnitude of the influence, the more weight the word will have in the final decision. Note that there are much more words with a negative influence. The explanation for that lies in the skewedness of the training data: the dictionary pictured in Figure 2 (on the left hand side) was trained with 197 documents about a given category and 9,406 documents not about that category. The 9,406 negative examples used for training were in fact the union of all 89 other categories. The skewedness also explains why the maximum positive amplitude is greater than the maximum negative one. Specialised words of the category in focus (word with a large positive influence score) are more likely to have a denser concentration in the positive documents than the specialised words of the other category (actually categories).

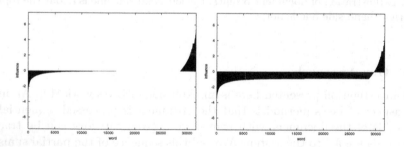

Fig. 2. Dictionary for one category of the Reuters dataset and a shifted version of the dictionary on the right.

The three following sub-sections will describe dictionary manipulations that proved to be beneficial, and explain the two ways propositional attributes are extracted from document profiles.

2.1   Shifting the Origin on the y-Axis

Figure 2 (right hand side) illustrates the result of shifting the origin on the y-axis in an attempt to equalise the maximum amplitudes. To shift the dictionary, we subtracted half the sum of the top positive value and the top negative value from all the words present in the dictionary or more formally:

$$\forall k \in d : \mathcal{I}'_{w_k} = \mathcal{I}_{w_k} - \frac{max(\mathcal{I}_w) + min(\mathcal{I}_w)}{2}$$

Where $\mathcal{I}_{w_k}$ and $\mathcal{I}'_{w_k}$ are respectively the influence of the word before and after shifting, $max(\mathcal{I}_w)$ the largest influence in the dictionary and $min(\mathcal{I}_w)$ the smallest. Note that a lot more words now have a negative influence, and that the magnitude of the positive influences has been reduced. This shift usually improves performance. A more sophisticated threshold selection method might fare even better.

Table 1. Dictionary sizes after cutting off at x% of the top influence value.

percent of the maximum positive/negative value cut off	# of words remaining in the dictionary
0% (whole dictionary)	31651
10%	4046
15%	2860
30%	1288
60%	166

2.2   Reducing the Size of the Dictionary

As mentioned earlier, the specialised words carry a large influence score, but their distribution is highly skewed: there are almost no specialised words for the negative class. On the other hand, words with a low influence score, are more evenly distributed between both the positive and negative class. Their low influence score causes them to play only a minor part in the final decision, but they can potentially add noise. We have therefore introduced a mechanism to prune words from the dictionary based on their influence score. The decision threshold is based on the maximum positive value and the maximum negative value (of the unshifted dictionary). The cut off value is determined as a percentage of the maximum values. A cut at 30% means that all the words with influence score between 0 and 30% of the maximum positive influence and between 0 and 30% of the maximum negative influence score will not be taken into account. Table 1 shows the non linear relation between the cut off value and the number of words left in the dictionary when applying this idea to the dictionary of Figure 2. This pruning effect is also illustrated in Figure 3 where the pruned dictionaries for the four different cut off values of 10%, 15%, 30% and 60% appear in clockwise order starting from the upper left corner. An additional advantage of pruning dictionaries is the potential speedup of the generation of document profiles.

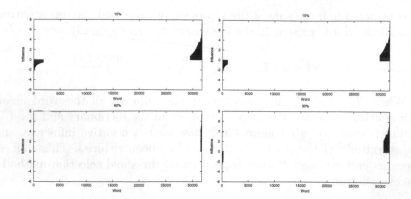

Fig. 3. Different cut off values applied to the dictionary of Figure 2; clockwise from the upper left corner the values are: 10%, 15%, 30% and 60%.

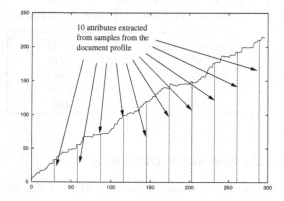

Fig. 4. Reading off attributes from a document profile.

2.3   Turning Document Profiles into Attributes

We have used two different methods to turn document profiles into a constant number of propositional attributes. We need to deal with the fact that the number of words in each document is different, therefore also the length of document profiles differs. The first methods solves that problem by simply reading off the value of the document profile after a certain percentage of the document has been read. Looking at Figure 4, we see that ten values are extracted, with equal-sized gaps in between. In a naive approach the maximum number of attributes that can be extracted in this manner is limited by the size of the smallest document. The second method for extracting attributes is even simpler, computing just some very high-level summary information about a document profile. Specifically, such a description comprises a mere seven attributes: the maximum and the minimum value encountered, the respective positions of these two extrema relative to the document length, a boolean indicator whether the maximum is

Table 2. The ten largest categories from the ModAPTE split.

category	# of training articles	# of test articles
earnings (earn)	2877	1087
corporate acquisitions (acq)	1650	719
money market (money-fx)	538	179
grain (grain)	433	149
crude oil (crude)	389	189
trade issues (trade)	369	117
interest (interest)	347	131
wheat (wheat)	212	71
shipping (ship)	197	89
corn (corn)	181	56

reached before the minimum, and the total number of words for and against the category (i.e. how many words carried a positive influence, how many carried a negative influence). Obviously this is just one of a few possible high-level summary descriptions, other potentially interesting attributes include document length or final value in the profile.

3   Experimental Setup

To investigate the performance of the two-step process described above we have conducted an empirical evaluation using the Reuters corpus[1]. We used the same train-test split as proposed in [13] where a total of 12,902 documents is split into a train-set of 9,603 documents and a test-set of 3,299 documents. We restricted our evaluation to the 10 most common categories, as presented in Table 2. The only pre-processing operation we did was to lower case the characters. We did not use any stemming nor stop word removal techniques.

For our evaluation we used the standard information-retrieval performance measures of precision and recall, as well as aggregate measures based on those two. The aggregate measures were F-measure and the precision-recall mean. The standard F-measure computes the harmonic mean of precision and recall and the precision-recall mean is the arithmetic mean (average) of those two measures. The four formulae are summarised in Table 3. Macroaveraging simply computes averages of either the F-measures or precision-recall means over several categories.

4   Experimental Results

We have conducted an extensive series of experiments to judge the performance of various standard classifiers using document profiles, and also to investigate

---

[1] The Reuters-21578 collection may be freely downloaded for experimentation purposes from *www.research.att.com/~lewis/reuters21578.html*

Table 3. Four information-retrieval evaluation measures: precision, recall, F-measure and precision-recall mean.

measure name	formula
precision	$\frac{tp}{tp+fp}$
recall	$\frac{tp}{tp+fn}$
F-measure	$\frac{2tp}{2tp+fp+fn}$
precision-recall mean	$\frac{tp(2tp+fn+fp)}{2(tp+fp)(tp+fn)}$

the effects of the dictionary tuning we have described above. For lack of space we will only concentrate some of the findings here, a complete report can be found in the forth-coming Master's thesis of the first author. We used the following classifiers from the Weka [14] package: J48 (C4.5, a decision tree algorithm [15]), OneR (rule based algorithm [16]), IBk (k-Nearest Neighbour (k-NN) [17]), SMO (Support Vector Machine [18]) and Naive Bayes (Naive Bayes algorithm [19]). We have also added a very simple classifier called *Polarity* that simply predicts the sign of the last value in the document profile. Polarity is closely related to *multinomial* Naive Bayes ([20]).

4.1   Which Classifier to Use?

Figure 5 shows the performance of the six different classifiers on the category *trade* per number of attributes taken from the profile. While J48, OneR, IBk and SMO show equivalent results—SMO shows an interesting behaviour, with recall rising at the expense of precision as the number of attributes increases past 100—*Naive Bayes* and Polarity do not seem to be as influenced as the afore-mentioned classifiers schemes by the number of attributes generated from the profile. They show impressive recall, but unfortunately also poor precision. Qualitatively speaking, graphs for other categories look similar.

Figure 6 depicts macroaveraged F-measures of the 6 classifiers on the 10 categories (described in Table 2) per number of samples. Two distinct clusters are noticeable: above 0.65 points F-measure with J48, IBk and OneR, and below with *Naive Bayes*, SMO and Polarity. Overall, J48 and IBk are clearly dominant, followed closely by OneR. The poor performance of SMO, *Naive Bayes* as well as Polarity is probably caused by the high correlation between the generated attributes. Summing up, J48 should be preferred to IBk for the slightly better results and the computationally expensive classification process of the lazy learning approach.

4.2   Pruning the Dictionary Plus Reading only Parts of a Document

In this section we only employ J48, because it performed well in the experiments reported in the last section, and it is fast. Figure 7 illustrates the effect of using a pruned dictionary (on the left hand side) and of only reading initial portions of documents (on the right hand side) using 150 attributes extracted from the

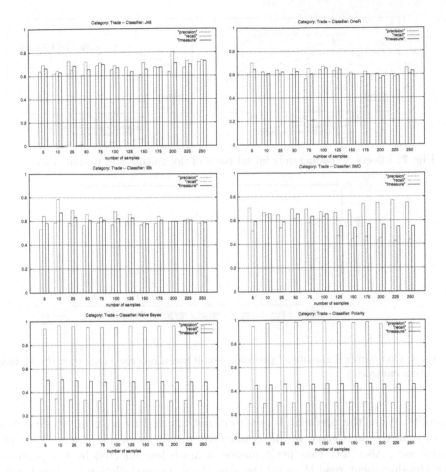

Fig. 5. Performances of six different classifiers on the Reuters category *Trade*.

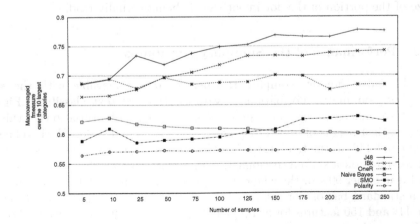

Fig. 6. Macroaverages for 6 classifiers plotted per number of attributes.

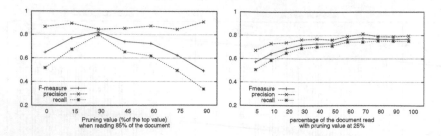

Fig. 7. Effects of reading only initial parts of documents and pruning dictionaries.

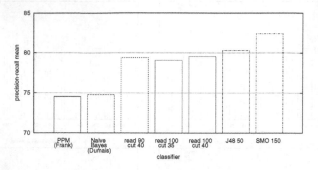

Fig. 8. Comparing 7 classifiers' precision-recall mean macroaverage over the 10 largest Reuters categories.

profile. While the precision is generally not affected by reducing the number of words in the dictionary, recall significantly decreases as the pruning percentage increases. Also note that performance is much less affected by the percentage of the document read than by the size of the pruned dictionary. Furthermore, precision appears to react more robustly than recall to the effect of reducing the size of the portion of the document that is being actually read.

4.3   Performance of the Summary Attributes

In this subsection we compare the performance achievable with the seven-attribute summary information to various standard text classifiers. Figure 8 shows the macroaverage of the precision-recall means of 3 variations of this classifier against results obtained by [3] for *Naive Bayes* and [9] for PPM and against SMO and J48 on the ten largest Reuters-21578 categories. The three variations all used J48 on a shifted dictionary, using either 35% or 40% as a cutoff value, and read either 90% or the whole document. Both SMO and J48 were run on top of a standard bag-of-words-based feature selection using info-gain. 50 features for J48 and 150 features for SMO yielded the best results. The results show that J48 using this tiny set of features outperforms *Naive Bayes* and PPM, closely approaches standard J48, but does not perform as well as SMO.

4.4   Complexity of the Attribute Generation

The algorithm's complexity is equivalent to the complexity of *Naive Bayes* in the sense that it is linear in the number of words and in the number of categories. The complexity of SMO, for comparison, is on average $n.log(n)$. Accessing the dictionary to retrieve influence values is generally $log(n)$, but if necessary, perfect hash functions could be used to reduce this dictionary access cost to a constant. Computing such hash functions will be easier for smaller dictionaries.

5   Conclusion

This paper has presented a text classification approach based on document profiles. Its predictive performance is comparable to more standard approaches, but the method is extremely simple, therefore fast and highly scalable. Our two-step approach effectively transforms a sparse learning problem into a dense one without having to explicitly select single features from the original representation.

The most promising direction for future work will be investigating combinations of the different sets of attributes available. Two approaches are possible: one can combine the high-level summary, the partial sums, as well as standard feature subsets into one larger single set of features. Secondly, single classifiers can be trained on these different feature sets in isolation and then be put together into ensembles. Another direction will be comparing our influence formula with the usual TFIDF document representation. A good starting point will be the thorough study carried out by Rennie et al. [21].

References

1. Yang, Y., Slattery, S., Ghani, R.: A study of approaches to hypertext categorization. Journal of Intelligent Info. Systems **18** (2002) 219–241
2. Roth, D.: Learning to resolve natural language ambiguities: a unified approach. In: Proc. of AAAI-98, 15th Conf. of the American Association for Artificial Intelligence, AAAI Press (1998) 806–813
3. Dumais, S., Platt, J., Heckerman, D., Sahami, M.: Inductive learning algorithms and representations for text categorization. In: Proc. of CIKM-98, 7th ACM Int. Conf. on Info. and Knowledge Management, ACM Press (1998) 148–155
4. Lewis, D.D.: Naive (Bayes) at forty: The independence assumption in information retrieval. In: Proc. of ECML-98, 10th European Conf. on Machine Learning, Springer (1998) 4–15
5. Joachims, T.: Text categorization with support vector machines: learning with many relevant features. In: Proc. of ECML-98, 10th European Conf. on Machine Learning, Springer (1998) 137–142
6. Rocchio, J.J.: Relevance feedback in information retrieval. The SMART Retrieval System: Experiments in automatic document processing (1971) 313–323
7. Yang, Y., Chute, C.G.: A linear least squares fit mapping method for information retrieval from natural language texts. In: 14th Int. Conf. on Computational Linguistics (COLING). (1992) 447–453

8. Yang, Y., Pedersen, J.O.: A comparative study on feature selection in text categorization. In: Proc. of ICML-97, 14th Int. Conf. on Machine Learning, Morgan Kaufmann (1997) 412–420
9. Frank, E., Chui, C., Witten, I.H.: Text categorization using compression models. In: Proc. of DCC-00, IEEE Data Compression Conf., IEEE Computer Society Press (2000) 200–209
10. Wiener, E.D., Pedersen, J.O., Weigend, A.S.: A neural network approach to topic spotting. In: Proc. of SDAIR-95, 4th Annual Symposium on Document Analysis and Info. Retrieval. (1995) 317–332
11. Scott, S., Matwin, S.: Text classification using WordNet hypernyms. In: Use of WordNet in Natural Language Processing Systems: Proceedings of the Conf. Association for Computational Linguistics (1998) 38–44
12. Lee, M.D.: Fast text classification using sequential sampling processes. In: Proc. of the 14th Australian Joint Conf. on Artificial Intelligence, Springer (2002) 309–320
13. Apté, C., Damerau, F., Weiss, S.M.: Automated learning of decision rules for text categorization. Information Systems **12** (1994) 233–251
14. Witten, I.H., Frank, E., Trigg, L., Hall, M., Holmes, G., Cunningham, S.J.: Weka: Practical machine learning tools and techniques with java implementations. In: Proc ICONIP/ANZIIS/ANNES'99 Int. Workshop: Emerging Knowledge Engineering and Connectionist-Based Info. Systems. (1999) 192–196
15. Quinlan, R.: C4.5: Programs for Machine Learning. Morgan Kaufmann (1993)
16. Holte, R.: Very simple classification rules perform well on most commonly used datasets. In: Machine Learning. Volume 11. (1993) 63–91
17. Aha, D., Kibler, D., Albert, M.: Instance-based learning algorithms. Machine Learning **6** (1991) 37–66
18. Platt, J.: Fast training of support vector machines using sequential minimal optimization. In: B. SchOlkopf, C. Burges, and A. Smola Advances in Kernel Methods - Support Vector Learning. (1998)
19. John, G.H., Langley, P.: Estimating continuous distributions in bayesian classifiers. In: Proc. of the Eleventh Conf. on Uncertainty in Artificial Intelligence, Morgan Kaufmann (1995) 338–345
20. McCallum, A., Nigam, K.: A comparison of event models for naive bayes text classification. In: AAAI-98 Workshop on Learning for Text Categorization. (1998)
21. Rennie, J., Shih, L., Teevan, J., Karger, D.: Tackling the poor assumptions of naive bayes text classifiers. In: Proc. of the 20th Int. Conf. on Machine Learning, Morgan Kaufmann (2003)

A Simple Algorithm
for Topic Identification in 0–1 Data

Jouni K. Seppänen, Ella Bingham, and Heikki Mannila

Laboratory of Computer and Information Science and HIIT Basic Research Unit
Helsinki University of Technology

Abstract. Topics in 0–1 datasets are sets of variables whose occurrences are positively connected together. Earlier, we described a simple generative topic model. In this paper we show that, given data produced by this model, the lift statistics of attributes can be described in matrix form. We use this result to obtain a simple algorithm for finding topics in 0–1 data. We also show that a problem related to the identification of topics is NP-hard. We give experimental results on the topic identification problem, both on generated and real data.

1 Introduction

Large collections of 0–1 data occur in many applications, such as information retrieval, web browsing, telecommunications, and market basket analysis. While the dimensionality of such data sets can be large, the variables (or attributes) are seldom completely independent. Rather, it is natural to assume that the attributes are organized into (possibly overlapping) *topics*, i.e., collections of variables whose occurrences are somehow connected to each other[1]. For example, in document data the topics correspond to topics of the document: e.g., phrases "data mining", "decision trees" and "association rules" probably are included in one topic, which might be called the "data mining" topic. In supermarket market basket data, the topics could correspond to classes of products such as soft drinks, vegetables, etc. In discretized gene expression data topics could correspond to groups of genes that are expressed in similar conditions or tissues.

Finding topics from data is by no means easy: the topics can be overlapping, and a particular topic is active only for a subset of documents. For example, simple frequent set based approaches are unable to find topics, as the attributes in a topic are seldom 1 together. There has been lots of work that searches for latent structure in 0–1 data (see, e.g., [1,2,3,4,5,6,7,8,9,10]). The approaches range from simple methods based on covariance-type statistics (e.g., [9]) to full probabilistic models (e.g., [4]) and to spectral approaches [10].

In order to discover topics from 0–1 data, one first has to specify the model for topics, and then give a method that finds topics corresponding to the model.

---

[1] Our usage of the word *topic* is similar but not identical to the meaning in information retrieval literature, where a topic is a probability distribution on the universe of terms, typically concentrating on a few terms.

N. Lavrač et al. (Eds.): PKDD 2003, LNAI 2838, pp. 423–434, 2003.

In this paper we describe a simple generative topic model, based on our previous work [11]. We prove some analytical results about the model by using the concept of lift [12]. We show that the lift statistics of individual attribute pairs can be described in matrix form as linear combinations of lift statistics of disjoint topics. Based on this observation, we give a simple algorithm for finding topics in 0–1 data. We also show that one form of the topic identification problem is NP-hard. We give experimental results on both generated and real data, showing that the algorithm works well in practice.

First we review some other methods for finding latent structure in binary data. Many of these generative models are quite powerful and are able to describe complex situations. On the other hand, finding exact solutions for them is computationally intractable, and it is difficult to get a clear picture of the quality of the obtained estimates. Many of the methods are also symmetric with respect to the data values 0 and 1; on the basis of the asymmetry in the data generating process, this can be viewed as a potential source of problems.

In nonnegative matrix factorization (NMF) [1], an observed data matrix is decomposed into a product of two unknown matrices. All three matrices have nonnegative entries. The observed data is regarded as a sum of latent variables. Lee and Seung give two algorithms for finding the unknown matrices; there is, however, no probabilistic interpretation of the results of NMF. Computationally, the methods seems very demanding and there are no clear results on the quality of the solutions [13].

The latent semantic analysis (LSA) method [2] uses singular-value decomposition to decompose an observed data matrix into a product of matrices. (In contrast to NMF, the matrices can have negative entries, too.) In a seminal paper by Papadimitriou et al. [3] some arguments were given to justify the performance of LSI by presenting a probabilistic corpus model. Their basic model is quite general and somewhat similar to ours.

Hofmann [4] has presented a probabilistic version of LSA, termed PLSA. His formal model is fairly close to ours and we will show comparative results on the models. For each observation vector, some topics are first selected according to some observation-specific topic probabilities; then, the topics generate attributes according to some topic-attribute probabilities. The attributes are conditionally independent given the topic. Hofmann's main interest is in good estimation of all the parameters using the EM algorithm, while we are interested in the structure of the data (that is, the probabilities of attributes belonging to topics) and also explaining why the methods would find topics.

Laten Dirichlet Allocation (LDA) [14,15,16] is a method in which the data model is closely similar to Hofmann's PLSA but the estimation of the parameters is computationally more demanding: a variational approximation to the data likelihood is needed prior to EM estimation of the parameters. Independent component analysis (ICA) ([8,17,18]) is a statistical method that expresses observed multidimensional sequences as combinations of unknown latent variables, that are statistically as independent as possible. The so called probe distances [19] of attributes can be used to find (possibly overlapping) sets of attributes that behave similarly with respect to other attributes; we studied this in an earlier paper [11]. Cooley and Clifton [9] compute the frequent sets in the data and cluster them using a hypergraph partitioning scheme, thus avoiding the problem of not having all attributes of a topic present in one data vector.

A popular method to analyze 0–1 data is the class of finite mixtures of multivariate Bernoulli distributions. However, for the Bernoulli models, the values 0 and 1 have symmetric status, while for our topic models defined in Section 2 this is not the case. Another important difference between Bernoulli (or any other) mixture model and our model is that in mixture models it is assumed that an observed 0–1 vector is only generated by one latent topic, although generation probabilities are given for all latent topics. In this paper we assume that a data vector is generated by the interaction of several latent topics. Binary generative topographic mapping [20,21] also assumes that the data vectors are generated by one latent topic at a time.

The rest of this paper is organized as follows. We describe our model and examine some of its analytical properties in Section 2. In Section 3 we study the lift statistic and describe the simple algorithm based on it. We give experimental results in Section 4, and conclude in Section 5.

2   Topic Models

In this section we present our concept of a topic model, give the likelihood function of the model, and discuss what kinds of parameter values are realistic. This form of the model was introduced earlier by us [11].

Let U be an n-element set of *attributes* (e.g., words). A k-*topic model* $\mathcal{T}$ arranges the n attributes into k *topics*. The model has the following parameters: a k-element vector $s = (s_1, \ldots, s_k)$ corresponding to the k topics, and a $k \times n$ matrix $\mathbf{Q}$ whose elements relate the topics to the attributes; the element corresponding to topic i and attribute A is denoted by $Q_{i,A}$. All elements of s and $\mathbf{Q}$ must be probabilities, i.e., reals in the range $[0, 1]$; however, neither s nor any row or column of $\mathbf{Q}$ is required to sum up to 1.

A data vector x (e.g., a document) is sampled from $\mathcal{T}$ as follows. First, the active topics are selected by sampling a k-element binary vector t whose every component t_i is 1 with probability s_i, independently of all other components. Second, the active topics generate the attributes. For each topic i, an n-element binary vector x_i is sampled so that the component corresponding to A is 1 with probability $t_i Q_{i,A}$, independently of all other components. The data vector x is then the logical *or* (i.e., maximum) of all the vectors x_i, $x = \bigvee_{i=1}^{k} x_i$.

It would be possible to add another layer on top of the topics, selecting the topic probabilities anew for each data vector from, e.g., a Dirichlet distribution. Many of our results could be generalized to such settings, which however fall outside the scope of this treatment. This type of approach has been taken in [3,4,14,15,16].

We next present the likelihood function of a k-topic model $\mathcal{T}$ with parameters $s, \mathbf{Q}$. The data D consists of vectors x, each considered independently of the others,

$$P(D \mid \mathcal{T}) = \prod_{x \in D} P(x \mid \mathcal{T}).$$

The probability of a single observation x is

$$P(x \mid \mathcal{T}) = \sum_{t} P(t \mid \mathcal{T}) P(x \mid t, \mathcal{T}).$$

The sum is taken over all k-element 0–1 vectors t, corresponding to all 2^k possible combinations of active topics. The probability of a topic combination depends on the parameters s only,

$$P(t \mid \mathcal{T}) = P(t \mid s) = \prod_{i=1}^{k} P(t_i \mid s_i) = \prod_{i=1}^{k} s_i^{t_i} (1 - s_i)^{1-t_i}.$$

The probability of an observation given the active topics depends on the parameters $\mathbf{Q}$ only,

$$P(x \mid t, \mathcal{T}) = P(x \mid t, \mathbf{Q}) = \prod_{A \in U} P(x_A \mid t, \mathbf{Q}),$$

where x_A denotes the element of x that corresponds to the attribute $A \in U$. A single attribute has a value of either zero or one, with distribution

$$P(x_A \mid t, \mathbf{Q}) = p_A^{x_A} (1 - p_A)^{1-x_A} = \begin{cases} 1 - p_A, & x_A = 0 \\ p_A, & x_A = 1, \end{cases}$$

where

$$p_A = 1 - \prod_{i=1}^{k} (1 - Q_{i,A})^{t_i}.$$

The likelihood function, if expanded fully, would have a large number of terms because of the sum over 2^k topic combinations t. This suggests a high computational complexity, and indeed the task of selecting the best t is difficult. This is illustrated by the following theorem, whose proof we defer to the Appendix.

Theorem 1. *The following problem is* NP*-complete: given a topic model $\mathcal{T}$, a single data vector x and a threshold ρ, decide whether there is a topic assignment t such that the probability of the data given the assignment exceeds the threshold, $P(x \mid t, \mathcal{T}) \geq \rho$.*

However, the models involved in the proof would best be described as contrived, so the result should not dissuade us from researching some reasonable subclass of topic models. But what kind of models are reasonable?

One assumption that we will make is that the topic probabilities s_i are small. This seems reasonable at least in the context of document data: if some words occur in a large fraction of all documents, in information retrieval they would be classified as stop words and not considered in searches; it is the less common words that distinguish interesting documents.

Another question is the amount of overlap between topics – if two topics consist of almost completely the same attributes, it does not seem easy to distinguish between them. In [11] we considered a class of "ε-separable" models, an idea similar to that in [3]. A model is ε-separable if every topic has a set of primary attributes and assigns at most a fraction ε of its attribute-activation weight to the non-primary attributes. However, the ε-separability property does not perfectly capture the idea of almost-disjoint topics, as the discussion in [11, before Lemma 3] notes: for example, several topics can "conspire" against another topic i by giving high weight to one of i's primary attributes. Even if every high weight is less than a fraction ε of the topic's total weight, it is possible that the majority of activations of that attribute come from the conspiring topics and not the primary topic.

This leads us to define a different separability concept: a model has θ-*bounded conspiracy* if every attribute A has a primary topic i such that

$$\sum_{j\neq i} Q_{j,A} \leq \theta Q_{i,A}.$$

We conjecture that a model is discoverable from data if it has low values of s_i and conspiracy bounded by some low θ.

3   Using the Lift Statistic

We now consider a statistic commonly called called *lift* or *interest* [12,22,23],

$$\text{lift}(A, B) = \frac{P(A \mid B)}{P(A)} = \frac{P(A, B)}{P(A)P(B)},$$

which is a kind of a relative risk factor: how much more common is it to observe A given that B is observed, compared to no information about B? Lift was chosen because it measures dependence, which is highly relevant to topic models – when two attributes belong strongly to the same topic, their co-occurrence should deviate significantly from the independence assumption. For independent A and B, $\text{lift}(A, B) = 1$, and the stronger the (positive) dependence, the higher the lift. Note that our model predicts $\text{lift}(A, B) \geq 1$ for all pairs $A, B \in U$; thus, one way of assessing whether the model fits a given data set is to see how $\text{lift}(A, B)$ is actually distributed.

Proposition 1. *Assume that attribute A is only generated by topic i. Then for any attribute B,*

$$\text{lift}(A, B) = \frac{P(t_i \mid B)}{P(t_i)} = \frac{P(t_i, B)}{P(t_i)P(B)}.$$

Proof. We factorize the probabilities: $P(A) = P(A, t_i) = P(t_i)P(A \mid t_i)$ and $P(A, B) = P(t_i, A, B) = P(t_i)P(B \mid t_i)P(A \mid t_i, B)$. Since A is only generated by topic i, $P(A \mid t_i, B) = P(A \mid t_i)$. Thus

$$\text{lift}(A, B) = \frac{P(A, B)}{P(A)P(B)} = \frac{P(t_i)P(A \mid t_i)P(B \mid t_i)}{P(t_i)P(A \mid t_i)P(B)}.$$

Using Bayes' theorem $P(B \mid t_i) = P(B)P(t_i \mid B)/P(t_i)$ and canceling terms we obtain the result.   □

What Proposition 1 says is that if A is a "core attribute" of topic i, i.e., an attribute generated by i only, then A represents i perfectly in lift calculations, even if $Q_{i,A} < 1$. Of course in practice, when the lift must be estimated from data, a small value of $Q_{i,A}$ can cause poor results. Another point to note is that the probability $P(B \mid t_i)$ appearing in the proof is *not* the model parameter $Q_{i,B}$. Instead, it is the probability that any topic will generate B conditioned on the fact that at least topic i is active. Proposition 1 has as immediate consequences two results that we used already in [11].

Corollary 1. *If attributes A and B are only generated by topic i, i.e., $Q_{j,A} = Q_{j,B} = 0$ for $j \neq i$, then $\mathrm{lift}(A, B) = s_i^{-1}$.*

Corollary 2. *If attribute A is only generated by topic i and attribute B is only generated by topic j, then $\mathrm{lift}(A, B) = 1$.*

Thus, the lift statistic between attributes belonging to one topic only is very simple. The interesting question is how lift behaves when an attribute belongs to several topics.

Assume that attribute A is only generated by topic i, and attribute B is generated by both topics i and j. Now $\mathrm{lift}(A, B)$ is, after simplification,

$$\frac{P(A, B)}{P(A)P(B)} = \frac{Q_{i,B} + s_j Q_{j,B} - Q_{i,B}s_j Q_{j,B}}{s_i Q_{i,B} + s_j Q_{j,B} - s_i s_j Q_{i,B}Q_{j,B}} \approx \frac{Q_{i,B} + s_j Q_{j,B}}{s_i Q_{i,B} + s_j Q_{j,B}}$$

where in the approximation we have assumed that $Q_{i,B}s_j Q_{j,B}$ and $s_i s_j Q_{i,B}Q_{j,B}$ are small compared to the other terms. The above formula generalizes to the case where B is generated by some other topics than i and j, too: before the approximation we then have several second order terms $s_\ell Q_{\ell,B}$ corresponding to all topics ℓ that generate B, and similarly several third order terms $s_\ell Q_{i,B}Q_{\ell,B}$ (in the numerator) or fourth order terms $s_i s_\ell Q_{i,B}Q_{\ell,B}$ (in the denominator).

Assume now that all the topic probabilities are (approximately) equal, i.e., $s_\ell \approx s$ for all topics ℓ. Then we can write the above formula as $\mathrm{lift}(A, B) \approx (s^{-1}Q_{i,B} + Q_{j,B})/(Q_{i,B} + Q_{j,B})$. Furthermore, let each topic ℓ have c_ℓ core attributes that are only generated by that topic. Then using Corollaries 1 and 2 we note that the lifts of A and all core attributes can be included in the formula as follows:

Observation. The lift between a core attribute A of topic i and an attribute B generated by topics i and j is

$$\mathrm{lift}(A, B) \approx \sum_{A'} \mathrm{lift}(A, A')c_i^{-1}\frac{Q_{i,B}}{Q_{i,B} + Q_{j,B}} + \sum_{D'} \mathrm{lift}(A, D')c_j^{-1}\frac{Q_{j,B}}{Q_{i,B} + Q_{j,B}}$$

where $\sum_{A'} \mathrm{lift}(A, A')c_i^{-1}$ is an averaged estimate of s^{-1}, $\sum_{D'} \mathrm{lift}(A, D')c_j^{-1} = 1$ and the two sums run over the core attributes A' and D' of topics i and j, respectively. Also, we may add a third summation including $\text{lift}(A, F')$ where F' is a core attribute belonging to topic l into which B does not belong to, as then $Q_{l,B} = 0$ and the whole term vanishes. This observation again generalizes to the case where B is generated by multiple topics.

The above reasoning included approximations in discarding high-order terms and the somewhat crude assumption that all s_i are equal. In any case, it does yield an idea of how to discover topics: for an attribute B that belongs to several topics, define a vector $\boldsymbol{\alpha}$ whose length is the total number of all core attributes. The element corresponding to A (a core attribute of topic i) is $\alpha_A = Q_{i,B}/(c_i \sum_j Q_{j,B})$. Then $\text{lift}(A, B) \approx \boldsymbol{\alpha}^T \text{lift}(A, \cdot)$ for all core attributes A, where we denote by $\text{lift}(A, \cdot)$ the vector of lifts between A and all core attributes (where $\text{lift}(A, A) = 0$). This gives us an algorithm for finding the topics in which the attributes belong, and also the parameters Q:

- Identify those attributes that belong to one topic only – this can be done by looking at the lift statistics, which are always either 1 or $1/s$ for those attributes.
- Cluster those attributes using some traditional clustering algorithm; at this stage the clusters do not overlap and do not cover all attributes – if an attribute B belongs to several topics, its lifts are intermediate between 1 and $1/s$, and so B is not clustered. For A belonging to one topic i only, $Q_{i,A} = P(AA')/P(A')$ which can be averaged over all A' belonging to the same topic i as A.
- For attributes B which are not clustered, find a decomposition $\text{lift}(B, \cdot) = \boldsymbol{\alpha}^T R$, where the square symmetric matrix R has the vectors $\text{lift}(A, \cdot)$ (of already clustered attributes A) as its columns. All of the lifts in this formula are known, so the vector $\boldsymbol{\alpha}$ can be estimated straightforwardly. The elements of $\boldsymbol{\alpha}$ are nonzero for those attributes that share a topic with B, and zero for others. Also, the elements are more or less constant within attributes of a given topic. Now $Q_{i,B} = \alpha_A c_i / \sum_j Q_{j,B}$ where α_A can be averaged over all A' belonging to topic i, c_i is known, and for small and equal s_j we can approximate $P(B) \approx s \sum_j Q_{j,B}$, which gives us $\sum_j Q_{j,B}$. We can also assume $\sum_j Q_{j,B} = 1$ and scale the estimated $Q_{i,B}$ accordingly.

4   Experimental Results

4.1   Generated Data

We designed experiments to see how the conspiracy statistic θ of a model affects our clustering results. The results corroborate our conjecture that low-conspiracy models are easier to discover. We constructed random models with θ-bounded conspiracy using the following recipe. The model has 10 topics and 100 attributes. The probability s_i of a topic was drawn uniformly at random from the interval $[0.01, 0.5]$. Each attribute was assigned a primary topic so that each topic was primary for 10 attributes.

To assign the within-topic attribute probabilities $Q_{i,A}$ so that the conspiracy parameter is θ, we first drew a number p uniformly from $[0, 1]$ and let $Q_{i,A} = p$ for the primary topic i. Then we distributed the mass θp to the non-primary topics in an uneven way. Each non-primary topic in random order received a fraction of ϕ of the remaining mass, where ϕ is chosen at random from $[0, 1]$, separately for each non-primary topic. The last topic received all remaining mass to make the mass sum up exactly to θp.

This way of generating a random model includes a number of somewhat arbitrary choices that we now justify. First, the topic probabilities s_i were chosen not from $[0, 1]$ but from a smaller interval. Some lower limit is necessary so that each topic is represented in a finite data sample; and an upper limit is needed by our algorithm, which distinguishes a topic by estimating its probability and cannot discover a topic that is almost always active. In a preliminary test (not shown), our algorithm's performance was best with low upper limits, and deteriorated rapidly when the upper limit approached 1. We chose 0.5 as the upper limit as a conservative approach: in document data, one would expect that individual topics have much smaller probabilities.

Second, we discuss the distribution of the within-topic attribute probabilities of non-primary topics. A more obvious strategy would be to draw the probabilities independently and then to normalize, but then the distribution would have become more even. With 9 non-primary topics, all the probabilities would center around $\theta/9$ times the primary probability, which makes the task far easier: none of the non-primary topics is likely to be confused with the primary topic. In contrast, our procedure typically results in a few non-primary topics with non-negligible topic-attribute probabilities for each attribute. We wish to mimic the behavior of true data sets, such as text document data: a term may have several meanings, perhaps a primary meaning and one or few secondary meanings, hence it belongs primarily to one topic of discussion and secondarily to a few other topics, but not to all possible topics.

In the experiment, we estimated the topic-attribute probabilities $\mathbf{Q}$ using the lift statistic, NMF, PLSA[2] and K-means. The NMF and PLSA methods estimate $\mathbf{Q}$ given the observed binary data. A naive alternative is the simple K-means algorithm which clusters the attributes into non-overlapping sets; we assume that $Q_{i,A}$ is equal for all attributes A of topic i and sums to 1 at each topic.

Figure 1 shows the mean squared errors (MSE's) of the estimated $\mathbf{Q}$, compared to the true probabilities used to generate the data. The conspiracy parameter θ runs from 0 to 1. At each θ, the topic probabilities s are sampled anew, so there is great variability in the data models. Originally, the topic-attribute probabilities estimated by the methods do not necessarily sum to 1 at each topic – they do in PLSA, but not either in the other methods or in the true data model – but we scale them accordingly, to be able to compare the MSE's.

In Figure 1 we see that at smaller θ, the Lift algorithm estimates the $\mathbf{Q}$ and thus the structure of the data very nicely. When θ grows very large, the data

---

[2] The PLSA method was kindly programmed by Mr. Teemu Hirsimäki.

model is more difficult to estimate. The behaviors of NMF and PLSA[3] do not depend on θ, which is natural: the methods are not primarily aimed for such θ-bounded data but instead are able to estimate the structure also when the topics are totally overlapping. The K-means algorithm estimates the structure of the data poorly for all θ.

4.2 Real Data

We performed experiments on bibliographical data on computer science available on the WWW[4]. We first tested the model's prediction that lift$(A, B) \geq 1$ for all A, B; while it does not hold perfectly because there are negative correlations between words, the vast majority of these negative correlations are statistically insignificant (details omitted). We preprocessed the data by removing a small set of stop words and all numbers, and then selected the 100 most frequent terms for further analysis.

We computed the lift statistics between all term pairs and used hierarchical average linkage clustering based on the inverses of lifts. Table 1 shows how the terms are clustered into topics. The number of clusters (21) was chosen based on the distance between clusters being merged in the process of hierarchical clustering: until these 21 clusters, the intercluster distances were quite small but distances between the final 21 clusters were large. The structure in Table 1 is immediately familiar to a theoretical computer scientist: the topics concentrate on different fields of the science.

We also performed topic finding on yeast gene expression data, using the same gene expression dataset as in [24] that combines the results of several different gene expression studies. The combined dataset measures the expression level of over six thousand genes in almost a hundred experiments; thus, we used the experiments as "attributes" and the genes as "measurements". The levels were discretized so that the top 5% expressed genes in each experiment were given the value 1. The results are not shown due to space constraints, but as a brief example, the discovered topics were seen to reflect cyclical behavior of the genes in the time-series experiments.

5 Concluding Remarks

We studied a simple generative topic model and showed that the lift statistics of attributes can be described in matrix form. Based on this, we obtained a simple algorithm for finding topics in 0–1 data. We also showed that a problem related to the identification of topics is NP-hard, and gave experimental results.

Several open problems remain. Our model is simple, and seems to yield good results; still, more complex models might do a better job at identifying, e.g., topics containing partly exclusive attributes. The identifiability of the model is another interesting issue: could one prove something about it? Further experimental studies are also needed.

---

[3] No simulated annealing was used in the EM algorithm of the PLSA.

[4] http://liinwww.ira.uka.de/bibliography/Theory/Seiferas/

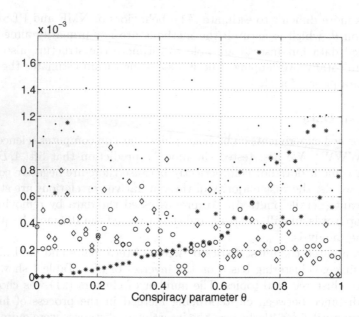

Fig. 1. Mean squared errors of **Q** at different conspiracy parameters θ. Lift $*$, NMF $\diamond$, PLSA $\circ$, K-means $\cdot$.

Table 1. Terms in different topics. (The order of the topics is not relevant).

topic	terms
1	algorithms approximation damath problems scheduling some tree two
2	analysis distributed libtr probabilistic systems
3	bounds communication complexity focs lower
4	algorithm efficient fast ipl matching problem set simple
5	design ieeetc network networks optimal parallel routing sorting
6	note tcs
7	finding graphs minimum planar polynomial sets sicomp time
8	graph number properties random tr
9	from information learning lncs theory
10	approach jacm linear new programming system
11	actainf binary search trees
12	abstract computation extended model stoc
13	automata finite languages mfcs
14	data dynamic infctrl logic programs structures using
15	applications icalp theorem
16	cacm computer computing science
17	crypto functions
18	jcss machines
19	algebraic beatcs computational geometry
20	de stacs van
21	codes dmath

References

1. Lee, D.D., Seung, H.S.: Learning the parts of objects by non-negative matrix factorization. Nature **401** (1999) 788–791
2. Deerwester, S.C., Dumais, S.T., Landauer, T.K., Furnas, G.W., Harshman, R.A.: Indexing by latent semantic analysis. Journal of the American Society of Information Science **41** (1990) 391–407
3. Papadimitriou, C.H., Raghavan, P., Tamaki, H., Vempala, S.: Latent semantic indexing: A probabilistic analysis. In: PODS '98. (1998) 159–168
4. Hofmann, T.: Probabilistic latent semantic indexing. In: SIGIR '99, Berkeley, CA (1999) 50–57
5. Carreira-Perpiñán, M.A., Renals, S.: Practical identifiability of finite mixtures of multivariate Bernoulli distributions. Neural Computation **12** (2000) 141–152
6. Gyllenberg, M., Koski, T., Reilink, E., Verlaan, M.: Non-uniqueness in probabilistic numerical identification of bacteria. J. Appl. Prob. **31** (1994) 542–548
7. Cadez, I.V., Smyth, P., Mannila, H.: Probabilistic modeling of transaction data with applications to profiling, visualization, and prediction. In Provost, F., Srikant, R., eds.: Proceedings of the Seventh ACM SIGKDD International Conference on Knowledge Discovery and Data Mining, San Fransisco, CA (2001) 37–46
8. Hyvärinen, A., Karhunen, J., Oja, E.: Independent Component Analysis. John Wiley & Sons (2001)
9. Clifton, C., Cooley, R.: TopCat: Data mining for topic identification in a text corpus. In: Principles of Data Mining and Knowledge Discovery. (1999) 174–183
10. Dhillon, I.S.: Co-clustering documents and words using bipartite spectral graph partitioning. In: Knowledge Discovery and Data Mining. (2001) 269–274
11. Bingham, E., Mannila, H., Seppänen, J.K.: Topics in 0-1 data. In Hand, D., Keim, D., Ng, R., eds.: Proceedings of the Eighth ACM SIGKDD International Conference on Knowledge Discovery and Data Mining (KDD–2002), Edmonton, Alberta, Canada (2002) 450–455
12. Castelo, R., Feelders, A., Siebes, A.: MAMBO: Discovering association rules based on conditional independencies. LNCS **2189** (2001) 289–298
13. Lee, D.D., Seung, H.S.: Algorithms for non-negative matrix factorization. In: Advances in Neural Information Processing Systems. (2000)
14. Blei, D.M., Ng, A.Y., Jordan, M.I.: Latent Dirichlet Allocation. In: Neural Information Processing Systems 14. (2001)
15. Minka, T., Lafferty, J.: Expectation-propagation for the generative aspect model. In: Proceedings of the 18th Conference on Uncertainty in Artificial Intelligence, Edmonton, Canada (2002)
16. Buntine, W.: Variational extensions to EM and multinomial PCA. In Elomaa, T., Mannila, H., Toivonen, H., eds.: Machine Learning: ECML 2002. Number LNAI 2430 in Lecture Notes in Artificial Intelligence. Springer-Verlag (2002) 23–34
17. Comon, P.: Independent component analysis — a new concept? Signal Processing **36** (1994) 287–314
18. Jutten, C., Herault, J.: Blind separation of sources, part I: An adaptive algorithm based on neuromimetic architecture. Signal Processing **24** (1991) 1–10
19. Das, G., Mannila, H., Ronkainen, P.: Similarity of attributes by external probes. In: Proceedings of the Fourth International Conference on Knowledge Discovery and Data Mining. (1998) 23–29
20. Pajunen, P., Karhunen, J.: A maximum likelihood approach to nonlinear blind source separation. In: Proc. Int. Conf. Artif. Neural Networks. (1997) 541–546

21. Girolami, M.: A generative model for sparse discrete bianry data with non-uniform categorical priors. In: Proc. European Symposium on Artificial Neural Networks, Bruges, Belgium (2000) 1–6
22. Silverstein, C., Brin, S., Motwani, R.: Beyond market baskets: Generalizing association rules to dependence rules. Data Mining and Knowledge Discovery **2** (1998) 39–68
23. Tan, P.N., Kumar, V.: Interestingness measures for association patterns: A perspective. Technical Report TR00-036, University of Minnesota (2000) (KDD 2000 Workshop on Postprocessing in Machine Learning and Data Mining).
24. Mannila, H., Patrikainen, A., Seppänen, J.K., Kere, J.: Long-range control of expression in yeast. Bioinformatics **18** (2002) 482–483

Appendix

Proof of Theorem 1. That the problem is in NP is simple to see: the certificate is the topic vector t, and the formula for $P(x \mid t, \mathcal{T})$ involves multiplying n numbers, each computable in $O(k)$ time.

To show NP-hardness, we reduce SAT to a topic assignment problem. Given a SAT instance of m clauses over n variables, we define a topic model with $2n$ topics and $n + m$ attributes. For each variable V_i, we create two topics T_i and T_i', and one attribute A_i. For each clause C_j, we create one attribute B_j. Each topic has probability 0.5, and each attribute has 0/1 within-topic probabilities as follows: attribute A_i has probability 1 in topics T_i and T_i' and probability 0 in other topics; attribute B_j has probability 1 in the topics T_i such that V_i appears positively in clause C_j and in the topics T_i' such that V_i appears negatively in clause C_j, and probability 0 in all other topics. We consider a data vector where all attributes have value 1.

Now, if the SAT problem has a satisfying truth assignment, it corresponds to a solution of the topic assignment problem where T_i is active if V_i is true and T_i' is active if V_i is false. This solution has likelihood 0.5^n, since exactly n topics are active, and the active topics explain all attributes A_i and B_j. Conversely, if a solution to the topic assignment problem exists such that the likelihood is at least 0.5^n, it must have at most n active topics. To explain attribute A_i, either T_i or T_i' must be active; thus the number of active topics is exactly n, and the solution corresponds to a truth assignment. Since the solution must also explain each attribute B_j, the truth assignment must satisfy the original problem. In summary, the SAT instance has a solution if and only if the topic assignment problem has a solution with likelihood at least 0.5^n. $\square$

Bottom-Up Learning of Logic Programs for Information Extraction from Hypertext Documents

Bernd Thomas

Universität Koblenz-Landau, Institut für Informatik
bthomas@uni-koblenz.de

Abstract. We present an inductive logic programming bottom-up learning algorithm (BFOIL) for synthesizing logic programs for multi-slot information extraction from hypertext documents. BFOIL learns from positive examples only and uses a logical representation for hypertext documents based on the document object model (DOM). We briefly discuss several BFOIL refinements and show very promising results of our IE system LIPX in comparison to state of the art IE systems.

1 Introduction

In the last decade several techniques and systems based on relational learning in the area of information extraction (IE) have been developed [10] . Though a handful approaches [1, 2, 5] exist which capture the idea of bottom-up and top-down rule learning inspired by inductive logic programming (ILP) [12], it is surprising that almost no system [8] tries to follow a pure logical ILP based approach. ILP in general offers broad varieties to be adapted to different problem domains by simply changing the problem representation and/or the hypothesis language. Our aim is to develop an algorithm for learning multi-slot wrappers for hypertext documents, based on logic programming and ILP concepts. This technology can easily be extended with additional information on the representational level (document pre-processing and hypothesis language) and algorithmic level (semantic least general generalization operators).

In Section 2 and 3 we introduce a DOM [4] based representation for hypertext documents and relational representation of text examples. Section 4 briefly explains the hypothesis language and derived example descriptions used for latter bottom-up learning. The **B**ottom-up **F**irst **O**rder **I**nductive **L**earning algorithm and results are presented in Section 5 and 6.

2 Document Representation

Throughout this paper we will focus on HTML documents. It should be noted that the approach presented in this paper is easily adaptable to XML or similar tag-based languages. In order to capture and model the syntactical and hierarchical aspects of HTML and XML documents we define the concept of TDOM-trees, which is strongly related to that of a *document object model* (DOM-tree). A node in a TDOM-tree consists of

N. Lavrač et al. (Eds.): PKDD 2003, LNAI 2838, pp. 435–446, 2003.

four features: a document reference (D_{id}), a node identifier (n_{id}), the corresponding to-ken t describing the document text denoted by the node and an ordered list of child node identifiers $([ch_1, \ldots, ch_n])$. Thus we represent a node in a TDOM-tree as a term $node(D_{id}, n_{id}, t, [ch_1, ch_2, \ldots, ch_n])$. The basic intention of tokens is, like in most other approaches, to group symbols from the text separated by white spaces or other separa-tors to typed words like integer, date, html-tags etc. Each token is represented as a term with a list of feature-value pairs, which is given by: $token([f_1, v_1], \ldots, [f_n, v_n])$, where f_i is an arbitrary feature name and v_i is an arbitrary feature value with $i = 1, n$. For exam-ple $Tok(<img\ src = "a.jpg">) = \{token([(ttype, html), (value,'<img\ src = "a.jpg">')$, $(spos, 0), (epos, 16), (tag, img), (src,' a.jpg')])\}$.

Node identifiers are terms representing a path from the root node to a node in the TDOM. To illustrate the idea of node identifiers assume every node in a tree is as-signed a unique number. The function $child : N_0 \times N_0 \to N_0$ computes for a given node number i and $n \in N_0$ the n-th unique child node of i. For example the term $child(child(child(root, 1), 0), 3)$ refers to the fourth child of the first child of the sec-ond child of the root node in the TDOM. For better readability and later handling we use a prolog list notation $[1, 0, 3]$, leaving out the root node, to denote node identifiers. Hence a *node identifier* is used to assign a unique term to each node in a TDOM. It also provides information about the position in the TDOM-tree. In fact, the notation of node identifiers is strongly related to the *Dewey-Notation* [18]. A leaf node in a DOM-tree represents text appearing at the "surface" of the hypertext document. For example a whole paragraph may be associated with one leaf node in a DOM-tree. In many cases, this representation is not accurate enough for IE tasks. We modify the concept of a DOM-tree such that a leaf node in a DOM-tree becomes many leaf nodes in a TDOM-tree. Each of these nodes represent one token from the text.

Given this notation, an arbitrary HTML document D can be represented as a set of ground unit clauses describing a TDOM model of D. $\mathcal{T}(D_i)$ denotes the TDOM of D with $D_{id} = i$. A $\mathcal{T}(D_0)$ representation for an example HTML page is shown in Figure 1. To be able to compare node identifiers we define the following order relation. A node identifier n_i is smaller than a node identifier n_j written $n_i < n_j$ iff $\exists x \in N_0 : n_j.x > n_i.x \wedge \forall y \in N_0 : y < x$ it holds that $n_j.y = n_i.x$ where $n_i.n$ denotes the n-th child number (starting from left) of a node identifier. Two node identifiers n_i and n_j are equal if they have the same length and $n_i \not< n_j \wedge n_j \not< n_i$. For example: $[0, 0, 3] < [0, 2]$.

Node identifiers have nice properties for wrapper-learning. Similar to expressions in the XPATH language [19] node identifier expressions can be used to refer to more than one node by the use of variables. The node identifier $[0, 1, 1, X]$ refers to every child node of the <ul> environment of Figure 1. For example, the term $[X, 3]$ refers to all child nodes of the root nodes with at least 3 child nodes. It is important to point out that vari-ables can only be substituted by one value and not by partial node identifier expression like $[0, 1]$. Furthermore additional constraints can be introduced by using one variable more than once (e.g. $[0, X, 2, X, 0]$) or more than one variable (e.g. $[Y, X, 2, X, Y]$. Then pattern variables with the same name are not treated disjunctively and thus have to be instantiated with the same value. In fact, in the XPATH query language such expres-sions can only be expressed by means of iterative programming language constructs like for-loops and thus are not as elegant and compact and easy to handle.

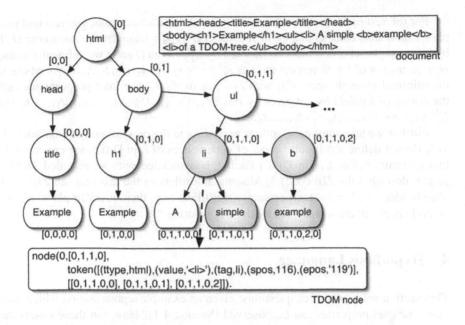

Fig. 1. HTML document, simplified TDOM-tree, TDOM node and span ([0,1,1,0],1,2)

This notation makes it easy to generalize on node identifiers by means of lgg operations [15]. Assume one text example is located in one document in node $[0,1,1,0,0]$ and in the other document in node $[0,1,1,1,0]$. A reasonable first step in learning an extraction rule is the assumption that all nodes described by the generalized node identifier $[0,1,1,X,0]$ are good extractions.

3 Example Representation

One essential concept of our approach is that of *span*. Informally spoken a span determines a subtree in a TDOM-tree. We pick up the idea mentioned by [3] where a span is defined as a triple consisting of a node identifier N and a left and right delimiter L,R. Delimiters determine the left and right boundaries of an interval of child nodes contained in a span. For example the span $([0,1,1,0],1,2)$ of the example TDOM (Figure 1) refers to the set of node identifiers $\{[0,1,1,0,1],[0,1,1,0,2],[0,1,1,0,2,0]\}$. More precise: a span $S = (N,R,L)$ is the set of all reachable descendant nodes starting at the i-th child node of node N with $i = R..L$. In general we assume a depth first traversal to enumerate all nodes of a span to ensure the left to right order of the text at the surface of a document.

A *minimal example span MS* for a given text T is the span with the least cardinality including the text T. For example let T be a text fragment from the document (Figure 1) like simple example and S_1 be a span with $([0,1,1],0,1)$ and S_2 be the span from our previous example. Clearly both S_1 and S_2 contain T but $card(S_1) > card(S_2)$ and therefore S_2 is the only existing minimal example span of T with respect to the example TDOM because: $\neg \exists S' : card(S') < card(S_2)$ where S' is a span including T.

For the rest of the paper we focus on multi-slot extraction tasks, where a text example t with n slots consists of a tuple of texts $<t_1,\ldots,t_n>$ taken from a document D. The initial example set of text tuples is denoted by E_T^D. Given D and t we define the *example representation* of t with respect to D as $e_t^D := <<s_1,\ldots,s_n>,<t_1,\ldots,t_n>>$ where s_i is the minimal example span of t_i with $i = 1..n$ in $\mathcal{T}(D)$. For later purposes we define the notion of a *validation set* given by $VS(E_T^D,p) = \{p(D_{id},[s_1,\ldots,s_n],[t_1,\ldots,t_n]) \mid t \in E_T^D \wedge e_t^D = <<s_1,\ldots,s_n>,<t_1,\ldots,t_n>>\}$

Further we take some assumptions according to the presentation of examples: 1) t_1 to t_n do not define a particular order of occurrences of t_i in D (i.e. we can not follow that t_i occurs before t_{i+1} in D). 2) Each t_i is associated with an intended semantics (e.g. t_i describes the ZIP code) 3) Missing slot fillers in the text (e.g. no ZIP field or placeholder stated in the text) or empty slot fillers (e.g. there is a ZIP placeholder but no code is given) are represented by the empty string "".

4   Hypothesis Language

This section will cover three questions: given an example representation which important relational properties can be observed (Section 4.1)? How can these observations be represented? How are these representations used to define a hypothesis language for inductive learning of extraction rules (Section 4.2)?

4.1   Observing Example Properties

We write $s.n$, $s.l$ and $s.r$ to refer to the components of a span $s := (n,l,r)$. Given an example representation e_t^D we investigate each tuple argument t_i and its span s_i according to the following four levels. Note, the following predicates can be exchanged by arbitrary other ones describing relational information regarding the training examples.

Structural Level: the position of a span s_i and its neighbor nodes are investigated: $xpath(D_{id},s,tl)$ holds if D_{id} is a document id (Section 2), s is a span and tl is the list of tokens associated with each node following the path from the root node to the node of s.
$xspan(D_{id},s,tl)$ holds if tl is the associated list of tokens of all nodes of span s.
$xright\_brother(D_{id},n,t_r)$ holds if n is a node identifier and t_r is the associated token of the right neighbor node of n. Analogously we define a left brother predicate.

Textual or Content Level: a relation between the example text, its tokens associated with the leaf nodes and its span is defined:
$span\_text\_and\_tokens(D_{id},s,t,tl)$ holds if tl is the list of tokens associated with all leaf nodes of span s for text t.

Delimiter Level: predicates to incorporate a widespread idea of IE approaches to learn right and left delimiters of relevant text parts are defined:
$start\_end\_nodes(D_{id},t,n_l,n_r)$ holds if n_l is the start node and n_r the end node of text t in $\mathcal{T}(D)$ referred to by D_{id}.

xpredecessor(D_{id}, n, n_i, tl) holds if the token list tl contains the tokens associated with the first n nodes we meet going backwards in a depth first search[1] to n_i. Analogously we define *xsuccessor* to collect all n successor tokens we meet by a depth first traversal after having met n_i. We call n the *context distance*.

```
extract(D, [[0,1,0,9,X,5] : 0 : R], [[EIER]]) :-
   xpath(D, [0,1,0,9,X,5],
         [token([(ttype,html), (value,'<html>'), (tag,html), (spos,'0'), (epos,'5')]),
          token([(ttype,html), (value,'<center>'), (tag,center), (spos,'136'), (epos,'143')]),
          token([(ttype,html), (value,'<nobr>'), (tag,nobr), (spos,'146'), (epos,'151')]),
          token([(ttype,html), (value,'<table border cellpadding,2>'), (tag,table), (border,''),
                 (cellpadding,'2'), (spos,'1090'), (epos,'1117')]),
          token([(ttype,html), (value,'<tr>'), (tag,tr), (spos,V19), (epos,V20)])]),
   unify(C1, 0), unify(C2, R), member(C1, [0]), member(C2, [0, 1, 2]),
   xspan(D, [0,1,0,9,X,5] : 0 : R,
         [token([(ttype,html), (value,V1), (tag,td), (align,V2), (spos,V3),(epos,V4)]), T1ITR1]),
   xleft_brother(D, [0,1,0,9,X,5],
         [token([(ttype,html), ((value,'<td align,right>')), (tag,td), (align,right), (spos,V5), (epos,V6)])]),
   xright_brother(D, [0,1,0,9,X,5],
         [token([(ttype,html), (value,V7), (tag,td), (align,V8), (spos,V9), (epos,V10)])]),
   span_text_and_tokens(D, [0,1,0,9,X,5] : 0 : R, [EIER], [T1ITR1]),
   start_end_nodes(D, [0,1,0,9,X,5] : 0 : R, [0,1,0,9,X,5,0], [0,1,0,9,X,5,R]),
   xpredecessor(D, 7, [0,1,0,9,X,5,0],
         [token([(ttype,html), value,V1, tag,td, align,V2, spos,V3, epos,V4),
          token([(ttype,html), (value,'<td align,right>'), (tag,td), (align,right), (spos,V5), (epos,V6)]),
          token([(ttype,html), (value,'<td align,right>'), (tag,td), (align,right), (spos,V11), (epos,V12)]),
          token([(ttype,html), (value,'<td align,right>'), (tag,td), (align,right), (spos,V13), (epos,V14)]),
          token([(ttype,html), (value,'<td align,left>'), (tag,td), (align,left), (spos,V15), (epos,V16)]),
          token([(ttype,html), (value,'<td>'), (tag,td), (spos,V17), (epos,V18)]),
          token([(ttype,html), (value,'<tr>'), (tag,tr), (spos,V19), (epos,V20)])]),
   xsuccessor(D, 7, [0,1,0,9,X,5,R],
         [T0, token([(ttype,V21), (value,V22), (spos,V23), (epos,V24)]), T2, T3, T4, T5, T6]),
   xsmallest_common_span(D, [[0,1,0,9,X,5] : 0 : R], [0,1,0,9,X,5] : 0 : R,
         [token([(ttype,html), (value,V1), (tag,td), (align,V2), (spos,V3), (epos,V4)])]).
```

Fig. 2. Learned single-slot rule for QS-vol

Relational Span Level: to figure out relations between spans we define:

xsame_span_node(D_{id}, s_i, s_j) holds if n_i and n_j of spans $s_i = (n_i, l_i, r_i)$ and $s_j = (n_j, l_j, r_j)$ are unifiable.

xnode_less$(D_{id}, n_i, n_j, dist)$ holds if $n_i < n_j$. Where $dist$ is the list of differences between the components of n_j and n_i (e.g. *xnode_less*$(0, [1,4,0], [2,3,0,2], [1,-1,0])$). Analogously we define *xnode_greater*.

overlapping_span$(D_{id}, s_i, tl_i, s_j, tl_j)$ holds if $(s_i.l < s_j.l) \wedge (s_i.r \geq s_j.l) \wedge (s_i.r \leq s_j.r)$ where tl_i and tl_j are the corresponding token lists of s_i and s_j.

span_in_span(D_{id}, s_i, s_j) holds if span s_i is a subtree of span s_j.

---

[1] This captures the idea to interpret the document as a sequence of tokens rather than a tree, and we investigate the n preceding tokens of the token associated with n_i.

$xsub\_related\_span(D_{id}, s_i, s_j)$ holds if $s_j.n$ is a prefix of $s_i.n$ (e.g. $[1,2]$ is a prefix of $[1,2,3]$).

$xsmallest\_common\_span(D_{id}, [s_1, \ldots, s_n], s_x, tk_x)$ holds if s_x is the smallest span (wrt. to its number of nodes) in D_{id} such that each span s_i with $i = 1..n$ is a subtree of s_x and tk_x is the associated token with $s_x.n$.

4.2 Clause Descriptions of Examples

Now that we have defined predicates for the description of text example properties based on the representation of a TDOM, we introduce the concept of a clause description $CD(e_t^D)$ for an example representation e_t^D. In terms of *extensional* and *intentional* *object languages* a clause description of an example is an *intentional object description*. Furthermore we use the same language for the description of objects and hypotheses.

Let $\mathcal{L}_H$ be the set of predicates introduced in Section 4.1. We call this the *hypothesis language* which is used later for construction of rules. This is in analogy to standard ILP algorithms like FOIL [16]. It should be noted that the hypothesis language can be freely chosen. Furthermore let us assume that a logic program $P_{\mathcal{L}_H}$ is given that implements the intended semantics of the predicates in $\mathcal{L}_H$. To denote the union of $P_{\mathcal{L}_H}$ and $\mathcal{T}(D_i)$ we write $P_{\mathcal{L}_H}^D$. Now we can define $CD(e_t^D) = \{l'_\sigma \mid P_{\mathcal{L}_H}^D \vdash l'_\sigma$ with $l \in \mathcal{L}_H$ and l' is l instantiated according to its given semantics with $(s_i, t_i) \in e_t^D$ and σ calculated answer substitution}. Here $\vdash$ denotes the logical derivation operator and we assume a standard logical calculus (e.g. SLD-Resolution [11]).

Finally we define E_u^+ to be the set of clause descriptions for a given set of examples as $E_u^+ = \bigcup_{t \in E_T^D} CD(e_t^D)$. Additionally we extend every $CD(e_t^D)$ with a special predicate, the *rule head* defined as $extract(D_{id}, [s_1, \ldots, s_n], [t_1, \ldots, t_n])$, where every s_i and t_i is instantiated with the associated argument from e_t^D. Then $CD(e_t^D)$ forms a ground instantiated rule of the form $extract(D_{id}, [s_1, \ldots, s_n], [t_1, \ldots, t_n]) \leftarrow l_1, \ldots, l_n$ with $l_i \in \mathcal{L}_H$. Since we focus only on learning non recursive horn clauses, we do not have to use negation operators for the body literals and consider a marked predicate (e.g. $extract$) to build the head and all other literals in $CD(e_t^D)$ to form the body of a rule. Thus every $CD(e_t^D)$ is one rule describing exactly one text example with respect to D and $\mathcal{L}_H$. Accordingly computing all answers to the query $P_{\mathcal{L}_H}^D \cup E_u^+ \vdash extract(D_{id}, [s_1, \ldots, s_n], [t_1, \ldots, t_n])$ provides the validation set $VS(E_T^D, extract)$ (Section 3).

5 BFOIL Algorithm

The central idea of BFOIL is to learn in a bottom-up fashion from positive examples only a set of rules by means of least general generalization techniques [15]. In contrast to the standard *Top-Down* learning approaches starting with the most general hypothesis BFOIL starts with a set of ground rules (clause descriptions) as initial hypothesis and tries to generalize these clause sets by means of lgg operations. The term *clause-lgg* denotes the lgg of two clauses C_1 and C_2 defined as $clause-lgg(C_1, C_2) = \{lgg(l, m) \mid l \in C_1 \wedge m \in C_2 \wedge lgg(l, m)$ is defined}. In general the clause-lgg of two clauses has to be reduced, in the sense that redundant literals under θ-subsumption have to be removed.

Algorithm 5.1 Basic BFOIL algorithm

Require: $P = $ logic program ; $E_T^D = $ positive examples
$\quad\quad E_{learn}^+ \subseteq E_T^D$; $E_u^+ = \bigcup_{t \in E_{learn}^+} CD(e_t^D)$
1: $LearnedRules \leftarrow \emptyset$
2: **while** $E_u^+ \neq \emptyset$ **do**
3: $\quad Rule \in E_u^+$
4: $\quad E_u^+ \leftarrow E_u^+ \setminus \{Rule\}$
5: $\quad ProblemSet \leftarrow \emptyset$
6: $\quad$ **while** $E_u^+ \neq \emptyset$ **do**
7: $\quad\quad X \in E_u^+$
8: $\quad\quad R \leftarrow clause\_lgg(Rule, X)$
9: $\quad\quad$ **if** $apply(R, P, E_T^D).fp > 0$ **then**
10: $\quad\quad\quad ProblemSet \leftarrow ProblemSet \cup \{X\}$
11: $\quad\quad$ **else**
12: $\quad\quad\quad Rule \leftarrow R$
13: $\quad\quad\quad E_u^+ \leftarrow E_u^+ \setminus \{X\}$
14: $\quad LearnedRules \leftarrow LearnedRules \cup \{Rule\}$

Since it is obvious that calculating the clause-lgg of E_u^+ results in one rule that over-generalizes with high probability, BFOIL inductively tries to partition E_u^+ into sets of clauses $C_i \subseteq E_u^+$ such that the clause-lgg of each C_i forms a new rule that does not produce any false positive predictions (extractions). Since we only learn from positive

---

Function 5.2 apply(R,P,V) with false positive calculation

Require: $R :=$ rule ; $P = $ logic program ; $V = $ examples
1: $A \leftarrow \{R_{head}\sigma \mid P \cup R \vdash R_{head}\sigma$ with σ answer subst.$\}$
2: $fp \leftarrow |A \setminus (VS(V, R_{head}) \cap A)|$

---

examples, standard techniques to determine false predictions during the learning phase (validation on negative example sets) are not applicable. To yield good rules anyhow, it is essential to estimate the correctness of rules during learning. Thus we assume that the set E_T^D is exhaustively enumerated. This means every intended extraction from D is contained in E_T^D. Then we can conclude that if a rule extracts a tuple t from D with $t \notin E_T^D$ it is false positive. This introduces a *closed world assumption* [17] similar view on extraction examples and the absence of negative training data.

This seems to be a very strong restriction which requires tedious labeling. But since our approach does not need many examples (5-30 training examples Section 6 Figure 3) only a small number of documents have to be labeled.

In general an IE learning task has to deal with multiple documents $D_1 \ldots D_n$ and examples drawn from $D_1 \ldots D_n$ then we define $E_T^D = \bigcup_{i=1}^n E_T^{D_i}$. Additionally we assume that the logic program P is an implementation of $\mathcal{L}_H \cup (\bigcup_{i=1}^n \mathcal{T}(D_i))$. Algorithm 5.1 shows the basic BFOIL algorithm and Function 5.2 the function *apply* for calculating false positives. In the best case basic BFOIL returns one rule, the clause-lgg of E_u^+. Experiments showed that this happens if examples are identical wrt. to their structural properties in a TDOM. In the worst case basic BFOIL just memorizes each clause in

E_u^+. This might happen if examples are too different wrt. to the expressiveness of $\mathcal{L}_H$ and the clause-lgg leads to over-generalized rules.

5.1 BFOIL Refinements

The results of using basic BFOIL to multi-slot extraction are not satisfying. Imagine a clause $C_1 = CD(e_t^D))$ with $\{xpath(0, [1,2], [\ldots]), \ldots, xpath(0, [1,3,1], [\ldots]), \ldots\}$. The intention of this clause is that the first literal describes path features of the first argument and the second literal describes path features of the second argument of an example.

Algorithm 5.3 Consistent BFOIL algorithm

Require: $P =$ logic program ; $E_T^D =$ positive examples

      $E_{learn}^+ \subseteq E_T^D$; $E_u^+ = \bigcup_{t \in E_{learn}^+} CD(e_t^D)$

 1: *LearnedRules* $\leftarrow \emptyset$

 2: **while** $E_u^+ \neq \emptyset$ **do**

 3:     *Rule* $\in E_u^+$

 4:     $E_u^+ \leftarrow E_u^+ \setminus \{Rule\}$

 5:     *ProblemSet* $\leftarrow \emptyset$

 6:     $C \leftarrow \emptyset$

 7:     **while** $E_u^+ \neq \emptyset$ **do**

 8:       $X \in E_u^+$

 9:       $R \leftarrow clause\_lgg(Rule, X)$

10:       **if** $(apply(R, P, E_T^D, E_{learn}^+, C).fp > 0)$

        or $(\text{not } apply(R, P, E_T^D, E_{learn}^+, C).consistent)$ **then**

11:          *ProblemSet* $\leftarrow$ *ProblemSet* $\cup \{X\}$

12:       **else**

13:          *Rule* $\leftarrow R$

14:          $C \leftarrow C \cup \{X\}$

15:          $E_u^+ \leftarrow E_u^+ \setminus \{X\}$

16:     *LearnedRules* $\leftarrow$ *LearnedRules* $\cup \{Rule\}$

Calculating the clause-lgg of C_1 and C_2 generalizes each *xpath* literal in C_1 with each *xpath* literal in C_2. This is not what we want. Only the lgg of *xpath* literals describing the same argument i should be calculated from both clauses. With a simple syntactic transformation before the calculation of an lgg and re-transformation before evaluation of a generalized clause (rule) we can still use the standard lgg operation for learning. Adding a prefix arg_i to every predicate symbol of each literal in E_u^+ prevents the lgg to generalize from non-intended literals. This *prefix protection* is more an issue of representation than a refinement of the BFOIL algorithm.

    The basic BFOIL algorithm is not consistent (e.g. learned rules may not cover examples from E_{learn}^+). Imagine two examples e_1 and e_2. The second argument of e_1 is empty. Due to P and $\mathcal{L}_H$ the clause description for e_1 would not contain literals for the description of argument 2 to reduce the complexity associated with *empty substitutions*. Because of the absence of these literals the clause-lgg eliminates the literals for argument 2 from clause two. It is possible that the new rule still covers e_1 and does not produce any false positives, but does not cover e_2 anymore. For this reason, we keep track of examples that had been used successfully for learning the current rule (line 14

Algorithm 5.3). Every rule refinement (line 9) must cover all examples that have been successfully used in previous learning steps (line 10). Function 5.4 implements this test.

Function 5.4 apply(R,P,V,L,C) with consistency check

Require: R :=rule ; P = logic program ;
   $L \subseteq V$ = examples ; C = example descriptions
1: $A \leftarrow \{R_{head}\sigma \mid P \cup R \vdash R_{head}\sigma$ with σ answer subst.$\}$
2: $fp \leftarrow |A \setminus (VS(V,R_{head}) \cap A)|$
3: *consistent* $\leftarrow$ *true*
4: **while** $C \neq \emptyset \wedge consistent$ **do**
5: $c_e \in C$
6: $e \in L \wedge e$ is described by c_e
7: **if** $e \notin A$ **then**
8:   *consistent* $\leftarrow$ *false*
9: **else**
10:   $C \leftarrow C \setminus \{c_e\}$

A third refinement of BFOIL is the modification of the clause-lgg operator. Therefore we introduce the concept of a *semantic lgg operator*. Semantic lgg operators are closely related to the chosen hypothesis language and example representation in general. The key idea is to guide the lgg operation by additional knowledge to prevent over-generalization. For example the lgg of spans and the generalization of *xspan* literals tend to blow up the search space. The lgg of $xspan(0,([1,2,3],3,6),[...])$ and $xspan(0,([1,2,3],1,10),[...])$ is $xspan(0,([1,2,3],X,Y),[...])$ which is obviously to general from a practical point of view. For this reason we define additional semantical lgg operators. These operators provide semantical based generalization by adding special literals to the lgg of two clauses. We denote a semantic lgg operator similar to an inference rule:

$$\frac{C1\setminus\{xspan(D_1,(N_1,L_1,R_1),TL_1)\} \quad C2\setminus\{xspan(D_2,(N_2,L_2,R_2),TL_2)\}}{\{member(L,[L_1,...,L_2]),member(R,[R_1,...,R_2])\} \cup CL}$$

with $CL = clause\_lgg(C1,C2)$ and $xspan(D,(N,L,R),TL) \in CL$.

Extending the standard clause-lgg with semantic lgg operators can reduce the search space significantly, resulting in faster learning and extraction times. Especially if spans in a document are huge, the insertion of the member predicates are of practical relevance. Instead of considering all possible instances for the left and right delimiter of the span, they are constrained to take only values between the smallest and the greatest value seen so far. All results presented in this paper have been generated by using only one semantic lgg operator, that is for the *xspan* literal.

6   Results and Conclusion

We tested the BFOIL algorithm with our extraction system LIPX on the RISE repository [13]. RISE contains document resources with an extraction task description taken from various IE research papers and projects. Most publications refer to these problem cases as kind of standard tests. Unfortunately not all approaches give a complete overview of their results with respect to *precision* and *recall* values. We focused on extraction tasks from HTML documents only and learned multi-slot extraction rules for HTML resources as described in the RISE repository.

All tests were ran using a fixed number of randomly drawn examples to perform 20 learning and test runs for each problem class. The settings are shown in the first table of Figure 3 where t = no. total tuples; e^+ = no. examples and r = average no. of learned rules. For each problem class the learning examples were randomly drawn from one half of the available documents. The testing set consisted of all documents, but only the data tuples not used for learning were considered. An extraction was counted as

LIPX	multi-slot			single-slot			
	t	e^+	r	slot	t	e^+	r
CS				name	1151	20	7.2
Bigbook	204	30	1.6				
IAF	84	30	13.3	altname	11	6	4.7
				org		53	20	8.7
LA	157	5	1.4	CC	144	10	2.4
Okra	3335	25	1				
Quote	25	14	6.8	date	21	6	1
				vol	25	10	1.6
Zagats	140	30	1.3	addr	140	30	1.3

Fig. 3. Test settings

correct when all of its slots where correctly extracted. Values for precision, recall and F1 are displayed in percentages, all other values in totals. For all tests we used the hypothesis language described in Section 4.1 with context distance $n = 7$. Figure 4 shows the best F1 (harmonic mean of precision and recall) values.

Comparing LIPX results with other multi-slot IE-systems is not straightforward, because almost all systems set up different evaluation scenarios with respect to the number of examples, their selection criteria and the number of test iterations. The first table of Figure 5 shows the results (median) for single and multi-slot learning in comparison[2]. to the systems SoftMealy [7], Stalker [14] and Wien [9]. Even though

LIPX multi-slot	t	e^+	r	Pre	Rec	F1
Bigbook	204	30	1	100	98.3	99.1
IAF	84	30	12	85	31.5	46
LA	157	5	2	100	25.1	40.2
Okra	3335	25	1	100	95.4	97.7
Quote	25	14	7	100	63.6	77.8
Zagats	140	15	1	98.2	97.3	97.7

Fig. 4. Best F1 multi-slot results

LIPX is developed for multi-slot tasks we tested it on single slot extraction tasks to provide a comparison to one state of the art single slot extraction approach (BWI) of [6]. These results are listed in the second part of Figure 5. While learning single slot wrappers supersedes the *relational span level* predicates the single slot learning results also underline the high precision values observed with multi-slot learning. In 5 out of 7 cases LIPX shows better or equal precision values than BWI and BWI HMM. This is not too surprising, because in the worst case BFOIL only memorizes the examples. This does not happened with these test cases, but in two cases the F1 rates due to the low recall rate are not acceptable. There are mainly two interacting reasons for this behavior, which build a general observation for multi and single-slot learning. First, BFOIL seems to yield bad recall rates if only a few examples are present and those differ strongly regarding their relational description (Quote, IAF-altname). Secondly, some tests where run with only a few training examples, because of the bad runtime behavior caused by

---

[2] All values for SoftMealy, Stalker and Wien are taken from [7].

BFOIL's naive answer set computing for each generalized rule. Consequently the recall rate was low (LA, LA-cc, CS-name).

The presented approach offers a wide variety for extensions by modifying the token representation of text units for richer semantic text pre-processing. This allows to incorporate linguistic or additional general semantic information. By modification of the underlying hypothesis language we can adapt the presented approach to other mark up languages or focus on different relationships than those stated in this paper. Using natural language tools (e.g. part of speech tagger) for the pre-processing of documents in combination with an XML representation of such pre-processed documents also allows us to apply our methods to natural language texts. By extending the BFOIL algorithm with additional semantic lgg operators the hypothesis search space can be constrained and runtime behavior improved. An additional modification to increase the recall rate is, to accept rules that cover a small number of false positives. This modification was not tested yet. But it is easily accomplished by incorporating a threshold (e.g. if the percentage of false positive extractions is below 0.03 % (algorithm 5.1 line 9)). These observations show that all results presented in this paper de-

multi-slot comparison	Pre	Rec	F1	Pre	Rec	F1
	BigBook			IAF		
LIPX	100	89.3	94.4	84	19.4	32
SoftMealy		100			41 to 99	
Stalker		97			85 to 100	
Wien	100	100	100	too hard		
	Quote			Okra		
LIPX	100	24	38.8	100	88.1	93.7
SoftMealy		85			100	
Stalker		79			97	
Wien	too hard				100	
	LA			Zagats		
LIPX	100	12.3	21.8	97.9	85.5	90.6
single-slot comparison	Pre	Rec	F1	Pre	Rec	F1
	CS name			LA cc		
LIPX	95.6	9.7	16.9	96.7	54.8	70.4
BWI	77,1	31.4	44.6	99.6	100	99.8
BWI HMM	41.3	65	50.5	98.5	100	99.2
	IAF altname			IAF org		
LIPX	100	20	33.3	77.5	37.7	51.3
BWI	90.9	43.5	58.8	77.5	45.9	57.7
BWI HMM	1.7	90	3.3	16.8	89.7	28.3
	QS date			QS vol		
LIPX	100	100	100	100	92.9	96.3
BWI	100	100	100	100	61.9	76.5
BWI HMM	36.3	100	53.3	18.4	96.2	30.9
	Zagats addr					
LIPX	98	93.6	95.4			
BWI	100	93.7	96.7			
BWI HMM	97.7	99.5	98.6			

Fig. 5. Median results and comparison

pend strongly on the chosen hypothesis language and the degree of additional information chosen for the representation of TDOM nodes. So far we only made experiments with the one mentioned in Section 4.1 without any fine tuning (e.g. context distance, sem-lgg). LIPX shows partially bad learning time results, which clearly stems from the combinatorial explosion while applying a rule that became too general during the learning process. In fact, evaluating each new rule by computing the answer set for it leads to this problem. Thus we are doing research on using more efficient proof procedures than SLD-Resolution, clustering of example description rules and extending BFOIL with specification operators to minimize this problem. To summarize the capabilities: LIPX can learn single and multi slot wrappers for HTML or XML documents. It can handle slot fillers occurring in varying orders in the texts and it can handle slots that may be empty, missing or nested. Though the presented approach shows very promising results its runtime behavior is a major subject for improvement. Nevertheless the pure logic programming motivated and based technique to learn multi-slot wrappers, the general method of lgg operations for learning and its independency of the application domain, are auspicious properties.

References

1. M. E. Califf. *Relational Learning Techniques for Natural Language Information Extraction*. PhD thesis, University of Texas at Austin, August 1998.
2. F. Ciravegna. Learning to Tag for Information Extraction from Text. In *Workshop Machine Learning for Information Extraction, European Conference on Artifical Intelligence ECCAI*, August 2000. Berlin, Germany.
3. W. Cohen, M. Hurst, and L. S. Jensen. A flexible learning system for wrapping tables and lists in html documents. In *The Eleventh International World Wide Web Conference WWW-2002*, 2002.
4. W3C, Document Object Model (DOM) Level 2 Core Specification , 2000.   Version 1.0 http://www.w3.org/TR/DOM-Level-2-Core/.
5. D. Freitag. *Machine Learning for Information Extraction in Informal Domains*. PhD thesis, Computer Science Department, Carnegie Mellon University, Pittsburgh, PA, November 1998.
6. D. Freitag and N. Kushmerick. Boosted Wrapper Induction. In *Proceedings of the Seventh National Conference on Artificial*, pages 577–583, July 30 - August 3 2000. Austin, Texas.
7. C.-N. Hsu and C.-C. Chang. Finite-State Transducers for Semi-Structured Text Mining. In *Workshop on Text Mining IJCAI 99*, 1999.
8. M. Junker, M. Sintek, and M. Rinck. Learning for Text Categorization and Information Extraction with ILP. In *Proc. Workshop on Learning Language in Logic*, June 1999. Bled, Slovenia.
9. N. Kushmerick. *Wrapper Induction for Information Extraction*. PhD thesis, University of Washington, 1997.
10. N. Kushmerick and B. Thomas.   *Intelligent Information Agents R&D in Europe: An AgentLink perspective*, chapter Adaptive Information Extraction: A Core Technology for Information Agents. Springer, 2002.
11. J. Lloyd. *Foundations of Logic Programming*. Springer-Verlag, 2 edition, 1987.
12. S. Muggleton and L. D. Raedt. Inductive logic programming: Theory and methods. *Journal of Logic Programming*, 1994.
13. I. Muslea. The RISE Repository, 1999.   http://www.isi.edu/~muslea/RISE/.
14. I. Muslea, S. Minton, and C. Knoblock. A hierarchical approach to wrapper induction. In O. Etzioni, J. P. Müller, and J. M. Bradshaw, editors, *Proceedings of the Third International Conference on Autonomous Agents (Agents'99)*, pages 190–197, Seattle, WA, USA, 1999. ACM Press.
15. G. Plotkin. A note on inductive generalization. *Machine Intelligence*, (5):153–163, 1970. Edinburgh Univ. Press.
16. J. R. Quinlan. Learning logical definitions from relations. *Machine Learning*, 5:239–266, 1990.
17. R. Reiter. On Closed World Data Bases. In H. Gallaire and J. Minker, editors, *Logic and Data Bases*. Plenum Press, New York, 1978.
18. M. L. Scott. *Dewey Decimal Classification: A Study Manual and Number Building Guide*. Libraries Unlimited, 1998.
19. W3C, *xpath specification*, 1999.   http://www.w3.org/TR/xpath.

Predicting Outliers

Luis Torgo[1] and Rita Ribeiro[2]

[1] LIACC-FEP, University of Porto, R. Campo Alegre, 823, 4150 Porto, Portugal
ltorgo@liacc.up.pt
http://www.liacc.up.pt/~ltorgo
[2] LIACC, University of Porto, R. Campo Alegre, 823, 4150 Porto, Portugal
rita@liacc.up.pt

Abstract. This paper describes a method designed for data mining applications where the main goal is to predict extreme and rare values of a continuous target variable, as well as to understand under which conditions these values occur. Our objective is to induce models that are accurate at predicting these outliers but are also interpretable from the user perspective. We describe a new splitting criterion for regression trees that enables the induction of trees achieving these goals. We evaluate our proposal on several real world problems and contrast the obtained models with standard regression trees. The results of this evaluation show the clear advantage of our proposal in terms of the evaluation statistics that are relevant for these applications.

1 Introduction

The work described in this paper addresses applications where the main objective is to model rare extreme values, usually known as outliers. Given that the target variable is continuous we are facing regression problems. However, the main difference to standard regression tasks is that our main interest is to predict accurately the occurrences of rare high or low values of the target variable. A typical real world application is the prediction of stock market returns, where small and highly frequent returns are irrelevant for investors, while large movements of the market are the key events where accurate prediction pays off. Our interest is not only to anticipate the occurrence of an extreme value but also to be accurate at predicting its concrete value, because the amplitude of the outlier is relevant for the user of these applications, as it may lead to differentiated actions. Another major requirement of our target applications is the interpretability of the models. This means that discovering the conditions that lead to these extreme values is also a major goal of our models.

Applications where the main modeling objective are rare events abound in recent data mining literature. Nevertheless, existing related work is mostly focused on discrete target variables (i.e. classification tasks). These works include topics like activity monitoring [4], prediction of rare events [17,18], anticipation of surprising patterns [7], novelty detection, anomaly detection, among others. Most of this research is also linked to applications where a data stream is being

N. Lavrač et al. (Eds.): PKDD 2003, LNAI 2838, pp. 447–458, 2003.
© Springer-Verlag Berlin Heidelberg 2003

monitored with the goal of anticipating rare events, that is time-dependent data. This research is usually focused in the task of distinguish between interesting cases and "normal" occurrences.

The importance and impact of rare cases has been the topic of research on small disjuncts (e.g. [6,19]). This research is again mainly focused on classification tasks and is also strongly related to the study of applications with unbalanced class distributions (e.g. [5]).

A frequent strategy to bias the models towards being accurate in particular types of cases is the use of differentiated misclassification costs (e.g. [16]). This is a common practice in classification tasks and was also used in solving regression problems through a classification approach [15].

All these classification approaches do not solve the problem of being able to accurately predict the specific value of outliers, and are particularly inadequate when these spread over a wide range of values. If the amplitude of the extreme values is relevant for the user, for instance for taking different actions, all these approaches based on classification are not applicable. Obviously, one could further divide the classes representing the extreme values into more specific classes to differentiate their importance but that would mean that we would partition an already low populated class into several classes, thus making our modeling task even more difficult. As such, for this kind of applications only a regression model can handle the problem properly.

Buja and Lee [2] have recently presented a series of new splitting criteria for both classification and regression trees that address related problems. Regarding regression, they propose two different splitting criteria with two objectives: identifying extreme buckets of the data; and identifying pure (low variance) buckets. The first objective is particularly related to ours. The goal of Buja and Lee is to identify areas of the regression surface where the target variable shows a high or low mean value. Although our goal is related to this, we are particularly interested in applications where these extreme values are rare, which demands for specific criteria.

We propose a new splitting criterion for regression trees which enables the induction of models that meet our application requirements. In Section 2 we formalize our target problems and propose evaluation criteria that should guide the search for the best models. Section 3 describes the details of our proposal. The experimental evaluation of this proposal is presented in Section 4. We finish with the conclusions of this work and future research directions.

2   Problem Formulation

In this section we present a general description of our problem. Let D be a data set, consisting of n cases $\{\langle \mathbf{x}_i, y_i \rangle\}_{i=1}^{n}$, where $\mathbf{x}_i$ is a vector of p discrete or continuous variables, and y_i is a continuous target variable value. As we have mentioned before, we are interested in models that are able to predict accurately rare extreme values of Y. To achieve this goal we need to formalize the notion of rare extreme values. We use the statistical notion of outlier with this purpose.

Box plots are visualization tools that are often used to identify outliers. Extreme values are defined in these plots as values above or below the so-called adjacent values [3]. Let r be the interquartile range defined as the difference between the 3rd and 1st quartiles of the target variable. The upper adjacent value, adj_H, is defined as the largest observation that is less or equal to the 3rd quartile plus $1.5r$. Equivalently, the lower adjacent value, adj_L, is defined as the smallest observation that is greater or equal to the 1st quartile minus $1.5r$. Given these two limits we can define our rare extreme values as,

$$O = \{y \in D \mid y > adj_H \vee y < adj_L\}$$
$$O_H = \{y \in D \mid y > adj_H\} \tag{1}$$
$$O_L = \{y \in D \mid y < adj_L\}$$

Depending on the application we may have either O_L or O_H empty[1]. Figure 1 shows the box plots of the targets in two applications where we have different types of outliers. These values are drawn with circles in these graphs.

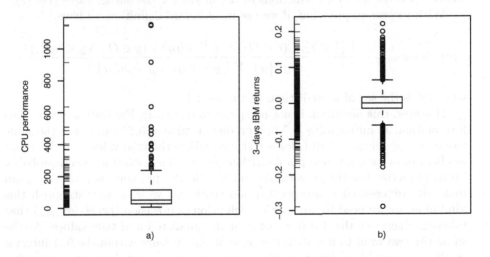

Fig. 1. Two example box plots with different types of extreme values: a) The relative performance of a set of CPUs; b) The 3-days returns of IBM closing prices.

Having described the main features of our target applications we need to define some evaluation criteria to guide the search for the best models. Typical performance measures used in regression settings, such as the mean squared error, are inadequate as they do not stress the fact that we are only interested in the performance in extreme values. This is the same kind of phenomenon as

---

[1] We will discard applications where both sets are empty as these are not relevant for this study.

the one reported regarding the use of classification accuracy on problems with unbalanced class distributions [8,10].

In the information retrieval literature (e.g. [9]) the notion of relevance seems particularly adequate to our needs. Relevance is defined as the value or utility of a system output as a result of a user search. Relevance is most of the times assessed using two measures: *precision* and *recall*. Precision is defined as the proportion of the cases predicted as target events that really are target events. Recall is defined as the proportion of existing target events that are captured by the model. Our proposal consists of adapting these two measures to our problem setup with the goal of developing a learning tool that maximizes the relevance of the induced model to our application goals.

We define recall in the context of our target applications as the proportion of outliers in our data that are predicted as such (i.e. covered) by our model,

$$recall = \frac{\mid \{\hat{y} \in \hat{Y}_O \mid (y \in O_H \wedge \hat{y} > adj_H) \vee (y \in O_L \wedge \hat{y} < adj_L)\} \mid}{\mid O \mid} \quad (2)$$

where $\hat{Y}_O$ is the set of $\hat{y}$ predictions of the model for the outlier cases (i.e. O).

With respect to precision, if we use its standard definition we have,

$$precision_{stand} = \frac{\mid \{\hat{y} \in \hat{Y}_O \mid (y \in O_H \wedge \hat{y} > adj_H) \vee (y \in O_L \wedge \hat{y} < adj_L)\} \mid}{\mid \{\hat{y} \in \hat{Y} \mid \hat{y} < adj_L \vee \hat{y} > adj_H\} \mid} \quad (3)$$

where $\hat{Y}$ is the set of $\hat{y}$ predictions of the model.

However, this definition is not adequate to our goals. For instance, with this formulation, assuming $adj_H = 5.6$, a predicted value of 5.8 would have the same value as a prediction of 10.1, for a test case where the true value is 10.5. In our applications this is not acceptable. Otherwise, the best solution would probably be to discretize the target variable and handle the problem as a classification task with differentiated misclassification costs. As we want to distinguish this kind of errors we need to use another definition of precision ($precision_{regr}$) that takes into account the distance between the predicted and true values. At the same time we want to maintain the scale of the measure within the 0..1 interval so that we are able to integrate recall and precision into a single measure using standard approaches. Our proposed definition of $precision_{regr}$ is the following,

$$precision_{regr} = 1 - NMSE_O \quad (4)$$

where $NMSE_O$ is the normalized squared error of the model for the outliers,

$$NMSE_O = \frac{\sum_{y_i \in O} (\hat{y}_i - y_i)^2}{\sum_{y_i \in O} (\bar{Y} - y_i)^2} \quad (5)$$

The value of $NMSE_O$ will usually be between 0 and 1. For the cases where this value goes above 1, which means that the model is performing worse than the naive average model, we consider that the precision of the model is 0.

Obtaining an overall evaluation measure from the values of recall and precision provides a global preference criterion that can be used to guide the search for the models. The F-measure [11] is among the most used measures and is defined as,

$$F = \frac{(\beta^2 + 1) \cdot precision \cdot recall}{\beta^2 \cdot precision + recall} \tag{6}$$

where β controls the relative importance of recall to precision. This is the definition we use replacing $precision$ by our proposed $precision_{regr}$.

3   An Approach Using Regression Trees

Regression trees are known for their computational efficiency, model interpretability and competitive accuracy. For these reasons we have decided to use these models as the base paradigm behind our proposal.

Standard regression trees are obtained using a procedure that minimizes the squared error. This means that the best splits for each tree node are chosen to minimize the weighed squared error between the two branches. As mentioned by Buja and Lee [2] this criterion is not adequate for several data mining applications. That is also the case of our target problems. Moreover, outliers can be a problem for standard regression trees as they may distort the selection of the best splits and may also have a large impact on the average values chosen for the leaves of the trees [14].

The main idea of our proposal to avoid the problems reported above is to use the F-measure presented in Equation (6) to guide the split selection procedure used to grow the trees. As such, the key distinguishing feature of our method is the criterion used to select the best test for each tree node. In our proposal the best split s^*, is chosen using the following criterion,

$$s^*(D_t) = \max_{s \in S} \max\left(F(D_{t_L}), F(D_{t_R})\right) \tag{7}$$

where S is the set of trial splits for the node t [2]; D_{t_L} is the subset of cases in t (D_t) that satisfy the test s (i.e. the left sub-branch of t), while D_{t_R} contains the remaining cases (i.e. $D_{t_R} = D_t - D_{t_L}$); and $F(D)$ is the F-measure for a set of cases.

In order to obtain the F-measure for the branches of a candidate split we need to obtain the values of precision and recall, which we do using the following formulas,

---

[2] That are the same as in a standard regression tree.

$$
precision_{regr_t} = \begin{cases} 1 - \dfrac{\displaystyle\sum_{y_i \in O_H(D_t)} (\bar{y}_t - y_i)^2}{\displaystyle\sum_{y_i \in O_H(D_t)} (\bar{Y} - y_i)^2} & \text{if } \bar{y}_t > adj_H \ \vee \ \bar{y}_t > \tilde{y}_t \\[4ex] 1 - \dfrac{\displaystyle\sum_{y_i \in O_L(D_t)} (\bar{y}_t - y_i)^2}{\displaystyle\sum_{y_i \in O_L(D_t)} (\bar{Y} - y_i)^2} & \text{if } \bar{y}_t < adj_L \ \vee \ \bar{y}_t \leq \tilde{y}_t \end{cases} \tag{8}
$$

where $O_H(D_t)$ $(O_L(D_t))$ is the set of cases of node t that belong to $O_H(O_L)$; $\bar{y}_t$ is the average Y value in the node; $\tilde{y}_t$ is the median Y of the node; and $\bar{Y}$ is the average Y in the training data.

This means that depending on the value of the node average we consider this branch as a tentative to predict high or low outliers, and calculate its precision accordingly. Even if the node average is not in the outlier range of values we still calculate the precision in the node, using the median as a threshold for deciding whether to calculate it with respect to high or low outliers.

Regarding recall we use,

$$
recall_t = \begin{cases} 0 & \text{if } \bar{y}_t \geq adj_L \ \wedge \ \bar{y}_t \leq adj_H \\[2ex] \dfrac{|\ y \in D_t \ \wedge \ y \in O_H \ |}{|\ O_H\ |} & \text{if } \bar{y}_t > adj_H \\[2ex] \dfrac{|\ y \in D_t \ \wedge \ y \in O_L \ |}{|\ O_L \ |} & \text{if } \bar{y}_t < adj_L \end{cases} \tag{9}
$$

When a trial split leads to a branch having an average target value that is not an outlier, the respective recall is zero. This would lead to an F value of zero according to Equation (6). This is a common situation particularly in top level nodes, where the partitions are still too big, and thus the average Y is seldom an outlier. Moreover, sometimes all trial splits for a node are in these circumstances. This means that we are not able to select the best split for these nodes as all splits have the same score, and thus the tree growth procedure would stop prematurely. These situations occur because in complex applications we seldom find a single split that is able to isolate extreme values in one of the branches so that the branch has an average target that is an outlier. This problem decreases as the tree grows because the number of cases in the nodes gets smaller and thus finding such splits is easier. Although these top level splits have zero recall we should still be able to establish a preference criterion to select one, because we can calculate their precision. In order to overcome this difficulty we have added a small threshold[3] to the value of recall in Equation (6) so that the value of F is not zero even when the recall is null.

Summarizing, our proposal consists of selecting the splits that are able to generate a branch (a subset of cases) with a high value of the F-measure. Notice, that we do not search for a weighted solution between the two branches.

---

[3] We have used the value of 0.001 in our experiments.

Even if one of the branches as a poor F score, as long as the other achieves a high F-measure we have a good candidate split. This strategy is similar to the one followed by Buja and Lee [2], which also do not search for splits with a good compromise between the left and right branches. These strategies lead to unbalanced trees. Still, we share the opinion of Buja and Lee that consider these trees more interpretable.

Another important question that needs to be addressed when developing a tree-based system, is the tree growth stopping criteria. This is a statistical estimation problem and most systems use a two-stages procedure consisting of growing an overly large tree (possibly overfitting the training data), and then use some statistical estimation procedure (e.g. cross validation) for post-pruning this tree[4]. Given that outliers are insignificant from a statistical perspective, these strategies are difficult to implement in our system because they are based on statistical significance. Because of this we have decided not to post-prune our trees. This is consistent with what is mentioned by Weiss and Hirsh [19] in the context of learning from small disjuncts. These authors mention that pruning is considered questionable when the learning objectives are small subsets of cases.

Our method obtains a tree model in a single stage, stopping the tree growth when one of the following conditions arise:

– The F-measure of the node is above a certain user-definable threshold,
– Or the node does not contain any extreme value (i.e. $D_t \cap O = \phi$).

In order to illustrate the effects of using the proposed splitting criteria as opposed to standard least squares methods, we describe a small example application. Due to space reasons we have chosen a dataset that leads to small trees. We have used the well-known CPU performance dataset. In this domain the task is to predict the relative performance of a set of CPUs given some hardware characteristics of these machines. The dataset has 23 high outlier values (values above 237, c.f. Figure 1a). Using a CART-alike regression tree[5] with a standard 1-SE cross validation pruning algorithm [1], we get the tree on the right-hand side of Figure 2. From the point of view of outliers this tree isolates two classes of outliers, both formed by machines with a maximum main memory size above 28000Kb: One class is less extreme in terms of performance (average performance of 299) and includes machines with cache size below 80Kb; and the other class contains machines with larger cache size that have higher performance (average of 667). According to this tree, all computers with less than 28000Kb memory have low performance. Still, there are three exceptions to this (the numbers between parentheses on each node are the number of outliers in that node) that are neglected by this tree. The solution of our model is given at the left hand-side of Figure 2. Our tree is much more specific in terms of describing the conditions leading to outliers. Moreover, it further distinguishes

---

[4] See [13] for an overview of pruning methods for regression trees.
[5] In this paper we have used as base implementation of regression trees the package *rpart* [12] of the open source statistical software R (www.r-project.org). This package is a close re-implementation of most of CART's [1] features.

454 Luis Torgo and Rita Ribeiro

the type of outliers. Namely, we can identify even more extreme performance machines that have a high main memory size (above 48000Kb). Our tree also describes the machines with an outlier performance that have less than 28000Kb memory. This tree is clearly more consistent with the distribution of the outliers (i.e. the type of machines with high performance), as it can be seen from the box plot of the target variable presented in Figure 1a. Although one may think that this tree could be simply overfitting the data, the fact is that as we will see on the results of our experiments for this domain, our models achieve a significantly higher precision, recall and F-value.

In summary, from the perspective of understanding the type of extreme values occurring in this domain, and also under which conditions these appear, we claim that our tree is more informative than a standard regression tree. Moreover, this higher interpretability is accompanied by better accuracy as it will be shown in Section 4.

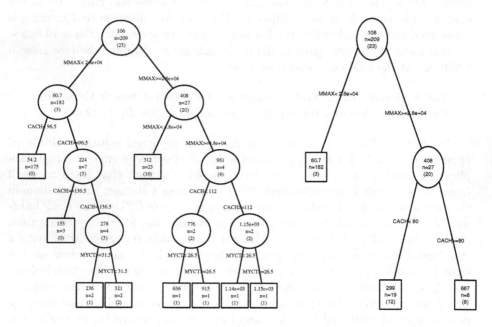

Fig. 2. Our regression tree vs the tree obtained by a CART-alike system, on the machine CPU dataset.

4 Experimental Results

In this section we perform an experimental analysis of the trees obtained with our method. Our analysis compares our proposal to its base paradigm, standard regression trees.

We have carried out a series of experiments using the datasets described in Table 1. These datasets include applications obtained from standard repositories as well as some commercial applications.

Table 1. Datasets description.

Datasets	# cases	continuous attr.	nominal attr.	# outliers	low outliers	high outliers
servo	167	0	4	30	0	30
triazines	186	60	0	9	9	0
algae1	200	8	3	12	0	12
algae2	200	8	3	10	0	10
algae3	200	8	3	22	0	22
algae4	200	8	3	16	0	16
algae5	200	8	3	13	0	13
algae6	200	8	3	19	0	19
algae7	200	8	3	21	0	21
machine_cpu	209	6	0	23	0	23
china	217	9	0	19	0	19
Boston	506	13	0	37	0	37
onekm	710	14	3	8	6	2
cw.drag	1449	12	2	52	1	51
co2.emission	1558	19	8	23	0	23
acceleration	1732	11	3	26	0	26
available.power	1802	7	8	121	0	121
bank8FM	4499	8	0	69	0	69
delta.ailerons	7129	5	0	107	41	66
ibm	8166	10	0	325	140	185
cpu.small	8192	12	0	430	430	0
delta.elevators	9517	6	0	132	60	72
cal.housing	20460	8	0	1071	0	1071
add	30000	10	0	63	0	63
fried.delve	40768	10	0	25	6	19

We have carried out 5 repetitions of 10-fold cross validation experiments using these datasets. These experiments were designed with the goal of estimating the average difference in precision, recall, and F-measure, between a standard regression tree and our proposed method. For the standard method we have used the package *rpart* of R, using cross validation error-complexity pruning with the 1-SE rule according to the method in [1]. Regarding our method we have used a F-value of 0.7 as threshold for deciding when to stop tree growth. The statistical significance of the observed differences was asserted through paired *t*-tests. Differences that are significant at the 95% level were marked with one sign, while differences significant at 99% have two signs. Plus (+) signs are used to mark differences favorable to standard regression trees, while minus (−) signs are used to indicate the significant wins of our method. Differences that are not significant at these confidence levels have no sign. The F-measure of each method was calculated with $\beta = 1$, meaning that the same weight was given to precision and recall (c.f. Equation (6)).

The results of our experiments are shown on Table 2. This table shows an overwhelming advantage of our method at least from the perspective of the F-measure, which was the criterion used to grow our trees. In effect, in the 25 datasets there were 18 significant wins, 3 insignificant wins and 4 insignificant losses of our proposal. The advantage is even more remarkable in terms of the proportion of outliers in the domain that are captured by the model (i.e. the recall). However, the results in terms of precision are not so interesting. We have tried to understand the reasons for this lack of precision in some domains. We have varied the F threshold that guides the criterion for stopping tree growth and have observed some variations on these results that seem to indicate that there is some space for improvement of our method by tunning this parameter. Apparently, tree growth may be stopping too soon for these large datasets, where our performance seems to be degrading. Still, if precision was the key objective we could also tune the β parameter of the F-measure that weights the preference between recall and precision when selecting the best splits of the trees[6].

Some of the results given in Table 2 deserve further explanations. Given the definition of precision we use (Equations (4) and (5)) it may seem strange to see some zero values in precision. These occur because some models have a NMSE at predicting the outliers equal or above one. Namely, for several datasets the CART tree is simply a single leaf node, which leads to a NMSE of one, and thus a precision of zero. The values of zero recall are consequence of models that do not predict any of the outliers as such, which occurs when a tree does not have any leaf with an average value that is an outlier.

Summarizing, the results of these experiments clearly show the advantage of our proposal in terms of predicting outliers. Nevertheless, we think some space is left for improvements particularly in terms of tunning the system by changing the stopping criterion as well as the weight between precision and recall. For large datasets, the best solution would probably be to keep a holdout set for proper tuning of these parameters.

5   Conclusions

We have described a new splitting criteria for regression trees with the goal of addressing a specific class of data mining applications. In these domains the main goal of modeling is to predict accurately outlier values in the target variable and also to understand under which conditions these values occur. Our proposal addresses these application goals by leading to regression trees designed to maximize both the number of outliers that are captured by the model and the precision at predicting their values.

The resulting trees were shown to achieve our goals in an extensive experimental comparison using 25 domains. In these experiments we have compared our approach to a standard regression tree and concluded that our proposal clearly outperforms these trees regarding the evaluation criteria that are adequate for this type of applications.

Regarding future work we plan to investigate more deeply the reasons for the failure of our models in terms of precision in some of the domains. Our current explanation lies on the tree growth stopping criteria and we intend to explore other alternatives to the current user settable threshold on the F-measure value.

---

[6] In our experiments we used equal weight.

Table 2. Regression trees vs our method in terms of Precision, Recall and F measure.

Datasets	Precision$_{regr}$			Recall			F measure		
	CART	Our method	Signif	CART	Our method	Signif	CART	Our method	Signif
servo	0.7616598	0.7829856		0.8713333	0.9800000	−−	0.8053263	0.8668980	−
triazines	0.1474294	0.3113946	−	0.0400000	0.2233333	−	0.0371048	0.2286451	−
algae1	0.2947144	0.4377268	−−	0.0000000	0.3266667	−−	0.0000000	0.3303920	−−
algae2	0.0000000	0.1394068	−−	0.0000000	0.0700000	−	0.0000000	0.0483434	
algae3	0.0000000	0.0994022	−	0.0000000	0.1820000	−−	0.0000000	0.0766102	−
algae4	0.0000000	0.1034622	−−	0.0000000	0.1416667	−−	0.0000000	0.0875694	−
algae5	0.0000000	0.1286299	−−	0.0000000	0.0673333	−	0.0000000	0.0569739	−
algae6	0.0000000	0.0481409	−	0.0000000	0.1600000	−−	0.0000000	0.0383576	−
algae7	0.0127059	0.0983871	−	0.0100000	0.1336667	−	0.0111917	0.0964774	−
machine.cpu	0.5517675	0.6879704	−−	0.8186667	0.8950000	−	0.6266528	0.7596690	−−
china	0.0000000	0.0740538	−	0.0000000	0.0706667	−	0.0000000	0.0621120	−
Boston	0.8243590	0.8225584		0.7580002	0.7595000		0.7675711	0.7361953	
onekm	0.4001506	0.3350779		0.0166667	0.2600000	−−	0.0059831	0.2701772	−−
cw.drag	0.9250750	0.8269861	++	0.8419906	0.9656190	−−	0.8734481	0.8864179	
co2.emission	0.8052664	0.8100812		0.4213333	0.5826667	−	0.4770925	0.6368632	−
acceleration	0.8964751	0.9010154		0.5600000	0.8210000	−−	0.6287861	0.8265599	−−
available.power	0.9668409	0.8567897	++	0.9091224	1.0000000	−−	0.9353684	0.9217586	
bank8FM	0.9781688	0.9529205	+	0.6532389	0.5876751		0.7592038	0.6971588	
delta.ailerons	0.6442916	0.5902405	++	0.1293077	0.1895810	−	0.1659074	0.2671672	−
ibm	0.0000000	0.0100769	−−	0.0000000	0.0055804	−	0.0000000	0.0038138	−
cpu.small	0.9939290	0.9856554	++	0.8519882	0.8620939		0.9165213	0.9189269	
delta.elevators	0.5980803	0.6054747		0.0419848	0.1734476	−−	0.0719734	0.2595248	−−
cal.housing	0.8425739	0.5854252	++	0.3140477	0.4403454	−−	0.4561053	0.5016178	−−
add	0.9320413	0.7925288	++	0.0000000	0.1496825	−−	0.0000000	0.2125705	−−
fried.delve	0.8816233	0.5351138	++	0.0735714	0.0700794		0.0959500	0.0913309	

References

1. L. Breiman, J. Friedman, R. Olshen, and C. Stone. *Classification and Regression Trees*. Statistics/Probability Series. Wadsworth & Brooks/Cole Advanced Books & Software, 1984.
2. A. Buja and Y.-S. Lee. Data mining criteria for tree-based regression and classification. In *Proceedings of ACM SIGKDD International Conference on Knowledge Discovery and Data Mining*, pages 27–36, 2001.
3. W. Cleveland. *Visualizing data*. Hobart Press, 1993.
4. T. Fawcett and F. Provost. Activity monitoring: Noticing interesting changes in behavior. In S. Chaudhuri and D. Madigan, editors, *Proceedings of the 5th ACM SIGKDD International Conference on Knowledge Discovery and Data Mining*, pages 53–62. ACM, 1999.
5. G.Weiss and F. Provost. The effect of class distribution on classifier learning: An empirical study. Technical Report Technical Report ML-TR-44, Department of Computer Science, Rutgers University, 2003.
6. R. Holte, L. Acker, and B. Porter. Concept learning and the problem of small disjuncts. In N. Sridharan, editor, *Proceedings of the 11th International Conference on Artificial Intelligence*. Morgan Kaufmann, 1989.

458   Luis Torgo and Rita Ribeiro

7. E. Keogh, S. Lonardi, and W. Chiu. Finding surprising patterns in a time series database in linear time and space. In *8th ACM SIGKDD International Conference on Knowledge Discovery and Data Mining*, pages 550–556, 2002.
8. I. Kononenko and I. Bratko. Information-based evaluation criterion for classifier's performance. *Machine Learning*, 6(1):67–80, 1991.
9. C. Meadow, B. Boyce, and D. Kraft. *Text Information Retrieval Systems*. Academic Press, 2nd edition, 2000.
10. F. Provost, T. Fawcett, and R. Kohavi. The case against accuracy estimation for comparing induction algorithms. In *Proc. 15th International Conf. on Machine Learning*, pages 445–453. Morgan Kaufmann, San Francisco, CA, 1998.
11. C. Van Rijsbergen. *Information Retrieval*. Dept. of Computer Science, University of Glasgow, 2nd edition, 1979.
12. T. Therneau and E. Atkinson. An introduction to recursive partitioning using rpart routines. Technical report, Mayo Foundation, 1997.
13. L. Torgo. A comparative study of reliable error estimators for pruning regression trees. In H. Coelho, editor, *Proceedings of the Iberoamericam Conference on AI (IBERAMIA-98)*, 1998.
14. L. Torgo. A study on end-cut preference in least squares regression trees. In P. Brazdil and A. Jorge, editors, *Proceedings of the Portuguese AI Conference (EPIA 2001)*, number 2258 in LNAI, pages 104–115. Springer, 2001.
15. L. Torgo and J. Gama. Regression using classification algorithms. *Intelligent Data Analysis*, 1(4), 1997.
16. P. Turney. Types of cost in inductive learning. In *Proceedings of the Workshop on cost-sensitive learning at the 17th ICML*, pages 15–21, 2000.
17. G. Weiss and H. Hirsh. Learning to predict rare events in event sequences. In R. Agrawal, P. Stolorz, and G. Piatetsky-Shapiro, editors, *Fourth International Conference on Knowledge Discovery and Data Mining (KDD'98)*, pages 359–363, New York, NY, 1998. AAAI Press, Menlo Park, CA.
18. G. Weiss and H. Hirsh. Learning to predict extremely rare events. In *AAAI Workshop on Learning from Imbalanced Data Sets*, pages 64–68. Technical Report WS-00-05, AAAI Press, 2000.
19. G. Weiss and H. Hirsh. A quantitative study of small disjuncts. In *Proceedings of AAAI/IAAI*, pages 665–670, 2000.

Mining Rules of Multi-level Diagnostic Procedure from Databases

Shusaku Tsumoto

Department of Medical Informatics, Shimane Medical University, School of Medicine
89-1 Enya-cho Izumo City, Shimane 693-8501 Japan
tsumoto@computer.org

Abstract. One of the most important features of expert reasoning is that each reasoning rule may be composed of several diagnostic steps, usually hierarchical differential diagnosis. For example, medical diagnosis include hierarchical diagnostic steps In this paper, the characteristics of experts' rules are closely examined from the viewpoint of hiearchical decision steps and a new approach to extract plausible rules is introduced, which consists of the following three procedures. First, the characterization of decision attributes (given classes) is extracted from databases and the concept hierarchy for given classes is calculated. Second, based on the hierarchy, rules for each hierarchical level are induced from data. Then, for each given class, rules for all the hierarchical levels are integrated into one rule. The proposed method was evaluated on medical databases, the experimental results of which show that induced rules correctly represent experts' decision processes.

1 Introduction

One of the most important problems in data mining is that extracted rules are not easy for domain experts to interpret. One of its reasons is that conventional rule induction methods[7] cannot extract rules, which plausibly represent experts' decision processes[9]: the description length of induced rules is too short, compared with the experts' rules. For example, rule induction methods, including AQ15[4] and PRIMEROSE[9], induce the following common rule for muscle contraction headache from databases on differential diagnosis of headache:

$[location = whole]$ $\wedge$[Jolt Headache $= no$] $\wedge$[Tenderness of M1 $= yes$]
$\rightarrow$ muscle contraction headache.

This rule is shorter than the following rule given by medical experts.

[Jolt Headache $= no$]
$\wedge$([Tenderness of M0 $= yes$] $\vee$[Tenderness of M1 $= yes$] $\vee$[Tenderness of M2 $= yes$])
$\wedge$[Tenderness of B1 $= no$] $\wedge$[Tenderness of B2 $= no$] $\wedge$[Tenderness of B3 $= no$]
$\wedge$[Tenderness of C1 $= no$] $\wedge$[Tenderness of C2 $= no$] $\wedge$[Tenderness of C3 $= no$]
    [Tenderness of C4 $= no$] $\rightarrow$ muscle contraction headache

where [Tenderness of B1 $= no$] and [Tenderness of C1 $= no$] are added.

N. Lavrač et al. (Eds.): PKDD 2003, LNAI 2838, pp. 459–470, 2003.

These results suggest that conventional rule induction methods do not reflect a mechanism of knowledge acquisition of medical experts.

In this paper, the characteristics of experts' rules are closely examined and a new approach to extract plausible rules is introduced, which consists of the following three procedures. First, the characterization of each decision attribute (a given class), a list of attribute-value pairs the supporting set of which covers all the samples of the class, is extracted from databases and the classes are classified into several groups with respect to the characterization. Then, two kinds of sub-rules, rules discriminating between each group and rules classifying each class in the group are induced. Finally, those two parts are integrated into one rule for each decision attribute.

The paper is organized as follows. Section 2 discusses the background of this study. Section 3 and 4 introduces rough sets and a characterization set. Section 5 gives an algorithm for rule induction. Section 6 shows an illustrative example and Section 7 discusses the results. Finally, Section 8 concludes this paper.

2   Background: Problems with Rule Induction

As shown in the introduction, rules acquired from medical experts are much longer than those induced from databases the decision attributes of which are given by the same experts. This is because rule induction methods generally search for shorter rules. One of the main reasons why rules are short is that these patterns are generated only by one criteria, such as high accuracy or high information gain. The comparative studies[9,10] suggest that experts should acquire rules not only by one criteria but by the usage of several measures. Those characteristics of medical experts' rules are fully examined not by comparing between those rules for the same class, but by comparing experts' rules with those for another class[9]. For example, the classification rule for muscle contraction headache given in Section 1 is very similar to the following classification rule for disease of cervical spine:

[Jolt Headache = no]
$\wedge$([Tenderness of M0 = yes] $\vee$[Tenderness of M1 = yes] $\vee$[Tenderness of M2 = yes])
$\wedge$([Tenderness of B1 = yes] $\vee$[Tenderness of B2 = yes] $\vee$[Tenderness of B3 = yes]
   $\vee$[Tenderness of C1 = yes] $\vee$[Tenderness of C2 = yes] $\vee$[Tenderness of C3 = yes]
   $\vee$[Tenderness of C4 = yes]) $\rightarrow$ disease of cervical spine

The differences between these two rules are attribute-value pairs, from tenderness of B1 to C4. Thus, these two rules can be simplified into the following form:

$$A_1 \wedge A_2 \wedge \neg A_3 \rightarrow muscle\ contraction\ headache$$
$$A_1 \wedge A_2 \wedge A_3 \rightarrow disease\ of\ cervical\ spine,$$

where A_1, A_2 and A_3 are given as the following formulae:
A_1 = [Jolt Headache = no], A_2 = [Tenderness of M0 = yes] $\vee$ [Tenderness of $M1$ = yes] $\vee$ [Tenderness of M2 = yes], and A_3 = [Tenderness of C1 = no] $\wedge$ [Tenderness of C2 = no] $\wedge$ [Tenderness of C3 = no] $\wedge$ [Tenderness of C4 = no].

The first two blocks (A_1 and A_2) and the third one (A_3) represent the different types of differential diagnosis. The first one A_1 shows the discrimination between muscular type and vascular type of headache. Then, the second part shows that between headache caused by neck and head muscles. Finally, the third formula A_3 is used to make a differential diagnosis between muscle contraction headache and disease of cervical spine. Thus, medical experts first select several diagnostic candidates, which are very similar to each other, from many diseases and then make a final diagnosis from those candidates.

This paper formalizes these procedures from the viewpoint of rough sets[5] and introduces a new approach to rule induction.

3 Rough Set Theory and Probabilistic Rules

In the following sections, we use the following notations introduced by Grzymala-Busse and Skowron[8], which are based on rough set theory[5]. These notations are illustrated by a small database shown in Table 1, collecting the patients who complained of headache.

Let U denote a nonempty, finite set called the universe and A denote a nonempty, finite set of attributes, i.e., $a : U \to V_a$ for $a \in A$, where V_a is called the domain of a, respectively. Then, a decision table is defined as an information system, $IS = (U, A \cup \{d\})$. For example, Table 1 is an information system with $U = \{1, 2, 3, 4, 5, 6\}$ and $A = \{age, location, nature, prodrome, nausea, M1\}$ and $d = class$. For $location \in A$, $V_{location}$ is defined as $\{occular, lateral, whole\}$.

The atomic formulae over $B \subseteq A \cup \{d\}$ and V are expressions of the form $[a = v]$, called descriptors over B, where $a \in B$ and $v \in V_a$. The set $F(B, V)$ of formulas over B is the least set containing all atomic formulas over B and closed with respect to disjunction, conjunction and negation. For example, $[location = occular]$ is a descriptor of B.

For each $f \in F(B, V)$, f_A denote the meaning of f in A, i.e., the set of all objects in U with property f, defined inductively as follows.

1. If f is of the form $[a = v]$ then, $f_A = \{s \in U | a(s) = v\}$
2. $(f \wedge g)_A = f_A \cap g_A$; $(f \vee g)_A = f_A \vee g_A$; $(\neg f)_A = U - f_a$

For example, $f = [location = occular]$ and $f_A = \{1, 5, 6, 7\}$. As an example of a conjunctive formula, $g = [location = occular] \wedge [nausea = no]$ is a descriptor of U and g_A is equal to $\{1, 5\}$.

It is also notable that d can be treated as a formula (or an attribute-value pair) because $B subseteq A$ is extended into $B subseteq A \cup d$ and d has the same nature as an atttribute $a \in A$: that is, since d is of the form $[d = class_i]$, $d_A = \{s \in U | d(s) = class_i\}$. For simplicity, d_A is denoted by D in subsequent sections.

By the use of the framework above, classification accuracy and coverage, or true positive rate is defined as follows.

Table 1. A small example of a database

No.	loc	nat	his	prod	jolt	nau	M1	M2	class
1	occular	per	per	0	0	0	1	1	m.c.h.
2	whole	per	per	0	0	0	1	1	m.c.h.
3	lateral	thr	par	0	1	1	0	0	common.
4	lateral	thr	par	1	1	1	0	0	classic.
5	occular	per	per	0	0	0	1	1	psycho.
6	occular	per	subacute	0	1	1	0	0	i.m.l.
7	occular	per	acute	0	1	1	0	0	psycho.
8	whole	per	chronic	0	0	0	0	0	i.m.l.
9	lateral	thr	per	0	1	1	0	0	common.
10	whole	per	per	0	0	0	1	1	m.c.h.

Definition. loc: location, nat: nature, his:history,
Definition. prod: prodrome, nau: nausea, jolt: Jolt headache,
M1, M2: tenderness of M1 and M2, 1: Yes, 0: No, per: persistent,
thr: throbbing, par: paroxysmal, m.c.h.: muscle contraction headache,
psycho.: psychogenic pain, i.m.l.: intracranial mass lesion, common.:
common migraine, and classic.: classical migraine.

Definition 1.
*Let R and D denote a formula in $F(B,V)$ and a meaning of a decision d.
Classification accuracy and coverage(true positive rate) for $R \to d$ is defined as:*

$$\alpha_R(D) = \frac{|R_A \cap D|}{|R_A|}, \ and \ \kappa_R(D) = \frac{|R_A \cap D|}{|D|},$$

where $|S|$, $\alpha_R(D)$, $\kappa_R(D)$ denote the cardinality of a set S, a classification accuracy of R as to classification of D and coverage (a true positive rate of R to D), respectively.

In the above example, when R and D are set to $[nau = 1]$ and $[class = common]$, $\alpha_R(D) = 2/5 = 0.4$ and $\kappa_R(D) = 2/2 = 1.0$.

It is notable that $\alpha_R(D)$ measures the degree of the sufficiency of a proposition, $R \to D$, and that $\kappa_R(D)$ measures the degree of its necessity. For example, if $\alpha_R(D)$ is equal to 1.0, then $R \to D$ is true. On the other hand, if $\kappa_R(D)$ is equal to 1.0, then $D \to R$ is true. Thus, if both measures are 1.0, then $R \leftrightarrow D$.

Finally, we define partial order of equivalence as follows:

Definition 2. *Let R_i and R_j be the formulae in $F(B,V)$ and let $A(R_i)$ denote a set whose elements are the attribute-value pairs of the form $[a,v]$ included in R_i. If $A(R_i) \subseteq A(R_j)$, then we represent this relation as:*

$$R_i \preceq R_j.$$

According to the definitions, probabilistic rules with high accuracy and coverage are defined as:

$$R \overset{\alpha,\kappa}{\to} d \ s.t. \ R = \vee_i R_i = \vee \wedge_j [a_j = v_k], \ \alpha_{R_i}(D) \geq \delta_\alpha \ and \ \kappa_{R_i}(D) \geq \delta_\kappa,$$

where δ_α and δ_κ denote given thresholds for accuracy and coverage, respectively.

4   Characterization Sets

4.1   Characterization Sets

In order to model medical reasoning, a statistical measure, coverage defined ins Section 2 plays an important role in modeling, which is equivalent to a conditional probability of a condition (R) under the decision (D): $P(R|D)$. Let us define a characterization set of D, denoted by $L(D)$ as a set, each element of which is an elementary attribute-value pair R with coverage being larger than a given threshold, δ_κ. That is,

Definition 3. *Let R denote a formula in* $F(B,V)$. *Characterization sets of a target concept (D) is defined as:*

$$L_{\delta_\kappa}(D) = \{R|\kappa_R(D) \geq \delta_\kappa\}$$

Then, three types of relations between characterization sets can be defined as follows:

$$\text{Independent type: } L_{\delta_\kappa}(D_i) \cap L_{\delta_\kappa}(D_j) = \phi,$$
$$\text{Boundary type: } \quad L_{\delta_\kappa}(D_i) \cap L_{\delta_\kappa}(D_j) \neq \phi, \text{ and}$$
$$\text{Subcategory type: } L_{\delta_\kappa}(D_i) \subseteq L_{\delta_\kappa}(D_j).$$

All three definitions correspond to the negative region, boundary region, and positive region, respectively, if a set of the whole elementary attribute-value pairs will be taken as the universe of discourse.

Tsumoto focuses on the subcategory type in [10] because D_i and D_j cannot be differentiated by using the characterization set of D_j, which suggests that D_i is a generalized disease of D_j. Then, Tsumoto generalizes the above rule induction method into the overlapped type, considering rough inclusion[11]. However, both studies assumes two-level diagnostic steps: focusing mechanism and differential diagnosis, where the former selects diagnostic candidates from the whole classes and the latter makes a differential diagnosis between the focused classes.

The proposed method below extends these methods into multi-level steps.

4.2   Characteristics

We consider the special case of characterization sets in which the thresholds of coverage is equal to 1.0. That is,

$$L_{1.0}(D) = \{R_i|\kappa_{R_i}(D) = 1.0\}$$

Then, we have several interesting characteristics.

Theorem 1. *Let* R_i *and* R_j *two formulae in* $L_{1.0}(D)$ *such that* $R_i \preceq R_j$. *Then,* $\alpha_{R_i} \leq \alpha_{R_j}$.

Thus, when we collect the formulae whose values of coverage are equal to 1.0, the sequence of conjunctive formulae corresponds to the sequence of increasing chain of accuracies.

For example, $[nat = per]$ and $[his = per]$ are elements of $L_{1.0}(m.c.h.)$ and those accuracies are: $3/7$ and $3/5$. Then, since the meaning of $([loc = occular] \vee [loc = whole]) \wedge [his = per]$ is equal to $[1, 2, 5, 10]$, the accuracy of $[nat = per] \wedge [his = per]$ is $3/4$.

Since $\kappa_R(D) = 1.0$ means that the meaning of R covers all the samples of D, its complement $U - R_A$, that is, $\neg R$ do not cover any samples of D. Especially, when R consists of the formulae with the same attributes, it can be viewed as the generation of the coarsest partitions. Thus,

Theorem 2. *Let R be a formula in $L_{1.0}(D)$ such that $R = \vee_j [a_i = v_j]$. Then, R and $\neg R$ gives the coarsest partition for a_i, whose R includes D.*

From the propositions 1 and 2, the next theorem holds.

Theorem 3. *Let A consist of $\{a_1, a_2, \cdots, a_n\}$ and R_i be a formula in $L_{1.0}(D)$ such that $R_i = \vee_j [a_i = v_j]$. Then, a sequence of a conjunctive formula $F(k) = \wedge_{i=1}^{k} R_i$ gives a sequence which increases the accuracy.* $\square$

5   Rule Induction with Grouping

As discussed in Section 2, When the coverage of R for a target concept D is equal to 1.0, R is a necessity condition of D. That is, a proposition $D \to R$ holds and its contrapositive $\neg R \to \neg D$ holds. Thus, if R is not observed, D cannot be a candidate of a target concept. Thus, if two target concepts have a common formula R whose coverage is equal to 1.0, then $\neg R$ supports the negation of two concepts, which means these two concepts belong to the same group. Furthermore, if two target concepts have similar formulae $R_i, R_j \in L_{1.0}(D)$, they are very close to each other with respect to the negation of two concepts. In this case, the attribute-value pairs in the intersection of $L_{1.0}(D_i)$ and $L_{1.0}(D_j)$ give a characterization set of the concept that unifies D_i and D_j, D_k. Then, compared with D_k and other target concepts, classification rules for D_k can be obtained. When we have a sequence of grouping, classification rules for a given target concepts are defined as a sequence of subrules. From these ideas, a rule induction algorithm with grouping target concepts can be described as Figure 1. This algorithm first calculates $L_{1.0}(D_i)$ for $\{D_1, D_2, \cdots, D_k\}$. Second, from the list of characterization sets, it calculates the intersection between $L_{1.0}(D_i)$ and $L_{1.0}(D_j)$ and stores it into L_{id}. Third, the procedure calculates the similarity (matching number) of the intersections and sorts L_{id} with respect of the similarities. Fourth, the algorithm chooses one intersection $(D_i \cap D_j)$ with maximum similarity (highest matching number) and group D_i and D_j into a concept DD_i. These procedures will be continued until all the grouping is considered (Fig. 2). Finally, rules for generated group and diseases are induced by using a rule induction algorithm shown in Fig. 3.

procedure *Total Process*;
  var inputs
    L_D : *List*; /* A list of Target Concepts */
  begin
    Calculate a set of characterization set L_c;
      Calculate a set of intersection L_{id};
        Calculate a list of similarity measures L_s;
          Calculate a list of grouping L_g; (Fig. 2)
      Induce a set of rules for L_g: L_r; (Fig. 3)
      Combine Rules in L_r for each D_i;
  end {*Total Process*}

Fig. 1. An Algorithm for Total Process

procedure *Grouping* ;
  var inputs
    L_c : *List*;
    /* A list of Characterization Sets */
    L_{id} : *List*;
    /* A list of Intersection */
    L_s : *List*;
    /* A list of Similarity */
  var outputs
    L_{gr} : *List*;
    /* A list of Grouping */
  var
    k : *integer*; L_g, L_{gr} : *List*;
  begin
    $L_g := \{\}$;
    $k := n$
    /* n: A number of Target Concepts*/
    Sort L_s with respect to similarities;
    Take a set of (D_i, D_j), L_{max}
      with maximum similarity values;
    k:= k+1;
    forall $(D_i, D_j) \in L_{max}$ **do**
      begin
        Group D_i and D_j into D_k;
        $L_c := L_c - \{(D_i, L_{1.0}(D_i)\}$;
          $L_c := L_c - \{(D_j, L_{1.0}(D_j)\}$;
          $L_c := L_c + \{(D_k, L_{1.0}(D_k)\}$;
        Update L_{id} for DD_k;
        Update L_s;
        $L_{gr} := ($
          Grouping for L_c, L_{id}, and L_s) ;
        $L_g := L_g + \{\{(D_k, D_i, D_j), L_g\}\}$;
      end
    return L_g;
  end {*Grouping*}

Fig. 2. An Algorithm for Grouping

procedure *RuleInduction* ;
  var inputs
    L_c : *List*;
    /* A list of Characterization Sets */
    L_{id} : *List*; /* A list of Intersection */
    L_g : *List*; /* A list of grouping*/
    /* $\{\{(D_{n+1}, D_i, D_j), \{(DD_{n+2},.)...\}\}\}$ */
    /* n: A number of Target Concepts */
  var
    Q, L_r : *List*;
  begin
    $Q := L_g$; $L_r := \{\}$;
    if $(Q \neq \emptyset)$ **then do**
      begin
        $Q := Q - first(Q)$;
        $L_r :=$ *Rule Induction* (L_c, L_{id}, Q);
      end
    $(DD_k, D_i, D_j) := first(Q)$;
    if $(D_i \in L_c$ and $D_j \in L_c)$ **then do**
      begin
        Induce a Rule r which discriminate
        between D_i and D_j;
        $r = \{R_i \to D_i, R_j \to D_j\}$;
      end
    else do
      begin
        Search for $L_{1.0}(D_i)$ from L_c;
        Search for $L_{1.0}(D_j)$ from L_c;
        if $(i < j)$ **then do**
          begin
            $r(D_i) := \vee_{R_l \in L_{1.0}(D_j)} \neg R_l \to \neg D_j$;
            $r(D_j) := \wedge_{R_l \in L_{1.0}(D_j)} R_l \to D_j$;
          end
        $r := \{r(D_i), r(D_j)\}$;
      end
    return $L_r := \{r, L_r\}$;
  end {*Rule Induction*}

Fig. 3. An Algorithm for Rule Induction

6   Example

Let us consider Table 1 as an example for rule induction. For a similarity function, we use a matching number[3] which is defined as the cardinality of the intersection of two the sets. Also, since Table 1 has five classes, k is set to 6.

6.1   Grouping

From this table, the characterization set for each concept is obtained as shown in Fig 4. Then, the intersection between two target concepts are calculated. Since *common* and *classic* have the maximum matching number, these two classes are grouped into one category, D_6. Then, teh characterization of D_6 is obtained as :
$D_6 = \{[loc = lateral], [nat = thr], [jolt = 1], [nau = 1], [M1 = 0], [M2 = 0]$ from Fig 5.

In the second iteration, the intersection of D_1 and others is considered as shown in Fig 6. From this matrix, we have two possibilities of grouping: one is to group *m.c.h.* and *i.m.l.* That is, these two diseases are grouped into D_7: $D_7 = \{([loc = occular] \vee [loc = whole]), [nat = per], [prod = 0]\}$ The other one is to group D_1 and *i.m.l.*, where $D_7 = \{[jolt = 1], [M1 = 0], [M2 = 0]\}$.

In the third iteration of the former case(3_a), the intersection is calculated as Fig 7 and D_2 and psycho are grouped into D_3: $D_{3a} = \{$ [nat=per], [prod=0] $\}$ In the latter case(3_b), it is calculated as Fig 8 and *m.c.h.* and *psycho* are grouped into D_8: $D_{8a} = \{$ [nat=per], [prod=0] $\}$. Fig 9 and 10 depicts the two results of grouping like a dendrogram in clustering analysis[3].

$L_{1.0}(m.c.h.) = \{([loc = occular] \vee [loc = whole]), [nat = per], [his = per],$
$[prod = 0], [jolt = 0], [nau = 0], [M1 = 1], [M2 = 1]\}$

$L_{1.0}(common) = \{[loc = lateral], [nat = thr], ([his = per] \vee [his = par]), [prod = 0],$
$[jolt = 1], [nau = 1], [M1 = 0], [M2 = 0]\}$

$L_{1.0}(classic) = \{[loc = lateral], [nat = thr], [his = par], [prod = 1],$
$[jolt = 1], [nau = 1], [M1 = 0], [M2 = 0]\}$

$L_{1.0}(i.m.l.) = \{([loc = occular] \vee [loc = whole]), [nat = per],$
$([his = subacute] \vee [his = chronic]), [prod = 0],$
$[jolt = 1], [M1 = 1], [M2 = 1]\}$

$L_{1.0}(psycho) = \{[loc = occular], [nat = per], ([his = per] \vee [his = acute]),$
$[prod = 0]\}$

Fig. 4. Characterization Sets for Table 1

6.2   Rule Induction

Due to the limitation of space, we focus on rule induction based on the first model. Figure 9 shows one candidate of the differential diagnosis. For the differential diagnosis of *common*. First, this model discriminate between D_6(*common*

	m.c.h.	common	classic	i.m.l.	psycho
m.c.h.	–	{[prod=0]}	∅	{([loc=occular]∨[loc=whole]), {[nat=per],[prod=0]}	{[nat=per],[prod=0]}
common	–	–	{[loc=lateral], [nat=thr],[jolt=1], [nau=1], [M1=0], [M2=0]}	{[prod=0],[jolt=1], [M1=0], [M2=0] }	{[prod=0]}
classic	–	–	–	{[jolt=1],[M1=0],[M2=0]}	{ }
i.m.l.	–	–	–	–	{[nat=per], [prod=0]}

Fig. 5. Intersection of Two Characterization Sets (Step 2)

	m.c.h. D_6	i.m.l.	psycho
m.c.h.	–	{}	{([loc=occular]∨[loc=whole]), {[nat=per],[prod=0]}	{[nat=per],[prod=0]}
D_6	–	–	{[jolt=1], [M1=0], [M2=0]}	{ }
i.m.l.	–	–	–	{[nat=per],[prod=0]}

Fig. 6. Intersection of Two Characterization Sets after the first Grouping (Step 3)

	D_6 D_7	psycho
D_6	– {}	{ }
D_7	– –	{[nat=per],[prod=0]}

	m.c.h. D_7	psycho
m.c.h.	– {}	{[nat=per],[prod=0] }
D_7	– {}	{ }

Fig. 7. Intersection of Two Characteriza-tion Sets after the first Grouping (1) (Step 4a)

Fig. 8. Intersection of Two Characteriza-tion Sets after the first Grouping (2) (Step 4b)

and *classic*) and D_8 (*m.c.h.*, *i.m.l.* and *psycho*). Then, *common* and *classic* within D_6 are differentiated. Thus, a classification rule for *common* is composed of two subrules: (discrimination between D_6 and D_8) and (discrimination within D_6). On the other hand, a classification rule for *m.c.h.* is composed of three subrules: (discrimination between D_6 and D_8), (discrimination between D_7 and psycho) and (discrimination within D_7).

Let us consider the first case. The first part can be obtained by the intersection in Figure 7. That is, $D_8 \rightarrow [nat = per] \wedge [prod = 0]$; $\neg[nat = per] \vee \neg[prod = 0] \rightarrow \neg D_8$. Then, since from Figure 4, the difference set between $L_{1.0}(common)$ and $L_{1.0}(classic)$ is $\{[prod = 1]\}$, for a classification rule for *common* within D_7 is: $[prod = 0] \rightarrow common$.

Combining these two parts, the classification rule for *common* is: $(\neg[nat = per] \vee \neg[prod = 0]) \wedge [prod = 0] \rightarrow common$. After its simplification, the rule is:

$$\neg[nat = per] \rightarrow \neg common,$$

whose accuracy is equal to 2/3. In the same way, the rule for *classic* is obtained as:

$$\neg[nat = per] \wedge [prod = 1] \rightarrow classic.$$

7   Experimental Results

The above rule induction algorithm was implemented in PRIMEROSE4.5 (Prob-abilistic Rule Induction Method based on Rough Sets Ver 5.0), and was applied

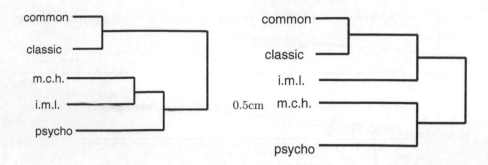

Fig. 9. Grouping by Characterization Sets (First Model)

Fig. 10. Grouping by Characterization Sets (Second Model)

to databases on differential diagnosis of headache, meningitis and cerebrovascular diseases (CVD), whose precise information is given in Table 2. In these experiments, δ_α and δ_κ were set to 0.75 and 0.5, respectively. Also, the threshold for grouping is set to 0.8 [1]. This system was compared with PRIMEROSE4.5[11], PRIMEROSE[9] C4.5[6], CN2[2], AQ15[4] with respect to the following points: length of rules, similarities between induced rules and expert's rules and performance of rules.

In this experiment, the length was measured by the number of attribute-value pairs used in an induced rule and Jaccard's coefficient was adopted as a similarity measure[3]. Concerning the performance of rules, ten-fold cross-validation was applied to estimate classification accuracy.

Table 2. Information about Databases

Domain	Samples	Classes	Attributes
Headache	52119	45	147
CVD	7620	22	285
Meningitis	141	4	41

Table 3 shows the experimental results, which suggest that PRIMEROSE5 outperforms PRIMEROSE4.5 (two-level) and the other four rule induction methods and induces rules very similar to medical experts' ones.

8   Discussion

The readers may wonder why lengthy rules perform better than short rules since lengthy rules suffer from overfitting to a given data. One reason is that a decision

---

[1] These values are given by medical experts as good thresholds for rules in these three domains.

Table 3. Experimental Results

Method	Length	Similarity	Accuracy
	Headache		
PRIMEROSE5.0	8.8 ± 0.27	0.95 ± 0.08	95.2 ± 2.7%
PRIMEROSE4.5	7.3 ± 0.35	0.74 ± 0.05	88.3 ± 3.6%
Experts	9.1 ± 0.33	1.00 ± 0.00	98.0 ± 1.9%
PRIMEROSE	5.3 ± 0.35	0.54 ± 0.05	88.3 ± 3.6%
C4.5	4.9 ± 0.39	0.53 ± 0.10	85.8 ± 1.9%
CN2	4.8 ± 0.34	0.51 ± 0.08	87.0 ± 3.1%
AQ15	4.7 ± 0.35	0.51 ± 0.09	86.2 ± 2.9%
	Meningitis		
PRIMEROSE5.0	2.6 ± 0.19	0.91 ± 0.08	82.0 ± 3.7%
PRIMEROSE4.5	2.8 ± 0.45	0.72 ± 0.25	81.1 ± 2.5%
Experts	3.1 ± 0.32	1.00 ± 0.00	85.0 ± 1.9%
PRIMEROSE	1.8 ± 0.45	0.64 ± 0.25	72.1 ± 2.5%
C4.5	1.9 ± 0.47	0.63 ± 0.20	73.8 ± 2.3%
CN2	1.8 ± 0.54	0.62 ± 0.36	75.0 ± 3.5%
AQ15	1.7 ± 0.44	0.65 ± 0.19	74.7 ± 3.3%
	CVD		
PRIMEROSE5.0	7.6 ± 0.37	0.89 ± 0.05	74.3 ± 3.2%
PRIMEROSE4.5	5.9 ± 0.35	0.71 ± 0.05	72.3 ± 3.1%
Experts	8.5 ± 0.43	1.00 ± 0.00	82.9 ± 2.8%
PRIMEROSE	4.3 ± 0.35	0.69 ± 0.05	74.3 ± 3.1%
C4.5	4.0 ± 0.49	0.65 ± 0.09	69.7 ± 2.9%
CN2	4.1 ± 0.44	0.64 ± 0.10	68.7 ± 3.4%
AQ15	4.2 ± 0.47	0.68 ± 0.08	68.9 ± 2.3%

attribute gives a partition of datasets: since the number of given classes are 4 to 45, some classes have very low support due to the prevalence of the corresponding diseases. Thus, the disease with the low frequency may not have short-length rules by using the conventional methods. However, since our method is not based on accuracy, but on coverage, we can support the disease of frequency. Another reason is that this method reflects the reasoning style of domain experts. One of the most important features of medical reasoning is that medical experts finally select one or two diagnostic candidates from many diseases, called focusing mechanism. For example, in differential diagnosis of headache, experts choose one from about 60 diseases. The proposed method models induction of rules which incorporates this mechanism, whose experimental evaluation show that induced rules correctly represent medical experts' rules.

This focusing mechanism is not only specific to medical domain. In a domain in which a few diagnostic conclusions should be selected from many candiates, this mechanism can be applied. For example, fault diagnosis of complicated electronic devices should focus on which components will cause a functional problem: the more complicated devices are, the more sophisticated focusing mechanism is required. In such domain, proposed rule induction method will be useful to induce correct rules from datasets.

9 Conclusion

In this paper, the characteristics of experts' rules are closely examined, whose empirical results suggest that grouping of diseases ais very important to realize automated acquisition of medical knowledge from clinical databases. Thus, we focus on the role of coverage in focusing mechanisms and propose an algorithm for grouping of diseases by using this measure. The above example shows that rule induction with this grouping generates rules, which are similar to medical experts' rules and they suggest that our proposed method should capture medical experts' reasoning. This research is a preliminary study on a rule induction method with grouping and it will be a basis for a future work to compare the proposed method with other rule induction methods by using real-world datasets.

Acknowledgments

The author thanks for three reviewers' insightful comments. This work was supported by the Grant-in-Aid for Scientific Research (13131208) on Priority Areas (No.759) "Implementation of Active Mining in the Era of Information Flood" by the Ministry of Education, Science, Culture, Sports, Science and Technology of Japan.

References

1. Aha, D. W., Kibler, D., and Albert, M. K., Instance-based learning algorithm. *Machine Learning*, 6, 37-66, 1991.
2. Clark, P. and Niblett, T., The CN2 Induction Algorithm. *Machine Learning*, 3, 261-283, 1989.
3. Everitt, B. S., *Cluster Analysis*, 3rd Edition, John Wiley & Son, London, 1996.
4. Michalski, R. S., Mozetic, I., Hong, J., and Lavrac, N., The Multi-Purpose Incremental Learning System AQ15 and its Testing Application to Three Medical Domains, in *Proceedings of the fifth National Conference on Artificial Intelligence*, 1041-1045, AAAI Press, Menlo Park, 1986.
5. Pawlak, Z., *Rough Sets*. Kluwer Academic Publishers, Dordrecht, 1991.
6. Quinlan, J.R., *C4.5 - Programs for Machine Learning*, Morgan Kaufmann, Palo Alto, 1993.
7. *Readings in Machine Learning*, (Shavlik, J. W. and Dietterich, T.G., eds.) Morgan Kaufmann, Palo Alto, 1990.
8. Skowron, A. and Grzymala-Busse, J. From rough set theory to evidence theory. In: Yager, R., Fedrizzi, M. and Kacprzyk, J.(eds.) *Advances in the Dempster-Shafer Theory of Evidence*, pp.193-236, John Wiley & Sons, New York, 1994.
9. Tsumoto, S., Automated Induction of Medical Expert System Rules from Clinical Databases based on Rough Set Theory. *Information Sciences* **112**, 67-84, 1998.
10. Tsumoto, S., Extraction of Experts' Decision Rules from Clinical Databases using Rough Set Model *Intelligent Data Analysis*, 2(3), 1998.
11. Tsumoto,S. Extraction of Hierarchical Decision Rules from Clinical Databases using Rough Sets. *Information Sciences*, 2003 (in print)

Learning Characteristic Rules Relying on Quantified Paths

Teddy Turmeaux[1], Ansaf Salleb[1], Christel Vrain[1], and Daniel Cassard[2]

[1] LIFO, Université d'Orléans, rue Léonard de Vinci
BP 6759, F-45067 Orléans Cedex 2, France
{Turmeaux,Salleb,Vrain}@lifo.univ-orleans.fr
[2] BRGM, 3, avenue Claude Guillemin, B.P. 6009
Orléans cedex 2, France
d.cassard@brgm.fr

Abstract. In this paper, we address the *characterization* task and we present a general framework for the *characterization* of a target set of objects by means of their own properties, but also the properties of objects linked to them. According to the kinds of objects, various links can be considered. For instance, in the case of relational databases, associations are the straightforward links between pairs of tables. We propose 𝔠𝔞𝔯𝔞𝔠𝔱𝔢𝔯𝔦𝔛, a new algorithm for mining characterization rules and we show how it can be used on multi-relational and spatial databases.

Keywords: Machine Learning, Inductive Logic Programming, Data Mining, Characteristic Rules, Relational Databases, Spatial Databases.

1 Introduction

Characterization is a descriptive data mining task which aims at mining concise and compact descriptions of a set of objects, called the *target set*. It consists in discovering properties that characterize these objects, taking into account their own properties but also properties of the objects linked to them.

In comparison to classification and discrimination, characterization is interesting since it does not require negative examples. This is an important feature for some real world applications where it is difficult to collect negative examples.

Several fields have contributed to this task. On the one hand, characterization has been treated as descriptive generalization in the field of Machine Learning [12]. Characterizing a set of objects has also been considered as computing the least general generalization (l.g.g.) in Inductive Logic Programming [14], but such an approach leads to complexity problems. An object oriented view for computing the l.g.g. called *structural matching* has been proposed in [8,17] and applied to air traffic control in [9]. On the other hand, in Data Mining, Han et al. [7,6] have introduced attribute oriented induction for data generalization, but in their framework, background knowledge such as taxonomies is needed for generalizing data, and objects are described in a single table, which limit the applicability of such a method.

N. Lavrač et al. (Eds.): PKDD 2003, LNAI 2838, pp. 471–482, 2003.

We can also consider that characterization is close to the task of mining frequent properties on the target set. This task has already long been studied [1, 11, 5, 16] , since in many systems, it is the first step for mining association rules. Nevertheless, most works suppose that data is stored in a single table, and few algorithms [3] really handle multi-relational databases. Moreover, the frequency (also called the support) is not sufficient to characterize the objects of the target set, because it is also important to determine whether a property is truly a characteristic feature by considering also the frequency of that property outside the target set.

The approach we propose handles multi-relational databases taking into account the structure of the database. It relies on the definition of a *Quantified Path* which is an expression that specifies how to take into account different kinds of objects and their relationships, starting from the target objects. For instance, considering as a target set the set of films produced by a given person Sp and denoted by $Movie_{(Sp)}$, the following expression:

$$Movie_{(Sp)} : \exists Award :: Award.kind\ in(Oscar, GoldenPalm)$$

is a characteristic rule which means that each movie produced by Sp has received at least one Oscar award or Golden Palm award. The expression $Movie_{(Sp)} : \exists Award$ is a quantified path. It specifies that we are interested in the properties satisfied by at least one award received by Sp'movies. On the other hand, considering the Quantified Path $Movie_{(Sp)} : \forall Award$ means that we are looking for properties satisfied by all the awards received by all Sp's movies.

At LIFO, we have developed $\mathfrak{CaracteriX}$, a levelwise[1] algorithm, for mining interesting characteristic rules. It starts with the most general Quantified Paths, exploring the search space, according to notion of generality between rules. Moreover, it uses two heuristics, *link-coverage* and *open-coverage*, to efficiently prune the search space. Another important feature of our approach is the form of the rules, which relies on quantified paths defining how to 'navigate' between sets of objects. As far as we know the form of rules we have introduced has not yet been used in that field.

The paper is organized as follows. Section 2 formalizes the problem of mining characteristic rules. In Section 3, we give definitions on which our approach relies: the notion of quantified paths, properties and characteristic rules, the notion of coverage and generality orders. Section 4 is devoted to the general algorithm and Section 5 to experiments.

2   Problem Statement

The characterization task we are interested in can be formulated as follows:

- given a set of types T_i, and attributes for describing objects of type T_i,
- given a set $\mathcal{E}$ of objects, $\mathcal{E} = \mathcal{E}_1 \cup \mathcal{E}_2 \cdots \cup \mathcal{E}_n$, where each $\mathcal{E}_i$ contains objects with the same type T_i,

---

[1] see [11, 13, 15] for a description of levelwise algorithms family.

- given a set $\mathcal{R}$ of binary relations (in the following, r_{ij} denotes a binary relation on $\mathcal{E}_i \times \mathcal{E}_j$)
- given a target set $\mathcal{E}_{target}$, such that there exists i, $\mathcal{E}_{target} \subseteq \mathcal{E}_i$,
→ find a set of characterization rules of $\mathcal{E}_{target}$.

The size of the search space for the characterization rules depends, among others, on the number of relations in $\mathcal{R}$ and on their cardinalities. Without restrictions on the possible forms of the rule, the search space may become so large that the learning task is intractable.

Example 1. Application to relational databases

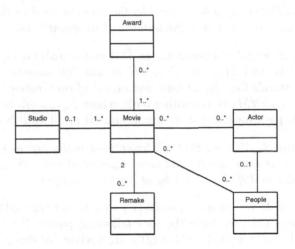

Fig. 1. Movies database

Our approach is illustrated throughout this paper by a running example *Movies*[2] given in Figure 1. This database is stored in a relational form composed of several files. There is information on actors, casts, directors, producers, studios,... The main file *Movie* is a list of movies described by their category, title, year, process, and so on. The actors are listed with their roles in another file *Casts*. More information about individual actors such as *name, date of birth, gender* and *origin* can be found in the file *Actors*. The file *People* gives more information about actors, directors, producers, writers, and cinematographers. *Remakes* links movies to their remakes, whereas *Awards* gives the different awards that can be won by a movie. Finally, *Studios* provides some information about each studio, such as the *location* and the *founder*.

For instance, we could be interested by characterizing the properties of comic movies, or the properties of movies produced by a given producer, and so on.

---

[2] inspired from `http://kdd.ics.uci.edu/databases/movies/movies.html`

3   General Framework

3.1   Quantified Path

Definition 1. *A **Quantified Path** (denoted in the following by $\mathcal{QP}$) on X_0 is a formula:*

$$Q_1\, X_1 \ldots Q_n\, X_n$$

where $n \geq 0$, X_0 represents the target objects, and for each $i \neq 0$, $Q_i = \forall$ or $\exists$, X_i is a type of objects, and there exists a relationship in $\mathcal{R}$ between X_{i-1} and X_i. When necessary, in order to remember the target set, it will prefixed by X_0 leading to $X_0 : Q_1\, X_1 \ldots Q_n\, X_n$.
Let us notice that when there exists several relationships between X_{i-1} and X_i, the quantifier Q_i may be indexed by the relation used in the $\mathcal{QP}$.
A $\mathcal{QP}$ has a size n that is the number of its quantifiers.

Example 2. • Links between movies (M) and awards (W) give two paths denoted by $M : \forall W$ and $M : \exists W$. $M : \forall W$ means *"all awards of each movie"*, while $M : \exists W$ stands for *"for at least one award of each movie"*.
• $P_{name=Hit} : \forall M \forall W$ is another path, where $P_{name=Hit}$ is a target set of people (P). This path means that we are interested in all awards of all Hit's movies.

Definition 2. *We say that two quantified paths are* variants *if they have the same size, if they involve the same type of objects, the same relations in the same order and if they differ by at least a quantifier.*

Example 3. If we consider people (P) as a target set and links between people and movies (M), we have the four following paths: $P : \forall M \forall W$, $P : \forall M \exists W$, $P : \exists M \exists W$, $P : \exists M \forall W$. These $\mathcal{QP}$s are variants of size 2.

Definition 3. *We say that a quantified path δ_1 is* more general *than a quantified path δ_2 (denoted by $\delta_1 \succeq \delta_2$) iff δ_1 and δ_2 are variants and for $1 \leq i \leq size(\delta_1)(= size(\delta_2))$, either:*
- $Q_i^1 \equiv Q_i^2$, or
- $Q_i^1 = \exists$ and $Q_i^2 = \forall$.

Example 4. For instance, we have $P : \exists M \exists W \succeq P : \forall M \exists W \succeq P : \forall M \forall W$ and also $P : \exists M \exists W \succeq P : \exists M \forall W \succeq P : \forall M \forall W$ but $P : \forall M \exists W \not\succeq P : \exists M \forall W$ and $P : \exists M \forall W \not\succeq P : \forall M \exists W$.

3.2   Properties

A set of properties is associated to each type of objects. We consider many kinds of properties such as: *attribute=value*, *attribute $\in \{value_1, \ldots, value_n\}$, attribute $\geq$ value, attribute $\leq$ value*, and even aggregates such as: count, min, max, … For a type T and a property p on T, we assume that there exists a boolean function $\mathcal{V}_p$, such that for each object o of type T, $\mathcal{V}_p(o) = true$ or $\mathcal{V}_p(o) = false$. It means that a property may be satisfied by an object o or not.

Definition 4. *We define two basic properties True and False such that for any object o, $\mathcal{V}_{True}(o) = true$ and $\mathcal{V}_{False}(o) = false$.*

Definition 5. *We say that a property p_1 is more general than a property p_2 (denoted by $p_1 \succeq p_2$) iff all objects that satisfy the property p_2 also verify the property p_1.*

Example 5. The property $W.kind \in \{Oscar, GoldenPalm\}$ where W represents the set of awards is more general than $W.kind \in \{GoldenPalm\}$.

3.3 Characteristic Rules

Definition 6. *We define a characteristic rule on a target set X_0 as the conjunction of a quantified path δ and a property p , denoted by: $X_0 : \delta :: p$.*

Definition 7. *We say that two characteristic rules r_1 ($T : \delta_1 :: p_1$) and r_2 ($T : \delta_2 :: p_2$) are variants if δ_1 and δ_2 are variants and $p_1 \equiv p_2$.*

Example 6. $P_{name=Hit} : \forall M :: M.category = Suspense$
is a characteristic rule, where $P_{name=Hit}$ is a target set of *People* whose name is Hit. This rule means that all Hit's movies belong to the *Suspense* category.

3.4 Coverage

The notion of coverage is defined for a property p relatively to a quantified path δ. It measures the number of objects that have this property. For a rule $r = X_0 : \delta :: p$ and an object $o \in X_o$, we define $\mathcal{V}_{\delta :: p}(o)$ recursively as follows:
- $\mathcal{V}_{\forall X. \delta' :: p}(o) = \mathcal{V}_{\delta' :: p}(o_1) \wedge \cdots \wedge \mathcal{V}_{\delta' :: p}(o_n)$ or *false* if there is no object linked to o
- $\mathcal{V}_{\exists X. \delta' :: p}(o) = \mathcal{V}_{\delta' :: p}(o_1) \vee \cdots \vee \mathcal{V}_{\delta' :: p}(o_n)$ or *false* if there is no object linked to o
- $\mathcal{V}_{\delta^{\emptyset} :: p}(o) = \mathcal{V}_p(o)$, that is *true* if o has the property p, *false* otherwise.
Where $o_1, \ldots o_n$ are the objects of type X linked to the object o, and $\delta^{\emptyset}$ is the empty path (size 0).

Example 7. Let us consider the rule: $P_D : \forall M \exists W :: w.kind \in \{Oscar, Golden\ palm\}$, where P_D denotes the directors in the relation people.
$\mathcal{V}_{\forall M \exists W :: w.kind \in \{Oscar, Goldenpalm\}}(Sp) =$
$\mathcal{V}_{\exists W :: w.kind \in \{Oscar, Goldenpalm\}}(film_1) \wedge \ldots$
$\wedge \mathcal{V}_{\exists W :: w.kind \in \{Oscar, Goldenpalm\}}(film_m)$
where $film_1, \ldots, film_m$ denote the movies directed by Sp.

Definition 8. *Coverage is given by the following:*
$$coverage(r, \mathcal{E}_{target}) = \frac{|\{o|o \in \mathcal{E}_{target}\ and\ v_r(o) = true\}|}{|\mathcal{E}_{target}|}$$

Example 8. Let us consider all the movies as the target set. The coverage of the rule $M : \exists A :: A.gender = female$ is equal to $\frac{2526}{11404}$, where 2526 is the number of movies with female actors and 11404 is the total number of movies. In the same way, we can calculate coverage($M : \exists A :: A.gender = animal, movies$)$= \frac{16}{11404}$.

3.5   Generality Order

Definition 9. *We say that a characteristic rule r_1 (δ_1::p_1) is more general than a rule r_2 (δ_2::p_2) (denoted by $r_1 \succeq r_2$) iff $\delta_1 \succeq \delta_2$ and $p_1 \succeq p_2$. We write $r_1 \succ r_2$, when $r_1 \succeq r_2$ and $\neg(r_2 \succeq r_1)$.*

Example 9. $M : \exists W :: W.kind\ in(Oscar, Golden\text{-}Palm) \succeq$
$M : \forall W :: W.kind\ in(Oscar)$.

Lemma 1. *Coverage is monotone with respect to the generality order, i.e., if $coverage(r_2, \mathcal{E}_{target}) \geq \epsilon$ and $r_1 \succeq r_2$ then $coverage(r_1, \mathcal{E}_{target}) \geq \epsilon$, or else if $\neg(coverage(r_1, \mathcal{E}_{target}) \geq \epsilon)$ and $r_1 \succeq r_2$ then $\neg(coverage(r_2, \mathcal{E}_{target}) \geq \epsilon)$.*

3.6   Specialization Operator

Definition 10. *We define the specialization operator ρ as a binary relation on the set of characteristic rules as follows:*
$\rho(\delta :: p) = \{\delta'::p | \delta'\ differs\ from\ \delta\ by\ one\ \exists\ quantifier\ set\ to\ \forall\} \cup \{\delta::p' | p \succ p'$ *and there is no p'' s.t. $p \succ p'' \succ p'\}$*
Let us notice that for all $r'' \in \rho(r)$, there is no $r' \notin \rho(r)$ such that $r \succ r'$ and $r' \succ r''$.

Example 10. Suppose that we consider only the following properties for Actors: {$Actor.gender = male, Actor.gender = female$}, and Movies as the target set. The complete search space starting with $\exists A :: True$ is given in Figure 2.

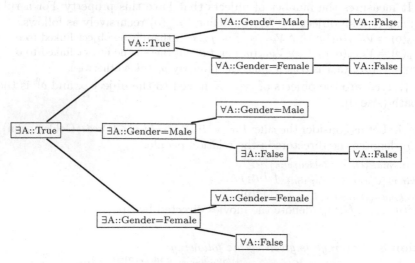

Fig. 2. Search space starting with the rule Movies: $\exists$ A::True

The definition of a specialization operator allows to define a top down, levelwise, search strategy, for mining characteristic rules. For pruning the search space, we define two notions: *open-coverage* and *link-coverage*.

3.7   Link-Coverage

We define *link-coverage* $(\delta::p, \mathcal{E}_{target})$ =coverage(open(δ)::$True, \mathcal{E}_{target}$) Intuitively, link coverage measures the number of target objects for which there exists at least an object linked to them through δ. This can be useful when there is a 0..* relation, which means that some objects can be linked to none objects by this relation.

3.8   Open-Coverage

We define *open-coverage* $(\delta::p, \mathcal{E}_{target}) = coverage(open(\delta)::p, \mathcal{E}_{target})$ where $open(\delta)$ is obtained by setting all the quantifiers of δ to $\exists$. Intuitively, open-coverage counts the number of target objects for which there is at least an object linked to them by δ and satisfying p.

3.9   Interesting Characteristic Rules

For a rule $\delta :: p$, coverage measures the number of objects in the target set having the property p. We would like to estimate whether this property is really characteristic of $\mathcal{E}_{target}$ or not. This can be achieved by verifying if the property covers *enough* objects in the target set, while covering *few* objects outside the target set. One should find a trade-off between these two conditions and estimate the quality of rules.

Furthermore, in descriptive data mining tasks, such as characterization, thousands of rules may be discovered, so making the rule filtering step as a necessary post processing step. In our framework, we define a function named *Interesting* that can filter the rules relying on such heuristics in order to keep only interesting ones. In [10], Lavrač et al. analyze some rule evaluation measures used in Machine Learning and Knowledge Discovery. They propose only a measure that can be considered as a measure of novelty, precision, accuracy, negative reliability, or sensitivity. In our experiments, we used their *novelty* measure: the novelty of a rule $H \longleftarrow B$ is given by: (P represents a probability)

$$Novelty(H \longleftarrow B) = P(HB) - P(H) * P(B)$$

For a characteristic rule r, for each object $o \in \mathcal{E}$, we can consider the facts $o \in \mathcal{E}_{target}$ and $\mathcal{V}_r(o) = true$. We are looking for a strong association between these two facts. This one can be estimated by the novelty measure. In our framework, the novelty of a rule can be estimated by:

$$Novelty(r) = \frac{|\{o|o \in \mathcal{E}_{target} \text{ and } v_r(o) = true\}|}{|\mathcal{E}|}$$
$$- \frac{|\mathcal{E}_{target}|}{|\mathcal{E}|} \cdot \frac{|\{o|o \in \mathcal{E} \text{ and } v_r(o) = true\}|}{|\mathcal{E}|}$$

According to [10], we have $-0.25 \leq Novelty(r) \leq 0.25$. A strongly positive value indicates a strong association between the two facts.

> **Function Interesting** $(r, \mathcal{E}_{target})$: **boolean**
> If Novelty$(r) \rightarrow 0.25$ then return True
> else return False

We can also use other measures such as entropy, purity, or Laplace estimate. See [4] for more details. In addition to the novelty we used in our experiments the Laplace estimate given by:

$$Laplace(r) = \frac{coverage(r, \mathcal{E}_{target})+1}{coverage(r, \mathcal{E}_{target})+coverage(r, \mathcal{E}-\mathcal{E}_{target})+2}$$

$0 \leq Laplace(r) \leq 1$. If a rule covers no examples, then Laplace is equal to 0.5.

4   Algorithm

We can use a variant of the *levelwise algorithm* [11] for mining all potentially interesting characteristic rules.

> **Caracterix Algorithm**
> **input** $\mathcal{C}_1 = \{r,$ such that there is no $r', r' \succ r \}$
> $i = 1$
> **while** $\mathcal{C}_i \neq \emptyset$
> 1. $\mathcal{F}_i$ $= \{r \in \mathcal{C}_i | link\text{-}coverage(r, \mathcal{E}_{target}) \geq \epsilon\}$
> 2. $\mathcal{F}'_i$ $= \{r \in \mathcal{F}_i | open\text{-}coverage(r, \mathcal{E}_{target}) \geq \epsilon\}$
> 3. $\mathcal{F}''_i$ $= \{r \in \mathcal{F}'_i | coverage(r, \mathcal{E}_{target}) \geq \epsilon\}$
> 4. $\mathcal{C}_{i+1} = (\bigcup \rho(r) | r \in \mathcal{F}''_i) \setminus \bigcup_{j \leq i} \mathcal{C}_j$
> 5. $i = i+1$
> **end while**
> **output** $\{r \in \bigcup_{j<i} \mathcal{F}''_j | Interesting(r, \mathcal{E}_{target})\}$

Caracterix starts with $\mathcal{C}_1$, the set of the most general characteristic rules given by the user. The algorithm then iterates coverage tests (lines 1,2,3) and generation of next candidate rules (line 4), taking care to discard previously considered rules. The iteration stops when it is not possible to generate further candidate rules. Pruning heuristics, link-coverage (line 1) and open-coverage (line 2) are used to reduce the number of coverage evaluations done in line 3. Open-coverage and, *a fortiori*, link-coverage are the same for variant rules. They are stored and retrieved as needed to avoid unnecessary computations. Let us notice that these pruning strategies only exclude characteristic rules that do not fulfill the minimum coverage requirement ϵ. The algorithm then outputs the set of all interesting rules.

Lemma 2. Caracterix *is correct and complete w.r.t.* $\mathcal{C}_1$.

Proof. The proof relies on the following inequality: $link\text{-}coverage(r, \mathcal{E}_{target}) \geq open\text{-}coverage(r, \mathcal{E}_{target}) \geq coverage(r, \mathcal{E}_{target})$.

5   Experiments

The model that we have proposed and the system Caracterix have been developed by the first three authors at LIFO, and experimented on a real geographic database provided by the BRGM[3]. The rules that have been learned have been evaluated by a geologist expert (the fourth author of the paper). For this purpose, we have extended our framework in order to take into account the spatial dimension, mainly the topological and distance information between geographic objects. In our experiments we have used a GIS [2], which handles many layers: geographic, geologic, seismic, volcanic, mineralogic, gravimetric,.... These layers store more than 70 thousands geographic objects. We aim at finding characterization rules for characterizing mineral ore deposits using geological information, faults, volcanos ... This task can be stated as follows:

- given a set $\mathcal{E}$ of geographic objects, $\mathcal{E} = \mathcal{E}_1 \cup \mathcal{E}_2 \cup \mathcal{E}_3 \cup \mathcal{E}_4 \cup \mathcal{E}_5$, where $\mathcal{E}_1$ contains *mineral deposits*, $\mathcal{E}_2$ represents the *geology*, $\mathcal{E}_3$ the *volcanoes*, $\mathcal{E}_4$ the *faults* and $\mathcal{E}_5$ the *seisms*;
- given a set $\mathcal{R}$ of binary relations based on spatial proximity;
- given a target set $\mathcal{E}_{target} = \{gold\ mines\} \subseteq \mathcal{E}_1$;
$\rightarrow$ find a set of characterization rules of $\{gold\ mines\}$

To take into account the distance between objects, we introduce a parameter λ and r_{ij}^{λ} represents a binary relation between objects in $\mathcal{E}_i$ and objects in $\mathcal{E}_j$ parameterized by λ. In the case of geographic objects, this parameter may denote the distance between objects. For instance $r_{1,3}^{100km}$ represents a binary relation between mineral deposits and volcanoes at a distance less or equal to 100 kms. As a consequence, the notion of quantified path described in section 3.1 has been extended, considering the parameter λ used in binary relations. For instance: $M : \forall_{10km}F\forall_{5km}V$ denotes all the volcanoes at less than 5 kilometers than faults at less than 10 kilometers than each mine. In order to handle distance information between objects, we construct growing buffers around target objects progressively, while checking for the properties satisfied by objects entering into the buffers. This notion is illustrated by Figure 3, where buffers are constructed around mineral deposits.

The Quantified path generality order defined in Section 3.1 can be extended to such parametrized quantified paths. In fact, in the case of characteristic rules with one parameter, we have: $\delta_\lambda \succeq \delta_{\lambda'}$ if ($\lambda \geq \lambda'$ and λ, λ' *indexes* a $\exists$) or ($\lambda \leq \lambda'$ and λ, λ' indexes a $\forall$). We have:

$$M : \forall_{3Km}F \succeq M : \forall_{5Km}F \succeq M : \forall_{10Km}F$$
$$M : \exists_{10Km}F \succeq M : \exists_{5Km}F \succeq M : \exists_{3Km}F$$

Intuitively, this means that if a property holds for all faults at a distance less than 10km from a mine, then this property also holds for all faults at less than 5km and 3km from this mine. Vice versa, if there exists a fault at less than 3km from a mine with a given property, than there exists a fault at less than 5km and less than 10km with the same property.

When we have more than one parameter, we can induce a *partial* order, by taking into account the relation $\delta_{\lambda_1,...,\lambda_n} \succeq \delta_{\lambda'_1,...,\lambda'_n}$ if $\forall i$, ($\lambda_i \geq \lambda'_i$ and λ_i, λ'_i indexes a $\exists$) or ($\lambda_i \leq \lambda'_i$ and λ_i, λ'_i indexes a $\forall$).

---

[3] French public institution based on Earth Sciences

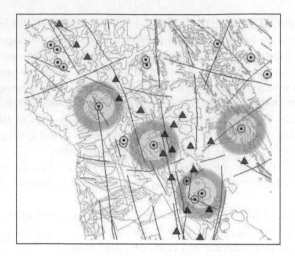

Fig. 3. Buffers around some target points in the GIS. Layers represented here are *geology, mineral deposits, fault and volcanoes*

Table 1. Some examples of tested rules

Rule	Coverage	Laplace	Novelty
M: M.Era ∈ {Mesozoic,Cretacious}	4.59%	0,750	0,0080
M: M.Era ∈ {Mesozoic, Jurassic, Cretacious}	6,42%	0,148	-0,0133
M: M.Lithology = sedimentary deposits	5,50%	0,070	-0,0413
M: M.Lithology=volcanic deposits	64,22%	0,266	0,0102
M: M.Distance_Benioff ∈ [170..175]	66,97%	0,365	0,0529
M: ∃$_{10km}$ G::G.Age=tertiary	86,24%	0,259	0,0086
M: ∃$_{5km}$ V::V.Age=recent	7,34%	0,310	0,0030
...	...	...	...

5.1   Results

Our system tested hundreds of rules. Some examples are given Table 1.

The following rule has been discovered and covers 60% of gold mines and rejects most of the other mines.

$M : ∃_{10km}$ G :: M.MainSubstance= au∧
    G.CodeGeology= $TertiaryVolcanic$∧
    M.BenioffDepth∈ [75..150]∧
    M.Distance_Benioff∈ [170..275]∧
    M.BenioffSlope ∈ [8°..16°]∧
    G.Age= $tertiary$∧
    M.Lithology= $volcanic$∧
    M.Gitology= $epithermal$∧
    M.Morphology= $veins$

This rule, considered as interesting by experts, expresses that for all gold mines, there exists a tertiary volcanic geology at a distance less than 10 km from this mine, and these mines are epithemal ones with a morphology of veins and are at a benioff depth between 75 and 150 km and at a slope benioff of 8° and 16°. According to geologist experts, this rule is interesting because it is

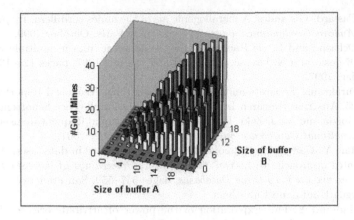

Fig. 4. Link Coverage of the rule $M : \exists_A F \; \exists_B V::$True

related to a natural phenomenon: the plate tectonics.

Figure 4 illustrates the notion of link-coverage and represents the number of gold mines that contain at least a fault in a buffer of size A around the mine and such that the fault contains at least a volcanoe in a buffer of size B around this fault.

6 Conclusion

In this paper, we have presented a new general approach for mining a new kind of characteristic rules in a target set of objects. These rules handle both properties and quantified paths. These latters specify how to take into account different kinds of objects and their relationships, in other words, how to go from objects to others without flattening the tables describing these objects. We propose 𝕮aracteri𝕏, a levelwise algorithm exploring the search space looking for characteristic rules, taking into account a generality relation between rules. Moreover, the notions of link-coverage and open-coverage are useful heuristics to prune the search space. We have experimented our approach on a geographic database and we have submitted our rules to geologists. They considered that these rules are interesting and give a good description of a set of chosen target objects. Quantified paths give a convivial way to look for the characteristics of the target objects according to the spatially linked objects. In the future, we aim at extending our framework on other kinds of databases, such as object oriented databases.

References

1. R. Agrawal, T. Imielinski, and A. N. Swami. Mining association rules between sets of items in large databases. *In Proc. of the ACM SIGMOD International Conference on Management of Data*, pages 207–213, 1993.

2. D. Cassard. Gis andes: A metallogenic gis of the andes cordillera. In *4th Int. Symp. on Andean Geodynamics*, pages 147–150. IRD Paris, October 1999.
3. L. Dehaspe and L. De Raedt. Mining association rules in multiple relations. In S. Džeroski and N. Lavrač, editors, *ILP97*, volume 1297, pages 125–132. Springer-Verlag, 1997.
4. J. Furnkranz. Separate-and-conquer rule learning. Technical Report OEFAI-TR-96-25, Austrian Research Institute for Artificial Intelligence Schottengasse, 1996.
5. K. Gouda and M. J. Zaki. Efficiently mining maximal frequent itemsets. *1st IEEE International Conference on Data Mining*, November 2001.
6. J. Han, Y. Cai, and N. Cercone. Knowledge discovery in databases: An attribute-oriented approach. In Li-Yan Yuan, editor, *Proceedings of the 18th International Conference on Very Large Databases*, pages 547–559, San Francisco, U.S.A., 1992. Morgan Kaufmann Publishers.
7. J. Han and Y. Fu. Exploration of the power of attribute-oriented induction in data mining. In Usama M. Fayyad, Gregory Piatetsky-Shapiro, Padhr Smyth, and Ramasamy Uthurusamy, editors, *Advances in Knowledge Discovery and Data Mining*. AIII Press/MIT Press, 1996.
8. Y. Kodratoff and J-G. Ganascia. *Machine Learning: An Artificial Intelligence Approach*, chapter Improving the Generalization Step in Learning. Morgan Kaufmann, 1986.
9. Y. Kodratoff and C. Vrain. Acquiring first order knowledge about air traffic control. *Knowledge Acquisition Journal, B.R. Gaines & J.H. Boose, (Eds.), Academic Press Limited*, pages 353–386, 1993.
10. N. Lavrač, P. Flach, and B. Zupan. Rule evaluation measures: A unifying view. In S. Džeroski and P. Flach, editors, *ILP99*, volume 1634 of *LNAI*, pages 174–185. Springer-Verlag, 1999.
11. H. Mannila and H. Toivonen. Levelwise search and borders of theories in knowledge discovery. *Data Mining and Knowledge Discovery*, 1(3):241–258, 1997.
12. R. S. Michalski. A theory and methodology of inductive learning. In *Machine Learning: An Artificial Intelligence Approach*, volume 2(4), pages 83–134, 1983.
13. T.M. Mitchell. Version spaces: A candidate elimination approach to rule learning. In *Proceedings of the 5th International Joint Conference on Artificial Intelligence IJCAI*, pages 305–310. Cambridge MA, August 1977.
14. G.D. Plotkin. A note on inductive generalization. In *Machine Intelligence*, volume 5, pages 153–163. Edinburgh University Press, 1970.
15. L. De Raedt and S. Kramer. The level-wise version space algorithm and its application to molecular fragment finding. In Bernhard Nebel, editor, *Proceedings of the Seventeenth International Joint Conference on Artificial Intelligence (IJCAI-01)*, pages 853–862. Morgan Kaufmann, 2001.
16. A. Salleb, Z. Maazouzi, and C. Vrain. Mining maximal frequent itemsets by a boolean approach. In IOS Press Amsterdam F. van Harmelen, editor, *ECAI'2002*, pages 385–389, Lyon, France, 2002.
17. C. Vrain. *Machine Learning, an Artificial Intelligence Approach*, volume 3, chapter OGUST: a system which learns using domain properties expressed as theorems, pages 360–382. Morgan Kaufman publisher, 1990.

Topic Learning from Few Examples

Huaiyu Zhu, Shivakumar Vaithyanathan, and Mahesh V. Joshi

IBM Almaden Research Center, 650 Harry Road, San Jose, CA, USA
{huaiyu,vaithyan,joshim}@us.ibm.com

Abstract. This paper describes a semi-supervised algorithm for single class learning with very few examples. The problem is formulated as a hierarchical latent variable model which is clipped to ignore classes not of interest. The model is trained using a multistage EM (msEM) algorithm. The msEM algorithm maximizes the likelihood of the joint distribution of the data and latent variables, under the constraint that the distribution of each layer is fixed in successive stages. We demonstrate that with very few positive examples, the algorithm performs better than training all layers in a single stage. We also show that the latter is equivalent to training a single layer model with corresponding parameters. The performance of the algorithm was verified on several real-world information extraction tasks.

1 Introduction

Several real world problems fall into the category of single class learning, where training data is available for only a single class. Examples of such problems include the identification of a certain class of web-pages - e.g., "personal home pages" or "call for papers" [10]. Building training data for such problems can be a particularly arduous task. A good sample of the positive class must involve all aspects that can lead to inclusion in the positive class. Constructing a negative class would require a uniform representation of the universal set excluding positive class [10].

Information extraction is another area where single class learning problems arise naturally. Information needs of users are too diverse and numerous to allow the creation of significant numbers of labeled examples. Consider, for example, an oil company's corporate reputation management group interested in monitoring articles about its and its competitor's image in the areas of diversity at work place, oil spill issues, environmental policies etc. Obtaining comprehensive examples for every one of these topics is almost impossible. Users are typically willing to provide only very few carefully crafted positive examples for each topic of interest.

The need for single class learning has been recognized and there have been a few previous efforts focusing on learning from positive examples. In [7], the algorithm maps the data using a kernel and then uses the origin as the negative class. In practice this was found to be very sensitive to parametric changes and some heuristic modifications were suggested to include more than just the origin into the negative class [4]. Recently [10] includes unlabeled examples in an iterative framework that identifies examples not sharing features with positive examples, which are then treated as negative examples for training a support vector machine. These approaches have concentrated on identifying negative examples and using them in a discriminative training framework. The motivation

N. Lavrač et al. (Eds.): PKDD 2003, LNAI 2838, pp. 483–494, 2003.

in these approaches has been towards building classifiers that do not degrade in accuracy with the growth in the size of labeled data [10].

Generative modeling approaches have also been applied to the problem of partially labeled data. Unsupervised approaches use joint distributions over the features to identify clusters in the data. In particular, finite mixture models trained using the popular Expectation-Maximization (EM) algorithm [2] have been used extensively. An interesting approach in [5] modifies the EM algorithm to allow the incorporation of labeled data. This approach can, in theory, be used with small amount of labeled data and [5] reported encouraging experiments on multi-class problems where labeled data are available for each class. A variant of this approach to the single class problem, but with larger amounts of labeled data, has been described in [3] with good results.

In this paper we focus on the single-class learning problems with the following two characteristics: (1) The topic of interest only constitutes a very small proportion of candidate data, and (2) The topic is specified by very few positive examples (*seeds*) which usually do not represent a fair sample of the topic. For such problems, single stage clustering algorithms do not perform well: The precision is low unless the number of clusters is large, while the recall is low unless the number of clusters is small. In order to overcome this, we use a hierarchical latent variable model trained with a novel multistage EM (msEM) algorithm. The algorithm concentrates on the class of interest, guided by the labeled examples. Experiments show that the algorithm generalizes well from small number of seeds that form skewed samples of the desired topic.

2   Latent Variable Models and Semi-supervised EM Algorithm

2.1   Latent Variable Model

One commonly used model for clustering is a mixture model of the form

$$p(z) = \sum_a p(z|a) \cdot p(a). \tag{1}$$

where the variable a is a latent variable and is interpreted as class label. Training of this model involves adjusting the parameters of the probability distributions $p(z|a)$ and $p(a)$. This model can be trained effectively using the EM algorithm [2].

Given a dataset $z = \{z_1, z_2, \ldots, z_n\}$ of individual observations of z, the EM algorithm is an iterative algorithm that maximizes the log-likelihood of the model,

$$\sum_i \log p(z_i) = \sum_i \log \sum_a p(z_i, a). \tag{2}$$

2.2   Semi-supervised EM Algorithm

The EM algorithm for maximizing (2) is an *unsupervised* algorithm. However, some *prior* information is available. E.g., in single class classification, we often have some *seed* information – a few labeled examples for the class of interest. Incorporating such *seed constraints* into the EM algorithm results in a *semi-supervised EM algorithm* (ssEM) [5]. We use the version as shown in Fig. 1.

- Set initial model parameters for the distributions $p(z|a)$ and $p(a)$.
- Iterate until convergence over the following two steps:

    E-Step: For each z_i, compute $q(a|z_i)$ by Bayes rule and seed constraint

    $$\begin{cases} q(a|z_i) = p(a|z_i) = p(z_i, a)/p(z_i), & z_i \notin \text{Seeds} \\ q(a = 1|z_i) = 1, & z_i \in \text{Seeds.} \end{cases} \tag{3}$$

    M-Step: Estimate new parameters for $p(z|a)$ and $p(a)$ by maximizing

    $$\sum_{i,a} q(a|z_i) \log p(z_i, a). \tag{4}$$

Fig. 1. Semi-supervised EM algorithm (ssEM)

However, with very few labeled examples, seed constraints alone are not sufficient to tackle the problem at hand. For the task of identifying a single class from multiple possibilities, there is a trade-off between the number of components in the mixture model (1) and the precision of the chosen class. If the number of components in the mixture model is small, the chosen component is likely to contain a large number of spurious datapoints. If the number of components is large, the desired class might be fragmented among many different components. We now proceed to describe more powerful models and algorithms to address this.

2.3   Hierarchical Latent Variable Models

A two level hierarchical model can be obtained by replacing the model (1) with

$$p(z) = \sum_{a_0, a_1} p(z|a_0, a_1) \cdot p(a_1|a_0) \cdot p(a_0), \tag{5}$$

where a_0 and a_1 are two levels of latent variables in the hierarchy.

A full-blown model of the form (5) can be expensive to train due to the combinatorial effect of the hierarchical hidden variables in the E-step. Since we are only interested in a single class, it is intuitively plausible that clipping off the branches in the hierarchical model not corresponding to the class of interest would reduce a substantial amount of computation without large impact on performance. This suggests using the following "clipped model" (assuming $a = 1$ corresponds to the class of interest)

$$p(z_i) = \sum_{a_0=1} p(a_0) \sum_{a_1} p(a_1|a_0) \cdot p(z_i|a_0, a_1) + \sum_{a_0 \neq 1} p(a_0)p(z_i|a_0). \tag{6}$$

However, as the following two lemmas show, training either of them with ssEM does not exhibit advantage over single layer models. The full potential of the clipped model can be realized by a new training algorithm, which we propose in Section 3.

Lemma 1. The hierarchical model (5) can be represented as a single layer model. They both behave identically under standard EM training algorithm.

Proof. The model (5) is a marginal distribution of

$$p(z_i, a_0, a_1) = p(z_i|a_0, a_1)p(a_0, a_1) \tag{7}$$

Suppose the combination (a_0, a_1) takes n distinct values. For example, if a_0 takes n_0 values and a_1 takes n_1 values, then $n = n_0 n_1$. Introduce a variable c with n distinct values, the distribution (5) is identical to

$$p(z_i) = \sum_c p(z_i|c)p(c). \tag{8}$$

The derivation of the training algorithm involves expressions of $p(z_i|a_0, a_1)$ and $p(a_0, a_1)$, which can be replaced with expressions of $p(z_i|c)$ and $p(c)$. The resulting training algorithm is therefore equivalent after simple renaming of parameters. □

Lemma 2. The clipped hierarchical model (6) can be represented as a single layer model. They both behave identically under standard EM training algorithm.

Proof. Since a_1 does not occur for $a_0 \neq 1$, we can arbitrarily set $a_1 = 1$ for $a_0 \neq 1$. Then (6) is a marginal of

$$p(z_i, a_0, a_1) = p(z_i|a_0, a_1)p(a_0, a_1) \tag{9}$$

where $p(z_i|a_0, a_1) = p(z_i|a_0)$ and $p(a_0, a_1) = p(a_0)$ for $a_1 = 1$. This case is analogous to that of Lemma 1, except that (a_0, a_1) would take value in a discrete set with $2n - 1$ elements. Hereafter the proof follows that of Lemma 1. □

3   Training Clipped Model with Multistage EM Algorithm

For problems involving large number of components in the mixture model, the semi-supervised EM can be successfully applied when labeled data are available for different classes [5]. However, a more powerful algorithm is needed when only a few labeled datapoints are available for just one class. We propose a multistage EM algorithm (msEM) to train the clipped model, which is more suitable for such problems.

3.1   Generalized Form of Likelihood

The log-likelihood (2) can be written in a more general form [1]

$$N \sum_i q(z_i) \log \sum_a p(z_i, a), \tag{10}$$

where $q(z_i) = 1/N$ is the *empirical distribution* of the data and N is the size (number of datapoints) of the dataset. Using (10), the M-step (4) of the EM algorithm can now be written as maximizing

$$\sum_{i,a} q(z_i)q(a|z_i) \log p(z_i, a). \tag{11}$$

The convergence properties of the EM algorithm still hold even when $q(z_i)$ is not the uniform distribution over the observed data z [1]. The EM algorithm can therefore be regarded as a mapping $q(z) \to q(z, a)$.

- For each layer $m = 0, 1, \ldots$, set initial model parameters for $p_m(z|a)$ and $p_m(a)$.
- Set the first layer empirical distribution $q_0(z_i) = q(z_i) = 1/N$ for all datapoint z_i.
- Iterate until convergence:
  - For each layer $m = 0, 1, \ldots$, carry out the following three steps
    * **E-Step:** For each z_i, compute $q_m(a_m|z_i)$ by Bayes rule and seed constraint

$$\begin{cases} q_m(a_m|z_i) = p_m(a_m|z_i) = p_m(z_i, a_m)/p_m(z_i), & z_i \notin \text{Seeds} \\ q_m(a_m = 1|z_i) = 1, & z_i \in \text{Seeds}. \end{cases} \quad (12)$$

    * **M-Step:** Estimate new parameters for $p_m(z|a)$ and $p_m(a)$ by maximizing

$$\sum_{i, a_m} q_m(z_i) q_m(a_m|z_i) \log p_m(z_i, a_m). \quad (13)$$

    * Set empirical distribution for the next layer $q_{m+1}(z_i) = q_m(z_i|a_m = 1)$ for all datapoint z_i using Bayes rule

$$q_m(z_i|a_m) = q_m(z_i) q_m(a_m|z_i)/q_m(a_m). \quad (14)$$

Fig. 2. Multistage semi-supervised EM algorithm (msEM)

3.2 Multistage Semi-supervised EM Algorithm

The msEM algorithm for the clipped model trains each layer successively by incorporating the empirical distribution from the previous layers (Fig. 2).

Comparing the E and M-steps of the above algorithm, (12) and (13), with those of the ssEM algorithm, (3) and (11), it can be seen that the computation for layer m implements the EM algorithm that maximizes the generalized log-likelihood

$$\sum_i q_m(z_i) \log \sum_{a_m} p_m(z_i, a_m), \quad (15)$$

where $q_m(z_i)$ is given by $q_m(z_i|a_m = 1)$. The intuition behind this algorithm is that, by weighting each datapoint with $q_m(z_i)$ instead of the uniform distribution, we are deemphasizing those z_i that are less likely to be in class 1 as predicted by layer $m - 1$. The discrimination in layer m could then conceivably concentrate more on the finer details not addressed at layer $m - 1$. Each layer acts as a regularizer to restrict the variability of the model in the next layer.

3.3 Deriving the Updated Empirical Distribution

In the msEM algorithm, the empirical distribution q_m of layer m is computed from the results of layer $m - 1$. The rule for computing q_m can be derived from a global optimization problem involving layers 0 through m. The objective function for layer m is

$$\sum_{i, a_0, \ldots, a_{m-1}} q(z_i, a_0, \ldots, a_{m-1}) \log \sum_{a_m} p(z_i, a_0, \ldots, a_m). \quad (16)$$

Maximizing (16) for successive m with ssEM implies that layer m is trained under the constraint that $q(z_i, a_0, \ldots, a_{m-1})$ is fixed. We now show that this is indeed equivalent to the msEM given in Fig. 2.

For layer 0, the objective function (16) reduces to

$$\sum_i q(z_i) \log \sum_{a_0} p(z_i, a_0). \tag{17}$$

The ssEM algorithm for maximizing (17) is the same as the msEM algorithm for layer 0, with the following substitutions,

$$q(z_i, a_0) = q_0(z_i, a_0), \qquad p(z_i, a_0) = p_0(z_i, a_0). \tag{18}$$

For layer 1, the objective function (16) reduces to

$$\sum_{i, a_0} q(z_i, a_0) \log \sum_{a_1} p(z_i, a_0, a_1). \tag{19}$$

The E- and M-steps corresponding to (3) and (11) are therefore

$$\begin{cases} q(a_1 | a_0, z_i) = p(a_1 | a_0, z_i) = p(z_i, a_1 | a_0)/p(z_i | a_0), & z_i \notin \text{Seeds} \\ q(a_1 = 1 | a_0, z_i) = 1, & z_i \in \text{Seeds}. \end{cases} \tag{20}$$

$$\text{maximize} \sum_{z_i, a_0, a_1} q(z_i, a_0) q(a_1 | z_i, a_0) \log p(z_i, a_0, a_1). \tag{21}$$

Since the model is clipped, it is clear that the E-step (20) is equivalent to (12) with the following substitutions,

$$p_1(a_1, z_i) = p(a_1, z_i | a_0 = 1), \qquad q_1(a_1, z_i) = q(a_1, z_i | a_0 = 1). \tag{22}$$

The M-step objective function in (21) can be expanded into three terms,

$$\sum_{a_0} q(a_0) \log p(a_0) + \sum_{a_0 \neq 1} \sum_{z_i} q(z_i, a_0) \log p(z_i | a_0)$$
$$+ \sum_{a_0 = 1} \sum_{z_i, a_1} q(a_0) q(z_i | a_0) q(a_1 | z_i, a_0) \log p(z_i, a_1 | a_0). \tag{23}$$

Maximizing the first two terms leads to $p(z_i, a_0)$ already calculated in (18). Using the definitions of p_1 and q_1 in (22) and the fact that the third term only involves $a_0 = 1$, it can be verified that maximizing the third term is equivalent to maximizing (13), provided that the empirical distribution is given as $q_1(z_i) = q_0(z_i | a_0 = 1)$. Therefore we have proved the equivalence for layer 1.

Similarly it can be shown that for any m, the E- and M-steps in the msEM are equivalent to the steps in an ssEM that maximizes (16), provided that $q_{m+1}(z_i) = q_m(z_i | a_m = 1)$.

3.4 Considerations on the Convergence of Multistage EM

In our experiments the msEM always converged with speed similar to ssEM. Here we outline an approach to prove the convergence. It is known that each iteration of the standard EM algorithm increases the likelihood, which converges to a stationary point [2]. Under certain conditions, this also results in the convergence of the probability distributions p and q to fixed points [8]. Under certain stronger conditions, the mapping $q(z) \rightarrow q(z, a)$ is continuous. Assuming that each layer of the clipped model satisfies all these conditions, the convergence of msEM can be proved by induction. Layer 0 implements ssEM so $q_0(z, a)$ converges. Suppose $q_{m-1}(z, a)$ converges. Then $q_m(z) = q_{m-1}(z|a = 1)$ also converges. The continuity of the mapping $q_m(z) \rightarrow q_m(z, a)$ then implies the convergence of $q_m(z, a)$.

3.5 Multistage EM Interpreted as Reverse Boosting

We note briefly the relationship between the msEM algorithm and boosted density estimation [6]. An extensive description of this relationship is available from [11].

Let $a_{0:m} = (a_0, \ldots, a_m)$ and denote by $a_{0:m} = 1$ the condition $a_0 = \cdots = a_{m-1} = 1$. The msEM algorithm can be regarded as building successively more complex models with weighted weak learners. At stage m the model built so far is

$$
P_m(z) = \sum_{a_0 \neq 1} p(a_0) p(z|a_0)
$$
$$
+ p(a_0 = 1) \sum_{a_1 \neq 1} p(a_1|a_0 = 1) p(z|a_0 = 1, a_1)
$$
$$
+ \ldots
$$
$$
+ \prod_{l=0}^{m-1} p(a_l = 1|a_{0:l-1} = 1) \sum_{a_m} p(a_m|a_{0:m-1} = 1) p(z|a_{0:m-1} = 1, a_m).
$$

$$(24)$$

The msEM algorithm attempts to improve the classification of a single class by getting successively better density estimates for that single class subject to the seed constraints. The weak learner chosen at layer m is a finite mixture model $p_m(z, a_m)$, trained with ssEM. In contrast to the boosted density estimation [6], which emphasizes regions with

---

1. Set the initial weights $w_i = 1/N$.
2. for m=1 to M
   (a) Use EM algorithm to compute $p_m(z_i|a_m)$ and $p_m(a_m)$ that maximizes $\sum_i w_i \log \sum_a p_m(z_i|a_m) p_m(a_m)$ subject to *seed constraints*, obtaining $q_m(a_m|z_i)$ in the process.
   (b) Set $w_i = q_m(z_i|a_m = 1)$, calculated with Bayes rule (14).
3. Output final model P_M

Fig. 3. Multistage EM interpreted as reverse boosting

Table 1. Experimental datasets and topics

Dataset	# Entities	# Datapoints	# Features	Half-window	Topics	# Seeds
TN	2151	87,251	24,808	400 characters	"chip", "web"	3
OP	10	30,000	77,762	75 words	"diversity"	2–8

> *Intel has not reduced its capital spending budget of $7.5 billion for the year, in part to accommodate the introduction of 300-millimeter wafer production. Chips produced on the new wafers will also be made with the more advanced 0.13-micron manufacturing process and contain copper wires.* **Intel** *currently makes its chips with the 0.18-micron manufacturing process and uses aluminum. The micron measurements refer to the size of features on the chip. The shift will result in smaller, cooler, faster and cheaper processors. "Intel expects chips produced on 300-millimeter wafers to cost 30 percent less than those made using the smaller wafers, " Tom Garrett, Intel's 300-millimeter program manager, said in a statement.*

Fig. 4. Example passage for the chip manufacturing topic

high uncertainty, the msEM emphasizes regions with high certainty of being in the class of interest, by increasing the weight of datapoints that perform well in the previous layer. The msEM algorithm in boosting framework is shown in Fig. 3, where the weight w_i corresponds to $q_{m+1}(z_i)$ in the msEM algorithm.

4   Experimental Setting

4.1   Data Sets

We experimented with msEM and several existing algorithms on two real-world document collections. The datasets are formed from passages in documents crawled from the Web. For each document, a set of proper names are identified as being of interest (named entities). A passage of a fixed window size surrounding each named entity is taken as its context. An example of such a passage is shown in Fig. 4. The context is tokenized into words (discarding punctuations), removing stop words (a list of 232 common words) and stemming (using Porter's Stemmer), resulting in a vector of feature counts. The named entity and the context feature counts together form a datapoint z_i. Note that each document can provide multiple datapoints.

Two test collections were made (Table 1). The collection TN was gathered from the Tech News section of CNet (www.cnet.com). A named entity tagger was used to identify organizational names (such as Intel, IBM, Microsoft etc.). The collection OP consists of web-pages discussing oil companies. A list of organizational names (each containing variations of the same organizational name) was obtained from industry experts. Some of these datasets will soon be made publicly available for other researchers to test their algorithms.

4.2   Topics and Seeds

The experiments were conducted on the following three topics:

chip Steps taken by semiconductor manufacturers to produce cheaper, faster and thermally more efficient microprocessors and microchips. (Dataset TN)
web Web Service protocols for business process integration. (Dataset TN).
diversity Issues related to diversity at work-place. (Dataset OP).

For the "chip" topic, three seeds occurring in two documents were identified from the corpus as relevant to the query. For the "web" topic, three seeds from three documents were selected as seeds. The "diversity" topic was used extensively to understand the behavior of the different algorithms. A series of experiments with several seed sets of different characteristics were used, as described below.

4.3   Algorithms, Parameters, and Evaluations

As comparison to msEM, the results are evaluated against several alternative algorithms: semi-supervised EM algorithm on single-layer latent variable models (ssEM)[5], a simple nearest neighbor algorithm (NN)[9] and a proximity pattern search (PPS).

For the ssEM algorithm, we used $p(z|a) = p(x|a)p(y|a)$ where x and y are the named entity and the context feature count vector, respectively, $p(x|a)$ is a discrete distribution, and $p(y|a)$ is the multinomial distribution. Laplace's smoothing is used in the M-step [5]. Let k be the numbers of components in the mixture model. The parameters of $p(z|a)$ are initialized by assigning $q(a = j|z_i) = 1/k + \epsilon, j = 1, \ldots, k$, for non-seed z_i and $q(a = 1|z_i) = 1$ for seed z_i, followed by an M-step, where ϵ is a small noise used for breaking symmetry among the components.

Each layer in the msEM algorithm uses the same model as in the ssEM algorithm with exactly two components ($k = 2$). The initializing of each layer is also the same as above, except that no symmetry breaking is necessary.

The NN algorithm is performed on the same tokenized context (without the named entity) using cosine similarity on the feature count vectors. A datapoint is deemed on topic if the similarity is higher than a cutoff value.

For the topics "chip" and "web" we also tested a type of proximity pattern search (PPS), based on patterns given by domain experts. A data point is deemed on topic if the pattern matches within a distance of the named entity in the original (not tokenized) context.

Each of these algorithms contains a parameter that controls the trade-off between precision and recall: number of layers for msEM, number of nodes for ssEM, similarity cutoff for NN and the distance for PPS. Results for different values of the control parameters are shown in the next section.

Results of the algorithms were manually evaluated by domain experts to establish a set of ground truth. Since the datasets are too large to be evaluated completely, only results from the high-precision versions of each algorithm are pooled together and evaluated. This allows us to calculate the precision of these algorithms and the number of correctly retrieved datapoints, but not the actual recall (which requires knowing the total number of on-topic datapoints in the whole corpus).

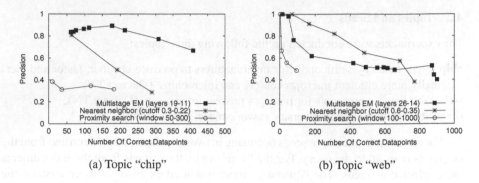

(a) Topic "chip"   (b) Topic "web"

Fig. 5. Precision vs number of correct datapoints

5   Results

Our experiments with ssEM [5] consistently returns very poor results. The algorithm was run for several different number of components ranging from 20 to 100 and the results were significantly worse than those of the other algorithms. We therefore do not report the actual numbers in this paper.

5.1   Results on the TN Collection

The results obtained by the algorithms on different parameter settings on the TN dataset are shown in 5(a) and 5(b). An immediate glance of these figures indicates that msEM is a clear winner for topic "chip", outperforming the other algorithms significantly. For topic "web" the NN algorithm is very competitive and the results drop off at 0.35. A look at the results of the pattern-matching proximity search helps shed more light on the relative performance between the msEM and nearest neighbor. For topic "chip" the best performance for proximity search is a precision of 0.4 which drops to a low of 0.3 with increasing recall – while for topic "web" the best performance is about 0.9 dropping to a low of just below 0.45. This seems to suggest that topic "web" is defined by simpler patterns than topic "chip". Closer examination of the results and discussion with the domain expert revealed that the "web" topic was particularly narrow that can be effectively defined by spotting a few keywords. In effect the smoothing/generalization effect provided by the msEM algorithm did not provide any advantages – instead worsened the results.

5.2   Effects of Seeds

To better understand the effect of seeds on the two algorithms msEM and NN, we conducted several experiments on the "diversity" topic using the dataset OP. This topic consists of several subtopics such as issues concerning minority, domestic partner benefit, gender equality, etc. The complexity of the task is increased due to the fact that these subtopics share very few words in common. We identified two types of seeds:

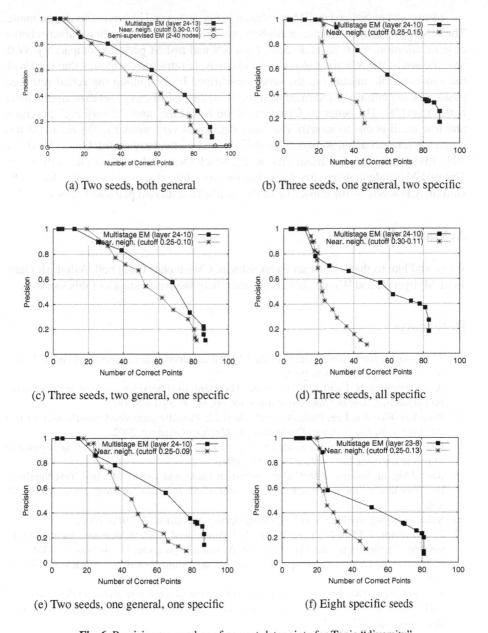

(a) Two seeds, both general

(b) Three seeds, one general, two specific

(c) Three seeds, two general, one specific

(d) Three seeds, all specific

(e) Two seeds, one general, one specific

(f) Eight specific seeds

Fig. 6. Precision vs number of correct datapoints for Topic "diversity"

- General seeds: passages discussing work place diversity policies in general.
- Specific seeds: passages discussing a specific instance of a company changing its policy on domestic partner benefits.

The results of NN and msEM are shown in Fig. 6. Several observations can be made. The NN is almost always better at highest precision. The msEM is almost always better at generalization. With general seeds, both NN and msEM perform comparably. With specific seeds, the NN algorithm almost exclusively retrieves datapoints that deal with the same specific instance as the seeds (confirmed by examining the actual retrieved passages). In contrast, at intermediate precision, the msEM can generalize significantly better than NN. The better performance of the NN at the range of very high precision and low recall is due to its retrieving only datapoints very similar to the seeds. At this range the generalization ability of the msEM is not particularly useful. On the hand, the NN fails to generalize for the specific seeds, which forms a skewed sample of the topic. The msEM is able to generalize better because its retrieval set is not entirely defined by similarity to seeds — the clustering of the unlabeled data also plays an important role.

Acknowledgments

We would like to thank Sreeram Balakrishnan, Christopher Campbell, Ashutosh Garg, Jussi Myllymaki and Wayne Niblack for their help in various stages of this work.

References

1. S.I. Amari. Information geometry of the EM and em algorithms for neural networks. *Neural Networks*, 8(9):1379–1408, 1995.
2. A. P. Dempster, N. Laird, and D. Rubin. Maximum likelihood for incomplete data via the EM algorithm. *J. of the Royal Statistical Society, ser. B*, 39:1–38, 1977.
3. Bing Liu, Wee Sun Lee, Philip Yu, and Xiaoli Li. Partially supervised classification of text documents. In *International Conference On Machine Learning*, 2002.
4. Larry Manevitz and Malik Yousef. One class SVMs for document classification. *Journal of Machine Learning Research*, 2:139–154, 2001.
5. Kamal Nigam, Andrew K. McCallum, Sebastian Thrun, and Tom M. Mitchell. Text classification from labeled and unlabeled documents using EM. *Machine Learning*, 39(2/3):103–134, 2000.
6. Saharon Rosset and Eran Segal. Boosting density estimation. In *NIPS*, 2002.
7. B. Scholkopf, J. Platt, J. Shawe-Taylor, A. J. Smola, and R. C. Williamson. Estimating the support of a high-dimensional distribution. *Neural Computation*, 13:1443–1472, 2001.
8. C. Wu. On the convergence properties of the EM algorithm. *Annals of Statistics*, 11:95–103, 1983.
9. Yiming Yang and Xin Liu. A re-examination of text categorization methods. In *Proceedings of SIGIR-99, 22nd ACM International Conference on Research and Development in Information Retrieval*, pages 42–49, 1999.
10. Hwanjo Yu, Jiawei Han, and Kevin Chen-Chuan Chang. PEBL: Positive example based learning for web page classification using SVM. In *Proceedings of 2002 SIGKDD Conference*, pages 239–248, 2002.
11. Huaiyu Zhu and Shivakumar Vaithyanathan. A multistage EM algorithm for training reduced hierarchical latent variable models. Technical Report RJ-10283, IBM Research Report, January 2003.

Arbogodaï, a New Approach for Decision Trees

Djamel A. Zighed[1], Gilbert Ritschard[2],
Walid Erray[1], and Vasile-Marian Scuturici[1]

[1] ERIC Laboratory, University of Lyon 2, C.P.11 F-69676 Bron Cedex, France
zighed@univ-lyon2.fr
[2] Dept of Econometrics, University of Geneva, CH-1211 Geneva 4, Switzerland
ritschard@themes.unige.ch

Abstract. Decision tree methods generally suppose that the number of
categories of the attribute to be predicted is fixed. Breiman et al., with
their Twoing criterion in CART, considered gathering the categories of
the predicted attribute into two superclasses. In this paper, we propose
an extension of this method. We try to merge the categories in an optimal
unspecified number of superclasses. Our method, called *Arbogodaï*, allows
during tree growing to group categories of the target variable as well as
categories of the predictive attributes. At the end, the user can chose to
generate either a set of single rules or or a set of multi-conclusion rules
that provide interval like predictions.

1 Introduction

Induction trees are among the most popular supervised methods proposed in
the literature. They are appreciated for the simplicity and the high efficacy of
the algorithms, for their ease of use and for the easily interpretable results they
provide. Hastie et al. [3], p. 313, designate them as the learning tool that comes
closest to the requirements of an "off-the-shelf" method.

Many induction trees methods have been proposed so far in the literature.
Some like ID3 [6], C4.5 [7] and CHAID [4,5] build n-ary trees, others like
CART [2] produce binary trees or, like SIPINA and Branching Programs [11,12],
latticed graphs that generalize trees by allowing the merging of nodes.

All these methods were originally intended for categorical attributes and require therefore that quantitative variables be discretized. This discretization can
be done at once before growing the tree. Most of the tree growing methods, nevertheless, handle quantitative variables in an automatic manner by dynamically
choosing the optimal discretization thresholds at each node. Some methods also
attempt to reduce the number of categories of nominal attributes by partitioning
them into a smaller number of classes. CART, for example, merges the categories
into two new superclasses at each new split. This has the advantage of avoiding to uselessly increase the number of nodes. Indeed, the higher the number of
nodes, the greater are the chances that some of them will have too few cases to
get reliable estimates of the response classes probabilities.

There are two main ways for partitioning the values of the predictive attributes. The first is for instance a characteristic feature of CHAID [4]. At each

N. Lavrač et al. (Eds.): PKDD 2003, LNAI 2838, pp. 495–506, 2003.

node, the local discriminating power of each categorical attribute is tested using all possible partitions of its values. Partitions in two or more groups are explored. Thus, for each split, a predictor is selected simultaneously with its locally best partition.

The second strategy is used for instance by Breiman et al. [2] in their CART method. At each node, CART looks only for the best bi-partition of each predictor. It generates thus only binary trees. With their Twoing criterion, the authors of CART propose however also a strategy that extends their principle to the response variable. When the response is multi-valued, using Twoing is equivalent to seek, for every predictor, simultaneously the best bi-partition of its values and the best bi-partition of the response values. The Twoing is the value of the Gini impurity for the best couple of bi-partitions and is used for selecting the split variable at each node.

In this paper, we extend the principle of a simultaneous search of a double bi-partition. We combine the CHAID and CART approaches. Like CHAID we look at each step for the best not necessarily binary partition of the attributes. Like CART with Twoing we explore also the partitioning of the values of the target variable. Unlike CART, we do not, however, restrict ourself to bi-partitions. At each step we look for the simultaneous grouping of the predictor values and of the target variable values that optimizes the chosen criterion. We make use here of results given in [9,10]. This gives rise to a new induction tree method that we call *Arbogodaï*. This kind of tree is characterized by a number of value classes of the target variable that varies from one node to the other. It is dynamically determined at each new split. When the majority class in a leaf contains several response values, the corresponding prediction rule becomes a multiple conclusion rule. For instance, we can get a rule like "a female customer aged between 30 and 40 with a monthly income ranging form 4000 to 5000 euros will chose a red or blue car". Indeed, we can easily compute which of the two colors is more frequent in the leaf. Hence, we can also derive classical simple rules. With *Arbogodaï*, the user has the possibility to chose the kind of rule that best suits her needs.

The paper is organized as follows. Section 2 introduces the notations and recalls the basic induction tree concepts. Section 3 discusses the optimal reduction of the crosstable that crosses at each node the target variable with the predictive attribute. Then Section 4 introduces the *Arbogodaï* algorithm that looks for such an optimal reduction when testing the attributes at each new node. In section 5, we specify the mutiple conclusion nature of the induced rules and propose adapted error rates for *Arbogodaï* trees. We report also experimentations that attempt to compare the generalization performances of *Arbogodaï* with other tree algorithms. Finally, in Section 6 we make some concluding remarks.

2   Principle of Induction Trees and Notations

Let Ω be the population concerned by the learning problem. The profile of any member ω of Ω is described by p variables, $X_1, \ldots, X_p$, called either exogenous variables, predictive attributes or predictors. These variables can be qualitative or quantitative. The set of values taken by X_j is denoted by $\mathcal{X}_j$. We consider also a target attribute C, sometimes called response, endogenous or dependent variable, and designate by $\mathcal{C}$ the set of response values. Like the X_j's, C can be qualitative or quantitative. Since the attributes X_j and the target variable C take only a finite number of different values in a given dataset, the sets $\mathcal{X}_j$ and $\mathcal{C}$ are finite. We denote by m_j the number of different values taken by the attribute X_j and by ℓ the number of different response values c_i. Thus, $\mathcal{C} = \{c_1, \ldots, c_\ell\}$. The goal of induction trees is then to generate a model $\phi(X_1, \ldots, X_p)$ in the form of a decision tree for predicting the value of C from the knowledge of the values taken by the predictive attributes. The tree ϕ is induced from a training sample $\Omega_L \subset \Omega$.

The growing process of the tree is quite simple. As illustrated in Figure 1, the set Ω_L is iteratively split by means of, at each step, one of the predictive attributes $X_1, \ldots, X_p$. The goal is to get distributions among the values of the target variable that are as different as possible. The leaves of the trees obtained at each step t of the growing process define a partition S_t of Ω_L that becomes finer and finer with t. The root of the tree corresponds to the trivial partition $S_0 = \{\Omega_L\}$. The tree given in Figure 1 partitions Ω_L in three subsets corresponding to the nodes s_2, s_3 and s_4. In leaf s_3 for example, we have the set of cases of Ω_L that take values $X_1 = $ male and $X_2 < 5000$. At step t, the partition S_t is derived from the previous one S_{t-1} by seeking the best leaf-attribute couple (s_k, X_j), i.e. that for which the splitting of $s_k \in S_{t-1}$ according to the values of X_j maximizes the gain of information on the target variable between S_{t-1} and S_t. The gain of information is usually measured as the reduction in uncertainty for the target variable or as the increase in the strength of association between the partition and the target variable. The growing process stops when the criterion can no longer be improved or when some stopping criterion is reached.

Let n be the grand total of cases, n_{ik} the number of cases with value c_i for the target variable in the class (leaf) s_k of the partition S, $n_{.k}$ the total number of cases in leaf s_k, $n_{i.}$ the total number of cases with value c_i. The corresponding observed frequencies are denoted respectively by f_{ik}, $f_{.k}$ and $f_{i.}$, and $f_{i|k} = n_{ik}/n_{.k}$ stands for the conditional frequency of value c_i in the leaf s_k.

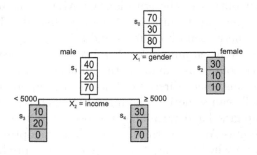

Fig. 1. An induced tree

Table 1. Contingency table defined by X_j at a node s

	x_{j1}	$\cdots$	x_{jk}	$\cdots$	x_{jm_j}	Total
c_1	n_{11}	$\cdots$	n_{1k}	$\cdots$	n_{1m_j}	$n_{1.}$
$\vdots$	$\vdots$	$\ddots$	$\vdots$		$\vdots$	$\vdots$
c_i	n_{i1}	$\cdots$	n_{ik}	$\cdots$	n_{im_j}	$n_{i.}$
$\vdots$	$\vdots$		$\vdots$	$\ddots$	$\vdots$	$\vdots$
c_ℓ	$n_{\ell 1}$	$\cdots$	$n_{\ell k}$	$\cdots$	$n_{\ell m_j}$	$n_{\ell.}$
Total	$n_{.1}$	$\cdots$	$n_{.k}$	$\cdots$	$n_{.m_j}$	n

At any node s of a tree, an attribute X_j defines a partition of the cases in s. This partition is described by the columns of the $\ell \times m_j$ contingency table (Table 1) that crosses the target variable (rows) with X_j (columns).

The criteria used to measure the gain of information brought by a split defined by X_j are computed from this table. For instance, some methods try to maximize the reduction in uncertainty as measured by entropies. In this case, the uncertainty after the split is defined as the weighted mean of the uncertainty of the columns of the contingency Table 1

$$I(S) = \sum_{k=1}^{m_j} \frac{n_{.k}}{n} h(f_{1|k}; \ldots; f_{i|k}; \ldots; f_{\ell|k}) \tag{1}$$

where $h()$ is, for example, the Shannon entropy, $-\sum_{i=1}^{\ell} f_{i|k} \log_2 f_{i|k}$, or the quadratic entropy, also known as the Gini diversity index, $\sum_{i=1}^{\ell} f_{i|k}(1 - f_{i|k})$. Alternatively, some methods like CHAID, optimize the strength or the statistical significance of the association between the resulting partition (columns of Table 1) and the target variable (rows of Table 1).

3   Optimal Reduction of a Contingency Table

Let us recall that CHAID tries, at each step, to merge the columns of crosstables like Table 1 to find the best grouping of values for each candidate attribute, i.e. the grouping that optimizes the criterion. CHAID makes no change, however, on the values of the target variable. *Arbogodaï*, like the Twoing approach in CART, considers merging both columns and rows. Unlike the Twoing rule that looks for the best solution among 2×2 tables only, we seek however the best cross-partition without constraints on the number of rows and columns. This section discusses this issue. First, we motivate the approach by showing that the best n-ary split can sometimes be missed by successive binary splits. Then, we examine strategies for determining the best simultaneous partition of the rows and the columns. We show that the search of the optimal solution becomes rapidly untractable when the number of values of the predictive and/or target variable exceeds 5 or 6. This leads us to use a heuristic to get a quasi-optimal solution.

3.1   Motivation of n-ary Partitions

Consider the crosstable of Table 2. The best bi-partition of its columns is $S_{\text{bin}} = \{\{a,b\},\{d,e\}\}$, whether we use the Gini, the Twoing, the significance of Pearson's Chi-Squares or an association measure like the t of Tschuprow. Now, the best 3 way partition is $S_{3\text{way}} = \{\{a\},\{b,d\},\{e\}\}$ with any of the criteria except Twoing which is not applicable. Clearly $S_{3\text{way}}$ cannot be obtained by splitting the classes of S_{bin}. This proves that multiple binary partitions are not equivalent to n-ary partitions and can sometime miss optimal solutions.

The merging of response values is different in nature from that of predictive attributes. Indeed, the partition of the response values does not translate into a split of the node. Considering such mergings in the optimization process merits therefore some further justification. This is given by simply extending the argument of Breiman et al. ([2], p. 105) who argue that searching for superclasses (the groups of the partitions of the response values) provides strategic information on the similarities of responses. When two or more responses, red car and blue car for example, are almost equally frequent it may be a better strategy to predict that the customer will buy a red or a blue car than explicitly a red one. Simultaneously, it may be useful to know that yellow and pink colors are much less improbable than all other non red and non blue proposed colors. There is thus no reason to limit the argument to two superclasses only. Multi-superclasses provide a more refined strategic information.

3.2   Optimal Reduction

Let $\mathbf{T}_s(C, X_j)$ denote the contingency table obtained by crossing the target variable C with the predictive attribute X_j at a given node s of the tree. From here on we shall drop the subscripts j and s when there is no ambiguity. Our objective is to seek the couple of partitions that produces the table $\mathbf{T}(C, X)$ with maximal row-column association θ. We write the association criterion as $\theta(\mathcal{C}, \mathcal{X})$ to make it clear that it varies with the partitioning of the values of C and X. Let us recall that $\mathcal{X}$ stands for the set of distinct values of X in the concerned population, and $\mathcal{C}$ for the set of distinct values of the target variable C.

We have to distinguish between ordered and unordered sets $\mathcal{X}$ and $\mathcal{C}$. In the unordered case, i.e. for nominal attributes, we denote respectively by $\mathcal{P}_x$ and $\mathcal{P}_c$ the sets of possible partitions of $\mathcal{X}$ and $\mathcal{C}$. In the ordered case, i.e. for ordinal or quantitative attributes, only the merging of adjacent categories is allowed. We denote by $\mathcal{A}_x$ and $\mathcal{A}_c$ the sets of such allowed partitions.

Table 2. A n-ary solution different from that of successive binary splits

	a	b	d	e	Total
c_1	200	100	10	1	311
c_2	10	150	150	10	320
c_3	1	10	100	200	311
Total	211	260	260	211	942

For the unordered case, assuming the criterion has to be maximized, we seek the best couple of partitions in $\mathcal{P}_c \times \mathcal{P}_x$. We have to solve 5 $\arg\max \theta(\mathcal{C}, \mathcal{X})$ for $(\mathcal{C}, \mathcal{X}) \in \mathcal{P}_c \times \mathcal{P}_x$. Finding the optimal solution requires thus to scan $|\mathcal{P}_c||\mathcal{P}_x|$ crosstables. For ordered variables, we replace $\mathcal{P}$ with $\mathcal{A}$.

The number of partitions $|\mathcal{P}|$ of a set of size m is given by Bell's formula, $B(m) = \sum_{k=0}^{m-1} \binom{m-1}{k} B(k)$ with $B(0) = 1$. In the ordinal case the number of partitions $|\mathcal{A}|$ is simply $\sum_{k=0}^{m-1} \binom{m-1}{k} = 2^{m-1}$. Thus, when both variables are nominal with $m = \ell = 10$ categories, we would have to test more than 13 billions of tables. This is indeed intractable.

3.3   Reduction Heuristic

To face the limitations mentioned above, two ways can be considered. The first, is to seek the optimal solution among reasonably sized tables only, i.e. by limiting partitions to a maximum of say 5 or 6 classes. The second approach consists in using a heuristic that would provide a solution close to the true optimum. We have considered this last case in our work on the optimal aggregation of contingency tables [10] in which we studied an algorithm first introduced in [9]. Our experiences with the heuristic led us to two main conclusions. First, the heuristic provides solutions that are most of the time very close to the true optimum. Secondly, the optimal solution is very unstable. It depends indeed strongly on the sample considered. In a learning framework, where the learned rules are intended to be applied outside the learning sample, it is then not crucial to know the exact learning optimum. A solution close to the optimum is largely sufficient. We decided therefore to adopt here the heuristic studied in [10] that we recall briefly.

The heuristic is a simple greedy algorithm. It iteratively merges two rows or two columns. At each step, it merges the couple of either rows or columns that provides the greatest improvement in the criterion. The algorithm reduces thus at each step the table by one row or one column. The process stops when any additional merging would deteriorate the criterion.

Let $\mathcal{C}^k$ and $\mathcal{X}^k$ be the partitions of the values of C and X after step k. For C for example, we denote by $\mathcal{P}_c^k$ the set of partitions that can be obtained in the nominal case by grouping two classes of $\mathcal{C}^k$. When $\mathcal{C}^k$ is an ordered set, we consider the set $\mathcal{A}_c^k$ of partitions that can be obtained by grouping adjacent classes of $\mathcal{C}^k$.

Assuming the criteria θ has to be maximized, the row-column configuration achieved after step k is, for the nominal case, the solution of

$$
\begin{cases}
\arg\max \theta(\mathcal{C}^k, \mathcal{X}^k) \\
\text{s.t.} \qquad \mathcal{C}^k = \mathcal{C}^{(k-1)} \quad \text{and} \quad \mathcal{X}^k \in \mathcal{P}_x^{(k-1)} \\
\qquad \text{or} \quad \mathcal{C}^k \in \mathcal{P}_c^{(k-1)} \quad \text{and} \quad \mathcal{X}^k = \mathcal{X}^{(k-1)}
\end{cases}
\tag{2}
$$

For ordinal variables, we replace $\mathcal{P}^{(k-1)}$ by $\mathcal{A}^{(k-1)}$.

The algorithm starts from the finest partitions $\mathcal{X}^0$ and $\mathcal{C}^0$ of the values observed at the concerned node s. It seeks iteratively, the tables $\mathbf{T}_k(C, X)$

($k = 1, 2, \ldots$) corresponding to the double partition solution of problem (2). The procedure is repeated as long as we have $\theta(C^k, X^k) \geq \theta(C^{(k-1)}, X^{(k-1)})$.

4   Arbogodaï Trees

We first explain the principle of the *Arbogodaï* algorithm and, then, describe how it works on an example.

4.1   Principle of the Algorithm

Arbogodaï follows the general principle of tree growing presented in Section 2. Its specificity is an additional preparatory step before testing the attributes at a node. This step consists in optimally reducing the size of the table that crosses the target variable with every attribute. The splitting criterion is then computed using the found partitions of both the attribute and the target variable values. The splitting of the selected node is done according to the found classes of values of the selected predictive attribute.

This additional step plays a role similar to discretization. The merging of values can indeed be assimilated to some sort of discretization that works also on nominal variables. Remember, however, that the merging is done here simultaneously at each step on the target and the predictive attribute.

The reduction of the table is that for which the row-column association θ is maximized. Indeed we use the heuristic of Section 3.3 and measure the association θ with the t of Tschuprow: $t = \{n^{-1}[(\ell - 1)(m - 1)]^{-1/2} \chi^2\}^{1/2}$, where $\chi^2 = \sum_{i=1}^{\ell} \sum_{k=1}^{m} (n n_{i.} n_{.k})^{-1} (n n_{ik} - n_{i.} n_{.k})^2$ is the Pearson Chi-Squares statistic. Unlike some other association measures, the t of Tschuprow may increase with the merging of either rows or columns (see [10].)

The splitting criterion is the reduction in uncertainty (gain in purity) achieved with the columns of the reduced table as compared to its margin. The uncertainty after the split is computed for every X_j by applying formula (1) on the optimal reduced table for X_j at the considered node s. Using the * to denote quantities derived from the reduced table, the gain in uncertainty reads, with the quadratic (Gini) entropy:

$$h(C^*) - h(C^*|X_j^*) = \sum_i \left[\left(\sum_k f_{.k}^* f_{i|k}^{*2} \right) - f_{i.}^{*2} \right] \tag{3}$$

In addition, we use Laplace estimates for the proportions, i.e. the f^*'s are computed by adding a constant λ to each cell of the reduced table and of its margins. This penalizes the gain of uncertainty obtained at nodes with small counts. With very small counts, i.e. when λ represents a significant proportion of the count, a split may even deteriorate the uncertainty criterion (see [12] p.76.)

4.2   Example

We now describe the *Arbogodaï* algorithm through an example. We consider the *Flags* dataset from the UCI repository [1]. The response variable C takes 6

Table 3. Step 1 optimal crosstable and Laplace estimates of column distributions

C / X_7	$\{c,d,e,h\}$	$\{a,b,g\}$	$\{f\}$
$\{c_1\}$	33	6	0
$\{c_2,c_4,c_5,c_6\}$	2	100	1
$\{c_3\}$	17	9	26
Total	52	115	27

$f^*_{i|k}$	$\{c,d,e,h\}$	$\{a,b,g\}$	$\{f\}$	$f^*_{i.}$
$\{c_1\}$	0.618	0.059	0.033	0.203
$\{c_2,c_4,c_5,c_6\}$	0.055	0.856	0.067	0.528
$\{c_3\}$	0.327	0.085	0.900	0.269
$f^*_{.k}$	0.271	0.581	0.148	1

nominal values $C = \{c_1, c_2, c_3, c_3, c_4, c_5, c_6\}$ and there are 29 mixed categorical and quantitative predictive attributes $X_1, \ldots, X_{29}$. The dataset contains 194 cases. Figure 2 shows an extract of the two first levels of the *Arbogodaï* tree for these data.

Step 1. At the root of the tree, we have the distribution of all 194 cases among the 6 values of the response C. The 29 predictive attributes are successively tested. For every attribute, we first determine the optimal reduced crosstable with the target variable. We then select the attribute for which the gain in uncertainty computed on the reduced table is maximal. The winner is X_7, which takes 8 values: $\mathcal{X}_7 = \{a, b, c, d, e, f, g, h\}$. The two simultaneous groupings found by the heuristic of Section 3.3 are $\mathcal{X}_7^* = \{\{c,d,e,h\}; \{a,b,g\}; \{f\}\}$ and $C^* = \{\{c_1\}; \{c_2, c_4, c_5, c_6\}; \{c_3\}\}$. The corresponding crosstable is shown in Table 3 together with the table of the derived conditional frequencies $f^*_{i|k}$. The latter have been computed by setting $\lambda = 1$. The marginal uncertainty is $h(C^*) = 1 - .20^2 - .53^2 - .27^2 = .61$ and the uncertainty after the split, which is the weighted average of the uncertainty of each column, is $h(C^*|X_3^*) = .31$. The gained information is thus .3. This is the maximal value achievable with any of the 29 attributes.

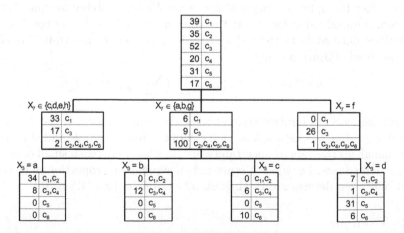

Fig. 2. Example of an Arbogodaï tree

Table 4. Crosstable for splitting the two leftmost leaves with X_{29}

	$\{a,b,d,e\}$	$\{f\}$
$\{c_1, c_2, c_4\}$	38	1
c_3	0	3
other	0	0

	$\{a,d,e\}$	$\{b,f\}$
c_3	2	2
c_4	8	0
other	0	0

Step 2. The process is repeated on every terminal node of the previously obtained tree. Notice that we try to merge the original set of values $\mathcal{X}$ and $\mathcal{C}$ and not the set of previously merged classes. In our example, the next best split occurs at the middle node ($X_7 \in \{a, b, g\}$). The attribute selected for splitting this node is X_3. The 6 values of the target C were merged to form 4 target classes. However, no merging of the attribute could improve the association between X_3 and the target C. The node is therefore split in 4 new classes corresponding to the 4 values of X_3. This leads to the tree with 6 leaves shown in Figure 2.

Following steps. In our example, the tree growing process is stopped after step 2. Without explicit stopping rules, the growing continues until the criterion can no longer be improved. At step 3, *Arbogodaï* would scan the 6 leaves of the previously grown tree.

Two further remarks should be made: (i) At a same level, nodes that do not result from a same parent may have different partitions of the set $\mathcal{C}$ of response values. (ii) When the same attribute is used as the splitting variable at more than one node, its values are not necessarily partitioned the same way for each split. For example, growing the tree of Figure 2 one level further leads to split each of the two left most leaves of level 2 with the same attribute X_{29}. The corresponding crosstables are given in Table 4. It can be seen that the values of C are once partitioned as $\{\{c_1, c_2, c_4\}, \{c_3\}, \{c_5, c_6\}\}$ and once as $\{\{c_3\}, \{c_4\}, \{c_1, c_2, c_5, c_6\}\}$. Likewise, attribute X_{29} is used once with the partition $\{\{a, b, d, e\}, \{f\}\}$ and once with $\{\{a, e, d\}, \{b, f\}\}$.

5   Induces Rules and Their Accuracy

Arbogodaï can generate two types of classification rules: (i) Classical rules by disregarding the merged classes of response values in the final leaves. (ii) Multiple conclusion rules for leaves with merged response values. This Section specifies the nature of these rules, defines error rates adapted for them and presents experimentation results.

We give hereafter the multiple conclusion rules generated by the tree of Figure 2. Each path joining the root to a leaf defines the premise of a rule. The conclusion is drawn from the distribution in the leaf, i.e. the rule predicts for cases falling in the leaf the modal value in the leaf, or modal class of values when some are merged. The tree has 6 leaves giving rise to the 6 following rules (the value between parentheses is the confidence of the rule for the training data). Clearly, when the majority class contains only one value we get classical rules.

Here, only R_3 and R_4 provide multiple conclusions in the form of "either c_1 or c_2."

R_1 : **If** $X_7 \in \{c, h, e, d\}$ **then** $C = c_1$	(33/52)
R_2 : **If** $X_7 = f$ **then** $C = c_3$	(26/27)
R_3 : **If** $X_7 \in \{a, b, g\}$ **and** $X_3 = a$ **then** $C \in \{c_1, c_2\}$	(34/42)
R_4 : **If** $X_7 \in \{a, b, g\}$ **and** $X_3 = b$ **then** $C \in \{c_3, c_4\}$	(12/12)
R_5 : **If** $X_7 \in \{a, b, g\}$ **and** $X_3 = c$ **then** $C = c_6$	(10/16)
R_6 : **If** $X_7 \in \{a, b, g\}$ **and** $X_3 = d$ **then** $C = c_5$	(31/45)

5.1 Error Rates

The accuracy of the learned rules is usually assessed with the misclassification error rate or equivalently with the classification success rate. For classical rules, the misclassification rate reads $err = 1 - \sum_{s \in S} f_s f_{\max|s}$ where f_s is the proportion of cases in leaf s and $f_{\max|s} = \max_i f_{i|s}$ is the frequency of the modal response in leaf s.

For multiple conclusion rules, two kinds of error rates can be defined:

$$\text{superclass error} \qquad serr = 1 - \sum_{s \in S} f_s f^*_{\max|s}$$

$$\text{weighted superclass error} \quad werr = 1 - \sum_{s \in S} f_s f^*_{\max|s} \Big(\sum_{i \in \mathcal{C}_{\max, s}} \hat{p}_{i|\max, s} \, f_{i|\max, s} \Big)$$

where $\mathcal{C}_{\max, s}$ is the set of response values in the modal superclass at leaf s, $f_{i|\max, s}$ the frequency of response c_i in that superclass and $\hat{p}_{i|\max, s}$ an estimation of the probability of c_i in the superclass. We get resubstitution error rates when the frequencies are those of the learning sample and generalization error rates when the frequencies are obtained from validation data. The estimations $\hat{p}_{i|\max, s}$'s are in any case computed on the training data. To get more reliable estimates, we use the marginal distribution at the parent node. This can be justified as follows. Values are merged when their distributions among the values of the split attribute are similar. Hence, their distributions inside the superclass are similar too and therefore similar to the marginal distribution.

The *superclass error*, *serr*, is computed as for classical rules but with the superclass frequencies $f^*_{i|s}$ instead of the single response frequencies $f_{i|s}$. Doing so, we do not care indeed of classification error inside the modal superclasses. This may have sense independently for each rule. We cannot compare, however, the error rate of a rule that predicts for instance c_1 with that of a rule that predicts c_1 or c_2. Hence, the global superclass error does not make much sense.

The *weighted superclass error*, *werr*, takes the uncertainty inside the majority class into account. It assumes that each case falling in a leaf is randomly assigned to a value in the modal superclass. The supposed random assignment is done according to the learned distribution inside $\mathcal{C}_{\max, f}$. For instance for our example tree, a case $(X_7 = a, X_3 = a, C = c_2)$ is correctly classified in the modal superclass of leaf 3. In that leaf, the estimated proportion of cases taking $C = c_2$

in the superclass is 85%. Thus, we weight this correct classification and count it as a .85 correct classification. In resubstitution, if we use $\hat{p}_{i|max,s} = f_{i|max,s}$, this is equivalent to weighting down the success rates with the Gini uncertainty of the distribution inside the superclass: $werr = \sum_s [1 - (1 - serr_s)\text{Gini}(C_{max,s})]$, where $serr_s$ is the superclass error for rule s.

It is well known that the learning error rate suffers from an optimistic bias. It underestimates the generalization error rate. For validation, it is then common to compute the classification error rate on a separate dataset not used for learning. Alternatively, and perhaps more frequently, a cross-validation error rate is computed. A 10 folds cross-validation (10CV), for instance, consists in splitting the learning sample into 10 approximately equally sized parts. Dropping each time a different part we get 10 learning datasets from which 10 trees are induced. For each of them we compute the error rate on the dropped out data. The cross-validation error rate is the mean values of the 10 resulting error rates.

5.2   Experimentation

We have experimented our approach on 8 benchmark datasets. Table 5 gives the cross-validation success rates obtained for each dataset with *Arbogodaï* and, for the sake of comparison, with CART and CHAID. For *Arbogodaï*, we give the rate derived from both the classical and the weighted superclass error. *Arbogodaï* ranks first for 5 of the 8 datasets whatever error is considered. Unsurprisingly, its superiority is mostly significant when the number of values of the target variable is large.

Table 5. Cross-Validation classification success rates (in percents)

Dataset	CART		ChAID		Arbogodaï			
	$1-err$	stdev	$1-err$	stdev	$1-err$	stdev	$1-werr$	stdev
Iris (3 cl.)	95.11	0.08	94.81	0.08	98.35	0.11	95.50	0.08
Flags (6 cl.)	75.14	0.40	75.21	0.40	78.83	0.41	83.37	0.34
Breast (2 cl.)	97.54	0.17	97.19	0.15	98.17	0.13	98.08	0.17
Car (4 cl.)	83.47	0.32	93.62	0.23	86.75	0.32	87.81	0.31
Ionosphere (2 cl.)	92.10	0.19	89.68	0.20	89.34	0.3	93.36	0.25
Pima (2 cl.)	84.44	0.38	83.55	0.38	81.39	0.38	81.20	0.40
Wine (3 cl.)	97.71	0.19	97.99	0.19	98.09	0.07	95.21	0.20
Zoo (7 cl.)	87.57	0.22	85.99	0.26	88.61	0.12	94.04	0.16

6   Conclusion

To conclude, we would like to point out that the *Arbogodaï* method is well suited for mixed nominal and ordinal multi-valued attributes since the merging of any or only adjacent values can be set on the fly. It is also able to handle similarly nominal and ordinal, hence quantitative, target variables. Thus, *Arbogodaï* could be seen as some sort of regression tree. The originality is that, unlike for instance

CART that generates point predictions for each leaf, *Arbogodaï* would provide interval predictions. The multi-conclusion of an *Arbogodaï* rule can hence be seen as a generalized interval for qualitative responses. Finally, let us mention that we are presently designing further experiments for comparing *Arbogodaï* with other tree methods and especially CHAID and CART. This aspect requires a careful investigation. Indeed, the parameterization of the trees (depth, pruning, stoping rules,...) plays a crucial role on the classification performance. We are trying, therefore, to set up rigorous conditions that would ensure more fair, hence more useful, comparison results. We also plan to investigate the relationship to the minimal description length (MDL) principle [8], as the optimally reduced tables can be seen as theories that best describe, locally at each node, the relevant knowledge about the relation between attributes and the target variable.

References

1. Blake, C., Merz, C.: UCI repository of machine learning databases. http://www.ics.uci.edu/~mlearn/MLRepository.html (1998)
2. Breiman, L., Friedman, J.H., Olshen, R.A., Stone, C.J.: *Classification And Regression Trees.* Chapman and Hall, New York (1984)
3. Hastie, T., Tibshirani, R., Friedman, J.: *The Elements of Statistical Learning.* Springer, New York (2001)
4. Kass, G.V.: An exploratory technique for investigating large quantities of categorical data. *Applied Statistics* **29** (1980) 119–127
5. Morgan, J.N., Sonquist, J.A.: Problems in the analysis of survey data, and a proposal. *Journal of the American Statistical Association* **58** (1963) 415–434
6. Quinlan, J.R.: Induction of decision trees. *Machine Learning* (1986) 81–106
7. Quinlan, J.R.: *C4.5: Programs for Machine Learning.* Morgan Kaufmann, San Mateo (1993)
8. Rissanen, J.: A universal prior for integers and estimation by minimum description length. *The Annals of Statistics* **11** (1983) 416–431
9. Ritschard, G., Nicoloyannis, N.: Aggregation and association in cross tables. In Zighed, Komorowski, Zytkow, eds.: *Principles of Data Mining and Knowledge Discovery.* Springer-Verlag, Berlin (2000) 593–598
10. Ritschard, G., Zighed, D.A., Nicoloyannis, N.: Maximisation de l'association par regroupement de lignes ou colonnes d'un tableau croisé. *Revue Mathématiques Sciences Humaines* **39** (2001) 81–97
11. Zighed, D.A., Auray, J.P., Duru, G.: *SIPINA : Méthode et logiciel.* Editions A. Lacassagne, Lyon (1992)
12. Zighed, D.A., Rakotomalala, R.: *Graphes d'induction: apprentissage et data mining.* Hermes Science Publications, Paris (2000)

Author Index

Lecture Notes in Artificial Intelligence (LNAI)

Lecture Notes in Computer Science